Wastewater Treatment Plant Design

Edited by
P. Aarne Vesilind
R. L. Rooke Professor of Engineering
Department of Civil and Environmental Engineering
Bucknell University, Lewisburg, Pennsylvania

Water Environment Federation

Improving Water Quality for 75 Years

Founded in 1928, the Water Environment Federation (WEF) is a not-for-profit technical and educational organization with members from varied disciplines who work toward the WEF vision of preservation and enhancement of the global water environment. The WEF network includes water quality professionals from 79 Member Associations in over 30 countries.

Water Environment Federation
601 Wythe Street, Alexandria, VA 22314-1994 USA
(703) 684-2400
http://www.wef.org

IMPORTANT NOTICE

The material presented in this publication has been prepared in accordance with generally recognized engineering principles and practices and is for general information only. This information should not be used without first securing competent advice with respect to its suitability for any general or specific application.

The contents of this publication are not intended to be a standard of the Water Environment Federation (WEF) and are not intended for use as a reference in purchase specifications, contracts, regulations, statutes, or any other legal document.

No reference made in this publication to any specific method, product, process, or service constitutes or implies an endorsement, recommendation, or warranty thereof by WEF.

WEF makes no representation or warranty of any kind, whether expressed or implied, concerning the accuracy, product, or process discussed in this publication and assumes no liability.

Anyone using this information assumes all liability arising from such use, including but not limited to infringement of any patent or patents.

Library of Congress Cataloging-in-Publication Data
Water Environment Federation.
  Wastewater treatment plant design / by Water Environment Federation.
     p. cm.
The primary source of material for this textbook was: Design of municipal wastewater treatment plants. 4th ed. 1998.
Includes bibliographical references and index.
     ISBN 1-57278-177-7 (hardcover)
  1. Sewage disposal plants—Design and construction. 2. Sewage—Purification. 3. Sewage sludge. I. Design of municipal wastewater treatment plants. II. Title.
TD746.W38 2003
628.3—dc21   2003007284

Co-published by IWA Publishing, Alliance House, 12 Caxton Street, London SW1H 0QS, UK
Telephone: +44 (0) 20 7654 5500; Fax: +44 (0) 20 7654 5555; e-mail: publications@iwap.co.uk
Web: www.iwapublishing.com
ISBN (book only): 1 84339 029 8 (IWA Publishing)
ISBN (2-vol. set with student workbook): 1 84339 052 3

Copyright © 2003 by the Water Environment Federation
Alexandria, VA 22314-1994 USA
ISBN 1-57278-177-7 (textbook only)
ISBN 1-57278-181-5 (2 vol. set, with student workbook)
Printed in the USA 2003

Contents

Preface xxvii
About the Editor xxix
Acknowledgments xxx

1 Fundamentals 1-1
Introduction 1-1
Water Quality 1-6
   Measures of Water Quality 1-6
      Dissolved Oxygen 1-7
      Biochemical Oxygen Demand 1-7
      Solids 1-9
      Nitrogen 1-10
      Phosphorus 1-11
      Bacteriological Measurements 1-11
   Water Quality Standards 1-13
Options for Municipal Wastewater Treatment 1-14
Materials Balances 1-16
Engineering Economics 1-19
   Calculating Annual Cost 1-20
   Calculating Present Worth 1-22
   Calculating Sinking Funds 1-22
   Calculating Capital Plus O&M Costs 1-23
Value Engineering 1-23
Conclusions 1-24
References 1-24
Symbols Used in this Chapter 1-25

2 The Design Process 2-1
Introduction 2-1
Project Participants 2-2
Project Sequence and Design Standards 2-3
Sources, Quantities, and Characteristics of Municipal Wastewater 2-4
   Population and Flow Projections 2-4
   Infiltration and Inflow 2-5
   Industrial, Commercial, and Institutional Wastewater Contributions 2-6
   Other Wastewater Contributions 2-6
   Community Water Supply 2-7
   Domestic Wastewater 2-8
   Industrial Wastewater 2-9
   Wastewater Variability 2-9
   Determining Design Flows 2-11
Occupational Safety 2-12
   Occupational Safety and Health Act and Federal Regulations 2-12
   State and Local Codes 2-13
   National Fire Protection Association Recommendations 2-13

Americans with Disabilities Act 2-13
Designing Safe Facilities 2-13
Site Selection. 2-14
Land Use 2-14
Receiving Water Location and Requirements 2-15
Area Requirements 2-16
Satellite Facilities. 2-17
Elevation and Topography 2-17
Geology, Hydrogeology, and Soils 2-18
Transportation and Site Access 2-18
Utility Services 2-19
Noise Control. 2-19
Air Quality Control. 2-19
Other Environmental Considerations 2-19
Permit Requirements. 2-19
Process Options and Selection 2-20
Plant Layout 2-21
Treatment Facilities. 2-21
Providing for Future Expansion. 2-23
Tank Geometry 2-23
Treatment Redundancy. 2-23
Hydraulics 2-24
Flexibility of Operation. 2-24
Continuing Treatment During Construction. 2-24
Maintenance. 2-25
Administration, Staff, and Support Facilities. 2-25
Monitoring and Control Facilities 2-26
Maintenance and Storage 2-26
Laboratories. 2-27
Administrative Offices. 2-27
Staff Facilities 2-27
Other Layout Considerations. 2-27
Environmental Impact 2-28
Conclusions 2-30
References. 2-30
Symbols Used in this Chapter 2-31

3   Plant Hydraulics 3-1
Introduction 3-1
Hydraulic Considerations 3-1
Hydraulic Profile. 3-1
Flowrates 3-2
Unit Process Liquid Levels. 3-4
Unit Process Redundancy 3-4
Flow Distribution 3-4
Plant Head Loss 3-5
Minimum and Maximum Velocity. 3-5

Fundamentals of Hydraulic Engineering..................................3-7
   Hydraulic Head...3-7
   Pipe Flow...3-7
   Open-Channel Flow..3-11
   Weir Control...3-12
   Effluent Launders..3-14
   Flow Distribution (Channels and Pipe Headers)....................3-14
Unit Process Hydraulics and Other Hydraulic Elements..................3-16
   Screens..3-17
   Grit Tanks...3-17
   Primary Settling Tanks...3-17
   Aeration Tanks...3-18
   Secondary Settling Tanks...3-18
   Disinfection Tanks...3-18
   Other Unit Processes...3-19
   Yard Piping..3-19
   Outfalls...3-19
Pumps and Pumping...3-19
   Pump Characteristics...3-19
   System Head Curves...3-21
   Pump Selection...3-23
Conclusions...3-25
References..3-26
Symbols Used in this Chapter..3-26

4 Preliminary Treatment 4-1

Introduction...4-1
Screening..4-1
   Purpose of Screening..4-1
   Bar Screens...4-2
      Manually Cleaned Bar Screens.................................4-3
      Mechanically Cleaned Screens.................................4-3
      Chain- or Cable-Driven Screens...............................4-5
      Reciprocating Rake Screen....................................4-5
      Catenary Screens...4-6
      Continuous Self-Cleaning Screens.............................4-7
   Comminutors and Grinders..4-8
      Comminutors..4-8
      Grinders...4-9
   Screening Quantities and Characteristics.........................4-10
   Design Practice..4-11
   Management of Screenings...4-13
Grit Removal..4-14
   Purpose of Grit Removal..4-14
   Methods of Grit Removal..4-15
      Aerated Grit Chambers.......................................4-15
      Vortex Grit Removal...4-17

Detritus Tanks ... 4-17
Horizontal-Flow Grit Chambers ... 4-17
Hydrocyclones ... 4-17
Grit Quantities and Characteristics ... 4-18
Design Practice ... 4-18
Removing Grit from the Chambers ... 4-19
Inclined Screw or Tubular Conveyors ... 4-19
Chain-and-Bucket Elevators ... 4-20
Clamshell Buckets ... 4-20
Pumping ... 4-20
Grit Washing ... 4-20
Grit Disposal ... 4-20
Septage ... 4-21
Flow Equalization ... 4-21
Conclusion ... 4-24
References ... 4-25
Symbols Used in this Chapter ... 4-25

5 Primary Treatment ... 5-1
Introduction ... 5-1
Primary Settling ... 5-2
Settling Theory ... 5-2
Types of Settling Tanks ... 5-5
Design Considerations ... 5-6
Depth ... 5-7
Hydraulic Residence Time ... 5-8
Weir Rate ... 5-8
Flow-Through Velocity ... 5-9
Surface Geometry ... 5-10
Inlets ... 5-10
Outlet Conditions ... 5-12
Maintenance Provisions ... 5-14
Enhanced Settling ... 5-13
Preaeration ... 5-14
Chemical Coagulation ... 5-15
Sludge Collection and Removal ... 5-16
Sludge Collection ... 5-16
Primary-Sludge Quantities and Properties ... 5-17
Scum Collection and Disposal ... 5-19
Scum Collection ... 5-19
Scum Management ... 5-19
Thickening with Waste Activated Sludge ... 5-20
Imhoff Tanks ... 5-20
Fine Screens ... 5-21
Conclusions ... 5-23
References ... 5-25
Symbols Used in this Chapter ... 5-25

6 Suspended-Growth Biological Treatment 6-1

Introduction 6-1
The Activated-Sludge Process 6-1
   Process Description 6-1
   Effect of Influent Load and Characteristics 6-3
   Historical Perspective 6-3
   The Activated-Sludge Environment 6-4
   Biological Growth and Substrate Oxidation 6-5
   Process Design for Carbon Oxidation 6-6
      Volume of Aeration Tanks 6-10
      Sludge Generated and Wasted 6-12
      Oxygen Demand 6-13
      Return Activated-Sludge Capacity Requirements 6-13
      Clarifier Sizing 6-13
   Process Design for Nitrification 6-13
   Other Environmental Effects 6-15
      Temperature 6-15
      Dissolved Oxygen 6-16
      Nutrients 6-16
      Toxic and Inhibitory Wastes 6-17
      pH 6-17
Process Configuration Alternatives 6-18
   Tank Shape 6-18
      Complete Mix 6-18
      Plug Flow 6-19
      Oxidation Ditch 6-20
      Aerated Lagoon Reactors 6-21
   Loading Rates 6-21
      Conventional 6-21
      Low Rate 6-22
      High Rate 6-23
   Feeding and Aeration Patterns 6-23
      Conventional 6-23
      Contact Stabilization 6-23
      Step Feed 6-24
      Tapered Aeration 6-24
   Other Variations 6-25
      Pure Oxygen 6-25
      Sequencing Batch Reactors 6-26
      Activated Carbon Addition 6-27
      Hybrid Systems 6-27
Oxygen Transfer 6-27
   Diffused Aeration 6-28
      Porous Diffuser Systems 6-28
      Nonporous Diffusers 6-29
      Other Diffused Aeration Systems 6-31
   Mechanical Surface Aeration 6-32

Secondary Settling 6-33
  General Design Considerations 6-33
  Process Design 6-34
    Solids Loading Rate 6-34
    Overflow Rate 6-35
    Depth. 6-35
    Influent Structure 6-36
    Scum Removal 6-36
    Outlet Structure 6-36
    Sludge Withdrawal 6-38
  Control Strategies 6-39
    Return Activated Sludge Flowrates 6-39
    Return Activated Sludge Flow Patterns 6-39
    Waste Activated Sludge Rate Control 6-40
    Type of Pumps and Systems 6-41
Conclusions 6-41
References. 6-42
Symbols Used in this Chapter 6-43

7 Attached-Growth Biological Treatment 7-1
Introduction 7-1
Trickling Filters. 7-2
  Theory of Operation 7-3
  Loading Terminology 7-4
  Factors Affecting Performance and Design 7-5
    Wastewater Composition 7-5
    Wastewater Treatability 7-5
    Pretreatment and Primary Treatment 7-5
    Trickling Filter Media Type 7-5
    Trickling Filter Depth 7-6
    Trickling Filter Hydraulics and Loadings 7-7
    Ventilation 7-8
    Odor, Vector, and Macroorganism Control 7-9
  Design Formulas 7-9
    National Research Council Formula 7-10
    Velz Formula 7-10
    Kinematic Wave—Velz Model. 7-11
    Kincannon and Stover Model 7-12
    Logan Model 7-13
  Nitrification 7-13
Rotating Biological Contactors 7-13
  Process Concepts and Principles. 7-15
  Factors Affecting Performance 7-15
    Organic and Hydraulic Loading 7-15
    Influent Wastewater Characteristics 7-16
    Wastewater Temperature 7-16
    Biofilm Control 7-16
    Dissolved Oxygen Levels. 7-16

  Process Design 7-16
    Denitrification 7-18
 Combined Biological Treatment. 7-18
  Trickling Filter–Solids Contact. 7-18
  Roughing Filter–Activated Sludge 7-18
  Biofilter–Activated Sludge 7-19
  Trickling Filter–Activated Sludge 7-19
 Developing Fixed-Film Processes 7-19
  Subsided Fixed-Bed Reactors. 7-19
  Floating Bed Aerated Filters. 7-19
 Secondary Clarification 7-20
 Conclusions 7-20
 References. 7-20
 Symbols Used in this Chapter 7-22

8 Biological Nutrient Removal 8-1
Introduction 8-1
Phosphorus Removal Processes 8-2
 Basic Theory. 8-2
 Design Options. 8-5
  Two-Stage Process 8-5
  Combined Chemical and Biological Process 8-5
  Sidestream Fermentation Processes 8-6
  Sequencing Batch Reactors 8-7
  Chemical Polishing 8-7
Nitrogen Removal Processes 8-7
 Basic Theory. 8-7
 Suspended-Growth Processes for Nitrogen Removal. 8-12
  Single-Sludge Processes 8-12
  Dual-Sludge Processes. 8-15
 Attached Growth Processes for Nitrogen Removal 8-17
  Denitrification Filters 8-17
  Fluidized Bed Denitrification. 8-17
Simultaneous Phosphorus and Nitrogen Removal. 8-18
 Modified Bardenpho 8-18
 A^2/O Process. 8-18
 University of Cape Town Process. 8-18
 PhoStrip II Process 8-19
 Fixed-Growth Reactor–Suspended-Growth Reactor 8-19
 Phased Isolation Ditches 8-21
Design Considerations. 8-21
 Design Procedures for Phosphorus Removal. 8-21
 Design Procedures for Nitrogen Removal. 8-22
Conclusions 8-25
References. 8-25
Symbols Used in this Chapter 8-27

9 Alternative Biological Treatment 9-1
Introduction 9-1
Soil Absorption Systems 9-1
Lagoon Systems 9-2
    Facultative Lagoons 9-3
        Treatment Performance 9-3
        Design Procedures 9-3
        Controlled Discharge Lagoons 9-4
        Hydrograph Controlled Release Lagoons 9-5
    Partially Mixed Aerated Lagoons 9-5
    Dual-Power Multicellular Aerated Lagoons 9-7
    Advance Integrated Lagoon Systems 9-7
Land Treatment Systems 9-7
    Preapplication Treatment 9-9
    Site Requirements 9-9
    Slow-Rate Systems 9-10
        Treatment Performance 9-10
        Design Objectives 9-11
    Overland Flow Systems 9-13
        Treatment Performance 9-13
        Design Factors 9-14
        Design Procedures 9-14
        Suspended Solids Loadings 9-14
        Biochemical Oxygen Demand Loadings 9-14
        Land Requirements 9-15
        Vegetation Selection 9-15
        Distribution System 9-15
        Monitoring Wells 9-15
    Rapid Infiltration Systems 9-15
        Treatment Performance 9-15
        Design Procedures 9-16
Floating Aquatic Systems 9-17
Constructed Wetlands 9-17
    Performance of Constructed Wetlands 9-18
    Land Requirements 9-19
Conclusions 9-19
References 9-20
Symbols Used in this Chapter 9-21

10 Physical–Chemical Processes 10-1
Introduction 10-1
Process Selection 10-2
Granular Media Filtration 10-2
    Process Description 10-2
    Design Objectives 10-2
    Filtration Theory 10-3
    Application of Granular Media Filtration to Wastewater Treatment 10-3
    Process Design 10-5

    Pretreatment to Enhance Filterability................................ 10-5
    Filter Type and Loading Rates...................................... 10-5
    Media Selection and Characteristics................................. 10-6
Automatic Backwashing Filters 10-7
    Moving Bed Filters 10-7
    Pulsed Bed Filters 10-8
    Operation.. 10-8
Activated Carbon Adsorption 10-10
    Process Description 10-10
    Application of Activated Carbon to Wastewater Treatment 10-11
    Design 10-11
        Carbon Characteristics 10-12
        Types of Carbon Adsorption Units 10-13
        Unit Sizing 10-14
        Backwashing.. 10-15
        Control of Biological Activity 10-16
        Carbon Transport... 10-16
        Carbon Regeneration 10-16
Chemical Treatment 10-17
    Phosphorus Precipitation... 10-17
        Phosphorus Removal Methods 10-18
        Precipitants.. 10-18
        Solids Considerations 10-21
    pH Adjustment 10-22
    Rapid Mixing... 10-23
        Impeller Mixers 10-23
        Other Mixing Devices... 10-23
        Fluid Regimes.. 10-23
        Design Considerations... 10-24
    Chemical Feed Systems 10-26
        Solution Feed 10-27
Membrane Processes 10-28
    Process Description 10-28
    Process Objectives... 10-29
    Pretreatment.. 10-29
    Membrane Systems 10-29
    Membrane Module Configuration 10-31
        Membrane Filtration.. 10-31
        Reverse Osmosis.. 10-31
    Reject and Brine Disposal 10-31
        Disposal to Surface Water..................................... 10-32
        Deep-Well Injection 10-32
        Evaporation Ponds.. 10-32
        Evaporation to Dryness and Crystallization 10-32
Conclusions 10-32
References.. 10-33
Symbols Used in this Chapter.. 10-33

11 Disinfection, Reoxygenation, and Odor Control 11-1
Disinfection..11-1
Disinfection Kinetics 11-2
Chlorine Disinfection..11-5
    Chemistry of Chlorine Disinfection 11-5
        Elemental Chlorine 11-5
        Hypochlorites..11-5
        Sulfur Dioxide 11-6
        Chemistry of Chlorine in Water.............................11-6
    Chlorine Toxicity and Effects on Higher Organisms...................11-7
    Aftergrowth 11-7
    Dechlorination 11-8
    Safety and Health 11-8
    Shipment and Handling...11-8
        Facility Design 11-8
    Design and Selection of Equipment 11-10
        Chlorinators and Sulfonators...............................11-10
        Chemical-Feed Pumps 11-10
        Sulfur Dioxide Feeders 11-11
    Feed Control Strategies 11-11
    Reactor Design 11-11
Ultraviolet Disinfection 11-12
    Mechanism of Ultraviolet Disinfection.............................11-13
    Ultraviolet Inactivation Kinetics...................................11-15
    Ultraviolet Disinfection System Design.............................11-16
        Wastewater Characteristics 11-16
    Ultraviolet Disinfection Equipment 11-17
Effluent Reoxygenation 11-17
    Cascade Reoxygenation..11-17
    Mechanical and Diffused Air Reoxygenation 11-19
    Relationship of Reoxygenation to Other Unit Processes 11-19
Odor Control 11-19
    Sources of Odor 11-20
    Odor Control 11-22
        Upstream Controls 11-22
        Chemical Additions.......................................11-22
        Adsorption Systems.......................................11-22
        Biological Systems..11-22
        Combustion Systems......................................11-23
        Ozonation 11-23
        Wet Scrubbers 11-23
Conclusions 11-23
References..11-24
Symbols Used in this Chapter..11-24

12 Production and Transport of Wastewater Sludge 12-1
Introduction 12-1
Sludge Quantities 12-2
    Estimating Sludge Quantities 12-3
    Primary Solids Production 12-3
    Secondary Solids Production 12-5
    Combined Solids Production 12-9
    Chemical Solids Production 12-11
Sludge Characteristics 12-11
    Primary Sludge 12-11
    Secondary Sludge 12-12
    Combined Sludge 12-13
    Chemical Sludge 12-13
Liquid Sludge Storage 12-13
Liquid Sludge Transport 12-13
    Flow and Head Loss Characteristics. 12-14
    Design Approach. 12-14
        Dilute Sludges. 12-14
        Thickened Sludges. 12-14
    Kinetic Pumps. 12-16
        Solids-Handling Centrifugal Pumps 12-17
        Recessed-Impeller Pumps. 12-17
        Screw–Combination Centrifugal Pumps 12-17
        Grinder Pumps 12-17
    Positive Displacement Pumps. 12-18
        Plunger Pumps 12-18
        Progressing Cavity Pumps 12-19
        Air-Operated Diaphragm Pumps 12-20
        Rotary Lobe Pumps 12-21
        Pneumatic Ejectors 12-21
        Peristaltic Hose Pumps 12-21
        Reciprocating Piston Pumps 12-21
    Other Pumps. 12-21
        Air-Lift Pumps 12-21
        Archimedes Screw Pumps 12-22
        Long-Distance Pipelines. 12-23
        Common Design Deficiencies in Pumps and Piping. 12-23
        Standby Capacity 12-24
Dewatered Sludge Storage 12-24
Dewatered Sludge Transport 12-24
    Pumps 12-24
        Progressing Cavity Pumps 12-25
        Hydraulically Driven Reciprocating Piston Pumps 12-25
    Conveyors 12-26
        Belt Conveyors 12-26
        Screw Conveyors 12-27

Dried Sludge Transport 12-27
   Belt Conveyors 12-27
   Screw Conveyors. 12-28
   Drag Conveyors 12-28
   Bucket Elevators 12-29
   Pneumatic Conveyors 12-29
Conclusions 12-30
References. 12-30
Symbols Used in this Chapter. 12-31

13 Sludge Conditioning 13-1
Introduction 13-1
Chemical Conditioning 13-2
   Inorganic Chemicals 13-2
      Lime. 13-2
      Ferric Chloride 13-3
      Aluminum Salts 13-3
   Organic Polymers 13-3
      Polymer Charge 13-4
      Polymer Molecular Weight 13-4
      Polymer Forms 13-5
   Dosage Optimization. 13-6
      Jar Test 13-8
      Capillary Suction Time Test 13-8
      Standard Shear Test 13-9
      Buchner Funnel Test 13-9
      Sludge Compactability 13-10
      Sludge Consistency 13-10
Thermal Conditioning. 13-11
Freeze–Thaw Conditioning 13-11
Conclusions 13-13
References. 13-13
Symbols Used in this Chapter. 13-14

14 Sludge Thickening, Dewatering, and Drying 14-1
Introduction 14-1
Gravity Thickening 14-2
   Theory of Gravity Thickening 14-3
   Design of Gravity Thickeners. 14-7
      Area Determination for Solids Thickening 14-7
      Other Design Considerations. 14-8
Dissolved Air Flotation Thickening 14-9
   Theory of Dissolved Air Flotation 14-11
   Design of Dissolved Air Flotation Thickeners 14-11
Gravity Belt Thickening. 14-12
   Theory of Gravity Belt Thickening 14-13
   Design of Gravity Belt Thickeners 14-15
Other Methods of Thickening 14-15

Rotary Drum Thickening.	14-15
Centrifuges	14-16
Centrifugal Dewatering	14-17
Theory of Centrifugation.	14-18
Design of Centrifugal Dewatering	14-21
Belt Filter Dewatering	14-23
Theory Of Belt Filter Dewatering.	14-23
Design Of Belt Filter Dewatering	14-23
Filter Press Dewatering	14-25
Theory Of Filter Press Dewatering.	14-25
Design Of Filter Press Dewatering	14-27
Drying Bed Dewatering	14-27
Sand Drying Beds	14-27
Other Types Of Drying Beds	14-30
Paved Drying Beds	14-30
Wedge-Wire Drying Beds	14-30
Vacuum-Assisted Drying Beds	14-30
Other Dewatering Methods	14-30
Reed Beds.	14-30
Lagoons	14-31
Vacuum Filters	14-32
Thermal Drying.	14-32
Theory of Thermal Drying.	14-33
Design of Thermal Drying Systems.	14-35
Direct Dryers	14-35
Indirect Dryers	14-35
Direct–Indirect Dryers.	14-36
Infrared Dryers	14-36
Conclusions	14-36
References.	14-38
Symbols Used in this Chapter.	14-39

15 Sludge Stabilization ... 15-1

Introduction	15-1
Anaerobic Digestion	15-2
Anaerobic Digestion Theory	15-2
Microbiology and Biochemistry.	15-2
Variables Affecting Anaerobic Digestion	15-3
Solids and Hydraulic Retention Times.	15-4
Temperature.	15-4
pH.	15-4
Toxic Materials	15-5
Applicability.	15-5
Process Variations	15-7
Low-Rate Digestion	15-7
High-Rate Digestion	15-8
Two-Stage Anaerobic Digestion.	15-9
Thermophilic Anaerobic Digestion.	15-9

- Design Criteria 15-10
  - Volatile Solids 15-10
  - Pathogen Reduction 15-10
  - Gas Quality and Quantity 15-11
- Anaerobic Digester Design 15-11
  - Design Data 15-11
  - Design Parameters 15-12
  - Volatile Solids Loading and Destruction 15-13
  - Gas Production 15-13
  - Gas Collection and Storage 15-14
  - Mixing 15-15
  - Heating 15-16
  - Chemical Requirements 15-18
  - Digester Covers 15-19
  - Tank Configuration and Construction 15-20
  - Effect of Digestion on Dewatering and Dewatering Recycle Streams 15-22
- Aerobic Digestion 15-24
  - Aerobic Digestion Theory 15-25
  - Aerobic Digestion Design 15-26
    - Reduction in Volatile Solids 15-26
    - Feed Quantities and Characteristics 15-27
    - Operating Temperature 15-27
    - Oxygen-Transfer and Mixing Requirements 15-27
    - Tank Volume and Retention Time Requirements 15-27
  - Process Variations 15-29
    - High-Purity Oxygen Aeration 15-29
    - Low-Temperature Aerobic Digestion 15-29
- Autothermal Thermophilic Aerobic Digestion 15-29
  - Theory of Autothermal Thermophilic Aerobic Digestion 15-30
  - Design of Autothermal Thermophilic Aerobic Digestion 15-30
  - Operation of Autothermal Digestion Plants 15-32
- Composting 15-33
  - Theory of Composting 15-34
    - Microbiology 15-34
    - Energy Balance 15-34
    - Carbon/Nitrogen Ratio 15-35
  - Process Objectives 15-35
    - Pathogen Reduction 15-35
    - Maturation 15-35
    - Drying 15-36
  - Design of Composting Systems 15-36
    - Process Alternatives 15-36
    - Bulking Agents and Amendments 15-36
    - Temperature Control and Aeration 15-38
    - Retention Time 15-38
- Alkaline Stabilization 15-38
  - Theory of Alkaline Stabilization 15-39
  - Design of Alkaline Stabilization Systems 15-40

  Liquid Lime Stabilization 15-40
  Dry Lime Stabilization 15-41
Sludge Combustion 15-41
 Sludge Heat Values 15-41
 Process Alternatives 15-42
  Multiple-Hearth Furnace 15-43
  Fluidized-Bed Furnace 15-44
 Design of Sludge Combustion Systems 15-46
 Emissions Control 15-46
  Pollution Control Technology 15-46
Conclusions 15-47
References 15-47
Symbols Used in this Chapter 15-48

16 Beneficial Use and Ultimate Disposal 16-1
Introduction 16-1
The 40 CFR Part 503 Regulations 16-2
 Land Application 16-2
  Pollutant Limits 16-4
  Pathogen and Vector-Attraction Reduction 16-4
  Management Practices 16-4
 Distribution and Marketing 16-5
 Composting 16-6
 Heat Drying 16-6
 Lime Stabilization 16-6
 Surface Disposal 16-6
 Pathogen and Vector-Attraction Reduction 16-7
  Class A Pathogen Requirements 16-8
  Class B Pathogen Requirements 16-9
  Pathogen Treatment Processes 16-10
  Vector-Attraction Reduction Requirements 16-11
 Combustion (Incineration) 16-12
  Pollutant Limits 16-12
Design of Land Application Operations 16-13
 Application Rates 16-14
 Methods of Application 16-14
 Odor Control 16-15
Design of Landfilling Operations 16-15
Design of Dedicated Land Disposal Operations 16-16
 Application Rates 16-17
 Methods of Application 16-17
Conclusions 16-18
References 16-18
Symbols Used in this Chapter 16-19

Index I-1

List of Figures

1.1	Classification of total solids.	1-9
1.2	Classification of suspended solids.	1-10
1.3	Classification of dissolved solids.	1-11
1.4	Arrangement of a typical large wastewater treatment facility.	1-15
1.5	Unit operations in a typical large wastewater treatment plant.	1-16
1.6	A black box for calculating materials balances.	1-17
1.7	Sludge thickening tank.	1-18
2.1	Hourly variation in flow strength of municipal wastewater.	2-8
2.2	Ratio of extreme flows to average daily flow in New England.	2-11
2.3	Flow variation at a wastewater treatment plant.	2-12
2.4	Plant site area versus plant capacity. The two lines represent a approximate range of plant site areas for a given capacity.	2-17
2.5	Unit (A) and functional (B) design plant layouts.	2-22
2.6	Functional diagram of administration building. Multiple connecting lines show areas of frequent communications and interrelationships.	2-26
3.1	Typical hydraulic profile for influent pumping and primary treatment. Water surface elevations (in feet) represent flow of 160,000 m^3/d (42 mgd).	3-3
3.2	Typical profile for an activated-sludge plant. Water surface elevations (in feet) represent flow of 160,000 m^3/d (42 mgd).	3-3
3.3	Typical hydraulic profile for a tricking filter plant. Water surface elevations (in feet) represent flow of 160,000 m^3/d (42 mgd).	3-3
3.4	Design examples of equal flow distribution among clarifiers: (a) symmetrical layout and (b) asymmetrical layout.	3-6
3.5	Bernoulli relationship for a pipe	3-8
3.6	Weir sections: (a) sharp-crested weir and (b) broad-crested weir.	3-12
3.7	Weir flow conditions: (a) free flow and (b) submerged flow.	3-13
3.8	A manifold or distribution channel.	3-15
3.9	Hydraulics for a typical clarifier.	3-16
3.10	Pump setup for establishing its characteristic curve.	3-20
3.11	The pump characteristic curve.	3-21
3.12	A series of characteristic curves and associated efficiencies for a family of pumps.	3-22
3.13	Pumping from reservoir A to reservoir B.	3-22
3.14	System head curve.	3-23
3.15	System head curve superimposed on pump characteristic curves to determine operating point.	3-24
3.16	Pumps in series.	3-24
3.17	Pumps in parallel.	3-25
4.1	Hand cleaned bar screen at a small wastewater treatment plant.	4-3
4.2	Reciprocating rake bar screen.	4-4
4.3	Catenary bar screen.	4-6
4.4	Continuous self-cleaning screen.	4-7
4.5	Comminutor used at a small treatment plant.	4-8
4.6	Wastewater grinders: (a) channel unit and (b) in-line unit.	4-9

4.7	Average volumes of coarse screenings as a function of openings between bars.	4-10
4.8	An aerated grit chamber.	4-16
4.9	A septage dumping station: (a) plan, (b) profile at pavement centerline, (c) section A-A, and (d) section B-B.	4-22
4.10	Two equalization schemes: (a) in-line equalization and (b) side-line equalization.	4-24
5.1	Zones of an ideal settling tank.	5-3
5.2	Typical rectangular primary settling tank.	5-5
5.3	Typical circular primary settling tank.	5-6
5.4	Typical stacked settling tank.	5-7
5.5	Primary settling tank suspended solid removal as estimated by the overflow rate.	5-8
5.6	Inlet configuration of a rectangular primary settling tank.	5-10
5.7	Various designs of conventional center-feed inlets: (a) side feed, (b) vertical pipe feed, and (c) slotted, vertical pipe feed.	5-11
5.8	Short-circuiting flow in a rectangular basin.	5-12
5.9	Outlet configuration using V-notch weirs of a rectangular primary settling tank.	5-13
5.10	Outlet configuration using submerged collection trough of a rectangular primary settling tank.	5-14
5.11	Possible scheme for coagulant addition for enhanced primary settling.	5-16
5.12	Typical traveling-bridge sludge collector in a primary rectangular settling tank.	5-18
5.13	Tilting trough scum collector.	5-20
5.14	Sloping beach scum collector: (a) partial plan and (b) section.	5-21
5.15	Typical Imhoff tanks with design details: (a) plan and (b) section. Use larger sludge storage allowances for garbage solids and communities with less than 5000 population.	5-22
5.16	Inclined, self-cleaning static screen.	5-23
5.17	Rotary drum screen.	5-24
6.1	The activated-sludge process.	6-2
6.2	Nomenclature for activated-sludge flow sheet.	6-6
6.3	Effect of solids retention time, θ_c, on the settling properties of activated sludge.	6-10
6.4	Typical operating range for activated-sludge systems.	6-12
6.5	Sludge production in activated-sludge systems, with and without primary settling.	6-14
6.6	Predicted relationship between process efficiency and temperature at several values of θ_c under aerobic conditions.	6-17
6.7	Completely mixed reactor type activated-sludge system.	6-18
6.8	Plug-flow reactor type activated-sludge system.	6-20
6.9	Oxidation ditch reactor activated-sludge system.	6-20
6.10	Step-feed process activated-sludge system.	6-24
6.11	Tapered aeration process activated-sludge system.	6-25
6.12	Pure-oxygen activated-sludge system.	6-26
6.13	Some types of porous diffusers: (a) disks, (b) dome diffusers, and (c) perforated membrane diffusers.	6-30

6.14	Submerged turbine aerators: (a) axial flow and (b) radial flow.	6-31
6.15	Some types of surface aerators: (a) standard surface aerator, (b) surface aerator with draft tube, and (c) surface aerator with impeller.	6-32
6.16	Design solids loading on a secondary clarifier.	6-35
6.17	Surface overflow rate and effluent suspended solids recovery for five cities in California.	6-36
6.18	Center-column, energy-dissipating inlet and hydraulic flocculating feed well in a circular clarifier.	6-37
6.19	Alternative peripheral baffle arrangements: (a) Stamford, (b) unnamed, (c) Lincoln, and (d) interior trough.	6-38
6.20	Alternative skimming devices and sludge collection mechanisms for circular clarifiers: (a) revolving skimmer and fixed scum trough and (b) rotary ducking skimmer.	6-40
7.1	A typical trickling filter.	7-2
7.2	Types of commonly used trickling filter media.	7-6
7.3	Flow diagrams for single- and two-stage trickling filter plants.	7-8
7.4	Comparison of trickling filter operating data with performance predicted by the National Research Council formula.	7-11
7.5	Rotating biological contactor.	7-14
8.1	Removal mechanisms for excess biological phosphorus.	8-3
8.2	Typical profile of soluble phosphorus concentrations in a biological nutrient removal process.	8-4
8.3	The A/O process.	8-5
8.4	PhoStrip process.	8-6
8.5	OWASA process.	8-7
8.6	Sequencing batch reactor for removal of phosphorus.	8-8
8.7	Wuhrmann process for nitrogen removal.	8-12
8.8	Ludzack–Ettinger process for nitrogen removal.	8-13
8.9	Modified Ludzack–Ettinger process for nitrogen removal.	8-13
8.10	Four-stage Bardenpho process for nitrogen removal.	8-14
8.11	Oxidation ditch process for nitrogen removal.	8-15
8.12	Dual-sludge process for nitrogen removal.	8-16
8.13	Modified Bardenpho process for phosphorus and nitrogen removal.	8-18
8.14	A^2/O process for phosphorus removal.	8-19
8.15	Modified University of Cape Town process for phosphorus and nitrogen removal.	8-19
8.16	PhoStrip II process for phosphorus and nitrogen removal.	8-20
8.17	Schematic of the fixed-growth reactor–suspended-growth reactor process.	8-21
9.1	Installation of a lagoon liner.	9-7
9.2	Riprap for erosion control on the sideslopes of a facultative lagoon.	9-8
9.3	Typical constructed wetland illustrating open water and emergent vegetation.	9-18
9.4	Inlet manifold to the freewater surface wetland.	9-20
10.1	Granular-media filters following (a) biological secondary treatment, (b) nitrifying biological treatment plus tertiary filtration for biochemical oxygen demand and suspended solids control, (c) biological secondary treatment plus tertiary treatment for ammonia reduction, and	

	(d) biological secondary and tertiary treatment for ammonia and nitrate reduction.	10-4
10.2	Typical flow diagrams for tertiary treatment with carbon adsorption.	10-12
10.3	Typical downflow activated carbon contactor.	10-15
10.4	Various schemes of single-stage lime precipitation for phosphorus removal.	10-19
10.5	Typical feed system for wet chemicals.	10-28
11.1	Flow-proportional chlorination.	11-12
11.2	A typical chlorine contact chamber.	11-13
11.3	Open-channel UV disinfection systems: horizontal lamp configurations (top) and vertical lamp configuration (bottom).	11-14
11.4	Graphical presentation of the terms in cascade reoxygenation equations.	11-18
12.1	Typical solids balance in a wastewater treatment plant.	12-4
12.2	Typical primary settling tank performance for a single settling tank over many days of operation.	12-5
12.3	Variability in the production of raw primary sludge in Ontario wastewater treatment plants.	12-6
12.4	Secondary sludge yield for an activated-sludge system in a plant with primary settling.	12-8
12.5	Secondary sludge yield in an activated-sludge plant without primary settling.	12-8
12.6	Variability in total sludge production in secondary wastewater treatment plants in Ontario.	12-10
12.7	Frictional head losses for a 6-inch diameter sludge force main.	12-15
12.8	Frictional head losses for an 8-inch diameter sludge force main.	12-16
12.9	Details of a positive displacement plunger pump.	12-18
12.10	Progressing cavity pump.	12-19
12.11	Air-operated diaphragm pump.	12-20
12.12	Archimedes screw pumps used for return activated sludge.	12-22
13.1	Typical results from a specific resistance to filtration test; establishing the slope b and calculating the specific resistance to filtration R using eq 13.2.	13-10
13.2	Flow diagram of a low-pressure oxidation process.	13-12
14.1	Typical gravity thickener.	14-3
14.2	Solids flux curve resulting from a settling test in a cylinder.	14-4
14.3	Total solids flux curve, with flux due to settling (Cv) and flux due to withdrawal from the bottom of the thickener (Cu).	14-5
14.4	Graphical method for determining limiting solids flux and underflow concentration.	14-8
14.5	Schematic of dissolved air flotation thickener.	14-9
14.6	Solid-bubble contacting mechanisms in dissolved air flotation.	14-10
14.7	Gravity belt thickener.	14-12
14.8	Gravity belt thickener schematic.	14-14
14.9	Movement of solids on gravity belt thickener to allow water to drain through fabric belt.	14-14
14.10	Rotary drum thickener (side panel removed).	14-16
14.11	Schematic of two solid-bowl centrifuge configurations: (a) countercurrent flow and (b) cocurrent flow.	14-18

14.12	Belt filter press.	14-22
14.13	Filter press.	14-26
14.14	Typical filter press recessed plate.	14-26
14.15	Schematic of a section of a typical sand drying bed.	14-28
14.16	Rotary dryer.	14-34
14.17	Flow diagram of a fluidized-bed sludge drying system.	14-36
14.18	Flow diagram of a paddle or hollow-flight sludge drying system.	14-37
15.1	Microbiological pathway of anaerobic digestion.	15-3
15.2	Low-rate anaerobic digestion.	15-7
15.3	High-rate anaerobic digestion.	15-8
15.4	Two-stage, high-rate anaerobic digestion.	15-10
15.5	Fixed digester cover.	15-19
15.6	Floating digester cover.	15-20
15.7	Cylindrical tank with Downes-type floating cover.	15-21
15.8	Floating gas-holder cover.	15-22
15.9	Egg-shaped anaerobic digester.	15-23
15.10	Mixing systems for egg-shaped anaerobic digesters.	15-23
15.11	Egg-shaped digester installation.	15-24
15.12	Typical autothermal thermophilic aerobic digestion system.	15-31
15.13	Aerated static pile composting system.	15-36
15.14	Vertical plug-flow composting reactor.	15-37
15.15	Typical multiple-hearth furnace.	15-44
15.16	Typical fluidized-bed furnace.	15-45

List of Tables

1.1	Excerpts from the Surface Water Standard for the State of North Carolina.	1-14
1.2	6.125% compound interest factors.	1-21
2.1	Per capita flows.	2-5
2.2	Average wastewater flows from commercial sources.	2-6
2.3	Typical primary pollutant composition of domestic wastewater.	2-10
2.4	Composition of organic materials in wastewater.	2-10
4.1	Typical design criteria for coarse screening equipment.	4-2
4.2	Typical design properties of coarse screenings.	4-11
4.3	Kirshmer's values of β.	4-13
4.4	Typical design criteria for aerated grit chambers.	4-16
4.5	Typical design criteria for horizontal-flow grit chambers.	4-18
4.6	Estimated grit quantities.	4-18
5.1	Summary of overflow rates and side water depths recommended by various entities for primary settling tanks.	5-9
6.1	Typical kinetic constants for suspended-growth reactors.	6-13
6.2	Operational characteristics of activated-sludge processes.	6-21
6.3	Design parameters for activated-sludge processes.	6-22
8.1	Monod kinetic coefficients.	8-10
8.2	Design parameters for biological nutrient removal processes.	8-24
9.1	Design criteria for facultative ponds.	9-4
9.2	Facultative pond biochemical oxygen demand loading rates.	9-4
9.3	Features of land treatment systems.	9-8
9.4	Site requirements for land treatment processes.	9-10
9.5	Ranges of f values (fraction of applied nitrogen lost) for municipal wastewaters.	9-13
10.1	Typical carbon dosages for various wastewater influents.	10-17
10.2	Values of K_T used for determining impeller power requirements.	10-25
11.1	Odorous compounds in wastewater treatment.	11-21
12.1	Typical range of values for sludge yield coefficients.	12-7
12.2	Primary sludge characteristics.	12-12
12.3	Secondary sludge characteristics.	12-12
13.1	Typical dosages of polymer for thickening municipal wastewater sludges.	13-7
13.2	Typical dosages of polymer for dewatering municipal wastewater sludges.	13-7
14.1	Typical gravity thickening data for various types of sludges.	14-6
14.2	Typical gravity belt thickener performance.	14-15
14.3	Typical rotary drum thickener performance.	14-16
14.4	Factors affecting centrifugal thickening and dewatering.	14-20
14.5	Typical performance of a belt filter.	14-24
14.6	Design criteria for sand drying beds, using anaerobically digested sludge without chemical conditioning.	14-29
15.1	Toxic and inhibitory concentrations of selected inorganic materials in anaerobic digestion.	15-5

15.2	Toxic and inhibitory concentrations of selected organic materials in anaerobic systems.	15-6
15.3	Typical gas production rates.	15-11
15.4	Typical design parameters for low-rate and high-rate digesters.	15-12
15.5	Estimated volatile solids destruction in anaerobic digesters.	15-13
15.6	Heat-transfer coefficients for various construction materials.	15-18
15.7	Heat-transfer coefficients for various tank components.	15-18
15.8	Lime dosage required for liquid lime stabilization.	15-40
15.9	Representative heat values of some wastewater residuals.	15-43
16.1	Heavy metal concentration in Ontario sludges.	16-3
16.2	Nutrient characteristics of Ontario sludges.	16-3
16.3	Land application pollutant limits (dry-weight basis).	16-4
16.4	Maximum allowable pollutant concentrations in wastewater sludge for disposal in active landfills without a liner and leachate collection systems.	16-8
16.5	Time and temperature guidelines.	16-9

Preface

This textbook is intended for use in university courses devoted to wastewater treatment plant design. *Wastewater Treatment Plant Design* includes the most important aspects of wastewater engineering and presents this material in a useful and organized manner. The primary purpose of the book is to provide an educational tool for the instructor in assisting students to better learn the theory and practice of wastewater treatment.

The primary source of material for this textbook was *Design of Municipal Wastewater Treatment Plants* (Manual of Practice No. 8, jointly published by the Water Environment Federation and the American Society of Civil Engineers in 1998). The 1998 three-volume set was edited for use in a classroom, reducing the detail in some areas while adding sufficient detail in others. The student edition of the textbook includes a separate student workbook containing instructional objectives, practical examples, and problems on tear-out sheets for each chapter as well as a detailed glossary of terms and list of conversion factors.

The assumption is that students using this book have had at least one course in environmental engineering and at least one course in fluid mechanics or hydraulics. Knowledge of calculus would be helpful but is not necessary. The student is expected to emerge from such a course having not only knowledge of the technology of wastewater treatment, but also a deeper understanding of the design process. The hope is that, upon completion of this course, the young engineer will be encouraged to always ask the "why" question and challenge the status quo.

About the Editor

P. Aarne Vesilind is a R. L. Rooke Professor of Engineering in the Department of Civil and Environmental Engineering at Bucknell University in Lewisburg, Pennsylvania.

Following his undergraduate degree in civil engineering from Lehigh University in Bethlehem, Pennsylvania, Vesilind received his Ph.D. in environmental engineering from the University of North Carolina, Chapel Hill, in 1968. He spent a post-doctoral year with the Norwegian Institute for Water Research in Oslo, during which he developed a laboratory test for estimating the performance of dewatering centrifuges, and a year as a research engineer with Bird Machine Company, South Walpole, Massachusetts, after which he joined the faculty at Duke University in Durham, North Carolina, as an assistant professor in 1970. In January 2000, he retired from Duke, and was appointed to the R. L. Rooke Chair of Engineering in the Societal and Historical Context at Bucknell University.

In 1976–1977, he was a Fulbright Fellow at the University of Waikato in Hamilton, New Zealand; in 1991–1992, he was a National Science Foundation Fellow at Dartmouth College, Hanover, New Hampshire. He served as the chair of the Department of Civil and Environmental Engineering at Duke University for seven years and was twice elected by the School of Engineering faculty to chair the Engineering Faculty Council. He is a former trustee of the American Academy of Environmental Engineers and a past-president of the Association of Environmental Engineering Professors. He serves on many technical and professional editorial boards and has written nine books on environmental engineering, solid waste management, education, and environmental ethics. *Introduction to Environmental Engineering* (1998), Brooks/Cole Publishing, incorporates ethics to an undergraduate environmental engineering course. *Engineering, Ethics and the Environment* (1999), co-authored with Alastair Gunn, is published by Cambridge University Press. *So You Want to Be a Professor?* (2000), a handbook for graduate students, is published by Sage Publications. His latest books are *Sludge Into Biosolids* (2002), a publication of the International Water Association, co-edited with Ludovico Spinosa; *Solid Waste Engineering* (2002), Brooks/Cole Publishing, co-authored with Bill Worrell and Debra Reinhart; and *Hold Paramount: The Engineer's Responsibility to Society* (2003) co-authored with Alastair Gunn, Wadsworth Publishing.

Acknowledgments

Reviewers selected by the publisher who were asked to comment on the draft included Glen T. Daigger of CH2M Hill in Englewood, Colorado; Arthur H. Fagerstrom, P.E., of USFilter Operating Services; Albert C. Gray of the National Society of Professional Engineers in Alexandria, Virginia; Harlan G. Kelly and Dr. Alan Gibb of Dayton-Knight in West Vancouver, British Columbia; E. Joe Middlebrooks, an Environmental Engineering Consultant in Lafayette, Colorado; Jan A. Oleszkiewicz of the Civil Engineering Department at the University of Manitoba in Winnipeg; Albert B. Pincince of CDM in Cambridge, Massachusetts; Dr. Norbert W. Schmidtke, P.Eng., Environmental Processing Engineering Specialist in Plattsville, Ontario, Canada; and Rudy J. Tekippe of Montgomery Watson Harza in Pasadena, California.

In addition to providing review of the text, E. Joe Middlebrooks prepared examples for the student workbook. Jamie Troxell, a graduate student at Bucknell University in Lewisburg, Pennsylvania, prepared problems for the student workbook and solution set for instructors.

A Personal Note from the Editor

There is an old story about the man who survived the Johnstown flood and who liked nothing better than to brag about how he did it. When he died, he went to heaven and St. Peter asked him if there was anything they might do to make his stay there more comfortable. The man asked if it was possible to get some people together so he could tell them about how he survived the Johnstown flood. "No problem," replied St. Peter, "but you have to realize that Noah will be in the audience."

Editing this book is like talking to Noah about the Johnstown flood. The men and women who originally wrote it are the best of the best—people who have wide knowledge and expertise in their field and who took the time and trouble to share this with their professional colleagues. And then here is the professor who comes along and edits their stuff. Some would liken this to editing Mozart.

My excuse, reason, or justification for what I have done is that, while the original Manual of Practice is a compendium of the latest and best existing wisdom in wastewater treatment, this edited version is *educational* in character. That is, the book is primarily intended as a textbook and is structured using principles of educational pedagogy.

As I was editing the book and trying to decide what to include, what to omit, and what to add, I used the question, "Why should my students need to know that?," as my gold standard. It was not a question of what is important to the field, but what should be taught to engineering students, that governed my decisions. No doubt some of the original authors will find fault with my selection, especially those whose material I did not use (or severely edited), and I want to apologize in advance for any hurt feelings. I ask the reader to keep the purpose of the book in mind and to understand the basis for my decisions.

Several other points should be made clear. I omitted a lot of secondary references, particularly to U.S. Environmental Protection Agency publications that are based on other original research. The Manual of Practice has an extensive bibliography, and those interested in the references are referred to this source. I also revised some of the symbols used to achieve more uniformity and thus the equations in the Manual and in this text do not always correspond. I tried to consciously omit jargon and contractions from the text. Students have a hard enough time understanding the concepts without having also to memorize what BNR, RAS, DAF, etc., stand for. I retained a few acronyms, such as BOD, for obvious reasons. I was also careful about including unsubstantiated opinion or personal preferences and often omitted such material. While opinion is useful for readers of the Manual of Practice, it is not appropriate for this text. Finally, I used what I hope is a rational vocabulary. For example, "total solids" is everything that remains on evaporation. Total solids is made up of "suspended solids" and "dissolved solids". It, therefore, makes no sense, and is in fact confusing to students, to speak of "total suspended solids". I know what practitioners mean by that term (fixed plus volatile), but it is confusing to students first trying to understand the concepts. In the "Notes on Terminology" sidebars, I try to explain these distinctions and hope that students who use this book in their classes will continue to use a more rational vocabulary for the remainder of their professional careers.

The most ethically troubling problem in what I did, however, is blending my own writing with the writing of the original authors. There is no indication in the text of who wrote what, and the original authors will see their own words followed by my own and the two are not identified. Who then is the author, and am I really "editing" if I include my own stuff? Who then gets the credit for the work?

Let's do it this way. Listed below are the people who wrote the original chapters in the Manual of Practice, and they are the "authors" of this work. It is their material that forms the technical heart of the text. Credit should be also given to Glen Daigger and his team of authors and reviewers, who no doubt greatly improved the original Manual. To all of these people, I owe a debt of gratitude and hope they will agree that my editing has produced a useful educational tool for our field.

Orris E. Albertson	(12)
Peter J. Barthuly	(5)
Paul A. Bizier	(18)
John C. Calmer	(14)
Ronald W. Crites	(13)
Alvin C. Firmin	(2)
Jane B. Forste	(24)
Daniel Graber	(23)
Shin-Joh Kang	(15)
Carl M. Koch	(17)
Charles J. Myers	(6)
Michael C. Mulbarger	(3)
Tom M. Pankratz	(Glossary)
Albert C. Petrasek, Jr.	(16)
Joseph C. Reichenberger	(4)
Robert A. Ryder	(8)
Donald E. Schwinn	(1)
Jeanette Brown	(22)
Thomas W. Sigmund	(19, 20, 21)
Richard Stratton	(9)
Rudy J. Tekippe	(11)
Carlos E. Vargas	(7)
Thomas E. Weiland	(10)

Authors of the 1998 manual also include

Jamie Bengoechea
Harold Blumenfeld
Brandon Braley
John Bratby
George V. Crawford
Allan J. DeLorme
James H. Fisher
Richard Frykman
John R. Harrison
Bill Holloway
Joseph F. Lagnese
James R. Laurila
Gary M. Lee
Peter S. Machno
Jane Madden

Cal Miller
Lynne Hersho Moss
Doug Noffsinger
Robert W. Okey
Edward R. Prestemon
Bob Probst
Gus Proietti
Timothy M. Ptak
William Shosho
Ron Sieger
Karen L. Van Aacken
Ted V. Ware
Thomas E. Wilson

So thanks to all of these authors but, in the end, the buck stops with me, and any mistakes, disagreements, or problems in this text are my responsibility.

Finally, several people have been outstandingly helpful in helping with the editing of this text, particularly Norbert Schmidtke, Harlan G. Kelly, Jan A. Oleszkiewicz, and Jim Nissen. There were also numerous "anonymous" editors (listed above) who contributed to the effort, and to all of them I want to express my appreciation. And finally, Lorna Ernst and her team at the Water Environment Federation were more than helpful and accommodating of my sometimes outrageous requests. Thank you for putting up with me.

P. Aarne Vesilind
Bath, New Hampshire

2003

Fundamentals

Introduction

The need for community wastewater collection and treatment systems in the United States has evolved over a period of nearly 200 years. This evolution has occurred in a similar fashion in many other countries, initially being driven by the need to reduce human disease, then to eliminate gross water pollution effects, and, finally, to achieve levels of water quality that allow native marine organisms to return to normal growth patterns and allow full human recreational use. In this section, a brief historical overview that focuses on the way in which municipal wastewater treatment has evolved in the United States in response to the public's changing perception of the sometimes conflicting needs of nature and society is provided.

In the early 1800s, the continental United States was a sparsely populated, underdeveloped country. By the late 1800s, the population had grown from 5 million to 76 million inhabitants, and the occupied land mass had increased from 2.3×10^6 to 7.8×10^6 km² (0.9×10^6 to 3×10^6 mi²). Then, as now, governing bodies responded to the public's negative reaction to sights and odors perceived to be unpleasant by improving waste management. Pit privies and open drainage ditches were replaced by buried sewers that conveyed not only wastes but also stormwater to other locations where it would have less effect on other users. The sewered population increased from 1 million (in 1860) to approximately 25 million by 1900 (Babbitt, 1947). This dramatic increase in the sewered population reflects the public awareness of the link between human disease and waste disposal practices.

During the 1800s, wastewater disposal strategies were primarily a local responsibility but, eventually, communities found that simple wastewater collection and dilution at the nearest watercourse were inadequate. By the end of the 1800s, several important wastewater management practices had been developed, including the following:

- In 1886, the development of standards for wastewater treatment began with the establishment of the Lawrence, Massachusetts, experimental station.
- In 1887, the first formal biological waste treatment unit, an intermittent sand filter, was tried at Medford, Massachusetts.
- Rudolph Hering's recommendation of 0.11 m^3/s (4 ft^3/s) per 1000 persons for the Chicago, Illinois, drainage canal was the first attempt to define assimilative capacity of a receiving water body for wastewater.
- The first specific federal water pollution control legislation, which, through the Rivers and Harbors Appropriations of 1899 (often called the Refuse Act), prohibited dumping of solids to navigational waters without a permit from the U.S. Army Corps of Engineers, was enacted (Galli, 1974; Longest, 1983).
- In 1885, William Dibdin, an engineer for the municipal district of London, recognizing the significance of using microorganisms to treat municipal wastewater, ushered in the era of biological treatment.

By the early 1900s, approximately 60 towns and cities, totaling approximately 1 million inhabitants, had some form of municipal wastewater treatment. An objective of early treatment was removal of visible settling and floating solids. The early 1900s saw an era of rapid urbanization and industrial development during which scores of people migrated from farmlands to the cities, thereby creating additional demands to construct sewers in the growing population centers. Until the 1930s, the rate of increase of total population kept pace with the rate of increase of the sewered population. After 1930, however, the increase in total population outstripped the sewered population, resulting in a decrease in the percentage of total population served by sewers. Though sewer service lagged in the early 1900s, wastewater treatment technologies such as the following continued to evolve:

- In 1901, the first trickling filter was placed in operation in Madison, Wisconsin.
- In 1909, the first Imhoff tank was installed in the United States.
- In 1914, liquid chlorine was first applied for plant-scale disinfection.
- In 1916, the first activated-sludge plant began operation in San Marcos, Texas.

Increased levels of treatment were intended to protect the oxygen resources of the receiving stream, because anoxic conditions would result in nuisance odors, unsightly growth, and loss of desirable aquatic life such as fish. Initially, the oxygen demands of both carbon-based and nitrogen-based materials were of concern, but in the 1930s the objective of wastewater treatment became the control of oxygen demand of biodegradable carbon. Control of oxygen-demanding nitrogenous substances did not reemerge as a practical treatment objective until the 1960s when the nitrification of rivers began to exert a significant demand for oxygen.

Increased levels of treatment also increased the volume of residuals for disposal, thereby creating additional handling and disposal problems. Residual management practices trailed the development of liquid-processing alternatives. In the 1920s, separate heated digestion and the use of digester gas became standard procedure. The first vacuum filters and centrifuges for mechanical sludge dewatering were used in Milwaukee, Wisconsin, in 1921. In the early 1930s, the first large-scale process for sludge dewatering, drying, and incineration was used in Chicago, Illinois (Lue-Hing et al., 1986).

The mathematical modeling of stream assimilative capacity was derived from the work of Phelps (1944) who developed what is now known as the "Streeter–Phelps dissolved oxygen sag curve", which made it possible to predict the oxygen balances in streams for given discharges.

Water pollution abatement programs gradually evolved from local public works and public health programs and remained the responsibility of local governing bodies during the early 1900s; later, the states began to oversee these programs. Federal involvement was inconsequential until the depression years of the 1930s. At that time, federal expenditures and recovery programs financed local public works efforts that built many of the significant interceptors still found in today's older cities. As a result of federally financed programs, the number of people receiving some form of wastewater treatment increased dramatically. Also, cooperative watershed-planning agencies that emerged at this time addressed problems and needs on a regional basis rather than on a local, state, or national basis.

During World War II, U.S. water pollution control efforts were put on hold, though urban and industrial growth continued unabated. After World War II, water quality again became the focus of national attention, resulting in the passage of the Federal Water Pollution Control Act of 1948, the first comprehensive federal program aimed at controlling water pollution. In 1952, the extension of the act and other legislation that provided more appropriate funding at nearly 5-year intervals for the next 15 years increased federal involvement and stimulated the construction of municipal wastewater treatment plants. The Clean Water Restoration Act of 1966 authorized expenditures of more than $1 billion, with a matching federal grant program of more than 30%.

At that time, most municipal plants consisted only of settling processes, or what now is known as primary treatment. Secondary treatment using biological reactors was the exception rather than the rule. Improvements in treatment plant performance were attained by chemical addition to primary settling tanks, but this soon gave way to biological treatment. Design standards, however, were often less stringent than those currently accepted for secondary treatment. By the late 1950s, nearly 50% of the U.S. population had the benefit of some form of wastewater treatment.

Biological treatment of municipal wastewater is susceptible to upset from industrial wastes. A number of ordinances were enacted that controlled the discharge of industrial wastes to municipal wastewater treatment plants. Increased use of secondary treatment produced more biological solids, which are more difficult to process. In the 1950s and 1960s, dissolved air flotation was advanced for both liquid–solids separation and thickening. Organic polyelectrolytes for conditioning and enhanced liquid–solids separation also evolved from the postwar chemical industry as an alternative to, and enhancement of, inorganic chemical-conditioning strategies in sludge dewatering.

In the 1960s, heightened public awareness of, and dissatisfaction with, the condition of the U.S. water resources resulted in the development of water conservation programs. By 1970, new liquid-processing technologies began to emerge for increased removals of suspended solids, organics, phosphorus and nitrogen, and solids-processing technologies using elevated temperatures and pressures to oxidize or condition waste solids. The ready availability of phosphorus from synthetic detergents and the accelerated eutrophication of many lakes led to the reformulation of detergents and development of nutrient removal technologies.

The Federal Water Pollution Control Act Amendments of 1972 (Public Law 92-500) spurred changes already taking place. The law provided ambitious goals for U.S. waterways, broadened enforcement powers under a National Pollutant Discharge Elimination System permit program, and authorized $18 billion for construction grants (in the form of 75% participating funding grants) for the first three years of the program. Almost every component of a community's physical wastewater management system qualified for funding under the law's definition of treatment works.

The U.S. Environmental Protection Agency (U.S. EPA), to interpret and implement the desires of Congress, was created by executive order. The U.S. EPA actively participated in all aspects of water pollution control projects, including research, planning, design, construction, and operation of wastewater collection sewers and treatment facilities.

During the next 15 years, the federal government spent approximately $50 billion for construction grants. Changes to the original legislation added, eliminated, expanded, or contracted deadlines, implementing regulations and guidelines, goals, and definitions. Funding authorizations and appropriations also changed. Many states that had developed sophisticated regulatory organizations took over management and execution of the federal program. Designs were influenced by greater public participation in the decision-making process, more stringent industrial pretreatment programs and sewer use ordinances, and delays in construction of improvements because of increased regulatory involvement in funding and technical review.

During the last several decades, the number of people receiving secondary or higher levels of treatment has more than doubled. Additionally, untreated wastewater releases from sewered population centers have almost been eliminated.

Recent technical advances in wastewater treatment have included

- Greater understanding and application of processing enhancements to control activated-sludge bulking and foaming;
- Processing improvements that promote biologically enhanced nitrogen and phosphorus removal;
- Increased understanding of limitations of secondary clarification and the need for integrated design of aeration and clarification functions of activated-sludge processes;
- Large-scale application of sequencing batch reactors;
- Renewed interest in attached-growth biological treatment technologies, with greater understanding of their expanded capabilities;
- Belt filters and high-efficiency centrifuges for thickening and dewatering applications;
- Application of high-pressure dewatering presses;
- Use of both composting and chemical stabilization to achieve residual solids that are more acceptable for beneficial recycling, drying, and pelletization;
- Odor control and reduction;
- High rate, low solids retention time activated-sludge processes for carbonaceous removal;
- Stacked clarifiers on constrained sites;

- Egg-shaped digesters for solids stabilization;
- Increased use of more effective computer control systems for process control and plant maintenance;
- Disinfection using UV irradiation;
- Hybrid biological systems using both attached-growth and suspended-growth reactors;
- Tertiary filtration through sand, anthracite coal, and fabric media;
- Membrane bioreactors;
- Liquid–solids separation using membranes;
- Introduction of vortex separators for grit removal; and
- Development of constructed wetlands and other natural treatment systems to achieve high effluent quality.

In the 1990s, the U.S. Congress changed focus away from the historic grants programs of the 1970s and 1980s to programs offering low-interest rates on loans and a newly defined state and federal partnership. The expected economic pressure and competition for limited public revenue has resulted in an increased burden on designers for solutions that offer greater economy and require more extensive consideration of alternative technologies and management practices. Demands will increase for new technological solutions that can be readily accessed by designers to solve their particular problems.

With the exception of a few countries that have nutrient removal regulations for ecological reasons, secondary treatment is the norm in most developed countries of the world. In Latin America, the most common treatment technology is based on oxidation ponds because they are simple, less expensive, and easy to operate and maintain, and because large areas of inexpensive land are often available. Most developing countries are still coping with issues that focus on human health and life expectancy, rather than on receiving-water quality aesthetics or adverse effects on flora and fauna. As standards of living are elevated, and industrial development progresses, vastly increased amounts of water will be required, creating much larger quantities of wastewater and exacerbating receiving-water quality problems. A significant body of knowledge is available to help meet these challenges while avoiding some missteps of the past. Before recommending "imported technologies", however, it is important that design professionals interpret local technologies in use and evaluate human resource capabilities to ensure that the plant can be operated.

Water conservation has become more common in water-limited areas. Water conservation needs will undoubtedly result in (if not force) more beneficial recycling of treated wastewater for cooling, irrigation, agriculture, drinking water, and certain classes of industrial use. As water becomes scarce, intentional recycling of wastewater for non-potable purposes will expand. There will also be greater pressure to use reclaimed water for drinking water supplies, although this will require more research into technologies and overcoming public attitudes. Industries and communities may continue to share the wastewater treatment plant for compatible waste control. These considerations suggest that wastewater received at treatment plants is likely to increase in strength. Increased efforts to control the discharge of toxins to U.S. waterways will continue.

The job of the design engineer is to incorporate new technology and innovations to the facilities design, but at the same time produce a plant that will have a high probability of performing as expected under the variable and unpredictable loadings. There is, therefore, a balance that the engineer has to maintain to be both innovative as well as conservative in the design. Designers aware of past experiences, but amenable to change, can meet future challenges of new standards and goals as defined by the public's changing perception of need and national priorities. Water pollution abatement resulting in unbalanced atmospheric deterioration and land degradation is not a viable solution. To control the cost of treatment and address environmental impacts, the designer should explore opportunities in the service area and its collection system. In designing a municipal wastewater treatment plant, the designer walks a careful line between providing facilities that can respond to uncertainties of the future and excessive overdesign, the latter of which may result in the misuse of public monies for superfluous facilities. Good designs also provide flexibility to allow modifications and additions to meet future, perhaps more stringent, treatment requirements as environmental regulations continue to be strengthened.

Water Quality

When is water dirty? More specifically, when is water sufficiently dirty that it needs to be treated in a wastewater treatment plant? The answer of course depends on what is meant by "dirty". What may be pollution to some may in fact be to others an absolutely necessary component in the water. For example, trace nutrients are necessary for algal growth (and hence for all of aquatic life), while fish require organics as a food source to survive. These same constituents, however, may be highly detrimental if the water is to be used for industrial cooling. In this chapter, various parameters used to measure water quality are discussed first and then the question of what is clean and dirty water is considered further.

Measures of Water Quality

The standard reference of the water quality industry in the measurement of chemical, physical, and biological characteristics of water is *Standard Methods for Examination of Water and Wastewater*, published jointly by several professional organizations, including the Water Environment Federation. This comprehensive book, now in its 20th edition (APHA et al., 1998), is the result of professional consensus and wide-scale testing using round robin techniques. These test methods have the weight of federal law behind them and should always be used unless good reason can be found to do otherwise (e.g., in research).

The latest edition of *Standard Methods* includes hundreds of tests for measuring water quality. This discussion, however, is restricted to the following, each of which is covered in great detail in the manual:

- *Dissolved Oxygen*, a significant determinant of water quality in streams, lakes, and other watercourses;
- *Biochemical Oxygen Demand*, a parameter indicating the pollutional potential a discharge would have on watercourses;
- *Solids*, including suspended solids, which are unsightly in natural waters, and total solids, which include dissolved solids, some of which could be detrimental to aquatic life or to people who drink the water;
- *Nitrogen*, a useful measure of water quality in streams and lakes and a source of oxygen demand in rivers and streams. Also a nutrient in lakes and estuaries, responsible in part for accelerated eutrophication;

- *Phosphorus,* a critical nutrient in lakes and, when in excess, responsible (along with nitrogen) for accelerated lake eutrophication; and
- *Bacteriological measurements,* which are necessary to determine the potential for the presence of infectious agents such as pathogenic bacteria and viruses and are usually indirect due to the problems of sampling for a literally infinite variety of microorganisms.

Dissolved Oxygen

The availability of oxygen in various treatment processes, and in the waters receiving the plant effluent, is perhaps the single most important measure of water quality. Wastewater in primary clarifiers that are devoid of oxygen will have odors and might result in floating sludge. Depressed oxygen levels in biological reactors can greatly disrupt biological treatment. And finally, low oxygen levels in receiving waters can cause anaerobic conditions, with the commensurate odor and loss of desirable aquatic life.

Dissolved oxygen is measured with a commercially available oxygen probe and meter. A semipermeable membrane allows the oxygen in the water sample to penetrate the membrane and enter into a reaction that produces electrical current, which is then measured with a microammeter. The amperes developed as a result of the reaction with oxygen is proportional to the concentration of oxygen in the water sample and the meter, after calibration, reads directly in milligrams per liter of dissolved oxygen.

Biochemical Oxygen Demand

Determining the biochemical oxygen demand (BOD) is perhaps even more important than determination of dissolved oxygen levels. The BOD measures the rate at which oxygen is used by microorganisms decomposing organic matter. This parameter reflects both the *rate* at which organic matter is assimilated by microorganisms, and the *quantity* of organic matter available to the microorganisms. If a discharge to a watercourse has a high BOD, the microorganisms in the water will rapidly use up the available oxygen. If then the reoxygenation from the atmosphere cannot keep pace with the use of oxygen, the dissolved oxygen level drops, perhaps going to zero and creating unacceptable conditions in the stream. One of the primary objectives of wastewater treatment plants is the removal of this demand for oxygen, or BOD.

The demand for oxygen in the decomposition of pure materials can be estimated from stoichiometry, assuming that all of the organic material will decompose to carbon dioxide and water. Unfortunately, wastewaters are seldom pure materials, and it is not possible to calculate the demand for oxygen from stoichiometry. It is, in fact, necessary to conduct a test in which the microorganisms that do the converting are actually employed and the use of oxygen by these microorganisms is measured.

The rate of oxygen use (that is, the demand for oxygen or the BOD) is not a measure of some specific pollutant. Rather, it is a measure of the amount of oxygen required by aerobic bacteria and other microorganisms while stabilizing decomposable organic matter. If the microorganisms are brought into contact with a food supply (such as human waste), oxygen is used by the microorganisms during the decomposition. A low rate of use would indicate either the absence of contamination or that the available microorganisms are unable to assimilate the available organics. A third possibility is that the microorganisms are dead or dying.

The standard BOD test is run in the dark at 20 °C for five days. This is defined as 5-day BOD, or BOD_5, or the oxygen used in the first five days. The temperature is specified because the rate of oxygen consumption is temperature dependent. The reaction must occur in the dark because algae may be present and, if light is available, may actually produce oxygen in the bottle.

Although the 5-day BOD is the standard, it is also possible to have a 2-day, 10-day, or any other day BOD. One form of BOD is the *ultimate* BOD, or the oxygen demand after a very long time, when the microorganisms have oxidized as much of the organics as they can. Ultimate BOD tests are typically run for 20 days, at which point little additional oxygen depletion will occur. Approximately 68% of the ultimate BOD is exerted in the first five days, or $BOD_5 \approx 0.68\ BOD_{ult}$.

When the BOD test is conducted, it is assumed that the waste sample has within it the microorganisms that will decompose the organic material and decrease the dissolved oxygen. It is also possible to measure the BOD of any organic material (e.g., sugar) and thus estimate its influence on a stream containing plenty of microorganisms that would easily assimilate the sugar as a food source. Pure sugar, however, does not contain the necessary microorganisms for decomposition. *Seeding* is a process in which the microorganisms that are responsible for the oxygen uptake are added to the BOD bottle with the sample (sugar) in order for the oxygen uptake to occur. Seed microorganisms are typically obtained from treatment plant effluents (before disinfection).

The BOD test is a measure of oxygen use, or potential use. An effluent with a high BOD can be harmful to a stream if the oxygen consumption is great enough to eventually cause anaerobic conditions (the dissolved oxygen sag curve approaches zero dissolved oxygen in the stream). Obviously, a small trickle of wastewater flowing into a large river probably will have a negligible effect, regardless of the BOD concentration. Similarly, a large flow into a small stream can seriously affect the stream even though the BOD might be low. Accordingly, BOD loading can be expressed as kilograms of BOD per day (kg BOD/d) and is calculated in the metric system as

$$\text{kg BOD/d} = (\text{mg/L BOD}) \times (\text{flow in m}^3/\text{d}) \times 10^{-3} \qquad (1.1)$$

In the United States, engineers commonly use what are called "customary units" and often talk of "pounds of BOD", calculated by multiplying the concentration by the flowrate with a conversion factor so that

$$\text{lb BOD/d} = (\text{mg/L BOD}) \times (\text{flow in mil. gal/d}) \times (8.34\ \text{lb/mil. gal})/(\text{mg/L}) \qquad (1.2)$$

The 8.34 is a conversion factor.

The BOD_5 of most domestic wastewater is between approximately 150 and 250 mg/L, although many industrial wastes may be as high as 30,000 mg/L. The potential detrimental effect of an untreated dairy waste that might have a BOD_5 of 25,000 mg/L is quite obvious, because it represents a 100 times greater effect on the oxygen levels in a stream as would untreated domestic wastewater.

Because the BOD test depends on the use of oxygen by microorganisms in the bottle, it has notoriously bad reproducibility. Alternative tests have therefore been used when more precise measurements of potential oxygen demand are needed. The most widely used of these BOD surrogates is the *chemical oxygen demand,* or COD, test. This test uses strong oxidants to oxidize the organic matter, and the assumption is that all of the organics will be destroyed. By measuring the residual of the oxidizing chemical, the amount of oxygen required can be calculated from stoichiometry.

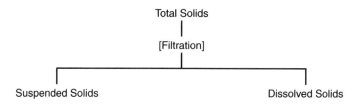

Figure 1.1 Classification of total solids.

Another test used to measure the organic strength of wastewater is the *total organic carbon*, or TOC, test that is conducted using a sophisticated total organic carbon meter. A sample of the wastewater is injected to a heated coil in which the organic material oxidizes to carbon dioxide gas, which is then detected and quantified. The amount of carbon dioxide produced is stoichiometrically related to the carbon in the sample.

While the precision of this test is excellent (it gives essentially the same answer every time the test is run with a given sample) it is no longer a measure of the original intent—a measure of the food supply for the microorganisms. For example, if a tiny particle of plastic is injected to the TOC monitor it will be oxidized to carbon dioxide, but this carbon is not available as a food source for the microorganisms. Thus, the test is precise, but not accurate. It does not really measure what is wanted. On the other hand, the BOD test is notoriously imprecise, but it accurately measures (although indirectly) the amount of organic material available to the microorganisms.

Solids

Solids can be a significant water pollutant and the separation of these solids from the water is one of the primary objectives of wastewater treatment. If high concentrations of solids are discharged to a watercourse, they can cause unsightly floating scum, or they can sink and form potentially hazardous "mud" banks in streams, rivers, or lakes. Most of the solids are also organic, so these particles will start to decompose and create a demand for oxygen. Finally, floating solids are a public health hazard because pathogenic organisms tend to "hide" on these solids and resist disinfection.

Strictly speaking, in wastewater anything other than water or gas is classified as a solid. The usual definition of solids, however, is the residue after evaporation at 103 °C (slightly higher than the boiling point of water). These solids are known as *total solids*. The test is conducted by placing a known volume of sample in a large evaporating dish and allowing the water to evaporate. The total solids are then calculated as the difference in the weights before and after evaporation, and expressed as milligrams per liter.

Total solids (Figure 1.1) can be divided into two fractions: the *dissolved solids* (DS) and the *suspended solids* (SS). If a teaspoonful of common table salt is placed in a glass of water, the salt dissolves. The appearance of water does not change, but the salt remains behind if the water is evaporated. A spoonful of sand, however, does not dissolve and remains as sand grains in the water. The salt is an example of dissolved solids while the sand would be measured as a suspended solid. Suspended solids are measured by running a specific volume of water through a fiberglass filter and weighing the residue retained by the filter. These become suspended solids, and are expressed in terms of milligrams per liter. The difference between total solids and suspended solids is dissolved solids (those solids that made it through the filter).

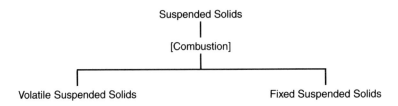

Figure 1.2 Classification of suspended solids.

Solids can be classified in another way: those that are volatilized at a high temperature and those that are not. The former are known as *volatile solids*, the latter as *fixed solids*. Although volatile solids are considered organic, at 600 °C, the temperature at which combustion takes place, some of the inorganics are decomposed and volatilized as well, but this is not considered a serious drawback. The term *inert solids* is often used to mean fixed solids.

Sometimes it is necessary to measure the volatile fraction of suspended material, because this is a quick (if gross) measure of the amount of microorganisms present. *Volatile suspended solids* (VSS) (Figure 1.2) are determined by simply combusting the fiberglass filter in a hot oven (600 °C), allowing the organic fraction to burn off, and reweighing the filter. The loss in weight is interpreted as volatile suspended solids.

The dissolved solids (Figure 1.3) can be classified in a similar fashion, although this is seldom necessary. *Fixed dissolved solids* is a measure of the inorganic dissolved chemicals in the water sample and is important when these chemicals become inhibitory to microbial growth.

Nitrogen

Nitrogen is an important element in biological reactions. In the organic and ammonia form, nitrogen demands oxygen for oxidation to nitrate, NO_3^-, and is partially responsible for depressing oxygen levels in streams. In the nitrate, or fully oxidized form, nitrogen is a necessary nutrient in lake eutrophication—accelerated metabolic activity in a lake that eventually results in the lake becoming filled with biomass and ending up as a peat bog. Accelerated eutrophication can also occur in estuaries, and even slow-moving rivers.

Nitrogen can be tied up in high-energy compounds such as amino acids and amines and in this form the nitrogen is known as organic nitrogen. One of the intermediate

Note on Terminology

The classification of solids is, as the above discussion shows, rational and precise. Unfortunately, in the past few years U.S. engineers and operators have started to confuse this classification by using the term *total suspended solids* (TSS) to mean the sum of the volatile suspended solids and the fixed suspended solids. Obviously, TSS is the same as suspended solids (SS) defined above.

In this text the term TSS, total suspended solids, is not used. Given the popularity of the term, however, especially among wastewater treatment plant operators, the student should be familiar with its meaning.

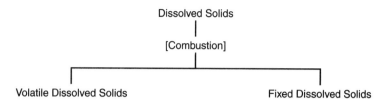

Figure 1.3 Classification of dissolved solids.

compounds formed during biological metabolism is ammonia-nitrogen. Together with organic nitrogen, ammonia is considered an indicator of recent pollution. These two forms of nitrogen are often combined in one measure, known as *Kjeldahl nitrogen*, named after the scientist who first suggested the analytical procedure.

Aerobic decomposition eventually leads to the nitrite (NO_2^-)- and finally nitrate (NO_3^-)-nitrogen forms. A high nitrate and low ammonia-nitrogen therefore suggests that pollution has occurred, but quite some time ago.

All of the above forms of nitrogen can be measured analytically by colorimetric techniques. The basic idea of colorimetry is that the ion in question combines with some compound resulting in a color. The compound is in excess, so that the intensity of the color is proportional to the original concentration of the ion being measured. For example, ammonia can be measured by adding a compound called Nessler reagent to the unknown sample. This reagent is a solution of potassium mercuric iodide, K_2HgI_4, and reacts with ammonium ions to form a yellow-brown colloid. Since Nessler reagent is in excess, the amount of colloid formed is proportional to the concentration of ammonia ions in the sample. The color is measured photometrically.

Phosphorus

Phosphorus is often the limiting nutrient in lake eutrophication. Because only a small quantity of phosphorus is available in a lake ecosystem, the metabolic activity is limited. Were it not for the limited quantity of phosphorus, the ecosystem metabolic activity could accelerate and eventually fill up with biomass because all other chemicals (and energy) are in plentiful supply. In order for the system to remain at homeostasis (steady state), some key component, in this case phosphorus, must limit the rate of metabolic activity by acting as a brake in the process.

Bacteriological Measurements

Bacteriological measurements are important in the protection of public health. It is one thing to declare that water must not be contaminated by pathogens (disease-causing organisms), and quite another to discover the existence of these organisms.

Seeking the presence of pathogens presents several problems. First, there are many kinds of pathogens. Each pathogen has a specific detection procedure and must be screened individually. Second, the concentration of these organisms can be so small as to make their detection virtually impossible. Looking for pathogens in most surface waters is a perfect example of looking for the proverbial needle in a haystack; yet only one or two organisms in the water might be sufficient to cause an infection if this water is consumed. In the United States, the pathogens of importance today include

Salmonella, Shigella, the hepatitis virus, *Entamoeba histolytica, Giardia lamblia, Cryptosporidium,* and the deadly *Escherichia coli H57* strain.

Salmonellosis is caused by various species of *Salmonella,* and the symptoms of salmonellosis include acute gastroenteritis, septicemia (blood poisoning), and fever. Gastroenteritis usually consists of severe stomach cramps, diarrhea, and fever, and although it makes for a horrible few days, is seldom fatal. Typhoid fever is caused by *S. typhi,* and this is a much more serious disease, lasting for weeks, and can be fatal if not treated properly. About 3% of victims become carriers of *S. typhi,* and although they exhibit no further symptoms, they pass on the bacteria to others primarily through the contamination of water.

Shigellosis, also called bacillary dysentery, is another gastrointestinal disease and has symptoms similar to salmonellosis. Infectious hepatitis, caused by the hepatitis virus, has been known to be transmitted through poorly treated water supplies. Symptoms include headache, back pains, fever, and eventually jaundiced skin color. While rarely fatal, it can cause severe debilitation. The hepatitis virus can pass through dirty sand filters in water plants and can survive for a long time outside the human body.

Amoebic dysentery, or amebiasis, is also a gastrointestinal disease, resulting in severe cramps and diarrhea. Although the normal habitat of *Entamoeba histolytica* that causes the disease is in the large intestine, it can produce cysts which pass to other persons through contaminated water and cause gastrointestinal infections. The cysts are resistant to disinfection and can survive for many days outside the intestine.

Originally known as "beaver disease" in the northeastern United States, giardiasis is caused by *Giardia lamblia,* a flagellated protozoan that usually resides in the small intestine. Out of the intestine, its cysts can inflict severe gastrointestinal problems, often lasting two to three months. Giardiasis was known as "beaver disease" because beavers can act as hosts, greatly magnifying the concentration of cysts in fresh water. Giardia cysts are not destroyed by usual chlorination levels but are effectively removed by sand filtration.

The latest public health problem has been the incidence of cryptosporidiosis, caused by *Cryptosporidium,* an enteric protozoan. The disease is debilitating, with diarrhea, vomiting, and abdominal pain, and lasts several weeks. There appears to be no treatment for cryptosporidiosis, and it can be fatal. In Milwaukee, in 1993, an outbreak of cryptosporidiosis affected more than 30,000 people and resulted in 47 deaths. The usual source seems to be agricultural runoff contaminating water supplies, but the cysts are resistant to the common methods of drinking water disinfection. Filtration provides the best barrier against bacterial water supply contamination.

There are many pathogenic organisms that can be carried by water and it is impossible to measure them all. Instead, indicator organisms are used to define bacteriological water quality. The indicator most often used is a group of microbes called *coliforms.* These microorganisms have five important attributes: they are

- Normal inhabitants of the digestive tracts of warm-blooded animals;
- Plentiful, hence not difficult to find;
- Easily detected with a simple test;
- Generally harmless except in unusual circumstances; and
- Hardy, surviving longer than most known pathogens.

Because of these five attributes, coliforms have become universal indicator organisms. But the presence of coliforms does not prove the presence of pathogens. If a large number of coliforms are present, there is a good chance of recent pollution by wastes from warm-blooded animals, and therefore the water may contain pathogenic organisms, but this is not proof of the presence of such pathogens.

The concentration of coliforms in water is measured by filtering a known quantity of water through a sterile filter, thus capturing any coliforms, and then placing the filter in a petri dish containing a sterile agar that soaks into the filter and promotes the growth of coliforms while inhibiting other organisms. After 24 or 48 hours of incubation, the number of shiny dark blue–green dots, indicating coliform colonies, is counted. If it is known how many milliliters of water were poured through the filter, the concentration of coliforms can be expressed as coliforms/100 mL.

Modern drinking water treatment practice is so effective in removing microorganisms that the incidence of coliform organisms is becoming increasingly rare. Instead of concentrations of coliforms, today's U.S. EPA drinking water standards are expressed in terms of the number of water samples that test positive for coliform organisms. To increase the sensitivity of bacterial measurements, the U.S. EPA is moving toward the use of *enterococci* as an indicator of contamination. Because of the huge numbers of coliforms in wastewater, there seems to be no logical reason to abandon this indicator for measuring the effectiveness of disinfection in wastewaters, but the question is still open and changes in testing methods are possible.

Water Quality Standards

Measurements of water quality are more useful in evaluating the quality of water with reference to some standards. There are three primary types of water quality standards—drinking water standards, effluent standards, and stream (or surface water quality) standards.

The U.S. EPA oversees, and states operate, programs designed to reduce the flow of pollutants into natural watercourses. All discharges are required to obtain a National Pollutant Discharge Elimination System (NPDES) permit. This permitting system has had a significant beneficial effect on the quality of surface waters and has been copied by most other developed nations. Typical effluent standards for a domestic wastewater treatment plant may range from 5 to 20 mg/L BOD_5, for example. The intent is to tighten these limits as required to enhance water quality.

Effluent standards are tied to the surface water quality because the objective is to protect these waterways. All surface waters in the United States are now classified according to a system of standards based on their greatest beneficial use. The highest classification is usually reserved for pristine waters, with the best use as a source of drinking water. The next highest includes waters that have had wastes discharged to them but that nevertheless exhibit high levels of quality. The categories continue in order of decreasing quality, while the lowest water quality is useful only for irrigation and transport.

The objective is to attempt to establish the highest possible classification for all surface waters and then to use the NPDES permits to convict polluters and enhance the water quality and increase the classification of the watercourse. Once at a higher classification, no discharge that would degrade the water to a lower quality level would be allowed. The objective is to eliminate pollution in all surface watercourses.

Stream standards for states are continually changing and evolving. An excerpt of surface water standards for North Carolina is shown in Table 1.1.

Table 1.1 Excerpts from the Surface Water Standard for the State of North Carolina.

Class	Best uses
Class C	Freshwater protected for secondary recreation, fishing, aquatic life (including propagation and survival), and wildlife. All freshwaters shall be classified to protect these uses at a minimum.
Class B	Freshwaters protected for primary recreation, which includes swimming on a frequent or organized basis, all of Class C uses.
Class WS-I	Waters protected as water supplied that are essential in natural and undeveloped watersheds.
Class WS-II	Waters protected as water supplies that are generally in predominantly undeveloped watersheds.
Class WS-III	Waters protected as water supplies that are generally in low to moderately developed watersheds.
Class WS-IV	Waters protected as water supplies that are generally in moderately to highly developed watersheds.
Class SC	Tidal waters.
Class T	Freshwaters protected for natural propagation and survival of stocked trout.
Class S	Waters that have characteristics of swamps.

Each class has some minimum water quality in terms of measurable characteristics. For example, Class C (the water classification with the lowest beneficial use) must meet these standards

- Dissolved oxygen >5 mg/L;
- Coliforms <200/100 mL;
- Normal pH;
- Temperature <2.8 °C (37 °F) above normal, and always <29 °C (84 °F);
- Turbidity <50 NTU; and
- Toxic substances
    —arsenic <50 µg/L
    —beryllium <6.5 µg/L
    —cadmium <2.0 µg/L
    —chromium <50 µg/L.

Other limits are for cyanide, fluoride, chlorine, lead, mercury, nickel, pesticides, selenium, toluene, copper, and many more.

Options for Municipal Wastewater Treatment

Although water can be polluted by many materials, the most common contaminants found in domestic wastewater that can cause damage to natural watercourses or create human health problems are

- Organic materials, as measured by the demand for oxygen (BOD);
- Nitrogen (N);
- Phosphorus (P);
- Suspended solids (SS);
- Pathogenic organisms (as estimated by coliforms); and
- Trace or persistent organics (such as chlorinated pesticides).

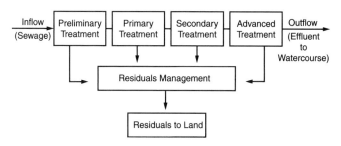

Figure 1.4 Arrangement of a typical large wastewater treatment facility.

Plants used for the treatment of municipal wastewater must be designed to produce effluents that will not adversely affect the watercourses to which the effluents are discharged.

Wastewater treatment can take many forms and the selection of the proper *process train* (series of *unit operations*) for a given location is one of the most important design decisions for the environmental engineer. In some locations, the most effective treatment system might consist of individual on-site treatment units (e.g., septic tanks and tile fields), whereas in others it might be lagoons or large, complex treatment plants. The ability of the community to pay for, operate, and maintain sophisticated treatment systems is also an important consideration. Good engineering involves providing the most trustworthy and effective solution to the problem of protecting the watercourses and human health at an affordable cost to the community.

Large wastewater treatment plants can vary considerably in terms of their design, but often take a general form as shown in Figure 1.4.

The typical wastewater treatment plant is divided into five main sections

- Preliminary treatment—removal of large solids and grit to prevent damage to the remainder of the unit operations.
- Primary treatment—removal of solids by settling. Primary treatment systems are usually physical processes, as opposed to being biological or chemical.
- Secondary treatment—removal of the demand for oxygen. These processes are commonly biological in nature.
- Advanced treatment—a name applied to any number of polishing or cleanup processes, one of which is the removal of nutrients such as phosphorus. These processes can be physical (e.g., filters), biological (e.g., oxidation ponds), or chemical (e.g., precipitation of phosphorus).
- Solids treatment and disposal—the collection, stabilization, and subsequent disposal of the solids removed by other processes.

As shown in Figure 1.5, a typical large wastewater treatment plant consists of *unit operations*, or individual treatment steps designed to accomplish a single task. For example, the function of the screen at the start of the process is to remove large materials that could harm the rest of the unit operations. It does not remove any of the gritty material, such as sand, which is the function of the grit chamber that often follows the screen in the treatment train. In some cases, a single unit operation will be designed to perform one task but will also help in the removal of other pollutants. For

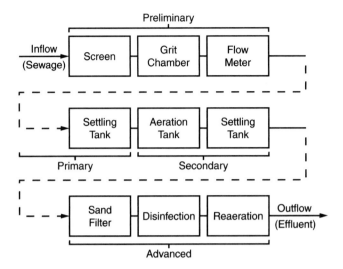

Figure 1.5 Unit operations in a typical large wastewater treatment plant.

example, the primary clarifier or settling tank is designed to remove suspended solids, but in so doing it also removes about 30% of the oxygen demand (BOD_5).

Not all wastewater treatment plants have these stages. A small plant might not have primary clarification, and the wastewater goes directly to a biological treatment step such as treatment in an oxidation ditch. Older plants have only preliminary and primary treatment, with no biological treatment at all. The simplest kind of plant is an oxidation pond that, if not overloaded, can perform all the functions of a full treatment plant. Although this text emphasizes plants that have the unit operations shown in Figure 1.5, it should be clear that this is not the only, or the best alternative, for all applications.

The overall objective of the treatment plant is to produce (1) an effluent that meets the regulatory requirements and (2) residuals that can be safely and efficiently disposed. The quality of the effluent is determined by the effluent requirements as set out by state and federal discharge permits. The options for residuals disposal is also dictated by local and federal regulations. While the plant is still on the drawing board or in the concept stage, the operation and performance of a wastewater treatment plant must be simulated. This can be facilitated by using the concept of materials balance.

Materials Balance

Materials balance is a premier tool for understanding and designing a plant. In conceptual design, a materials balance should be prepared to yield preliminary guidance concerning design quantities and significant differences between processing alternatives. In detailed design, the materials balance should be the first of the reference project documents created to ensure a commonality of understanding, consistent use of major design criteria and quantities, and a standard frame of reference and logic for the project team. The materials balance also provides the basis for the control logic of the subsequently prepared process and instrumentation diagram.

Consider a black box into which some material is flowing. All flows into the box are called *influents* and are represented by the letter X. If the flow is described as mass per unit

Figure 1.6 A black box for calculating materials balances.

time, X_{in} is a mass per unit time flowing into the box. Similarly, X_{out} is the outflow, or *effluent*, if no processes are occurring inside the box that will either make more of the material or destroy any, and if the flow is assumed not to vary with time (to be at *steady state*). If there are several flows into and out of the black box, the balance can be written as

$$[X_1 + X_2 + X_3 \cdots + X_m]_{in} = [X_1 + X_2 + X_3 \cdots + X_n]_{out} \quad (1.3)$$

where X = mass flowrate. The subscripts in the above equation refer to the locations of the flows in or out of the black box.

If one stream enters the black box, a flow (mass of X_0 per unit time) and the box splits the flow into two exit streams, 1 and 2, the flow from each is X_1 and X_2. A black box with one mass flow going in and two mass flows exiting is shown in Figure 1.6.

The two assumptions used in such materials balance equations are that the flows are in steady state (they do not change with time) and that no material is being destroyed (consumed) or created (produced), provided the density does not change. If these two assumptions are removed, and the amount of X can change with time (unsteady state) and this material can either be consumed or created, then the mass balance is

$$[X]_{accumulated} = [X]_{in} - [X]_{out} + [X]_{produced} - [X]_{consumed} \quad (1.4)$$

If the quantities do not change with time, the flows at one moment are exactly like the flows later, nothing can be accumulating in the black box, either positively (material builds up in the box) or negatively (material is flushed out), and the steady-state assumption holds and $[X]_{accumulated} = 0$. If nothing is being produced or consumed, $[X]_{produced} = [X]_{consumed} = 0$, and $[X]_{in} = [X]_{out}$.

In addition to mass balance, a volume balance must be maintained at steady state, or

$$[Q]_{in} = [Q]_{out}$$

where Q = volume flowrate.

The above equations hold if only one material is considered in the materials balance. Materials balances become considerably more complicated (and useful) when several materials flow through the system. Mass and volume balances can be developed with multiple materials flowing in a single system. In some cases, the process is one of mixing, where several inflow streams are combined to produce a single outflow stream, while in other cases a single inflow stream is split into several outflow streams according to some material characteristics. Because the mass balance and volume balance equations are actually the same equation, it is not possible to develop more than one materials balance equation for a black box, unless there is more than one material involved in the flow.

Chapter 1 Fundamentals

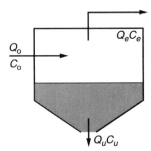

Figure 1.7 Sludge thickening tank.

This concept can be illustrated using a materials separator such as a sludge thickening tank, shown in Figure 1.7. The flow going into the tank is Q_0 at a solids concentration of C_0. The underflow concentration is C_u and the flowrate is Q_u, while the effluent flow and solids concentrations are Q_e and C_e, respectively. Two materials balances can be calculated: the volume balance and the mass balance. The volume balance is

$$Q_0 = Q_u + Q_e \tag{1.5}$$

That is, the flow (e.g., gallons per day) going into the thickener has to equal the flow out of the thickener.

The mass balance is calculated by recognizing that concentration times flowrate yields mass per unit time, or mass flowrate. For example, if the concentration is kg/m^3 and the flowrate is m^3/d, $C \times Q$ = kg/d. The mass balance is then

$$Q_0 C_0 = Q_u C_u + Q_e C_e \tag{1.6}$$

Some unit operations in wastewater treatment separate one material from another. The effectiveness of this separation is known as *recovery*, calculated as

$$R = \frac{X_1}{X_0} \times 100 \tag{1.7}$$

where R = recovery of X (mass/time) from exit stream 1 with a feed stream of X_0 (mass/time), in percent.

Because concentration can be converted to mass flow by multiplying by the flowrate, this equation can also be written as

$$R = \frac{C_1 Q_1}{C_0 Q_0} \times 100 \tag{1.8}$$

where
- Q_0 = liquid flowrate of material entering the separator;
- C_0 = concentration of material entering the separator;
- Q_1 = liquid flowrate of material recovered or exiting the separator in the desired exit stream (exit stream 1); and
- C_1 = concentration of material recovered.

The units are not important as long as they are consistent.

The above equations apply to situations where the system is in steady state and there is no production or destruction of the material of interest. Consider next a system where the material is being destroyed or produced in a reactor, but in which the steady-state assumption is maintained. That is, the system does not change with time so that if the flows are sampled at any given moment, the results will always be the same.

An anaerobic digester destroys solids (converting them to gas and liquid), so that if a mass balance were to be used to describe the digester, the equation would be (in steady state)

$$Q_0 C_0 = Q_s C_s + Q_d C_d + Z \qquad (1.9)$$

where
Q_0 = inflow of raw sludge (m³/d);
C_0 = solids concentration of raw sludge (kg/m³);
Q_s = outflow of supernatant (m³/d);
C_s = solids concentration of supernatant (kg/m³);
Q_d = flow of digester sludge out of the digester (m³/d);
C_d = solids concentration of digested sludge (kg/m³); and
Z = destruction (digestion) of solids (kg/d).

Engineering Economics

Wastewater treatment plants are designed to perform well using limited resources. The owners of the plants, usually the municipality or the waste treatment authority, want to minimize cost at a given level of performance. Because wastewater treatment facilities are most often long-term investments, the time value of money is important and engineering economics plays a significant role in deciding what kind of facility will be constructed. Funding of wastewater treatment plants is similar to funding of other utilities such as solid waste landfills and water treatment plants. Budgets consist of two general components, revenues and costs. Depending on the specifics of the planned facility, such as ownership and contractual arrangements, the complexity of the financial system can vary. Some systems are complex and require financial experts while others may only require a simple accounting system.

Wastewater treatment facilities almost always receive revenue from the sale of the service. When the home or business pays its monthly wastewater treatment bill (the so-called "sewer charge"), these funds provide the revenue to cover the costs of operating the facility. In a publicly owned system, the revenue should equal the cost. In a privately owned system, the revenue should exceed the cost, resulting in a profit.

A central idea in engineering economics is the time value of money: A dollar today does not have the same value as will a dollar a year from now. Ignoring inflation, a dollar invested today in an interest-bearing account will be worth a dollar plus the interest earned a year from now. Thus, a dollar today and a dollar a year from now cannot be summed to get two dollars. The values of these dollars are as different as apples and oranges, and they cannot be added.

The initial cost of the facility is very important to the municipality or agency. Such a cost is known as a *capital cost* and is a one-time investment. Capital costs are paid from the proceeds of bank loans, general obligation bonds, and revenue bonds. It is common for a municipality or other wastewater treatment authority to arrange for a bank loan to buy a major piece of equipment, a system upgrade, or even an entire treatment plant. The term of the loan should be shorter than the life of the equipment or facility. The interest rate on the loan is based on the risk the lender perceives and includes such variables as the financial health of the municipality and its wealth. A poor community that is already fully extended in its borrowing will have difficulty selling municipal bonds for further expenses and will have to pay high interest rates to obtain the necessary funds.

Municipal governments can use *general obligation bonds* to finance capital projects. These bonds are based on the full faith and credit of the government. In addition, the interest paid to the bond holder is usually tax exempt and thus the interest rate on these bonds is low. One problem with these types of bonds is that they may require a vote of the citizens before they can be issued.

Municipal governments can also use *revenue bonds* to finance capital projects. A revenue bond is guaranteed by the project. For example, a wastewater treatment facility may be funded by revenue bonds and the revenue from sewer-use fees would be used to pay off the bond. Unlike a general obligation bond, these bonds have more risk because only the project revenue is pledged—not the full faith and credit of the municipal government. Thus, these bonds have a higher interest rate and result in a higher cost to the municipality. The interest rate is generally still tax free and thus the rates are lower than typical bank loans.

One variation to use of revenue bonds is the ability of private companies to access revenue bond financing through the government. This allows a private company to own the capital project while at the same time use tax-exempt financing and receive a low interest rate. In addition, the private company can depreciate the assets and reduce its tax bill. Many waste-to-energy facilities have been built using these bonds. In addition, the companies also receive accelerated depreciation and investment tax credits for the project. In 1984, the federal government limited the amount of private projects that could be financed using tax-exempt bonds and now wastewater treatment plants must compete with other beneficial projects such as low-cost housing for this limited amount of funding.

Another complication is that public (and private) facilities require not only the capital cost for their construction, but also a yearly cost for their operation and maintenance (O&M). These *O&M costs* include such items as salaries, replacement parts, service, fuel, and the many other costs incurred by the owners of the facilities to keep them performing as designed so as to protect the initial capital investment.

Because of the time value of money, estimating the real cost of municipal facilities, such as wastewater collection and treatment operations, can be a complex problem. Economists and engineers use two techniques to "normalize" the dollars so that a true estimate of the cost of multiyear investment can be calculated. The first technique is to compare the costs of alternatives on the basis of *annual cost*, while the second is to compare the cost of alternatives on the basis of their calculated *present worth*. Both give the same answer but use a different calculation, and both include the capital cost plus the O&M cost.

Calculating Annual Cost
The capital costs of competing facilities can be estimated by calculating the cost that the municipality or agency would incur if it were to pay interest on a loan of that amount.

Table 1.2 6.125% compound interest factors.

Year	Capital recovery factor	Present worth factor	Sinking fund factor
1	1.06125	0.942	1.00000
2	0.54639	1.830	0.48514
3	0.37497	2.666	0.31372
4	0.28941	3.455	0.22816
5	0.23820	4.198	0.17695
6	0.20416	4.898	0.14219
7	0.17993	5.557	0.11868
8	0.16183	6.179	0.10058
9	0.14782	6.764	0.08657
10	0.13667	7.316	0.07542
11	0.12760	7.836	0.06635
12	0.12009	8.326	0.05884
13	0.11378	8.788	0.05253
14	0.10841	9.223	0.04716
15	0.10380	9.633	0.04255

Calculating the annual cost of a capital investment is exactly like calculating the annual cost of a mortgage on a house. The owner (municipality or agency) borrows the money and then pays it back in a number of equal installments. If the owner borrows $X and intends to pay back the loan in n installments at an interest rate of i, each installment can be calculated as

$$\$Y = \left[\frac{i(1+i)^n}{(1+i)^{(n-1)}} \right] \$X \qquad (1.10)$$

where
$\$Y$ = the installment cost ($);
i = annual interest rate, as a fraction;
n = number of installments; and
$\$X$ = the amount borrowed.

The expression

$$\left[\frac{i(1+i)^n}{(1+i)^{(n-1)}} \right] \qquad (1.11)$$

is known as the *capital recovery factor* or CRF. The capital recovery factor does not have to be calculated because it can be found in interest tables or is programmed into hand-held calculators. Table 1.2 is a table showing capital recovery factors if the interest rate is 6.125%.

Looking at Table 1.2, if the number of annual installments $n = 10$ and if the interest rate is 6.125%, the capital recovery factor is 0.13667. If $10 million is borrowed at the above terms, the annual payment would be $1,366,700 for each of the next ten years.

Calculating Present Worth

An alternative method of estimating the actual cost of a capital investment is to figure how much money one would have to invest right now, at a specified interest rate to have a given amount available every year for a fixed number of years. The present worth is used for this purpose and is calculated as the amount of money, Y, that must be invested for n years at interest rate i to have X available every year, or

$$\$Y = \left[\frac{1}{(1+i)^n}\right]\$X \quad (1.12)$$

where
  Y = the amount that has to be invested ($);
  i = annual interest rate, as a fraction;
  n = number of years; and
  X = amount available every year ($).

The term

$$\left[\frac{1}{(1+i)^n}\right] \quad (1.13)$$

is called the *present worth factor*, or PWF.

From Table 1.2, the present worth factor for 10 years is 7.316. If again $10 million is required, the funds needed every year would be

$$\frac{\$10,000,000}{7.316} = \$1,366,70$$

as before.

Calculating Sinking Funds

In some situations the municipality or agency must save money by investing it so that at a later date it will have a reserve available. In environmental engineering, this most often occurs when the landfill owner must invest money during the active life of the landfill so that when the landfill is full there are sufficient funds available to place the final cover on the landfill. Such funds are known as *sinking funds*. Sinking funds may also be established for expanding treatment plants without having to issue new bonds. To determine sinking funds, it is necessary to calculate the funds, Y, necessary to be invested in an account at i interest rate so that at the end of n years the fund has X dollars accumulated. The calculation is

$$\$Y = \left[\frac{1}{(1+i)^n - 1}\right]\$X \quad (1.14)$$

where
  Y = the amount that has to be invested ($);
  i = annual interest rate, as a fraction;
  n = number of years; and
  X = accumulated amount.

and the term

$$\left[\frac{1}{(1+i)^n - 1}\right] \qquad (1.15)$$

is known as the *sinking fund factor* (SFF).

Calculating Capital Plus O&M Costs

Capital investments require not only the payment of the loan in regular installments, but also the upkeep and repair of the facility. The total cost to the community is the sum of the annual payback of the capital cost plus the O&M cost, or [Total annual cost] = [O&M cost] + [Annualized capital cost].

Value Engineering

When design engineers are working on complex, intricate design projects such as that of a wastewater treatment plant, they might become so mired in the details that they forget to ask the most important and overarching questions about the design. *Value engineering* is the process of using a team of interdisciplinary individuals to challenge and provoke the design engineers and get them to justify (or change) their design, all to the benefit of the owners—often the public. Value engineering can be done within a firm, or the reviewers can be from outside of the firm. The job of the value engineering team is to ask hard questions, such as

- What is being designed and what is it supposed to do?
- How much is it expected to cost?
- What is it worth?
- How else might this be done?
- How much would alternatives cost?
- Are there good reasons to consider the alternatives?

The most important question has to be with regard to the function of the facility. A description of what the facility is supposed to accomplish is mandatory. Then, the question of cost is considered. The result is that the function is defined and the treatment process is recommended, but this process has to be at the least cost. Clearly difficult decisions have to be made, but every time a process is specified or a piece of equipment is selected, there has to be good reasons for doing so.

When the value engineering team tackles a project, it does so in phases. First the team seeks *information*—everything about the project, including public opinion and history of its inception. The second phase is called *speculation*, where the questions of basic functions are discussed and difficult questions asked. For example, the objective might be to reduce the polluted water in a harbor. How might this be done? Certainly a treatment plant is an option, but so are (1) a long outfall to the deep ocean, (2) a long interceptor sewer to some location outside the harbor, (3) relocation of the entire population of the city, and so forth. The objective in this phase is to not be practical, but to be innovative.

The third phase in the value engineering process is *evaluation* in which numerical rankings are given to the alternatives. In this phase, the deep outfalls and long sewers would be eliminated, leaving the only feasible alternative of a wastewater treatment plant. At this point the team enters the *recommendation* phase in which it develops some sense of what is and is not feasible, given alternatives. And finally, the recommendations are presented to the design staff in the *resolution* phase, where the ideas of the value engineering team are incorporated to the design.

The value engineering process can take place at any point in the design process. The further into the design, the more disruptive the results of the recommendations might be, and thus there is great advantage in initiating the process early in the design life.

Conclusions

Historically, the objective of wastewater treatment has been to protect human health and to preserve water quality for the enjoyment and pleasure of people. While there is little to argue against such an objective, it remains incomplete. Increasingly, engineers are asked to protect the "safety" of the environment: the streams, rivers, lakes, and even oceans. This is no longer a question of public health, but the well-being of the entire ecosystem and its inhabitants. We have begun to recognize that nonhuman nature—the global ecosystem—is absolutely necessary for the survival of the human species. The destruction of ecosystems will eventually lead to self-destruction. Thus our "environmental ethic" is most clearly based on selfish motives—wanting to preserve a quality environment for ourselves and our progeny.

But do we have room for a different kind of environmental ethic; one that assigns value to nonhuman nature for its own sake, and not just what good it is to humans? Should we offer nonhuman species and organisms protection from harm (such as from effluent discharges), not because we enjoy and need these species and organisms (as food, as enjoyment, or even as necessary members of an ecosystem that supports our lives), but as a part of the environment that has its own purpose? One of the rewards of a successful design of a wastewater treatment plant is that not only is it beneficial to the people who own and use the treatment plant, but that the nonhuman environment is protected as well.

References

American Public Health Association; American Water Works Association; Water Environment Federation (1998) *Standard Methods for the Examination of Water and Wastewater*, 20th ed.; Washington, D.C.

Babbitt, H. E. (1947) *Sewerage and Sewage Treatment*, 6th ed.; Wiley & Sons: New York.

Galli, G. (1974) 100 Years of Construction News. *Eng. News-Rec.*, **192**, 433.

Longest, H. L. (1983) The Past is Prologue: Looking Back, Looking Ahead at the U.S. EPA's Construction Giants Program. Paper presented at the 9th U.S. Japanese Conference on Sewage Treatment Technology, Tokyo.

Lue-Hing, C.; Zenz, D. R.; Kuchenrither, R. (1986) Treatment Technologies and Effluent Quality for the Future, 2000+. *Proceedings of Public Waste Management: Ocean Choice*, Massachusetts Institute of Technology, Cambridge.

Phelps, E. B. (1944) *Stream Sanitation*; Wiley & Sons: New York.

Symbols Used in this Chapter

- C = concentration of material (mass/volume)
- i = annual interest rate, as a fraction
- n = number of installments
- X = mass flow or material (mass/time)
- Q = volume flow (volume/time)
- R = recovery
- $\$X$ = the amount borrowed ($)
- $\$Y$ = the installment cost ($)
- Z = rate of material destruction (mass/time)

2

The Design Process

Introduction

Much of heavy industry during the turn of the 20th century was concentrated in the Ohio River Valley and the water quality of this great river was dismal and dangerous. In 1905, Pennsylvania passed a law requiring all communities to treat their wastewater before discharge. Pittsburgh was required to submit a plan for separating its combined sewers and for treating the wastewater prior to discharge. The city decided to fight this requirement and hired two prominent engineers of that time, Allen Hazen and George Whipple, to plead their case. After a thorough investigation, the engineers estimated that it would be too expensive for Pittsburgh to construct the new sewers and the wastewater treatment plant, and that it was "expedient" for the city to discharge its waste to the river and to allow the river to treat it. They calculated that it was less expensive for towns downstream from Pittsburgh that relied on the Ohio River for water supply to treat their drinking water than it was for Pittsburgh to treat its wastewater. Engineering opinion overwhelmingly supported Hazen and Whipple and the State Board of Health finally backed down. The Ohio River remained an open sewer well through the middle of the century.

Much has happened in the intervening 100 years. It is now understood that because the quality of the water in a stream is important, it is indeed the responsibility of each community to treat its wastewater before discharge to the watercourse. The duty of the engineer is to develop conceptually a plan for the treatment of a community's wastewater, and to do this at a low cost while protecting the environment. This is a high calling, and quite a responsibility.

In this chapter we discuss the preliminary considerations that an engineer must address before the detailed design begins. Often it is these considerations (or the lack of consideration) that doom a plant to failure and thus these cannot be overemphasized.

Project Participants

Decision making for any municipal wastewater treatment plant project involves many participants: the public, regulator, legal counsel, owner, engineer, financier or investment banker, operator, and contractor. Each role varies by project and is particularly affected by the procurement procedures used for construction of the proposed facilities.

All projects begin with the identification of a need by the regulator, public, legal counsel, or owner. Projects then proceed through five phases: facilities planning, design, construction, start-up, and operation. Roles of each participant vary with the project phases. The planning process should include community participation, not only in plant siting and aesthetics but in final effluent and residual disposal planning. Federal and state requirements will largely dictate effluent quality targets, but good-neighbor considerations will be critical in finding an effluent discharge site and configuration that wins public support, or at least acceptance. Public involvement is limited to the planning phase, although projects that fail to meet the public's needs, particularly with regard to good-neighbor status, will receive considerable public attention following implementation. Experience has shown that it is nearly always easier to upgrade or expand an existing plant than to site a new one. Often the public remains silent until some fieldwork starts and the realization sets in that a new plant is to be constructed.

Regulators (local, state, and federal) are involved throughout the project, approving planning and design documents, performing periodic reviews during construction, and reviewing grant and revolving loan fund applications, payment requests, and project closeout documentation. Regulators are also involved in outfall siting, effluent dispersion, and residual disposal planning issues. Reviews by regulators continue after project implementation with receipt and review of discharge compliance reports, operator training, and periodic inspection of the operating facilities.

Participation from financial or investment bankers is required at the time the finances are arranged, such as the sale of bonds. Many municipalities use short-term construction borrowing, with final bonds acquired at the end of construction. Larger projects may benefit from financial and bonding advice during the planning period or early in the design phase. Recently, private investments have had a greater role in providing municipal services.

The engineer serves as the owner's agent during facilities planning, design, construction, and start-up. However, this role is changing in projects where the facility is designed and built by a private firm and then is turned over to the owner after the plant is operational. In some cases the private firm stays on to operate the plant under contract.

The contractor is selected through a public bidding process. The contractor's role varies depending on the procurement approach, as discussed later in this chapter. However, the contractor is always responsible for construction of the facilities in accordance with contract specifications. In addition, the contractor may have design responsibilities, operational responsibilities, or both.

Often, operator involvement has unfortunately begun at the start-up phase of the wastewater treatment plant. Operator involvement in the planning and design process results in the operators feeling that they are running "their" plant rather than one that was just turned over to them by a designer. Project reviews, starting with alternative development and evaluations during facilities planning through various design submittals, are most constructive if operational input is included.

The owner oversees all project activities. With the exception of the public, which assists in identification of the owner's needs, all other project participants are responding to the needs of the owner. Representatives of the owner, other than plant operators, should be contacted at various points during planning, design, and construction to establish compliance with local requirements. On larger projects, the owner has staff engineers designated as project managers. Their job is to direct the in-house engineering and to interface with the consulting firm.

Many projects, especially larger ones with significant public funds involved, will at this juncture use *value engineering* techniques to evaluate the preliminary design (see Chapter 1). The recommendations of the value engineering team would be shared with all of the project participants, including representatives of the public. The design engineer is challenged to justify that the design to this point is the best (perhaps not the least cost) solution to the problem.

Project Sequence and Design Standards

Design practices for municipal wastewater treatment plants reflect rulemaking and guidance documents developed by the U.S. Environmental Protection Agency (U.S. EPA) during the Construction Grants Program of the 1970s and 1980s. Design consists of two phases: facilities planning (feasibility studies and basis-of-design reports, sometimes called a pre-design report) and actual design culminating in the preparation of documents for bidding and construction. Design–build and privatization projects alter the typical design sequence.

General design criteria should be firmly established during the early stages of the facilities planning process. These criteria include

- The planning period for the facilities (often about 20 years),
- Tide levels and flood protection,
- Standby electrical power for essential facilities,
- Equipment and tankage redundancy requirements,
- Methods and time frames for unit isolation and dewatering,
- General means of flow distribution and interprocess conveyance,
- Flow projections,
- Influent characteristics,
- Treatment performance requirements,
- Liquid stream process alternatives and evaluation of alternatives,
- Residual solids processing alternatives and their evaluation, and
- Selection of recommended treatment processes to satisfy performance requirements.

Contractual documents for bidding and construction of the treatment facilities are the product of the design phase. Contractual documents consist of the bid proposal, instructions to bidders, construction contract, addenda, general conditions of the contract, project specifications, and drawings. Design activities can be separated into five sequential phases: process criteria, concept, schematic, preliminary, and final design. All decisions relating to the size of the facility, construction materials, and major equipment selection types must be finalized before proceeding to detailed design.

Sources, Quantities, and Characteristics of Municipal Wastewater

With the exception of infiltration and inflow and industrial discharge conditions, the amount of wastewater is not subject to alteration by the engineering planning required for the wastewater treatment plant project. In general, the plant designer determines the wastewater characteristics and develops an "end-of-pipe" solution responsive to compliance standards and other wastewater management objectives. The selection of design flows is a critical part of plant design. This section outlines the variables that would affect the expected flow to the wastewater treatment plant.

Populations and Flow Projections

Population and flow projections for areas served by a wastewater treatment plant should be made before sizing of treatment processes and piping. Designs are based on a 20-year design period for any one phase of construction. Ultimate development of the collection area should be assessed to determine how the layout for the plant may look if the area is fully developed. The population projections should take into account nonpermanent residents and seasonal changes in populations (that is, heavy tourist areas or commercial areas). The sum of the nonpermanent and the permanent population is considered the functional population and is the basis for design flow. Although seasonal population may be present in a service area for only a portion of the year (for example, the summer months or the winter months), it may have a significant effect on the wastewater flow treated by the plant. If large fluctuations in flow are expected, the design must include provisions for isolating parts of the plant during periods of low flow and bringing these back into operation as the need arises during periods of high flow or organic loading.

Consideration of service area characteristics should be included when estimating flow per capita. When available, water consumption for an area should be used to estimate wastewater flow generation. At least 60 to 90% of the water consumption reaches the sewer system (the lower percentage is applicable in semiarid regions). If water consumption data are not available or an area is undeveloped, an estimation of flow per capita can be used to generate expected wastewater flow. Flow per capita ranges from approximately 230 to 420 L/cap·d (60 to 110 gpd/cap). Domestic flow contributions should be based on not less than approximately 270 L/cap·d (70 gpd/cap) nor more than approximately 380 L/cap·d (100 gpd/cap), unless historical data indicate values outside of this range. In some affluent areas with no water shortage, these values can be substantially higher. Design engineers should look at trends to estimate future per capita flows. Table 2.1 presents estimates for use in determining flowrates from sewered areas.

Commercial areas may significantly contribute to flow, yet not have a significant population count. Therefore, the type of industry or commercial business should be

Table 2.1 Per capita flows (from Metcalf & Eddy, Inc. *Wastewater Engineering*. Copyright © 1979, McGraw-Hill, Inc.: New York, with permission).

	Flow, gpd/cap[a]	
Source	Range	Typical
Apartment	53–90	69
Hotel	40–58	50
Dwelling		
Average	50–92	74
Above average	66–106	82
Luxury	79–145	100
Trailer park	32–53	40

[a] gpd/cap × 3.785 = L/cap·d.

identified whenever possible. If the area is zoned for commercial use but is not developed, an estimate based on the wastewater generation from similar businesses in the surrounding area offers a starting base for flow projections. Table 2.2 presents representative flows from commercial sources; however, site-specific flows should be determined whenever possible.

Infiltration and Inflow

Some of the most significant components of wastewater received at a treatment plant include *infiltration*, which refers to water seepage through collection system pipes and vaults, and *inflow*, which refers to surface and subsurface stormwater entering the collection system (mirroring the character of the precipitation event). Inflow can be immediate or delayed. *Delayed inflow* refers to the runoff associated with the melting of an accumulated snow cover.

Infiltration, a function of regional groundwater levels, will vary seasonally and annually. Old sewerage systems that have not been rehabilitated can substantially contribute to infiltration problems. Poor construction practices in new sewers can also leave them vulnerable to infiltration.

Recommended Standards for Wastewater Facilities (Great Lakes, 1997) defines an allowable infiltration or exfiltration rate of 19,000 (mL/m·d)/m (200 gpd/in. dia/mile) for new pipe construction. Acceptable infiltration values before replacement or rehabilitation becomes appropriate in older, existing sewers can be 10 or more times higher. This determination depends on a case-specific economic analysis of each sewerage system.

Inflow is high in communities with combined or older sewer systems. Although combined-sewer service may only represent a small fraction of a plant's service area, inflow derived from the combined-sewer service area will often dominate the design and operation of the treatment works. Inflow will reflect low-buffered, often acidic rainwater and the additional pollutants derived from rooftops, roadways, and land use of the service area.

A designer faces special issues when a plant serves a combined-sewerage service area because oversized combined sewers and interceptors serve as traps for sediment and settleable solids. Often, elevated masses of influent screenings, grit, and suspended solids received at the treatment plant during or following a storm reflects the extent of past accumulations in the sewers and pollutants introduced with the stormwater. This is also

Table 2.2 Average wastewater flows from commercial sources (from Metcalf & Eddy, Inc., *Wastewater Engineering*, 4th ed. Copyright © 2003, McGraw-Hill, Inc.: New York, with permission).

Source	Unit	Flowrate, gal/unit·d Range	Flowrate, gal/unit·d Typical	Flowrate, L/unit·d Range	Flowrate, L/unit·d Typical
Airport	Passenger	3–5	4	11–19	15
Apartment	Bedroom	100–150	120	380–570	450
Automobile service	Vehicle served	8–15	10	30–57	40
station	Employee	9–15	13	34–57	50
Bar/cocktail lounge	Seat	12–25	20	45–95	80
	Employee	10–16	13	38–60	50
Boarding house	Person	25–65	45	95–250	170
Conference center	Person	6–10	8	40–60	30
Department store	Toilet room	350–600	400	1300–2300	1500
	Employee	8–15	10	30–57	40
Hotel	Guest	65–75	70	150–230	190
	Employee	8–15	10	30–57	40
Industrial building (sanitary waste only)	Employee	15–35	20	57–130	75
Laundry (self-service)	Machine	400–550	450	1500–2100	1700
	Customer	45–55	50	170–210	190
Mobile home park	Unit	125–150	140	470–570	530
Motel (with kitchen)	Guest	55–90	60	210–340	230
Motel (without kitchen)	Guest	50–75	55	190–290	210
Office	Employee	7–16	13	26–60	50
Public lavatory	User	3–5	4	11–19	15
Restaurant:					
Conventional	Customer	7–10	8	26–40	35
With bar/cocktail lounge	Customer	9–12	10	34–45	40
Shopping center	Employee	7–13	10	26–50	40
	Parking space	1–3	2	4–11	8
Theater (indoor)	Seat	2–4	3	8–15	10

not uncommon in older sanitary sewerage systems that receive high inflow when a rainstorm occurs after an extended dry period.

Industrial, Commercial, and Institutional Wastewater Contributions

Quantities of industrial, commercial, and institutional components of municipal wastewater are sometimes difficult to estimate for design purposes, particularly for projected future contributions. Occasionally, the industrial or institutional component can dominate the wastewater treatment plant design.

Industrial contributions in any municipal wastewater may range from insignificant to many times the domestic contribution. Daily, weekly, holiday, and seasonal variations of industrial releases should be expected unless information to the contrary exists. The type of waste—simple carbohydrates or complex proteins and fats, soluble or particulate matter, organics or inorganics, and nutrient enriched or nutrient poor matter—can influence the selection of treatment processes and the performance of the treatment plant.

Other Wastewater Contributions

A municipal wastewater treatment plant often receives *septage* (the waste from septic tanks) generated in the surrounding, unsewered areas as well as solids from sewer cleanings. Landfill leachate, water treatment residuals, and, in some instances, groundwater contaminated with hazardous materials can also be discharged to wastewater treatment plants. Because septage includes solid residue that may have stabilized somewhat during several years of storage, residual organics may exhibit low biodegradability. Landfill leachate characteristics reflect the character and age of the material placed in the landfill and the amount of water that infiltrates the landfill from ground and surface sources.

Waste solids from a water treatment plant can be expected to exhibit the characteristics of suspended solids in the raw water supply and any solid (for example, powdered activated carbon) or solid-forming material (for example, alum and the resultant hydroxide precipitate) added during the course of treatment. The soluble pollutant phase of these waste solids reflects the organics removed from the raw water supply and the time of storage at the water treatment plant. Aluminum or iron salts added during water treatment enhance phosphorus removal at the wastewater treatment plant, although not as much as freshly added chemicals. Smaller plants may experience some problems related to flow and solids surging unless water plant discharges, particularly filter-backwashing wastes, are hydraulically equalized.

An understanding of the characteristics of the wastewater is just as important as knowing the quantities of wastewater expected to enter a treatment plant. There are a number of variables that define the characteristics of municipal wastewater.

Community Water Supply

The nonconsumptive portion of water used in a service area constitutes most of the wastewater routinely received at a plant. This component of wastewater reflects the character of the raw water supply, the water treatment processes, and the history of beneficial water use.

Groundwater contains scaling compounds and is highly buffered. Conversely, surface waters are often slightly mineralized and contain little or no buffer. Softening, demineralization, or both may be practiced with or without accompanying changes in background alkalinity. Simple, raw water coagulation and clarification with aluminum or iron salts will add anions and deplete alkalinity. Soft, unstabilized waters will aggressively solubilize metal from the water system and customer-distribution piping. Copper, for example, may affect biosolids quality and iron may adversely affect phosphorus removals.

The magnitude of the available buffer (alkalinity) is important when designing one or more of the following processes: nitrification, metal salt or lime addition for phosphorus removal, pH adjustment, and closed-system oxidation. The scaling nature of both the water supply and the wastewater can impair equipment if processes include boilers,

Note on Terminology

Although it is common to find confusion among the three words *sewer*, *sewage*, and *sewerage*, the definitions for these words are quite explicit. A *sewer* is a pipe in which *wastewater* (*sewage*) flows, and a whole system of these pipes is a *sewerage* system. Thus we have sewerage systems that consist of collection sewers (or collectors) and interceptor sewers (or interceptors) and in these pipes flows wastewater (and other flows such as infiltration and inflow). The combination of all the flows is commonly called wastewater because it is often impossible to determine the original source.

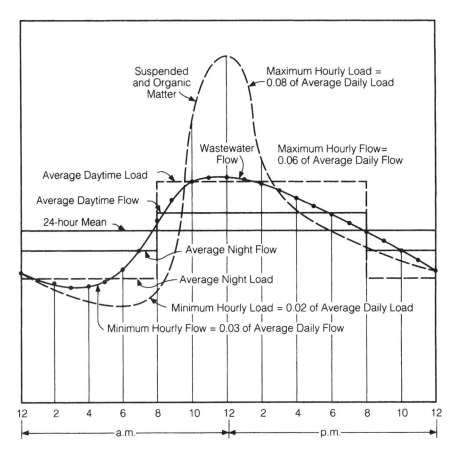

Figure 2.1 Hourly variation in flow strength of municipal wastewater (from Fair, G. M. et al. *Elements of Water Supply and Waste Disposal,* 2nd ed. Copyright © 1971, Wiley & Sons: New York. This material is used by permission of John Wiley & Sons, Inc.).

steam, cooling, or water seals. Chloride, sulfate, sodium, and other inorganics pass through wastewater treatment plants and can affect some disposal strategies. High chlorides also influence material selections for elevated temperature-processing schemes. During anaerobic conditions, high sulfates can result in concrete corrosion, odors, and toxic air (due to hydrogen sulfide emission). Magnesium can affect the return of nitrogen and phosphorus in sludge processing.

Domestic Wastewater

The most consistent component of municipal wastewater is the character of its domestic wastes. These wastes reflect the character and practices of the served population. As the collection system and service population base expand, the effect of the domestic population becomes less pronounced in terms of peak/average and minimum/average ratios of both flow and pollutant masses.

Domestic wastewater flow variations follow daily and weekly lifestyle patterns of the serviced residential customers. Figure 2.1 illustrates the classic variation of wastewater flows and pollutants. Minimum domestic flows and pollutant loadings are observed during the early morning hours; peak flows and pollutant loadings are experienced in the late morning or early afternoon. Fifty percent of the plant's domestic pollutant load during a weekday could arrive at the plant during an 8-hour period; the plant's maximum hourly load can be more than double its average hourly load.

In general, peak nitrogen concentrations will precede the flow peak, whereas peak phosphorus concentrations will coincide with or lag behind the flow peak. In residential and college communities, for example, phosphorus concentrations will reflect the weekend washing habits of the users.

Recommended Standards for Wastewater Facilities (Great Lakes, 1997) states that new wastewater treatment plants should be designed for a domestic load contribution of at least 0.077 kg/cap·d (0.17 lb/cap/d) of BOD_5 and 0.09 kg/cap·d of suspended solids unless available information justifies other design criteria. Historically, the 0.077 kg/cap·d value has been used to define the population equivalent of industrial wastes. Further, the U.S. EPA's standards also recommend that if garbage grinders are used in the service area, the design domestic loads can increase to 0.09 kg/cap·d (0.20 lb/cap/d) of BOD_5 and 0.104 kg/cap·d (0.23 lb/cap/d) of suspended solids (SS). Total nitrogen and total phosphorus loads can be based on averages of 0.018 and 0.003 kg/cap·d (0.04 and 0.006 lb/cap/d), respectively.

Domestic loads from an area's transient population should also be considered in designing a plant if the service area includes transients who work and visit the area during the day but maintain their permanent residence elsewhere. Information from New York suggests a nominal worker and transient flow contribution of 57 to 110 L/cap·d (15 to 30 gpd/cap). Corresponding pollutant (5-day biochemical oxygen demands [BOD_5] and SS) load contributions from these sources range from 0.009 to 0.023 kg/cap·d (0.02 to 0.05 lb/cap/d). Plumbing codes, architectural standards, or state criteria can be used to develop site-specific estimates for restaurants and hotels.

Table 2.3 shows the composition of domestic wastewater. Table 2.4 shows size classification, organic constituents and distribution, and biological oxidation rates.

Industrial Wastewater

Wastewater from industries can vary widely in strength and characteristics. Treatment plants in the past had to be designed to accommodate sometimes highly unusual characteristics from such diverse industries as meat processing plants, dairies, and steel mills. Today, industrial discharges are almost always treated at the industrial plants and the discharges are similar in strength to domestic wastewater. Industrial wastewaters that are not treated by *pretreatment* are subject to *surcharges*, or fees charged for excess wastewater strength. For example, a community may charge an industry so much per liter or million gallons discharged to the community sewerage system, but limit the discharges to BOD_5 less than 200 mg/L. If an industry produces wastewater that has a significantly higher BOD_5, it might be paying high surcharges, and it would probably be advantageous to the industry to pretreat the wastewater. Such pretreatment is now common in most communities, with the sewer charges and surcharges set so that it is always more economical for the industry to treat instead of paying the bill for discharging high-strength wastes.

Wastewater Variability

Influent flows and loads to a wastewater treatment facility vary on an hourly, daily, and seasonal basis. These variations in flow and load should be accounted for to ensure that

Table 2.3 Typical primary pollutant composition of domestic wastewater.

Parameter	Concentration by phase, mg/L[a]		
	Soluble[b]	Particulate	Total
Suspended solids[c]			
Volatile[d]			190
Inert			50
Total			240
5-day BOD[e]	65	135	200
Chemical oxygen demand[f]	130	260	400
Total nitrogen[g]	20	10	30
Total phosphorus[h]	5	2	7

[a] Based on assumed flow of 378 L/cap·d (100 gpd/cap).
[b] As typically defined with coarse filter, not necessarily soluble.
[c] Based on assumed 0.09 kg/cap·d (0.20 lb/cap/d) of suspended solids.
[d] 80% volatile.
[e] Based on assumed 0.08 kg/cap·d (0.17 lb/cap/d) of BOD_5 and 30% soluble.
[f] Twice the BOD.
[g] 0.02 kg/cap·d (0.04 lb/cap/d) of total nitrogen and 65% soluble, with ammonia-nitrogen composing most of the soluble content.
[h] 0.003 kg/cap·d (0.006 lb/cap/d) of phosphorus and 65% soluble.

the facility can adequately treat peak flows and loads and economically operate under minimum flow conditions.

Flow variation is a function of the size and type of service area and the geographical location of facilities. Where historical data are available, they should be analyzed to develop design criteria. In the absence of historical data, generalized curves such as Figure 2.2 can be used to estimate ratios of extreme to average daily flows.

Pollutant loads also vary, although not necessarily in sequence with the flow variation. Review of load variability at a number of wastewater treatment plants indicates that the variability is fairly consistent from plant to plant when expressed as a ratio to the average annual load. Probability in the range of 95 to 98% is an acceptable peaking factor to use for design purposes in wastewater treatment. Values in excess of these probabilities are likely a result of data error or other aberrations. Typical peaking

Table 2.4 Composition of organic materials in wastewater (Levine et al., 1985).

Item	Classification			
	Soluble	Colloidal	Supracolloidal	Settleable
Size range, μm	<0.08	0.08–1.0	1–100	>100
COD, % of total	25	15	26	34
TOC, % of total	31	14	24	31
Organic compostion, % of total solids				
Grease	12	51	24	19
Protein	4	25	45	25
Carbohydrates	58	7	11	24
Biochemical oxidation rate k, d^{-1} (log base 10)	0.39	0.22	0.09	0.08

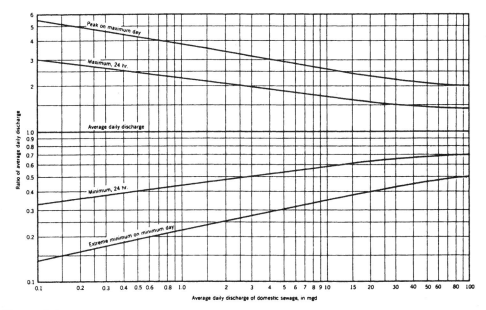

Figure 2.2 Ratio of extreme flows to average daily flow in New England (mgd × 3785 = m³/d) (WPCF, 1982).

factors for suspended solids and biochemical oxygen demand loads based on a 98th percentile are as follows:

- For suspended solids: maximum day = 2.1, maximum 3 day = 1.8, maximum 7 day = 1.6, and maximum 30 day = 1.3.
- For BOD_5: maximum day = 1.8, maximum 3 day = 1.4, maximum 7 day = 1.3, and maximum 30 day = 1.14 to 1.2.

Determining Design Flows

Engineers often speak in terms of "average flows" or "maximum flows per given time," such as the *maximum day*. This concept is best illustrated with the help of a graph such as Figure 2.3. The first curve of this figure shows the variation of wastewater flows as it might occur during a year, and the average of this is the *average annual flow*. If the flows for each week are calculated, the greatest flow that occurs during any one week is the *maximum week* flow. More than likely, during this week, there will be one day when the flow will the highest and this is known as the maximum day. Finally, during that day, there will be one hour when the flow will be the greatest it has ever been during any hour during that year, and this is known as the *maximum hour*. (The maximum hour does not have to occur on the maximum day.)

As illustrated above, these flows can vary considerably, and the ratios between the average annual flows and the maximum days and hours is highly dependent on such variables as population and infiltration and inflow to the sewerage system. Given this variability, determining design flows for a plant is difficult.

There are actually two design flows that must be established before a plant is designed. The first is the *design process flow*, used to design the processes such as

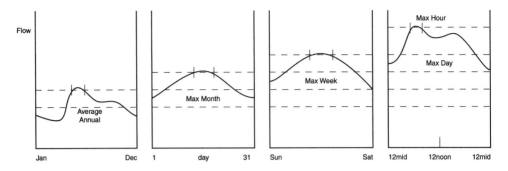

Figure 2.3 Flow variation at a wastewater treatment plant.

activated sludge and clarification. The other design flow is the *design hydraulic flow*, which is used to design the hydraulics of the treatment plant. The latter, the hydraulic flow, is always significantly higher than the design process flow. The engineer has to be sure that the plant will at the very least be able to pass the water through and not be flooded out during those times when the flow is the greatest, such as during hurricanes. At these times, the process is of secondary importance. The operator could be busy attempting to keep the plant from overflowing the tanks and have little concern for the process. During dry weather, however, the design process flow governs operation of the plant. Plants that are advertised as being, for example, 0.4 m^3/s (10 mgd) plants are able to treat 0.4 m^3/s (10 mgd) to the required effluent level. But their design hydraulic flow might be as high as 0.9 m^3/s (20 mgd) or more. The ratio between design process flow and design hydraulic flow depends on many variables such as the size of the community, the sources of the wastewater stream, and the type and condition of the wastewater collection system.

Occupational Safety

Governments at all levels establish regulations, requirements, and codes that contain minimum occupational health and safety standards. Laws, regulations, codes, and agency guidelines require or recommend occupational health and safety provisions. These provide the facility owner and designer with minimum requirements and design and operation guidance.

Occupational Safety and Health Act and Federal Regulations

The most significant safety statute is the Occupational Safety and Health Act (29 *CFR* 1910 and 1926). The goal of this act is to eliminate unsafe working conditions during construction and in general industry. Although these standards do not directly apply to publicly owned treatment facilities, many states have adopted, by reference, standards set in regulations of the U.S. Occupational Safety and Health Administration (OSHA). It is common practice for public facilities to meet these regulatory requirements because they apply directly to construction and operation of privately owned treatment plants that treat municipal or industrial wastes.

State and Local Codes
Most state environmental agencies have developed manuals for staff to use when reviewing plans and specifications. These contain safety items that should be incorporated into design and construction. Therefore, engineers should contact state environmental agencies early in the design stage.

Many requirements of existing and proposed fire codes are more detailed and stringent than OSHA regulatory requirements. In some instances, fire codes dictate layouts, construction materials, and safety equipment. Local fire departments often have the authority to establish requirements for site layouts to allow access for emergency response equipment and firefighters; establish the number, location, and type of fire hydrants; and set other requirements.

National Fire Protection Association Recommendations
A National Fire Protection Association (NFPA) committee has established recommendations for fire protection practices for wastewater treatment and collection facilities in *Standard for Fire Protection in Wastewater Treatment and Collection Facilities* (NFPA, 1999). This publication provides guidance for protection against fire and explosion hazards in wastewater treatment plants, including hazard classification of specific areas and processes. The publication calls for a specific fire-risk evaluation as part of the project design process. In addition, the publication lists locations, descriptions, electrical classifications, fire and explosion hazards, recommended materials of construction, recommended ventilation practices, and suggested fire protection measures associated with liquid stream treatment and solids treatment processes from municipal collection systems. Detailed information on sources of ignition, sources of hazards, and mitigation measures is also presented.

Americans with Disabilities Act
Facilities under design, renovation, or both must comply with the federal Americans with Disabilities Act regulations. This act covers accessibility issues, such as handrails, ramps, parking facilities, and lavatories, that must be considered during design.

Designing Safe Facilities
Many decisions that have an effect on occupational health and safety, such as costs, operability, layout, operator skill, and convenience, are made by representatives of an owner. A designer is required by convention and, in many cases, law to comply with appropriate codes, regulations, and industrial standards. Registered professional engineers are required by registration laws to protect the public health and welfare.

In their plans and specifications for fixed facilities and other facilities, designers provide for items such as portable fire extinguishers, equipment replacement parts, portable maintenance equipment, vehicles, laboratory equipment and supplies, and operating manuals. It is unlikely for a designer to specify such items as maintenance tools, personal safety gear, protective clothing, or housekeeping requirements. Plant managers will often conduct training sessions on safety, emergency response planning, employee screening, or other plant activities.

In addition to designing around existing codes, standards, regulations, and so on, a designer can facilitate the reduction of hazards by creating an inherently safer environment. Examples of this include using fewer hazardous chemicals (e.g., chlorine) in process designs, designing processes to operate closer to ambient conditions, designing equipment and facility layouts (such as easy access to valves) to accommodate confined space entry and lockout–tagout procedures, and using fencing and other security measures to reduce liability and chance of trespassers hurting themselves.

Site Selection

Ideally, a new site for a wastewater treatment plant is one that can be developed economically without unnecessarily stressing the environment. Although numerous considerations, such as social and environmental factors, are difficult to quantify, cost effectiveness and design and construction requirements can be quantified when choosing among potential sites.

Wastewater treatment facilities are almost universally perceived by the general public to be unacceptable neighbors. Public opposition is strong; however, with early involvement of the public in the planning process and a sincere desire by designers to listen and mitigate the public's concerns, much of the opposition can be minimized. Public involvement, or participation, is one of the most important elements in selecting and evaluating alternative sites. This is discussed in more detail later in this chapter.

Land Use

The previous uses and possible contamination of soil and groundwater at a wastewater treatment plant site often affect the cost to develop the site. One of the first steps in the site-selection process is to perform a *Phase I Environmental Site Assessment* to identify the presence of contaminated soil; buried, leaking fuel tanks; buried herbicide and pesticide containers; and other illegal dumping activities. These activities are not limited to urban areas; they are often found in rural areas too. When considering the demolition of existing structures, the presence of asbestos insulation, flooring, and roofing materials should be identified because these will have a significant effect on the site-development cost. Phase I environmental site assessments are inexpensive and can be performed quickly by experienced individuals before land purchase.

A determination of the effects of a wastewater treatment plant on the surrounding land in the area should consider zoning regulations, impacts on adjacent property values, and compatibility with activities on neighboring properties. Generally, constructing a plant in an industrial neighborhood rather than a residential area is more acceptable and less expensive (unless the former industrial site is contaminated with hazardous materials). Wastewater treatment facilities located near airports may require U.S. Federal Aviation Administration approval because airplane glide slopes may control the height of some structures. Lagoons and ponds attract birds, which also affect air-traffic operations.

If the selected site is surrounded by residences, measures should be taken to ensure that the plant is a good neighbor during both construction and operation. Such measures include minimizing noise, odors, aerosols, air particulates, chemical hazards, insects, and intrusive lighting. Attractive architecture and landscaping and proper consideration of prevailing winds help make wastewater treatment plants more acceptable neighbors.

In general, sites with special natural features should be avoided. The development of areas designated as wild, scenic, or recreational under the Wild and Scenic Rivers Act may be prohibited. Similarly, shorelines are often reserved for public use; this is especially important in urban areas where a shortage of open shoreline access exists. Facilities proposed along coasts should comply with the Coastal Zone Management Act.

If the site is within a "viewshed", a low-profile, well-screened facility will be needed. Three-dimensional computerized images using actual photographs of the site—enhanced to show the postconstruction condition—can be used to evaluate mitigation measures and provide the public with some assurance that the view will be protected.

Occasionally, a site may have some paleontological significance. In these cases, qualified experts in the area should be consulted. If there is reason to believe site development will expose fossils, the construction specifications should require an on-site paleontologist during excavation. Local libraries, museums, historical and archeological groups, Native American groups, and universities specializing in archeology should be contacted during the site-selection phase to determine whether the site has historical or cultural significance.

The amount of isolation and buffer area needed between plant processes and other property owners influences the amount of land required for the treatment facility. State and local design standards for wastewater treatment plants and building and zoning codes frequently contain requirements for buffer zones between the treatment facility and surrounding land uses. Wetlands are excellent buffer zones.

In the absence of specified local regulations, the buffer in residential areas should be at least 50 to 75 m (150 to 250 ft) (Arizona Department of Health Services, 1978). Facilities using facultative ponds or lagoons should have at least a 300-m (1000-ft) setback (Arizona Department of Health Services, 1978). Additional buffer zones are recommended to reduce odor and noise intrusion to the surrounding community.

Receiving Water Location and Requirements

If treated effluent is to be discharged to a lake, river, or stream, receiving water characteristics and water quality standards will affect the degree of treatment required. *National Pollutant Discharge Elimination System* (NPDES) permits and state and local health departments determine the discharge requirements. Selecting a site that will require advanced treatment to comply with stringent water quality standards may not be the most cost-effective choice.

Downstream uses of receiving water may affect the cost of the treatment plant. Stream uses such as potable water supplies, shellfish waters, or water-contact sports require higher reliability standards than streams not used for such purposes (U.S. EPA, 1974). Downstream impoundments may require more stringent nutrient removal limits. Downstream uses also determine levels of redundancy in treatment units, process equipment, and power supply. Similarly, coastal plants that release effluent within bathing, shellfishing, and tidal wetland zones require a greater degree of reliability than those that discharge a great distance from the shore and at great depths.

Plant personnel are better able to monitor effluent and its effects on water quality and aquatic life if a site is close to the receiving water. Sampling upstream and downstream of the plant outfall is easier, and a convenient supply of upstream water is available for preparing various dilutions for bioassay testing.

The bioassay test gaining importance in wastewater treatment plant design and operation is the *whole effluent toxicity* (WET) test. This test uses a standard species of stream life such as the flathead minnow or daphnia as a surrogate to measure the effect of the effluent on the receiving stream. The laboratory methods used to calculate WET vary greatly from state to state and it is important to know the detailed requirements at each jurisdiction.

The WET tests are used to set two kinds of design limits, the *acute limit* and the *chronic limit*. The acute limit is measured on the basis of lethal dose $(LD)_{50}$ or that dose at which 50% of the organisms die in some predetermined time, such as 24 hours. The chronic limit is based on detecting *no observable effect* during some longer period such as two weeks. The no observable effect concentration (NOEC) is that mixture of effluent and stream water where none of the organisms die. (Or more correctly, no more organisms die than would die in a control test with zero effluent.) The design limits are set based on the condition of the stream and relate to the sensitivity of the aquatic ecosystem. For example, suppose a mixture of plant effluent and stream water is used to measure the survival of flathead minnows. The stated design limit may be "more than 50% will survive in mixtures of 20% or less of effluent." If for the mixture of 20% effluent and 80% stream water, more than half of the minnows survive, then the effluent meets the acute WET design limit. The chronic limit may be stated as "the NOEC will occur at a mixture of no more than 5% effluent." In this case, if for a mixture of 5% effluent and 95% stream water no discernable effect was noted, then the effluent meets the chronic WET design limit.

Obviously, if a plant is designed and the effluent does not meet the required WET limits, then there is something in the wastewater that is affecting the aquatic ecosystem. This "something" might be quite difficult to detect, but a thorough analysis of the wastewater would then be necessary, and either pretreatment at the source or remedial action at the plant must be initiated.

Area Requirements

A site should accommodate present and anticipated future requirements. As growth occurs in the service area and treatment requirements increase, the plant will likely require additional space. The potential for such demands should be considered when selecting a site.

The area required for a plant of a particular capacity depends on the following considerations: degree of treatment required; processes to be used; degree of redundancy necessary; space requirements of ancillary and support facilities; and space required for access, circulation, and maintenance.

The layout and shape of process units can drastically affect the land areal requirements. Stacking of process units is expensive and complicates operation and maintenance (O&M), but has been done on occasion to accommodate restrictive sites. Likewise, square or rectangular tanks using common wall construction save considerable space when compared with circular tanks of like volume, but may add to maintenance costs and may not be operationally satisfactory for some processes.

Space requirements for maintenance, administration, storage, laboratory, and staff services influence total space requirements. In some cases, related activities, such as maintenance crews and the equipment for the sewerage system or the agency's industrial waste monitoring division, may be housed at the site.

Figure 2.4 presents the approximate amount of area required by activated-sludge plants of various sizes. It includes areas used for both process and administrative purposes. Dedicated solids disposal sites, sludge application, and buffer areas are not included. As may be expected, data vary considerably. Therefore, the figure should only be used as a preliminary guide to plant areal requirements.

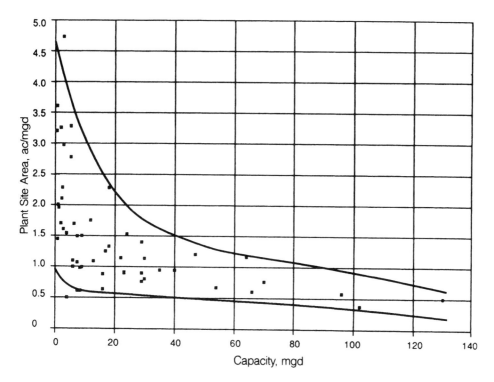

Figure 2.4 Plant site area versus plant capacity (mgd × 3785 = m³/d, ac × 0.4047 = ha). The two lines represent an approximate range of plant site areas for a given capacity.

Satellite Facilities

If one site cannot satisfy the space requirements of the entire treatment facility, a second site is needed. Smaller plants upstream from the main plant may be used to provide hydraulic relief, or in some cases the liquid stream is treated at one location while sludge is treated or disposed of at another location.

Water reclamation is more economically attractive when the reclamation facility is close to the reuse sites. This eliminates the need for extensive and costly reclaimed water conveyance systems and should be a prime factor in selecting a site for a reclamation facility.

Many agencies operate sludge handling, land-application systems, and composting facilities remotely from the liquid treatment facility. The selection of sites for sludge handling is similar to that for liquid stream treatment facilities; however, other factors such as soil type and location of the groundwater table must be considered. Increased truck traffic to and from facilities such as composting sites should be addressed in anticipation of public concerns. Because of the increased potential for odor at these sites, more buffer space and odor control measures will be required.

Elevation and Topography

A low-lying site eases the flow of wastewater from the service area by gravity and minimizes the number of pumping stations in the collection system. However, such a site may also require flood protection. Adequate protection may be provided by building earthen dikes around the perimeter of the site as long as the dikes do not create an obstruction in the waterway. Constructing the tops of process tanks, buildings, and pipe gallery entrances and building finished floor elevations above the expected high-water levels also provide flood protection; these methods can be costly and may actually negate the advantages of selecting a low-lying site.

Protection against damage to plant structures and equipment from a flood or tidal surge with a 100-year recurrence interval is customarily provided. Protection against even greater floods may be warranted in some cases. At a minimum, the plant should remain operational during a 25-year flood; however, plants are designed to remain operational during larger floods. State environmental regulatory agencies often require flood protection to a stated level, such as 1.2 m (4 ft) above the 100-year event. Effluent pumping or storage during episodes of high tides or high water may be required.

Geology, Hydrogeology, and Soils

Site geology, hydrogeology, and soil types significantly affect construction costs. Design problems are more complex when a site is underlain by more than one type of soil. The allowable bearing capacity of the soils and the permissible differential settlement may dictate the need to support tanks and structures on piles or caissons. Unfortunately, fine-grained alluvial soils and other unfavorable subsurface formations are often found near rivers and streams on otherwise ideal sites. Corrosive soils, which are detrimental to buried metals, may dictate the use of more expensive coatings, nonmetallic substitutes, or a cathodic protection system. Bedrock depth may dictate the hydraulic profile of the plant and inflate construction costs if rock excavation is required.

A high-groundwater condition also increases construction costs because dewatering requirements are more extensive during construction. Because of machine vibrations, dewatering may cause soil subsidence and damage adjacent structures. It may also be more difficult to keep basements, deep dry wells, and tunnels dry. Waterproofing membranes are recommended on the exterior (earth side) of these areas. A designer should also consider the potential of floating empty tanks and other buried structures and providing sufficient ballast or pressure-relief drains to prevent flotation during the highest possible groundwater conditions. The use of pressure-relief drains allows groundwater to enter empty tanks or buried chambers, including utility tunnels. However, relief valves can malfunction; therefore, more positive provisions may be needed, such as pumping groundwater from empty tanks during maintenance work. Groundwater in a tank during cold weather may freeze, and equipment such as fixed-air diffusion systems and waste sludge collectors may be damaged. Some facilities have elected to install permanent, continuous, groundwater-pumping systems to minimize the problem. The groundwater could be discharged to a receiving water if it conforms to permit requirements. However, the use of permanent groundwater-pumping systems will still not eliminate the need for relief valves.

Transportation and Site Access

It is important that a wastewater treatment plant site be accessible to personnel and delivery persons at all times. Access roads that have the same degree of flood protection as the treatment plant ensure the safe ingress and egress of operating personnel. Plants located close to all-weather highways facilitate the delivery of equipment and chemicals and off-site disposal of grit, screenings, and sludge. Access to fire and other emergency vehicles must be available.

Utility Services

A treatment plant should have sources of potable water, reliable electrical power, and telephone communication. In addition, it is usually beneficial to locate a site near a supply of natural gas. The availability of such utilities in sufficient capacity is an important consideration when selecting a site. Large plants are able to either draw power from two sources in the power grid, or have standby power (diesel motors coupled to generators).

Noise Control

Noise management is an important consideration in layout and design. Consider two aspects: first, the transmission of noise beyond plant boundaries and, second, the effect of excessive noise on the health and welfare of plant personnel. The latter is minimized through proper specification of equipment and sound-absorbing enclosures or isolation. Maximum noise levels for working areas are regulated under the federal Occupational Safety and Health Act (Occupational Noise Exposure, 1971).

Air Quality Control

Wastewater treatment plants can be sources of odors, chemical emissions, particulates, and aerosols, all of which must be controlled. In addition to the federal Clean Air Act requirements, some states and local jurisdictions require that permits be obtained from air-quality management agencies. These agencies often require approved permits before construction starts and another permit before operation, the latter of which is granted only after successful demonstration that the particular equipment or process complies. For major discharges, permits are often required for scrubbers, engines, compressors, and gas flares. The permit process can be time consuming and often requires special studies such as dispersion modeling and health-risk assessments. In the site-selection process, the ability to easily secure the required air-quality control permits is an important consideration.

Other Environmental Considerations

In addition to air emissions, there are other effects on the environment that should be considered. For example, locating a plant in an area of special habitat, such as wetlands or other critical ecosystems, requires extensive mitigation or may be totally prohibited. The federal Endangered Species Act protects threatened and endangered species and their habitats from adverse effects of proposed developments.

If the proposed site is of natural, ecological, or aesthetic significance or is home to threatened or rare and endangered species, site development will be expensive. In many cases, these issues can be mitigated by purchasing additional land that would be permanently dedicated to an artificial wetlands area or special habitat. In some areas, construction may be shut down during nesting seasons if the construction activities are expected to interfere with the wildlife. Areas of historical significance, such as ancient burial grounds, must also be respected. An archeological study is always recommended.

Permit Requirements

Numerous permits are required for a wastewater treatment plant, although these vary from state to state. It is imperative that all of the permits be identified during the planning process and a timeline and list of requirements for each be established. This will become the roadmap for project implementation. Some negotiation of permit limits and conditions may be appropriate to ensure that requirements are reasonable, technically feasible, and scientifically sound. Many permits depend on securing other approvals and, hence, are not all on parallel tracks. It is important to identify the sequence of the permits and direct the preliminary design to follow that sequence so that information is available when needed to support a particular permit. Failure to obtain the required permits on a timely basis can scuttle the best of projects.

Process Options and Selection

A plant designer should favor facilities with low maintenance, tolerance, and ample capacity to reflect the uncertainty of staffing, maintenance, and remedial action in a public marketplace and with political and public oversight where funding of significant capital improvements depends on public indebtedness. The designer should carefully balance the need to provide tolerant facilities that can respond to a multitude of future uncertainties with an overdesign that results in the misuse of public monies for superfluous facilities and capacity.

Procedurally, wastewater treatment consists of the serial application of reactors and separators. Reactors oxidize, reduce, solubilize, immobilize, or physically condition their contents and create gaseous products. Separators create two products depending on their separation objectives; for example, low- and high-suspended solids-concentration product streams are derived from a solids separator.

Reactor and separator processes can be *passive* or *active*. As a metaphor, human transportation can be active or passive, in that riding on a bus is passive transportation (for the passenger) while driving the car is active transportation. The driver of a car must constantly be able to make small adjustments in speed and direction, and to look out for danger ahead. Riding in a bus is passive in that one can sit back and let someone else (the bus driver) worry about these decisions. There are certain advantages to both passive and active unit operations in wastewater treatment.

An example of a passive unit operation in a wastewater treatment plant is the bar screen. The screen is constructed and is expected to perform under any flow condition. The performance of passive unit operations such as the bar screen is not amenable to operational manipulation once the entire process comes online. For such a process, the designer should err on the conservative side. Well-designed, passive unit processes are the classic systems chosen for plants that will have little attention and low personnel commitment. Generally, these processes require the highest capital investment and are the most unresponsive to new or unanticipated treatment requirements or inadequacies. Once upset, these processes may require the longest period to recover.

Active processes involve the opposite design considerations. An example of an active process is the activated-sludge system, which requires constant supervision and testing. The higher the activity of any process, the easier it is both to upset the process and to restore it from an upset condition. Active processes allow the plant and the processes to be optimized at the plant and typically offer opportunities to realize savings of both first costs and operating costs, especially early in the design period or when the facility must respond to seasonal treatment objectives.

A process' activity should not be confused with its reliability. The most demanding (and least reliable) unit processes for the designer and the operator are those that have a multitude of support systems and moving parts and operate under elevated temperatures, pressures, or both. Most of these processes exist in the solids-processing train of the plant, especially at large facilities.

When large volumes of residuals—in comparison with the average daily production volume (such as with a digestion process)—are handled by the solids-processing train, the designer should provide additional process redundancy as an allowance for

- Anticipated failures,
- Differing feed-solids characteristics, and
- Maintenance or repairs of the principal or supporting unit processes to ensure that the "rated" capacity is realized (with either new or old equipment).

The success of the entire plant and all of its unit processes may ultimately depend on the ability to remove solids from the plant. A single or undersized dewatering unit or residue management plan without a backup processing sequence or disposal outlet does not represent good practice.

Those processes that unavoidably solubilize pollutants create special design and operating issues that vary with technologies and operating strategies used at the plant. With the exception of the thermal-conditioning system, thermal processes achieve some beneficial destruction of solids. The engineer should use special care in evaluating and selecting processes that will reintroduce pollutants to the liquid-processing train, especially when the plant's design objective includes control of reintroduced pollutants.

Plant Layout

The layout of a wastewater treatment plant primarily considers the layout of treatment processes. Once the staff, administration, and ancillary facilities are sited, they are placed (based on function, hydraulics, and operation) for efficient, economical operation and the support and comfort of the staff.

Treatment Facilities

The arrangement of treatment processes on a plant site affects operating and construction costs. Before locating any facilities on a new site, a designer should develop a rough, preliminary hydraulic profile to establish both tops and bottoms of principal structures.

In general, there are two types of plant layouts: *unit design* and *functional design*. The unit design, illustrated in Figure 2.5, takes advantage of common walls and arranges the unit operation into a compact unit. Functional design, on the other hand, spreads the unit operations around so as to take full advantage of the topography. The advantages of unit design over functional design are

- Lower first cost because the construction is simplified;
- Ease of operations because there are few decisions to be made by the operator;
- Small footprint, requiring less land and resulting in larger buffers around the plant;
- Ease of expansion because the entire plant can be doubled if needed to double the flow, as shown in Figure 2.5;
- Reduced piping and few valves;

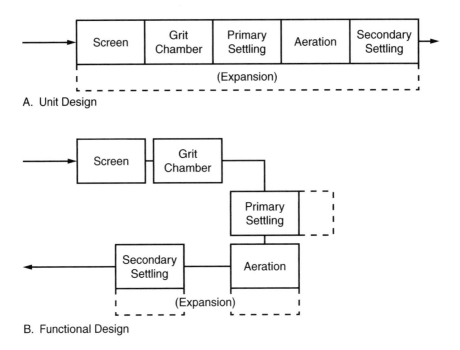

Figure 2.5 Unit (A) and functional (B) design plant layouts.

- Reduced head losses through the plant; and
- Common walls that speed construction.

On the other hand, the advantages of the functional design are

- Flexibility of operation,
- Ability to bypass units if necessary,
- Easy to expand any one of the units without having to expand the entire plant,
- Easy to add intermediate processes,
- Easy to add chemicals,
- Short chemical lines if chemicals are added at different points,
- Ability to use topography to best advantage,
- Short distance for operators to move between various unit operations, and
- Easy access to unit operations.

Many of these advantages and disadvantages are contradictory and the designer has to achieve a balance between the simplicity of the unit operation plant layout and the functional design. Commonly, the designer begins by laying out the plant according to the functional design and then moves unit operations to take advantage of some of the aspects of unit design. The unit design layout is most appropriate for smaller plants while the functional design becomes necessary for larger facilities.

Where possible, a designer should group similar unit processes to ease operation, minimize piping, and allow for expansion (Kawamura, 1991). Such groupings include solids thickening, digestion, dewatering, and disposal; influent pumping and preliminary treatment units such as screening, grit removal, and grit dewatering; and primary and secondary treatment units.

Arranging treatment processes to follow site contours helps maximize use of natural topography and reduces both pumping requirements and excavation costs for new structures.

Providing for Future Expansion
Provisions for future expansion must be considered because most wastewater projects upgrade, refurbish, or expand existing treatment plants. During the layout of a wastewater treatment project it is necessary to provide for future expansion or upgrading of treatment facilities. Provisions made to facilitate future expansion at the expense of short-term efficiency and convenience should be made only after due consideration. Maintaining adequate space during the arrangement of treatment units eases future construction. Providing adequate space where deep foundations or tanks will be constructed adjacent to existing shallow foundations or tanks avoids undermining in the future. A good guideline for planning and layout purposes is to keep clear of a zone extending downward and outward 45 deg from the bottom of a shallow foundation.

A designer should determine the hydraulics of the plant with future expansion in mind. Hydraulic channels, connectors, and splitters might be oversized so that future process improvements can be installed. For example, the hydraulic load on plastic pack trickling filters is far higher than that on rock filters, and yet they have the same footprint. In addition, a designer should provide sufficient elevation between process stages to permit increases in flowrate and permit adequate distribution of flow among multiple tanks. Space should also be allocated for junction and splitter boxes. It is not prudent design to construct a new plant that is hydraulically limited. Such plants are difficult and costly to expand or upgrade. Also, the existing plant should be able to operate while it is being expanded. Placing masonry plugs and stop-plank grooves in chambers and channels to which future connections must be made will help. The installation of a valve upstream of a blind flange or other line plug enables the plugged line to be connected without shutting down the entire line.

Tank Geometry
Tank geometry plays an important part in the plant layout. Circular, square, and rectangular tanks are used for primary and secondary sedimentation. Circular tanks are commonly used for primary and secondary clarifiers for plants where land area is not restrictive. They offer good performance with relatively simple mechanical equipment. Square tanks with center mechanisms require use of corner sweeps on the rake arms and more complex provisions for scum removal. These additions to the normal circular mechanism require more maintenance to achieve acceptable performance. Using common-wall construction for square and rectangular tanks minimizes space requirements and offers an opportunity to reduce construction costs.

Treatment Redundancy
Treatment redundancy determines to a great degree the plant reliability. Multiple units are required for all critical components of the treatment plant. Pumps needed to move

flow through the plant and blowers providing air to an activated-sludge system should have adequate capacity for peak conditions with the largest single unit out of service. Depending on the reliability classification of the plant, portions of the power distribution system should be duplicated. If the power distribution system cannot be duplicated, connection for a portable power generator should be considered.

It should be clear how the plant would operate if any one unit operation was shut down. If the bar screen mechanism broke, for example, is there an alternative (passive) bar screen through which the flow can be diverted? For some critical components such as secondary clarifiers, multiple units of the same size, make, and model number should be installed to facilitate maintenance and reduce inventories of spare parts. Some facilities may be allowed to function at a reduced level of service during an emergency such as that following a damaging earthquake or other natural disaster.

Hydraulics
Hydraulics are important because the plant layout should take advantage of the natural slope of the site to minimize site-development costs. The ideal plant site has gravity flow into the facility and gravity flow through the outfall to the receiving water. It is better to have gravity inflow to the facility with the treated effluent pumped out because effluent pumps are more efficient than raw wastewater pumps and less likely to clog. This consideration must be traded off with the costs for excessively deep structures. When laying out a site, it is important to analyze the hydraulics carefully to ensure that, under normal conditions, only the influent or the effluent is pumped—not both.

Flexibility of Operation
Flexibility of operation permits various operating modes and allows the operator to accommodate changes in wastewater characteristics and other conditions. The operations staff should be consulted early in the design process to address their operating strategies and procedures relative to the need for unit process drains, bypasses, redundancy, and so on. Routine maintenance, equipment breakdowns, loss of a utility (such as natural gas), or damage caused by an earthquake or other natural disaster require flexibility to provide alternate modes of operation.

Increased flexibility also complicates the associated instrumentation and control system. Piping arrangements, for example, can become both costly and complicated, leading to operator errors and, in the case of sludge piping, line plugging. Therefore, provisions for flushing the drain lines are necessary.

Continuing Treatment During Construction
Continuing treatment during construction often constrains the design. Planning the construction sequence carefully ensures that new facilities can be built without undue interruption of treatment. The site layout should recognize the future expansion. Pipelines and conduits should be located sufficiently far away from future structure excavations to avoid costly sheeting and shoring for support. Likewise, foundations for structures and tanks should be sufficiently deep to avoid undermining when there is adjacent excavation.

Future expansions will have less of an impact on plant operations if they occur in an area removed from normal operations traffic. Generally, this means away from the center of the site and more towards the property boundary. Construction on the boundary side, with its associated noise and dust, may affect neighbors. In cases where the plant site is abutted by occupied land uses it may be better to expand inward or away from the existing development. The existing treatment processes and structures will provide screening.

In terms of construction, project specifications should include the number of process units that can be taken out of service at any one time, allowable electrical power shutdown periods, and temporary diversion schemes. The specifications should state whether the contractor or the owner of the treatment plant is responsible for draining and cleaning process tankage and piping.

The engineer should consult with the owner's operations staff to design the specific scenario for the sequence of events that should occur during construction when continued operation of various plant facilities is required. This detailed process description and sequencing should be included in the contract documents.

Maintenance

Maintenance considerations can be important especially in cold-weather areas where the most critical equipment should be placed indoors. This is often not necessary in warmer areas where year-round outdoor maintenance is feasible. However, a roof over equipment even in warm- or hot-weather areas shields the equipment from direct solar radiation and provides a dry work area during rainy weather. In some cases, motor control and instrument enclosures and variable-speed drive control panels may require self-contained air-conditioning units.

Proper maintenance requires convenient access to equipment. Buildings housing large process equipment and piping should be sized to allow sufficient space for repairs. Bridge cranes are appropriate in areas where heavy lifts are required. In other areas, the installation of monorails, hoists, or outside jib cranes may be appropriate. Cast-in inserts capable of accepting a rig for lifting can be placed in concrete slabs above equipment and piping where infrequent use is anticipated. Floor doors and pass doors for the movement of components from one floor to another and to the exterior must be designed.

Adequate electrical outlets should be placed throughout the plant. Outlets of the proper voltage and phase must be located appropriately. Specialty outlets such as those used for welding may be placed in some locations. Compressed air is often piped to various areas for equipment maintenance. Hose bibs should be located throughout the plant. Hot water or steam is useful in galleries containing scum or sludge piping or in other solids-handling facilities.

Confined spaces must be adequately ventilated. The ventilation should be sized for adequate air changes based on the volume of the confined space in the absence of any liquid. Local codes for ventilation rates must be met.

Piping and electrical and instrumentation conduits should be routed in galleries or covered pipe trenches—especially pump rooms—to provide ease of access for maintenance and expansion or modification. Color-coding of piping is helpful and can reduce accidents.

Administration, Staff, and Support Facilities

A useful tool in the layout of administration, staff, and support facilities is a functional diagram, or "bubble" diagram, which is shown in Figure 2.6. A functional diagram identifies significant functional areas and interrelationships between these areas. In the layout of these facilities, it is important to conform with the Americans with Disabilities Act.

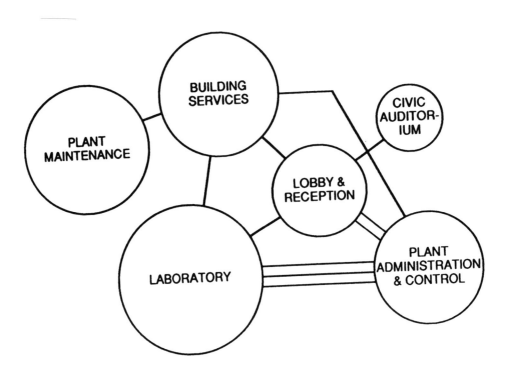

Figure 2.6 Functional diagram of administration building. Multiple connecting lines show areas of frequent communications and interrelationships.

Monitoring and Control Facilities

Monitoring and control facilities should be provided for personnel in charge of operating the plant or recording data and making calculations. These stations should be located near an equipment-intensive area of the plant or within view of an alarm panel or major control panel. The area should be relatively quiet, clean, well lighted, and well ventilated. In warmer climates, the area should also be air-conditioned.

Sampling points should be located in a well-mixed area of the stream to be sampled. The effect of recycle streams must be considered. In addition, sampling of plant influent should be located upstream of all recycle streams. Likewise, a secondary influent sample should not include return activated sludge or recirculation flows. Junction chambers, access manholes, or pipe taps are commonly provided for sampling.

Maintenance and Storage

Maintenance and storage facilities are best located near the center of a plant or in the most equipment-intensive area of the plant and should contain adequate, well-lighted bench space. The size of the maintenance staff and the services to be performed in-house determine the size of the facility and the tools to be provided. A shop located at ground level best facilitates the handling of large equipment. If welding is to be done in the area, attention should be given to ventilation, isolation, and power supply. Plants with a high degree of instrumentation require a separate instrumentation shop and maintenance staff.

The storage area for the plant's spare parts and maintenance supply inventory is best located near the shop area and should be large enough to accommodate an array of shelves, bins, and drawers. To ease deliveries, the storage area, which is secured, should be located adjacent to the roadway of the main plant. Paints, lubricants, pesticides, herbicides, and similar toxic, flammable, and hazardous materials should be stored in an isolated, secure area with adequate ventilation.

Laboratories
Laboratories provide required influent monitoring and analyses of process control parameters. The size of the plant, type of treatment provided, and extent of laboratory analyses to be performed on-site determine the size and layout of the laboratory. In larger plants, a receiving area for samples and other packages is useful. Refrigerators should be available at those locations for short-term sample storage. For many small plants, it is cost effective to contract all but the simplest analyses to outside laboratories. The type and anticipated frequency of tests and the analytical equipment required to conduct them must be considered in the design. Future operating and monitoring requirements should be evaluated and factored into the layout.

Administrative Offices
Administrative offices should be located near the front entrance of the plant so visitors may find them easily. A private office sized to hold meetings of three to six people provides a plant manager the privacy needed to handle personnel issues, as well as easy access to plant operating records, personnel files, cost records, and O&M manuals. These documents have to be protected from flooding or other damage.

If a plant has distinct organizational groups or departments, each manager may need a private office. Grouping the individual offices of a management team promotes communication among the various groups. In larger plants, an assembly room large enough to accommodate plant personnel meetings, on-site training, visitors, and public meetings is desirable.

Staff Facilities
Staff facilities such as toilets and showers should be provided for all personnel throughout the plant, especially in areas where employees are stationed and in locker rooms. The Occupational Safety and Health Administration (OSHA) requires a minimum of one toilet for every 15 to 20 people a facility will employ. Unless local codes require otherwise, a plant should have one shower for every 10 persons expected to complete work at the same time. Separate lockers for street clothes and work clothes may also be required.

Providing lunchrooms and break rooms for all personnel discourages the consumption of food and beverages in unsanitary areas of the plant. Restrooms or separate washrooms should be located close to the rooms where food or beverages will be consumed. An essential part of ensuring a safe and efficient plant is providing a training room, which should contain a large blackboard or other erasable board and a large wall suitable for displaying diagrams and other training aids

Other Layout Considerations
The design of a plant's roadways is important to the plant's efficiency. Roadways must provide access to all points where deliveries will be made or where materials will be loaded onto trucks and transported off-site. The turning radius of driveways must be large enough to handle large trucks.

Access to the site must be controlled. A perimeter fence and lockable gates should be provided. In larger plants, closed-circuit television may be used to control plant access and maintain security. In designing security measures, however, access to the site by outside emergency response teams should be considered. No visitor should have access to process units or open-flow conduits without either obtaining a guide or at least checking in with the plant operator.

Stormwater from developed areas of a site sometimes cannot be discharged to receiving waters without a permit. State and local codes concerning stormwater handling must be consulted.

A wastewater treatment plant is a diverse facility that receives a wide variety of materials. Chemicals, lubricants, disinfectants, spare parts, laboratory supplies, and various liquid wastes are among the materials frequently delivered to most plants. In addition, many plants ship out a wide variety of materials including sludge, biosolids, empty chemical containers, and solid wastes.

Odors represent one of the most offensive conditions to people living near or working in a wastewater treatment plant. The presence of odors and their effect on public relations should not be underestimated. Even if odor control facilities are not installed at a plant, the layout and design should allow for the addition of odor control systems should the need arise.

Parking should be provided for all personnel, disabled employees, and visitors. Visitor parking should be marked and placed close to the administration building so visitors can park and register without having to drive through the site.

Winter weather in some parts of the country may affect all types and sizes of treatment processes in plants. The most frequent source of winter problems is settling tanks, particularly scum-removal mechanisms. Whenever possible, abovegrade tanks, which maximize exposure to cold weather, should be avoided. If other conditions dictate the use of abovegrade tanks, tank walls should be insulated and perhaps covered. Earth mounding is also an effective method of insulating exposed tank walls. Cold weather significantly affects solids-management facilities. Disposal or reuse sites may not be available for extended periods of time because of weather conditions, and storage may be required. Some states have requirements for storing solids.

Pleasing architecture and attractive landscaping greatly improve a facility's image within the community and provide a pleasant atmosphere for the staff. This is particularly important if the plant is located in a scenic area or in a residential neighborhood. In such cases, additional costs of special architectural treatment and landscaping are justified. In residential areas, the buildings and site should be designed to integrate with the surrounding neighborhood. In older industrial areas, a new, aesthetically pleasing treatment plant can form the nucleus for the revitalization of the surrounding area.

Environmental Impact

Projects receiving funding from the U.S. government require compliance with the National Environmental Policy Act (NEPA) of 1969 and the Council on Environmental Quality Guidelines (Canter, 1996). The first step toward compliance is to determine whether the action is categorically excluded from NEPA. The types of actions that are categorically excluded by federal agencies vary. Construction of a new wastewater treatment plant or expansion of an existing one is not excluded.

Compliance with NEPA requires public participation throughout the process, starting with initial meetings, public meetings and presentations during the assessment process, and final public hearings. The process begins with an initial session involving interested federal, state, and local agencies. The environmental assessment follows with the public involved. The key output of the assessment is a determination of whether or not the federal action will significantly affect the quality of the human environment. If the results indicate that the action has no significant impact or the impacts can be mitigated, then a Finding of No Significant Impact (FONSI) is issued. This is akin to a "negative declaration". The public has 30 days to review the FONSI. There is a public hearing during the review process and comments are received and formal responses are prepared. If the FONSI is still deemed the appropriate action after the hearings and the comments, then the lead agency formally approves the document, allowing the action (or project) to go forward.

If the FONSI is appropriate, this does not mean that the project will not be scrutinized for environmental impact. It will likely have to go through some form of environmental assessment for the state. Environmental assessments are most common for treatment plant upgrades and other modifications.

If the FONSI is not appropriate, which is the case for almost all wastewater treatment plants, an environmental impact statement (EIS) must be prepared. This process starts with the filing of a notice of intent that is published by U.S. EPA in the *Federal Register*. Although there are no strict guidelines on how the EIS is to be prepared, the following discussion is a description of several alternatives within a general framework. The EIS should be in three parts: *inventory, assessment,* and *evaluation*.

The first duty in the writing of any EIS is the gathering of data (such as hydrological, meteorological, and biological information, etc.). A listing of the species of plants and animals in the area of concern, for example, is included in the inventory. There are no decisions made at this stage because everything properly belongs in the inventory.

The second stage is the analysis part commonly called the assessment. This is the mechanical part of the EIS in that the data gathered in the inventory are fed to the assessment mechanism and the numbers are examined accordingly. Numerous assessment methodologies have been suggested; one of which is the quantified check list.

The *quantified check list* is possibly the simplest quantitative method of comparing alternatives. It involves first the listing of those areas of the environment that might be affected by the proposed project, and then an estimation of

a) the *importance* of the impact,
b) the *magnitude* of the impact, and
c) the *nature* of the impact (whether negative or positive).

Commonly, the importance is given a rating such as 0 to 5, where 0 means no importance whatsoever, while 5 implies extreme importance. A similar scale is used for magnitude, while the nature is expressed as simply −1 for negative (adverse) impact and +1 for positive (beneficial) impact. The environmental impact (EI) is then calculated as

$$\text{Environmental impact} = \sum_{i=1}^{n} (I_i M_i N_i)$$

where
I_i = importance of ith impact;
M_i = magnitude of ith impact;

N_i = nature of ith impact, so that N = +1 if beneficial and N = −1 if detrimental; and
n = total number of areas of concern.

The environmental impact, which can be negative (detrimental) or positive (beneficial), is calculated for each alternative and tabulated.

The last step in the EIS is the evaluation of the calculated impacts. It is important to recognize that the previous two steps, inventory and assessment, are simple and straightforward procedures compared with the final step, which requires judgment and common sense.

Once the draft EIS has been published in the *Federal Register*, a formal comment period begins that leads to the preparation of the Final Environmental Impact Statement, another comment period, and, finally, a Record of Decision. This process is time consuming, particularly if the project is controversial or the environmental documentation is challenged, and additional time needs to be factored into the project implementation schedule. Also, many permits are contingent on having an approved environmental document (for example, an NPDES permit); this can affect the entire project timeline because the preparation and approval permits themselves are time intensive.

Conclusions

The objective of treatment plant design is to make the plant "operator friendly". It is necessary to make it easy for the operator to do the right thing, and very difficult for him or her to do the wrong thing. The designer has to mentally think of how the operation of the plant will take place. Where will the operators go, and will they be able to get there when it is raining or snowing? Are all the valves reachable? Are all the pipes color coded so that clean water is blue, sludge is brown, etc.? Are the railings in the right places? These and an infinite number of other concerns dog the design engineer. When the plant is built, and the operator finds that something is badly arranged, it is the fault of the design engineer. The engineer has nobody else to blame. Their job is to make the plant operator-friendly and sufficiently flexible so that the operator can make adjustments to the treatment scheme as the conditions, regulations, and wastewater characteristics change. There is no sense in spending millions on the construction of a plant only to have it perform poorly because a) it is impossible to operate it properly, or b) the operators do not understand what is to be done, or c) do the wrong thing because it is convenient to do so. Nobody said that design engineering was a soft job.

References

Arizona Department of Health Services (1978) *Minimum Requirements for Design, Submission of Plans and Specifications of Sewage Works;* Engineering Bulletin No. 11; Phoenix.

Canter, L. A. (1996) *Environmental Impact Assessment;* McGraw-Hill: New York.

Fair, G. M.; Geyer, J. C.; Okun, J. W. (1971) *Elements of Water Supply and Waste Disposal*, 2nd ed.; Wiley & Sons: New York.

Great Lakes—Upper Mississippi River Board of State Sanitary Engineering Health Education Services, Inc. (1997) *Recommended Standards for Wastewater Facilities;* Albany, New York.

Kawamura, S. (1991) *Integrated Design of Water Treatment Facilities*; Wiley & Sons: New York.

Levine, A. D.; et al. (1985) Characterization of the Size Distribution of Contaminants in Wastewater: Treatment and Reuse Implications. *J. Water Pollut. Control Fed.*, **57**, 805.

Metcalf & Eddy, Inc. (1979) *Wastewater Engineering*; McGraw-Hill: New York.

Metcalf & Eddy, Inc. (2003) *Wastewater Engineering*, 4th ed.; McGraw-Hill: New York.

National Fire Protection Association (1999) *Standard for Fire Protection in Wastewater Treatment and Collection Facilities;* NFPA 820; National Fire Protection Association: Quincy, Massachusetts.

Occupational Noise Exposure (1971) 36 CFR 105, 1910.95, *Federal Register,* May 29.

U.S. Environmental Protection Agency (1974) *Design Criteria for Mechanical, Electric, and Fluid System and Component Reliability*; EPA-430/9-74-001; Office of Water Program Operations: Washington, D.C.

Water Pollution Control Federation (1982) *Gravity Sanitary Sewer Design and Construction*; Manual of Practice No. FD-5; Washington, D.C.

Symbols Used in this Chapter

E = effect terms
I_i = importance of ith impact
M_i = magnitude of ith impact
n = total number of environmental areas of concern considered
N_i = nature of ith impact, so that $N = +1$ if beneficial and $N = -1$ if detrimental
W = weighing factors

Plant Hydraulics

Introduction

After a treatment concept has been selected, the next step is to determine the hydraulic profile, or hydraulic grade line, needed for the wastewater to flow through the wastewater treatment plant and its unit operations. The objective is to provide optimum hydraulic head to drive the plant. This may include making pipes and channels large enough to meet future expansion beyond the capacity required during the design year. Sufficient hydraulic head should be provided to permit good distribution of the plant flow to all treatment processes over the range of expected conditions without being excessive. During calculation of the hydraulic profile, the economics of building deeper plant structures should be considered in relation to building pumping stations. Pumping stations are an important consideration because their use leads to higher operation and maintenance costs and reduced reliability.

Hydraulic Considerations

Hydraulic Profile

The hydraulic profile describes the water level required for the wastewater to flow through the plant. A hydraulic profile is normally prepared for the main liquid flow path that extends from the plant inlet to the receiving water, but can also be prepared for ancillary flow trains such as sludge treatment and disposal facilities. The hydraulic profile is necessary for establishing water surface elevations and hydraulic structure elevations and should include water surface elevations, hydraulic control devices such as control valves and weirs, and critical elevations of process structures, channels, and pipelines. The profile may also include ground surface elevations, pipeline sizes, and other special features that will enhance operation of the treatment plant. Figures 3.1, 3.2, and 3.3 present examples of hydraulic profiles for three commonly used wastewater treatment processes.

Hydraulic calculations should begin at a control point where the plant discharges to a receiving watercourse, body of water, or the wet well of an effluent pumping station. A plant's water surface profile and pumping requirements are established from this control point. For example, if the receiving body is a river or stream, the controlling elevation is the required flood elevation as calculated by accepted hydrological methods or as obtained from an agency such as the U.S. Army Corps of Engineers. The level of flood protection (for example, the 100-year flood elevation) is established by the environmental regulatory agency that governs the respective area's plant design. With a storage basin or pond installed for 100% reuse applications, the controlling water level is the pond-overflow elevation. For a larger body of water such as a lake, wind setup should be considered in addition to the highest water surface elevation. With discharge to an ocean, the controlling water level is the high-tide elevation based on the selected design storm. When discharge is to an ocean, the higher specific gravity of seawater (1.025) compared with wastewater (1.00) is also a consideration. For every 12 m (40 ft) of depth below high tide, an additional 0.3 m (1 ft) of head should be provided to overcome the heavier weight of seawater.

In instances where plant effluent can flow to the receiving water body by gravity most of the time, except during high-flood stages, effluent pumping may be economically justified. In such cases, the control point of the hydraulic profile for the treatment plant will be the maximum wet-well elevation at the effluent pumping station.

The hydraulic profile is then calculated through each unit process up to the influent sewer. The calculation of the hydraulic profile is more involved than simply adding the head losses of various components and should take into account the need for equal flow distribution to and among all treatment processes. The hydraulic profile should also account for the interaction between water depth and velocity head as it relates to the hydraulic and energy grade line.

If the available hydraulic head at the plant inlet is not sufficient to meet the required head based on the hydraulic profile, revisions to the sizes and elevations of the hydraulic structures are necessary. If revisions do not produce the desired results, pumping should be used at the plant inlet or elsewhere along the flow path through the treatment plant. A cost-effectiveness analysis should be performed to determine the appropriate pumping location.

Flowrates

Peak flow is used for hydraulic design to establish maximum water levels and reveal minimum freeboards at tanks and channels. The average flow during the month in which the maximum water level occurs is commonly used for treatment process design. Minimum flow is used to show minimum levels throughout the plant to reveal maximum heights of free-falling water at weirs and channels and minimum submer-

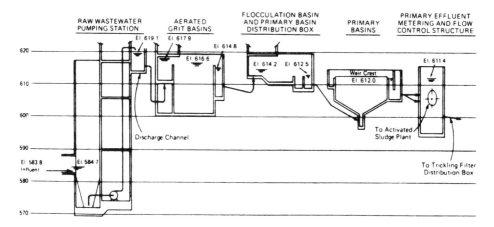

Figure 3.1 Typical hydraulic profile for influent pumping and primary treatment. Water surface elevations (in feet) represent flow of 160,000 m^3/d (42 mgd).

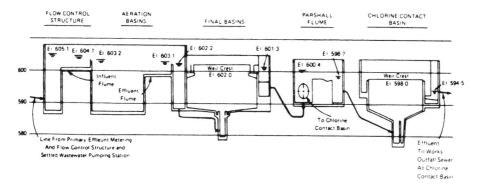

Figure 3.2 Typical profile for an activated-sludge plant. Water surface elevations (in feet) represent flow of 160,000 m^3/d (42 mgd).

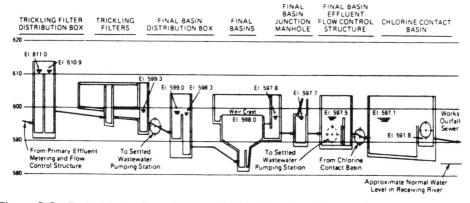

Figure 3.3 Typical hydraulic profile for a tricking filter plant. Water surface elevations (in feet) represent flow of 160,000 m^3/d (42 mgd).

gences on equipment. The unit processes should convey the maximum flow unless this flow would cause a hydraulic washout of the treatment plant. Equalization basins to minimize any negative effect on the treatment process may be necessary.

To calculate the hydraulic profile through the secondary facilities, pertinent channels and piping should be designed to accommodate the return activated-sludge flow, other types of process recycle flows, and the main plant flow. When they are substantial, process sidestream flows can also be included in the calculation of the hydraulic profile for the entire plant. The hydraulic profile associated with the peak flow should address the probability of treatment units being out of service. For this reason, unit process redundancy should be considered. Future plant expansion should be accounted for in sizing of piping and channels. Minimum flow should be included in the hydraulic design to avoid low velocities and potential solids deposition in pipes and channels.

Unit Process Liquid Levels

Each unit process should be hydraulically designed to prevent liquid from overtopping the walls of structures under all conditions. The top-of-structure elevation is set so that the freeboard is maintained above the high-water elevation. Depending on the degree of expected surface disturbance and the relative frequency of the condition, typical freeboards can range from an extreme low of 150 mm (6 in.) to 1 m (3.3 ft) or more. However, the typical minimum freeboard for maximum water levels is approximately 0.3 m (12 in.). Common hydraulic design allows for little or no submergence of the various weirs throughout the main process flow path. Some submergence is permitted when the peak-flow condition is judged to be an infrequent occurrence. Any effects on flow distribution as it affects treatment capacities should be considered when including submergence in the plant's hydraulic design. In addition to considering the maximum water surface elevation in relation to structure walls, a designer should address conditions such as surface waves, foaming, splashing (such as occurs with hydraulic jumps), and aeration (air bulking of liquid).

Unit Process Redundancy

The design treatment capacity of a plant should include unit process redundancy. Redundancy generally means that each of the unit processes, depending on the mechanical equipment, can have one or more tanks out of service at average flow without adversely affecting effluent quality. However, this may affect the hydraulic design. With one tank out of service, the flow to the unit process may exceed the maximum flow for which it was designed and cause a hydraulic overload. In addition to average-flow conditions, peak-flow conditions with individual unit processes out of service should be evaluated. A common approach is to provide hydraulic capacity to accommodate having one of each primary process unit out of service during peak flow, along the worst-case flow path relative to distance and flow. The effect on treatment process during such conditions should be reviewed. However, temporary overload of unit processes may be acceptable for short periods or during an emergency; in any case, the structure walls must not be overtopped.

Flow Distribution

To ensure proper treatment, it is essential to achieve equal flow distribution to each of the tanks that make up a unit process. Because weirs provide good flow distribution without the need for mechanical equipment, such as rate-controlling valves, and are less subject to clogging, the best technical solution for flow distribution is to provide an upstream weir control structure to split flow evenly to the tanks of each unit process. Figure 3.4a illustrates this type of layout for two clarifiers. The use of symmetry alone will not ensure equal flow distribution because flow over long weirs is particularly sensitive to head. For example, a small change in head on the order of 13 mm (0.5 in.) is often enough to adversely affect flow distribution to a group of processes having long effluent weirs as the means of level control. Flow distribution to circular clarifiers is a case in point. Because of the long weirs associated with circular clarifiers, even slight differential settling of the clarifier structure will impair flow distribution. Figure 3.4b illustrates an example of flow-control structures for plants with asymmetrical layouts. All weirs must be adjustable after construction.

An inlet header or distribution channel or pipe can also be used to distribute flow to a unit process. However, in such cases the inlet ports or gates to each tank should be designed with adequate head loss to ensure good distribution. This is described in more detail later in this chapter.

As with flow distribution, equal distribution of sludge to treatment units should be maintained. Unless provided for in the design, equal sludge distribution may not occur coincidentally with equal flow distribution. This is especially common where flow is distributed by use of channels. Where channels are used, such as upstream of grit tanks, the wastewater flow should be well mixed to ensure that the sludge distributes evenly with the flow.

Plant Head Loss

Head loss through plants varies widely. The specific plant designs are a function of the available head needed for pumping and site conditions. Equal distribution of flows to treatment units requires the expenditure of head (head losses); therefore, plants with lower than normal hydraulic head may be an indication that equal flow distribution is not occurring. Although plants may be designed to function well hydraulically outside typical norms, the total heads commonly found for secondary treatment plants range from 4.3 to 5.5 m (14 to 18 ft). This range of total head applies to treatment plants with facilities for flow measurement, pretreatment, and disinfection. Given the variability of this information, the hydraulics of each plant in relation to site and configuration must be evaluated and assumptions based solely on what are thought to be typical head losses associated with various types of unit processes should be avoided.

Minimum and Maximum Velocity

The appropriate velocity in the pipe or channel must be maintained. A minimum velocity of 0.6 to 0.76 m/s (2 to 2.5 ft/s) for raw wastewater is required to prevent solids deposition in channels and pipelines. Flushing velocities for raw wastewater are 1.5 to 1.8 m/s (5 to 6 ft/s). In some cases, the minimum velocity cannot be maintained because of specific process requirements. For example, the entrance velocity to a clarifier is reduced to 0.3 m/s (1 ft/s) to improve the settling characteristics and prevent disturbance of the sludge blanket. For channels where low velocities occur and sludge settling can be a problem, mixing or aeration is provided.

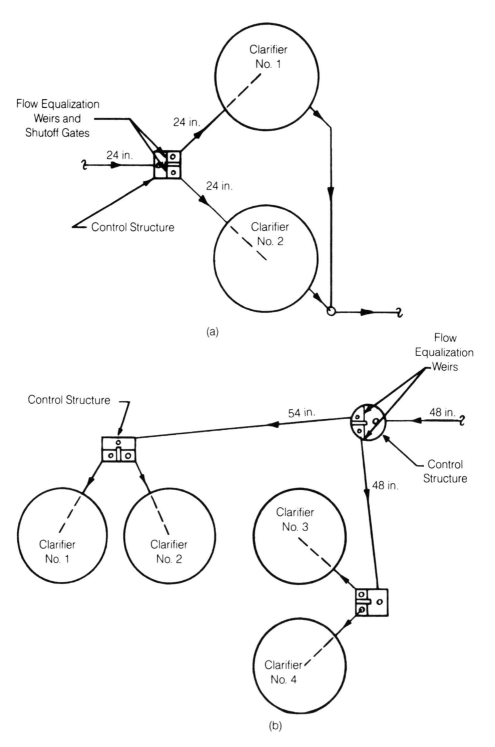

Figure 3.4 Design examples of equal flow distribution among clarifiers: (a) symmetrical layout and (b) asymmetrical layout (in. × 25.4 = mm).

Fundamentals of Hydraulic Engineering

Hydraulic Head

Hydraulic-head calculations can be classified as three types: static head, friction (including turbulence), and velocity head. Static head represents the potential energy measured by the difference in elevation between the static levels (that is, the level that would be achieved if no flow were to occur) in process units and receiving waters. For pumping applications, the static head represents the difference in elevation between the water surface elevation in the wet well and the water level at the discharge of the force main. Friction- and turbulence-head losses are defined as the loss in energy and the subsequent loss in hydraulic grade required to convey liquid through pipes, channels, fittings, inlets, outlets, reducers or increasers, or other items that create resistance to flow. Velocity head represents the dynamic or kinetic energy resulting from the movement of the liquid. When the flow velocity is reduced or stopped, this energy converts to static head. When the flow velocity increases, a portion of the static head converts to velocity head. The relationship of static head and velocity head is expressed by the Bernoulli equation for open-channel flow, as follows:

$$Y_1 + Z_1 + v_1^2/2g = \text{constant} \tag{3.1}$$

where
  Y = water depth, m (ft);
  Z = baseline elevation (for example, channel invert), m (ft);
  v = average velocity, m/s (ft/s); and
  g = acceleration due to gravity, 9.8 m/s² (32.2 ft/s²).

At any hydraulic section, the total energy represented by the terms in the Bernoulli equation is constant and consists of the water elevation ($Y_1 + Z_1$) plus the velocity head ($v_1^2/2g$). The water elevation is associated with the hydraulic grade line, or hydraulic profile, whereas, the total of the terms in the equation is associated with the energy grade line. Accounting for head loss occurring between downstream and upstream sections (1 and 2, respectively), the energy equation is

$$Y_1 + Z_1 + v_1^2/2g = Y_2 + Z_2 + v_2^2/2g + h_L \tag{3.2}$$

where h_L = head loss between sections 1 and 2, m (ft).

Along any channel, the energy equation is used to calculate the upstream depth (Y) given the other parameters. The head loss (h_L) is expressed as a direct value or in terms of velocity head for downstream and upstream sections ($v_1^2/2g$ or $v_2^2/2g$, respectively). Where negligible or no velocity occurs, such as in tanks, the hydraulic grade line elevation equals the energy grade line elevation.

Pipe Flow

Pipe flow, or *closed-conduit flow*, occurs when the fluid completely fills the pipe and places it under pressure. If the pipe is not full, no pressure is exerted on the pipe and the flow is characterized as *open-channel flow*, discussed later. This section is specifically about pipes that are full and under pressure.

For closed-conduit flow, the pressure head, $P/\rho g$, is substituted for the depth, Y, in the Bernoulli equation, where P is pressure in kilopascals or kilonewtons per square meter (pounds per square foot) and ρ is the fluid density in kilograms per cubic meter (lb s^2/ft^4). For the English system, ρg is more familiarly expressed as the specific weight, γ (N/m^3 [in. lb/ft^3]), and the Bernoulli equation is thus

$$Z_1 + (P/\gamma)_1 + (v^2/2g)_1 = Z_2 + (P/\gamma)_2 + (v^2/2g)_2 + h_L \tag{3.3}$$

where Z becomes the elevation. These variables are illustrated in Figure 3.5.

Head loss in closed-conduit flow occurs as a result of friction between the fluid and the pipe walls. The head loss can be measured empirically by running the fluid through a pipe and measuring the pressure at points 1 and 2. If both ends of the pipe are at the same elevation, $Z_1 = Z_2$, and because the velocity does not change, $(v^2/2g)_1 = (v^2/2g)_2$, then

$$(P/\gamma)_1 = (P/\gamma)_2 + h_L \tag{3.4}$$

$$h_L = (P/\gamma)_1 - (P/\gamma)_2 = \Delta P \tag{3.5}$$

The relationship between h_L and ΔP is not a simple one. At very low velocities, the flow in the pipe will be *laminar* (i.e., nonturbulent), exhibiting special flow characteristics and resulting in a laminar relationship between flow and pressure drop. Then, as the velocity increases, the flow will suddenly jump to *turbulent* flow. In addition, the type of fluid used and the roughness of the pipe will affect the pressure drop. In short, predicting the pressure drop from head loss in a pipe is not an easy task and requires a semiempirical approach.

The most widely used equation for estimating the head loss between two points in a closed conduit under pressure is the Darcy–Weisbach equation. This equation has the

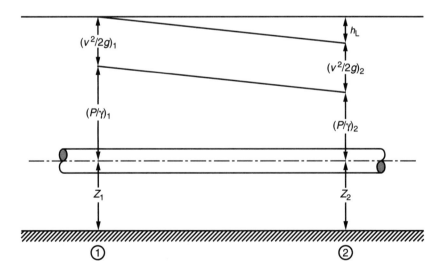

Figure 3.5 Bernoulli relationship for a pipe.

advantage of being dimensionally homogeneous and being applicable to all fluids and all pipes. The Darcy–Weisbach equation is written as

$$h_L = f\,(L/d)(v^2/2g) \tag{3.6}$$

where
h_L = head loss due to friction, m (ft) of pipe;
L = length of pipe, m (ft);
f = friction factor, dimensionless;
d = inside diameter of pipe, m (ft);
v = average velocity, m/s (ft/s); and
g = acceleration due to gravity, 9.8 m/s² (32.2 ft/s²).

All of the variables in the Darcy–Weisbach equation can be controlled or estimated except the friction factor, f. This term is estimated using a plot of f versus the Reynolds number, defined as

$$Re = dv\rho/\mu \tag{3.7}$$

where
d = pipe diameter, m (ft);
v = velocity, m/s (ft/s);
ρ = fluid density, kg/m³ (lb-s²/ft⁴); and
μ = fluid viscosity, kg·m/s (lb-s/ft²).

The plot of the friction factor versus the Reynolds number is commonly called the *Moody diagram*, found in many basic hydraulics and fluid mechanics books.

Because the velocity, v, appears in both the Reynolds number and the Darcy–Weisbach equation, solutions using the Moody diagram is by trial and error. The calculation is simplified in the case of water because the density and viscosity of water are known. Solving the Darcy–Weisbach equation for velocity, as

$$v^2 = \frac{h_L d^2 g}{fL} \tag{3.8}$$

$$v = \left[\frac{2g}{f}\right]^{1/2} d^{1/2} \left[\frac{h_L}{L}\right]^{1/2} = K d^{0.5}\, S^{0.5} \tag{3.9}$$

where
d = pipe diameter, m (ft);
K = a constant, function of the pipe roughness; and
S = slope of the hydraulic gradient, or h_L/L.

The most popular equation used in U.S. engineering practice for estimating head losses in closed conduits is the Hazen–Williams equation, which is based directly on the above modification of the Darcy–Weisbach equation. The Hazen–Williams equation was derived by conducting numerous experiments with different pipes and back-calculating the constant K. The Hazen–Williams equation for customary U.S. units is written as

$$v = 1.318\ C\ R^{0.63} S^{0.54} \tag{3.10}$$

where
- v = velocity (ft/s);
- R = hydraulic radius (ft);
- = $d/4$ for circular pipes, where d = pipe inside diameter (ft);
- S = slope of the energy gradient, (h_L/L);
- L = pipe length (ft); and
- C = the Hazen–Williams friction coefficient.

Note that the powers, 0.63 and 0.54 in eq 3.8, are very close to 0.5 as in the Darcy–Weisbach equation.

Solving the Hazen–Williams equation for head loss,
(international system of units [SI])

$$h_L = [v/(0.85C(R)^{0.63})]^{1.85} \times L \tag{3.11}$$

(customary U.S. units)

$$h_L = [v/(1.318C(R)^{0.63})]^{1.85} \times L$$

where
- h_L = friction head loss (m or ft);
- v = mean velocity of flow (m/s or ft/s);
- R = hydraulic radius, area/wetted perimeter (m or ft);
- C = surface-roughness coefficient (dimensionless); and
- L = pipe length (m or ft).

The surface-roughness coefficient varies with the pipe's type, age, and condition. Older ductile or cast iron pipes can have a surface-roughness coefficient value of 100; new polyvinyl chloride pipes can have a surface-roughness coefficient value of 140. When using new pipes and determining pump heads, the designer should consider two surface-roughness coefficient values: one corresponding to the new condition of the pipe and the other to the design life of the pipe.

A useful relationship can be derived from the Hazen–Williams equation by recognizing first that the equation can be solved for flowrate if the continuity equation, $Q = Av$, is used where Q is the flowrate (m³/s or ft³/s) and A is the area through which the flow occurs (m² or ft²), which is equal to $\pi d^2/4$. The Hazen–Williams equation (common U.S. units) can then be written as

$$Q = 0.285\ Cd^{2.63}\ S^{0.54} \tag{3.12}$$

For a given pipe, C (roughness coefficient) and d (diameter) are the same. The relationship between any one flowrate, Q_1, and any other flowrate, Q_2, through the same pipe is then

$$\frac{Q_1}{Q_2} = \left[\frac{S_1}{S_2}\right]^{0.54} \tag{3.13}$$

and because L, the length of pipe, is the same, and $S = h_L/L$,

$$\frac{Q_1}{Q_2} = \left[\frac{h_{L1}}{h_{L2}}\right]^{0.54} \tag{3.14}$$

This equation is especially useful when system head curves are to be constructed for pumped systems being designed. This is discussed further below.

Head losses caused by pipe fittings such as valves, bends, entrances, exits, and sudden enlargements or constrictions are caused by friction and turbulence as flow passes through the devices. Almost all of the head loss through fittings, however, is attributed to turbulence as the flow strives to return to hydraulic equilibrium after being distorted. It is better to calculate fitting losses using a constant, K, for the type of fitting, multiplied by the velocity head pertaining to the fitting. However, many engineers use the equivalent-length method to calculate head losses associated with fittings. Either method will produce satisfactory results; however, for fluids with viscosities different from that of water, the equivalent-length approach will be incorrect. The K values for commonly used fittings can be found in hydraulics handbooks. Using the K-value method, head loss for pipe fittings can be calculated using the following equation:

$$h_L = K(v^2/2g) \tag{3.15}$$

where
h_L = head loss, m (ft);
K = constant for type of fitting (dimensionless);
v = velocity in pipe, m/s (ft/s);
g = acceleration due to gravity, 9.8 m/s² (32.2 ft/s²).

Open-Channel Flow

When there is a free surface in a pipe or channel, the flow is *open-channel flow*. As the name implies, open-channel flow is associated with channels. Open-channel flow may also occur in unsurcharged influent sanitary sewers or in effluent outfall lines. The Manning equation is commonly used to calculate head loss for open-channel flow; in SI units, it is

$$h_L = L\left[\frac{vn}{R^{0.67}}\right]^2 \tag{3.16}$$

and in common U.S. units it is

$$h_L = L\left[\frac{vn}{1.49R^{0.67}}\right]^2 \tag{3.17}$$

where
h_L = head loss, m (ft);
R = hydraulic radius (cross-sectional area divided by wetted perimeter), m (ft);
n = roughness coefficient (dimensionless);
v = velocity, m/s (ft/s); and
L = channel length, m (ft).

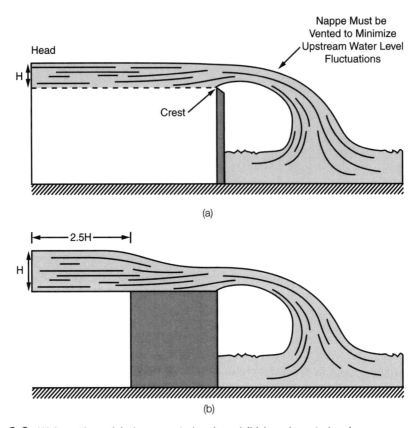

Figure 3.6 Weir sections: (a) sharp-crested weir and (b) broad-crested weir.

The roughness-coefficient values depend on the roughness of the conveyance channel and range from 0.009 for glass to more than 0.06 for natural channels. The most commonly used materials for channels in treatment plants are concrete and steel. New concrete and steel can have roughness-coefficient values as low as 0.010; as the materials deteriorate, roughness-coefficient values ranging from 0.013 to 0.016 can be expected. *Gravity Sanitary Sewer Design and Construction* (WPCF, 1982) provides a listing of widely accepted roughness-coefficient values.

In addition to head loss in channels, pipes, fittings, and valves, head loss also occurs with in-line equipment, flow meters and flumes, and weirs. In-line equipment and flow meter head loss data can be obtained from equipment manufacturers. Flumes, such as Parshall or Palmer–Bowlus flumes, are used for open-channel flow measurement. Head loss for flow through weirs, launders, ports, and distribution headers is described in the following sections.

Weir Control

Although weirs are sometimes used for flow measurement in plants, they more commonly serve as control devices to maintain a required water surface elevation in a unit process. Weirs are classified in accordance with the shape of the notch. Weir types include rectangular, V-notch, trapezoidal, proportional, and parabolic. The upper edge of the weir or weir plate is referred to as the *crest* of the weir. The depth of the water over the crest is the *head*. The head, or depth, over the weir is measured as the difference between the upstream water surface elevation above the point where weir drawdown occurs and the weir crest elevation (see Figure 3.6).

Weirs are either sharp crested or broad crested as shown in Figure 3.6. The most commonly used weirs in treatment plants are *rectangular, straight-edge,* and *V-notch*. Weir plates are bolted to a wall or trough so that they can be leveled independently of the structure. V-notch weirs are best suited for long weirs, such as those found in circular settling tanks.

V-notch weir angles range from 22.5 to 120°. The 90° V-notch is the most commonly used notched weir, particularly for circular settling tanks. Under free-flow (unsubmerged) conditions, as shown in Figure 3.7a, the head over a sharp-crested V-notch weir can be calculated with the following equation:

$$H = [Q/(C \tan \phi/2)]^{0.4} \quad (3.18)$$

where
- Q = flow, m³/s (ft³/s);
- ϕ = angle of the notch (deg);
- H = head over the crest, m (ft); and
- C = weir coefficient, 1.38 for SI units (2.5 for U.S. units), for a 90° weir.

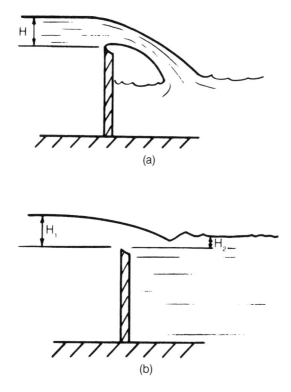

Figure 3.7 Weir flow conditions: (a) free flow and (b) submerged flow.

Head over a sharp-crested rectangular weir under free-flow conditions can be calculated using the following equation:

$$H = [Q/(C_w L)]^{0.67} \qquad (3.19)$$

where
- Q = flow, m³/s (ft³/s);
- L = length of weir, m (ft);
- H = head over the crest, m (ft); and
- C_w = coefficient that accounts for the approach velocity in terms of the ratio of the weir plate depth to the head over the crest. The values range widely depending on the approach velocity. The most commonly used value, 1.82 for SI units (3.3 for U.S. units), accurately represents deep water upstream, such as that found in tanks, with an extremely low approach velocity toward a sharp-crested rectangular weir.

Often, a rectangular weir has end contractions (that is, the ends of the weir project inward from the sides of the channel). End contractions cause the nappe to be aerated, thereby preventing pulsations of the upstream water surface. The shape of V-notch weirs also permits nappe aeration. The effects of the end contractions for rectangular weirs are accounted for by subtracting $0.1H$, where H is the weir head over the crest, from L, the length of the weir, for each end contraction. The equation for a rectangular weir with a contraction on each end is as follows:

$$H = [Q/(C_w[L - 0.2H])]^{0.67} \qquad (3.20)$$

These equations apply under free-flow conditions. When the downstream water elevation rises above the weir crest, the weir becomes submerged, as shown in Figure 3.7b. Submerged weirs are not used for flow measurement because the weir equations do not directly apply to submerged conditions, and thus produce sizable errors. The head over the crest can be calculated from the equations; however, the discharge flow must be corrected by using curves that have been developed experimentally (King and Brater, 1963). The head created upstream of a submerged weir should be adjusted accordingly.

Effluent Launders

Effluent launders are found at the outlet end of unit processes, such as settling tanks. These launders, or channels, serve to collect flow along its length as the flow moves downstream toward the outlet end. Conveying flow along the launders requires head to overcome the friction loss along the channel and the exchange of momentum as the water falls perpendicular to the flow stream and accelerates along the direction of flow. The *side-overflow formula*, described in detail in *Open-Channel Hydraulics* (Chow, 1959), is a differential equation for flow with increasing discharge. Because of the low average value of the Reynolds number along the entire length of the launder, use of a higher-than-normal friction factor is required (for example, 50% higher).

Flow Distribution (Channel and Pipe Headers)

Ports, or gates, are sometimes used to equally distribute flow from influent channels or pipe headers to unit processes, such as shown in Figure 3.8. The sensitivity of the

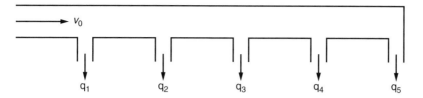

Figure 3.8 A manifold or distribution channel.

process dictates the need for equal flow distribution to each parallel tank and also within each tank. The head loss required to equally distribute the flow must overcome the hydraulic gradient along the channel or pipe. The hydraulic relationship between the header and gates in which the outflow is continuous or nearly continuous along the length of the header is (Camp and Graber, 1968; Fair et al., 1968)

$$h_{Lp} = \Delta h/(1 - m^2) \tag{3.21}$$

where

h_{Lp} = head loss through port, m (ft);
Δh = hydraulic grade differential along the channel or pipe, m (ft); and
m = ratio of the lowest flow and highest flow though the respective ports.

Energy in the form of head loss is required to cause the equal distribution of flow from the header to the process tank. This equation is used to determine the head loss that must be induced by the inlet gates or ports to ensure uniform distribution to all tanks from a common inlet header or distribution channel. For example, to keep the flow to port 1 in Figure 3.8 within 5% ($q_1/q_5 < 1.05$) of that to any other port such as port 5, the head loss through the gate should be approximately 10 times the hydraulic grade level differential over the entire length of the distribution channel or pipe header.

To determine the actual head loss through a gate or port, the following equation applies:

$$h_{Lg} = (Q/CA)^2/2g \tag{3.22}$$

where

h_{Lg} = head loss through the gate, m (ft);
Q = flow, m³/s (ft³/s);
C = gate or orifice coefficient (dimensionless);
A = area of gate or port opening, m² (ft²); and
g = acceleration due to gravity, 9.8 m/s² (32.2 ft/s²).

Values of the gate, or orifice, coefficient for calculating head loss through ports under varying conditions can be found in *Handbook of Applied Hydraulics* (Davis and Sorensen, 1969).

The hydraulic gradient differential is influenced by velocity head and head loss along the upstream header pipe (Camp and Graber, 1968)

$$\Delta h = (v_0^2/2g) - h_{Lh} \tag{3.23}$$

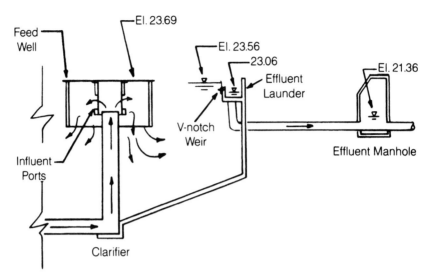

Figure 3.9 Hydraulics for a typical clarifier.

where
    Δh = hydraulic grade differential, m (ft);
    v_0 = header inlet velocity, m/s (ft/sec) ;
    g = acceleration due to gravity, 9.8 m/s² (32.2 ft/sec²); and
    h_{Lh} = header head loss, m (ft).

When the outlet flow from a pressurized header approximates a uniform continuous outflow along the length of the header, the head loss along the header can be estimated by calculating the head loss as if the inlet flow to the header were to be conveyed along the entire length of header and dividing the result by 3. This estimation procedure is valid for the ratio of the lowest flow and highest flow through the respective ports greater than 0.9.

Unit Process Hydraulics and Other Hydraulic Elements

Within each unit operation, devices are used to distribute flow, maintain a certain water depth, and control the flow. Typical devices include shutoff gates, weir gates, valves, ports, weirs, baffles, orifices, launders, and underdrains. Each of these devices imposes a head loss on the system and should be considered in the hydraulic calculations. For example, Figure 3.9 shows that head loss must be calculated between the manhole and clarifier, the side-weir equation is needed for calculating the water elevation in the clarifier effluent launder, and the V-notch weir equation is needed for calculating the head over the weir crest.

If the flow is to be equally distributed to multiple tanks or within an individual tank, distribution boxes, channels, and header pipes can be used for this purpose. Although modulating ports, gates, or valves can provide equal distribution, they require control systems that share the disadvantages inherent to any electrical and mechanical system (for example, failure and high maintenance). Where possible to install, static devices are always better suited for the distribution of flow.

Screens

Some processes can impose varying heads. One such process is the influent screen. The wastewater flowing through the bars of the screen creates head loss. The head loss increases if the screen becomes partially or wholly clogged by debris. As the screen is cleaned by mechanical rakes or other means, the head loss decreases. The typical head loss for bar screens ranges from 0.15 to 0.76 m (0.5 to 2.5 ft). The equation for calculating head losses through screens is as follows:

$$h_{Ls} = k(v_B^2 - v^2)/2g \tag{3.24}$$

where
- h_{Ls} = head loss through the screen, m (ft);
- v_B = velocity through bar screen, m/s (ft/s);
- v = velocity upstream of bar screen, m/s (ft/s);
- g = acceleration due to gravity, 9.8 m/s² (32.2 ft/s²); and
- k = head loss coefficient (1.0 to 0.7).

Expressing the upstream velocity in terms of multiples of the velocity through the screen (clean [full flow area] or partially clogged [decreased flow area]) simplifies the calculation of head loss. Knowing the downstream water depth, the Bernoulli equation is used to determine the upstream water depth. The highest head loss through the screens is based on the maximum clogging of the screen that can be tolerated. For mechanically cleaned screens, the head loss or change of head through the screens is regulated by a timer or, in some cases, a differential level sensor. Best design is to use a timer with an upstream head override. Head differential override is also used by many engineers.

Grit Tanks

Regardless of the type of grit tank used (gravity, aerated, etc.), equal distribution of flow and grit to the tanks is important. Although the flow may be equally distributed, equal distribution of grit may not occur. This is especially true when inlet header channels are used. Flow separation points (such as those found at sharp corners) along the inlet header channels can cause dead zones where grit drops out. Once settled, the grit tends to eventually migrate to the closest tank or tanks. Other than eliminating the possibility of dead zones, which is sometimes a difficult task, the grit-laden inlet flow can be aerated to keep the flow well mixed and the grit in suspension. The use of air to keep the grit suspended and well mixed may require odor control facilities to treat the offgases.

Primary Settling Tanks

Much of the head loss found in wastewater treatment plants is associated with inlet piping to primary settling tanks. One of the largest losses can be caused by the inlet ports to circular settling tanks. The size of the ports at the top of the inlet columns, and the associated head losses, often cannot be predicted because the required information depends on the tank manufacturer that is selected. To cover this uncertainty, tank manufacturers should be consulted for the appropriate head loss value during the design phase of a treatment plant project. In addition, the head loss caused by the inlet ports must be determined. The inlet column and inlet port head loss at the maximum design flowrate should be included in the equipment specifications for the settling tank.

In addition to the equal flow splitting required for each settling tank, equal distribution of flow is required within each tank itself. A rectangular settling tank, for example, requires good distribution of flow along its inlet conduit and through its ports or inlet gates. Best design seems to be the use of ports with diffusers or baffles.

Aeration Tanks

As with any unit process, equal distribution of flow to individual aeration tanks and within each tank is important. Head loss is needed to ensure equal distribution of flow. This generally occurs in the form of head loss caused by inlet gates. Although not as effective hydraulically, inlet weirs can also be used to distribute flows. The inlet gates should be sized to create sufficient head loss to overcome the head differential along the inlet header. To reduce excessive head losses through the inlet gates, large header channels are used to keep the head differential along the channel low. The channels may be aerated to keep sludge in suspension. Aeration of the channels is not an option in cases where the treatment process would be affected (i.e., anoxic tanks).

The mode of operation for the aeration tank should be addressed during calculation of the hydraulic profile. Aeration tanks can be operated as completely mixed flow, plug flow, or stepped flow (feed) and its variations. The calculation of the hydraulic grade line should anticipate unplanned changes to the operating mode because of future operational requirements. Regardless of the mode, equal distribution of flow to each tank and its cells should be adequately addressed during the hydraulic design. To prevent excessive head losses, multiple ports of different sizes—each dedicated to a particular operating mode—may be required. In plants that experience large diurnal flow variations, the parallel gates in the distribution channels must be frequently adjusted if the flow to various tanks is to be equalized. The effluent collector channel for aeration tanks can have zones of low velocity. To prevent sludge deposition, however, aeration of the effluent channel may be necessary.

Secondary Settling Tanks

Head loss caused by the inlet column and ports should be estimated and included in the equipment specifications for circular settling tanks. Often, return activated sludge is drawn directly from the settling tank hopper bottom. For some activated-sludge treatment plants, the recirculated activated sludge is drawn off via a pipe situated within the inlet column. For such piping configurations, the return activated sludge is withdrawn from above the inlet ports to the settling tank. Head losses affected by the obstructing return activated sludge withdrawal pipe should be included in the calculation of the column and port head losses.

Disinfection Tanks

Equal distribution of flow to and within disinfection or contact tanks is important. Because sludge settling is not as much of a concern, the tanks are designed to operate at low velocities. This permits head loss to be low. Head loss is kept at a minimum and is concentrated at the inlet to cause equal flow distribution and at the effluent weir to maintain proper water depth. To minimize weir head and to allow for proper operation of scum-removal equipment intolerant to wide water level fluctuations, finger weirs are commonly used. Finger weirs consist of multiple channels fitted with weirs that extend out into the tank at its outlet end. Baffles are often used to minimize short circuiting.

Other Unit Processes

Other unit treatment processes, such as trickling filters, rotating biological contactors, and gravity filters, have hydraulic requirements, some of which are peculiar to their design. For trickling filters to operate, sufficient water surface elevation must exist at the inlet structure to push the water through the distributor and keep it rotating over the media. Rotating biological contactors have minor head losses because water only has to flow through the trough. For effluent gravity filters, sufficient water surface elevation must exist over the filter media to convey the liquid through the media and the underdrains. Therefore, the manufacturers of these and other treatment units should be consulted to gather information on the water depth, velocity, and head loss requirements of the respective equipment.

Yard Piping

Yard piping is the system of pipes connecting the various processes. The construction material for yard pipe is almost always ductile iron pipe because of its strength and ease of disassembly and repair. Joints are commonly flanged for ease of repair. Piping should never be buried under concrete slabs or other tanks.

Outfalls

For ocean discharge, the higher density of seawater (1027 kg/m^3 [1.99 lb-s^2/ft^4]) must be taken into account when computing the static head at the outfall outlet. (For the U.S. system, the more familiar specific weight of 64 lb/ft^3 is used to calculate the static head of seawater.) Although distribution along an outfall's diffuser outlets is based on the same basic principles as those discussed previously in this chapter, proper distribution of effluent through the outlet diffusers depends on many other complicated parameters and interactions. Often, a detailed analysis of the currents in receiving waters is required before the hydraulic design of the outfall can be initiated. Outfalls should be sloped slightly *upward*, toward the effluent end, and not be level or be sloped downward. This allows bubbles to flow downstream along the crown and avoid bubble buildup and air binding in the pipe.

Additional considerations for the proper hydraulic design of outfalls include seawater purging for deep-water tunneled diffusers with long riser pipes; air in the outfall as it affects flow capacity, buoyancy, and related structural failure; and surging and related water hammer (Grace, 1978; Gunnerson and French, 1996).

Pumps and Pumping

Pump Characteristics

Pumps are used to add energy to fluids such as water and wastewater. In closed conduits, centrifugal pumps with various types of impellers are the most popular because they are efficient and dependable. In some cases, for example, in the pumping of viscous materials such as sludge, positive displacement pumps are useful. Dewatering devices are often fed by progressing cavity pumps that squeeze the sludge through pipes without surges. In open channels, Archimedes screws are the most popular devices for efficiently and effectively lifting wastewater moderate elevations. Sludge pumps are discussed further in Chapter 12. The discussion that follows applies to rotating impeller pumps, the most popular pumps in wastewater treatment plants.

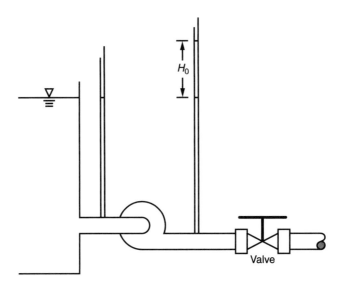

Figure 3.10 Pump setup for establishing its characteristic curve.

The ability of a pump to function in a given application is described by its *characteristic curves*, defined as the head developed by the pump versus the flowrate produced. The head (H) versus flowrate (Q) curve is constructed for a given pump using a setup such as pictured in Figure 3.10. A pump is installed in the line from a reservoir with positive suction pressure (the water level in the reservoir is above the elevation of the pump). The valve at the pump-discharge side is closed. Pressure in the pipeline can be measured immediately before and after the pump.

If the valve is closed and the pump is not on, the pressure at both points (shown as manometer tubes in Figure 3.10) are at the same level as the water in the reservoir. If now the pump is turned on, the pressure at the intake side remains the same, but the pump causes the pressure in the discharge side to increase by some value H_0, measured in meters (or feet) of water in the manometer tube. This value is plotted on an H versus Q plot as shown in Figure 3.11.

If the valve on the discharge side is now opened slightly, some flow will result, but total energy has to be equal so that the height of water in the manometer tube will drop slightly, indicating a lower pressure in the discharge line, the pressure head being converted to velocity head. These two values of H and its corresponding Q, call them H_1 and Q_1, are then plotted in Figure 3.11. As the valve is opened wider and wider, H continues to drop as Q increases, resulting in a plot known as the characteristic curve for this pump. Often a given style of pump (rotational speed, impeller width, angle of impellers, etc.) is manufactured in different diameters. If each pump is tested, a family of characteristic curves will result: each curve is specific to that pump impeller.

The amount of energy used to move the water and develop the pressure head is known as the *brake horsepower*, or BHP. The energy output from the pump is known as *water horsepower*, WHP, and defined as

$$\text{WHP} = QH \qquad (3.25)$$

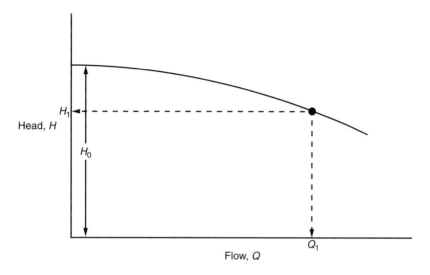

Figure 3.11 The pump characteristic curve.

where
Q = flowrate, m³/s (ft³/s);
H = head developed, N/m² (lb/ft²); and
WHP = water horsepower, N·m/s (lb-ft/s).

Thus the units of WHP are N·m/s, or power.

The efficiency of a pump is measured as the ratio of WHP (energy in) over BHP (energy out), as

$$\text{Efficiency} = \text{WHP/BHP} \tag{3.26}$$

For any run with the pump as discussed above, the energy input to the pump is measured as the kilowatts used, and the resulting output is calculated by multiplying the flowrate at the corresponding head produced. Thus for each flowrate, an energy efficiency can be calculated, and this can be plotted on the H versus Q plot. If the family of pump impellers of a given design (impeller diameters d_1, d_2, d_3, ...) are similarly tested and the efficiencies are calculated, a series of isoefficiency curves will result, as shown in a typical plot in Figure 3.12. This plot is then a complete picture of the characteristics of this pump and used in the selection of a pump for a given application.

System Head Curves

An application of a pump requires that the job the pump is supposed to do be defined. This is done by drawing a *system head curve*.

Suppose a specific application requires that water be pumped from the elevation in tank A to the elevation in tank B, as shown in Figure 3.13. The length of the pipe and all of its appurtenances such as valves, elbows, and other sources of minor losses, are well defined. The static head, or the difference in elevations, is H_0, and this is plotted on an H versus Q plot, as shown in Figure 3.14.

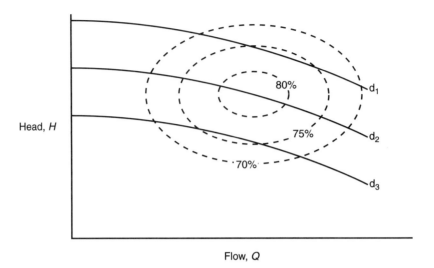

Figure 3.12 A series of characteristic curves and associated efficiencies for a family of pumps (d = impeller diameter).

An arbitrary flow is selected and the head loss at that flow is calculated, using equations such as Darcy–Weisbach or Hazen–Williams. For some flowrate, Q_1, the head loss is then calculated as H_1, including losses due to valves, pipe fittings, meters, and so on. This value of H_1 is then plotted at the corresponding Q_1. To complete the curve,

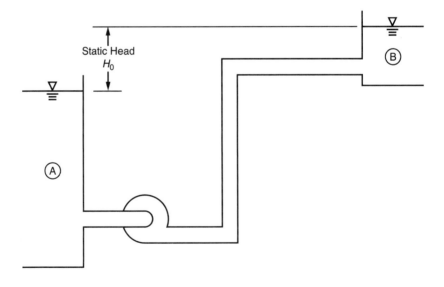

Figure 3.13 Pumping from reservoir A to reservoir B.

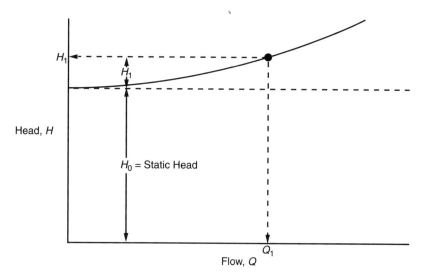

Figure 3.14 System head curve.

other values of Q must be chosen and the corresponding head losses calculated. This is, however, a tedious task, and the usefulness of the equation,

$$\frac{Q_1}{Q_2} = \left[\frac{h_{L1}}{h_{L2}}\right]^{0.54} \quad (3.27)$$

now becomes obvious. If the head loss is known for any flow, the head loss for any other flow can be calculated using this simple ratio. The full system head curve is then drawn.

Pump Selection

For a given application, as described by the system head curve, a pump is selected by superimposing the system head curve on the characteristic curves and selecting a pump that has a high efficiency. Figure 3.15 shows how the system head curve from Figure 3.14 is superimposed on the pump characteristic curve, Figure 3.12. The point at which these two curves, the characteristic curve and the system head curve, intersect, is known as the *operating point*. If this pump is applied to the system of pipes and appurtenances for which the system head curve was drawn and the pump is turned on, it should discharge the amount of water defined by the operating point.

Sometimes the head to be developed is quite high, and a single pump will not be sufficient to produce the necessary energy. In that case, two pumps can be placed in series, as shown in Figure 3.16. The H versus Q curves are added (vertically) to produce the characteristic curve for the two pumps in series. In other cases, any single pump may be inadequate to provide the required flowrate, and several pumps are then placed in parallel, as shown in Figure 3.17. In this case the system head curves are added horizontally. Such an arrangement is used most frequently where a significant variation in flow is expected, such as influent pumping stations.

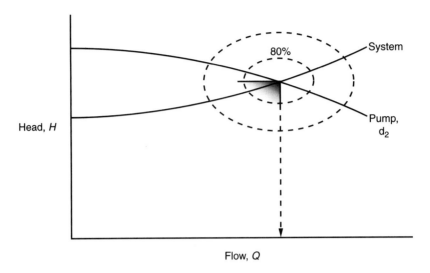

Figure 3.15 System head curve superimposed on pump characteristic curves to determine operating point.

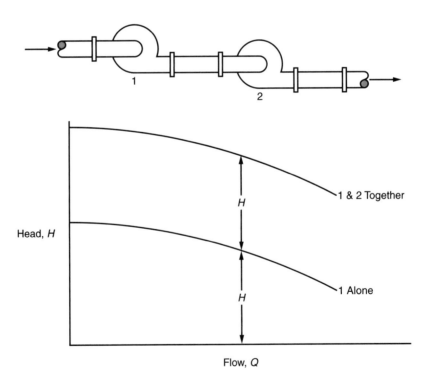

Figure 3.16 Pumps in series.

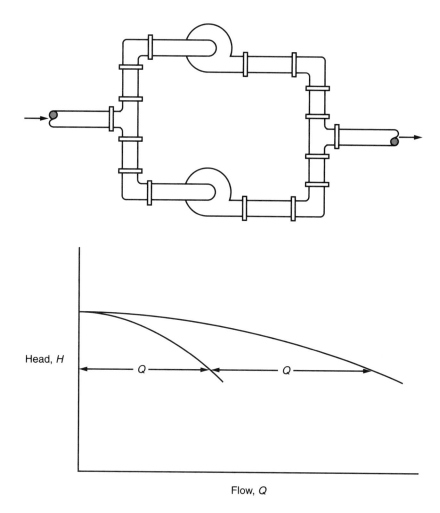

Figure 3.17 Pumps in parallel.

Conclusions

Humans have an inexplicable affinity for water. There is something mighty, clean, and beautiful about water in nature and we are drawn to it, be it a small puddle in the street or the magnificent Niagara Falls. This is all very Darwinian, of course. Our very survival depends on the availability of water for drinking, transport, power production, manufacturing, recreation, and yes, for the disposal of wastes. Being able to understand how water behaves in the macro sense is one of the great pleasures of engineering. Everyone knows that water flows downhill, but only engineers know how to make water run uphill. Hydraulic engineering, the topic of this chapter, is fun. But it is also serious. In the design of wastewater treatment plants, hydraulics can make the difference between success and failure. Overflowing clarifiers, pumps that do not have enough suction head, poorly operating distribution boxes—these and thousands of other hydraulic elements in a wastewater treatment plant—deserve careful engineering design and consideration. Because of space limitations, this chapter covers only the bare essentials. There is much more serious business, and fun, to be found in hydraulic engineering.

References

Camp, T. R.; Graber, S. D. (1968) Dispersion Conduits. *J. Sanit. Eng. Div., Proc. Am. Soc. Civ. Eng.*, **94**.

Chow, V. T. (1959) *Open-Channel Hydraulics*; McGraw-Hill: New York.

Davis, C.; Sorensen, K. (1969) *Handbook of Applied Hydraulics*; McGraw-Hill: New York.

Fair, G. M.; Geyer, J. Ch.; Okun, D. A. (1968) *Water and Wastewater Engineering: Water Purification and Wastewater Treatment and Disposal*; Wiley & Sons: New York; Vol. 2.

Grace, R. A. (1978) *Marine Outfall Systems: Planning, Design and Construction*; Prentice-Hall: Upper Saddle River, New Jersey.

Gunnerson, C. G.; French, J. A. (1996) *Wastewater Management for Coastal Cities: The Ocean Disposal Option*; Springer-Verlag: New York.

King, H. W.; Brater, E. F. (1963) *Handbook of Hydraulics*; McGraw-Hill: New York.

Water Pollution Control Federation (1982) *Gravity Sanitary Sewer Design and Construction*; Manual of Practice No. FD-5; Washington, D.C.

Symbols Used in this Chapter

A = area of gate or port opening, m² (ft²)
BHP = brake horsepower, N·m/s (lb-ft/s)
C = gate or orifice coefficient (dimensionless)
C = Hazen–Williams friction coefficient
C = surface-roughness coefficient (dimensionless)
C = weir coefficient
d = inside diameter of pipe, m (ft)
f = friction factor (dimensionless)
g = acceleration due to gravity, 9.8 m/s² (32.2 ft/s²)
h_L = friction head loss, m (ft)
h_{Lg} = head loss through gate, m (ft)
h_{Lp} = head loss through port, m (ft)
h_{Ls} = head loss through screen, m (ft)
H = head developed, N/m² (lb/ft²)
H = head over crest, m (ft)
k = head loss coefficient
K = a constant, function of the pipe roughness
L = pipe or channel, m (ft)
m = ratio of the lowest flow and highest flow through the respective ports
n = roughness coefficient (dimensionless)
P = pressure, N/m² (lb/in.²)
Q = flow, m³/s (ft²/s)
R = hydraulic radius (cross-sectional area divided by wetted perimeter), m (ft)
S = slope of the energy gradient (h_L/L)
v = velocity, m/s (ft/s)
v = velocity upstream of bar screen, m/s (ft/s)
v_B = velocity through bar screen, m/s (ft/s)

WHP = water horsepower, N·m/s (lb-ft/s)
Y = water depth, m (ft)
Z = baseline elevation (e.g., channel invert), m (ft)
Δh = hydraulic grade differential along the channel or pipe, m (ft)
γ = specific weight, N/m³ (in. lb/ft³)
μ = fluid viscosity, kg·m/s (lb-s/ft²)
ϕ = angle of the notch (deg)
ρ = fluid density, kg/m³ (lb s²/ft⁴)

Preliminary Treatment

Introduction

Preliminary treatment prepares wastewater influent for further treatment by reducing or eliminating problems that could otherwise impede operation or unduly increase maintenance of downstream processes and equipment. Typical problems include large solids and rags; abrasive grit; odors; and, in some cases, unacceptably high peak hydraulic or organic loadings. This chapter presents descriptions of and design considerations for preliminary treatment processes, including screening, grit removal, septage handling, and flow equalization.

Screening

Purpose of Screening

Screening, one of the oldest treatment processes, removes gross pollutants from the wastewater stream to protect downstream operations and equipment from damage. Screening, which removes oversized solids, is distinguished from comminution and grinding, which reduce the size of solids. Comminution and grinding, discussed later in this section, are used for preliminary treatment where handling of screenings would be impractical.

Some modern wastewater treatment plants use both coarse and fine screens. Coarse screens, with 6-mm (0.25-in.) and larger openings, remove large solids, rags, and debris

Table 4.1 Typical design criteria for coarse screening equipment.

Item	Range[a]	Comment
Trash rack		
Openings	38–150 mm	Commonly used on combined systems—opening size depends on equipment being protected
Manual screen		
Openings	25–50 mm	Used in small plants or in bypass channels
Approach velocity	0.3–0.6 m/s	
Mechanically cleaned bar screen		
Openings	6–38 mm	18-mm opening considered satisfactory for protection of downstream equipment
Approach velocity (maximum)	0.6–1.2 m/s	
Minimum velocity	0.3–0.6 m/s	Necessary to prevent grit accumulation
Continuous screen		
Openings	6–38 mm	This type of screen effective in the 6- to 18-mm range
Approach velocity (maximum)	0.6–1.2 m/s	
Minimum velocity	0.3–0.6 m/s	
Allowable head loss	0.15–0.6 m	
Comminutor (size reduction only)		
Openings	6–13 mm	Opening a function of the hydraulic capacity of unit
Grinder (size reduction only)		
Openings	6–13 mm	
Typical head loss	300–450 mm	In open channel

[a] mm × 0.03937 = in.; m/s × 3.281 = ft/s; m × 3.281 = ft.

from wastewater. Fine screens, with 1.5- to 6-mm (0.059- to 0.25-in.) openings, can be used to remove materials and may significantly increase operation and maintenance in downstream liquid and sludge-handling processes, particularly in systems without primary treatment. Very fine screens, with openings of 0.2 to 1.5 mm (0.008 to 0.059 in.), following coarse or fine screens can reduce suspended solids to near the primary treatment level. These very fine screens plus microscreens, with openings of 1 µm to 0.3 mm (3.9×10^{-8} to 1.2×10^{-2} in.), can be used for "effluent polishing" to upgrade secondary effluent to tertiary standards. This chapter covers coarse screens and comminution or grinding while fine screening is discussed later under primary treatment.

Bar Screens

Coarse screens, often called *bar screens*, are the first unit treatment process in a wastewater treatment plant. Bar screens remove solids and trash that could otherwise damage or interfere with downstream operations such as pumps, valves, mechanical aerators, and biological filters. Coarse screens consist of vertical or inclined steel bars spaced at equal intervals across a channel through which wastewater flows and are either manually or mechanically cleaned. Criteria used in their design include bar size, spacing, and angle from the vertical, channel width, and wastewater approach velocity. The most commonly used coarse screens are mechanically cleaned bar screens. This section also includes comminutors and grinders because of their purpose and position ahead of primary treatment. Table 4.1 summarizes typical size ranges and design criteria for coarse screening equipment.

Figure 4.1 Hand cleaned bar screen at a small wastewater treatment plant.

Trash racks are bar screens with large openings, 38 to 150 mm (1.5 to 6 in.), designed to prevent logs, timber, stumps, and other large, heavy debris from entering treatment processes. Trash racks, followed by bar screens with smaller openings, are principally used in combined sewerage systems that carry large amounts of large debris, especially during storms. Trash racks may be mechanically or manually cleaned. Where space is limited, plants sometimes have basket-type trash racks that are manually hoisted and cleaned.

Manually Cleaned Bar Screens
Manually cleaned bar screens have 30- to 50-mm (1- to 2-in.) openings with the bars set 30 to 45° from the vertical to ease cleaning (Figure 4.1). Screenings are manually raked from the screen onto a perforated plate where they drain before removal for disposal. If screens are cleaned infrequently, the backwater caused by the buildup of debris between cleanings may cause flow surges when the screens are cleaned. These high-velocity surges can reduce the solids-capture efficiency of downstream units.

Although manually cleaned screens require little or no equipment maintenance, they demand frequent raking to avoid clogging. Manually cleaned screens exist primarily in older, small (<4000 m³/d [1 mgd]) treatment facilities and in bypasses of comminutors and mechanically cleaned screens.

Mechanically Cleaned Screens
Mechanically cleaned screens have openings ranging from 6 to 38 mm (0.25 to 1.5 in.), with bars set from 0 to 30° from the vertical. Mechanical cleaning, compared with manual cleaning, tends to reduce labor cost; improve flow conditions and screening capture; reduce nuisances; and, in combined systems, better handle large quantities of stormwater debris and screenings. A mechanically cleaned screen is, therefore, almost

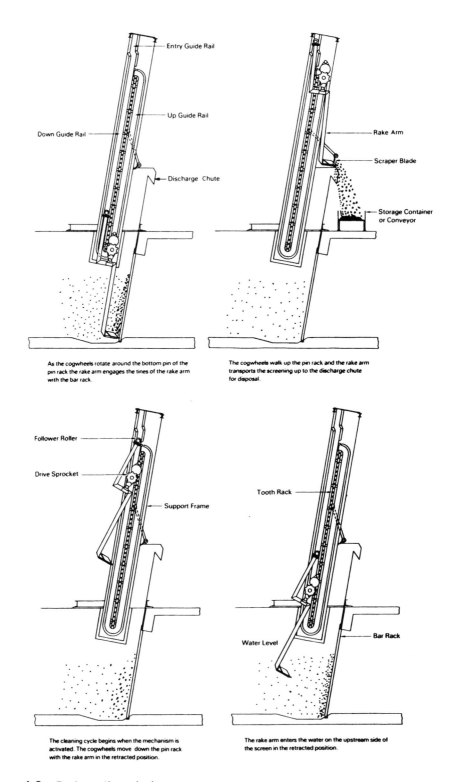

Figure 4.2 Reciprocating rake bar screen.

always specified for new plants of all sizes. Many types of mechanically cleaned bar screens are manufactured, including, but not limited to, chain or cable driven with front or back cleaning, reciprocating rake, catenary, and continuous.

Chain- or Cable-Driven Screens

Chain- or cable-driven screens are mechanical bar screens with a chain or cable mechanism to move the rake teeth through the screen openings. These screens are the oldest mechanized screening devices. These types of screens are manufactured in several configurations: front clean–front return, front clean–rear return, and back (or through) clean–rear return. The front clean–front return type most efficiently retains captured screenings by minimizing carryover. Such screens are used extensively in typical municipal wastewater applications. Front clean–rear return screens are used for heavy-duty applications. A disadvantage of a rear-return configuration is the possible return to the screened waste stream of debris that is not dislodged by the rake-cleaning mechanism. For any of the front-cleaned configurations, bars should be trapezoidal in cross section, thus providing a tapered opening to prevent wedging and trapping of solid material between the bars. There is the possibility of bottom jamming by unusual deposits of trash, particularly if no trash racks are used. Also, the chain- or cable-driven raking mechanism consisting of submerged sprockets or other mechanical devices is subject to fouling by grit and rags. Frequent inspection and maintenance of the drive mechanism are required, and, depending on the screen design, channel dewatering may be necessary.

With the back-cleaned, or through-cleaned screening configuration, the bars protect the rake and operating mechanisms from damage caused by large objects in the wastewater stream. The back-cleaning action pushes debris away from the bars, thus preventing these materials from being compressed and jammed between the bars. However, the longer rake teeth protruding through the bars are more susceptible to bending and breakage than are the teeth of front-cleaned screens.

Reciprocating Rake Screen

The reciprocating rake screen can be equipped with a back-clean–back-return mechanism or with a front-clean–front-return mechanism that minimizes solids carryover (Figure 4.2). The up-and-down reciprocating motion of the rake, similar to that of a person raking a manual bar screen, minimizes the possibility of jamming. These screens remain popular because their lack of submerged moving parts allows easy, trouble-free inspection and maintenance without dewatering of the channel.

Headroom requirements for reciprocating rake screens are greater than those for other types of screens. The estimated headroom requirement can be determined by adding the vertical depth of the screen to the screenings discharge height above the floor. Headroom deserves special attention from the design engineer.

Although many drive mechanisms are available—chain and cable, hydraulic, and screw operated—the most successful designs include a cogwheel-type drive. For these designs, the entire cleaning rake assembly, including the gear motor, is carriage-mounted on cog wheels that travel on a fixed pin or gear rack. The drive mechanism is designed to allow the rake to ride over obstructions encountered during the cleaning stroke. In the unlikely event that the rake becomes completely jammed, a limit switch is activated to turn off the drive motor. A disadvantage of reciprocating screens is the single rake, which limits their capacity to handle extreme loads, especially for deep applications where cycle times are long. Also, these systems require high overhead clearance.

Figure 4.3 Catenary bar screen.

Catenary Screens
Catenary screens consist of heavy tooth rakes held against the screen by the weight of its chain (Figure 4.3). The term *catenary* stems from the catenary curve formed by the operating chain ahead of the screen. A curved transition piece at the base of the screen

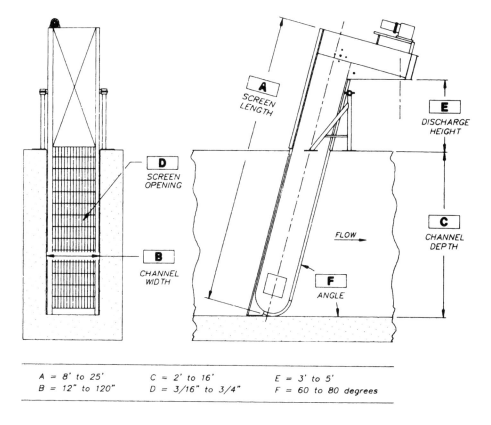

Figure 4.4 Continuous self-cleaning screen (in. × 24.5 = mm; ft × 0.3048 = m).

allows for efficient removal of solids captured at the bottom. Like reciprocating rake screens, all sprockets, shafts, and bearings are located out of the flow stream to reduce wear and corrosion and ease required maintenance. Because the cleaning rake is held against the bars primarily by just the weight of the chains, the rake can be pulled over large rags or solids, which are tightly held against the screen.

Continuous Self-Cleaning Screens

Continuous self-cleaning screens consist of a continuous "belt" of plastic or stainless steel elements that are pulled through the wastewater to provide screening along the entire length of the screen (Figures 4.4). Screen openings are designed with both horizontal and vertical limiting dimensions; the vertical spacing is slightly larger than the horizontal spacing. Continuous screens may have openings as small as 1 mm (0.039 in.), ranging up to more than 80 mm (3 in.). The greater solids-handling capacity of these screens allows smaller openings to be used, which results in greater capture of solids from the waste stream. Also, the continuous screening action of these units allows efficient removal of large quantities of solids. Continuous screens have either a lower gear sprocket or a guide rail at the channel bottom to support the screen elements. Construction of a recessed notch or step in the channel at the screen bottom is a good practice that can help prevent buildup of grit and debris ahead of the unit. Continuous

Figure 4.5 Comminutor used at a small treatment plant.

screens are designed to be pivoted up and out of the channel for maintenance. Some screen systems include spray bars and brushes to improve cleaning of screen elements. A disadvantage of continuous screens is possible solids carryover resulting from the front-clean–back-return design.

One problem with fine screens is that they are prone to clogging during peak-flow conditions where the high velocities in sewers scour grit and other fine particles from the bottom of sewers. This overwhelms the capabilities of the screens, and good design requires a bypass system and a way of dealing with the debris in the bypass flow. The bypass must also have screens to protect downstream operations. Such bypass screens have 6 mm (0.25 in.) openings.

Comminutors and Grinders

Comminutors
Comminutors are installed in the wastewater flow channel to screen and shred material to sizes from 6 to 19 mm (0.25 to 0.75 in.) without removing particles from the flow (Figure 4.5). Many small plants ($<1.9 \times 10^4$ m^3/d [<5 mgd]) continue to use comminutors. The use of such a device is intended to reduce odors, flies, and unsightliness often associated with screenings that are handled by other means and essentially eliminate screenings removal and handling. Unfortunately, solids from comminutors and screenings grinders have caused downstream problems, including deposits of plastics in digestion tanks and rag accumulations on air diffusers. Pulverized synthetic materials will not decompose in the digestion process and, if not removed, they could reduce public acceptance of biosolids for reuse as a soil amendment. Comminutors may create "ropes" or "balls" of material (particularly rags) that can clog treatment equipment (for example, mechanical aerators, mixers, pump impellers, pipelines, and heat exchangers).

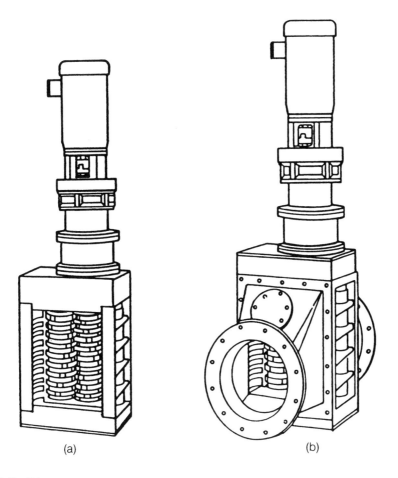

Figure 4.6 Wastewater grinders: (a) channel unit and (b) in-line unit.

As experience with comminutors in the United States and elsewhere has been generally unsatisfactory, they are avoided in new designs and are being removed from many existing plants.

Grinders

Grinders, shown in Figure 4.6, consist of two sets of counter-rotating, intermeshing cutters that trap and shear wastewater solids into a consistent particle size of 6 to 9 mm (0.25 to 0.38 in.). The cutters are stacked on two steel or stainless steel drive shafts with intermediate spacers. The shafts counter-rotate at different speeds to produce a self-cleaning action of the cutters.

Because of their design, grinders aggressively chop, rather than shred, wastewater solids. The chopping action helps reduce, but not eliminate, the formation of rag balls and rag ropes produced by comminutors. Where wastewater contains large quantities of solids and rags, grinders are sometimes installed downstream of coarse screens to help prevent frequent jamming and excessive wear. The grinders' restricted opening area, compared with screens, tends to cause greater head losses, which must be accounted for

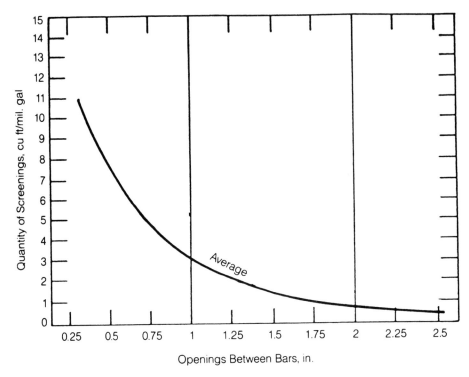

Figure 4.7 Average volumes of coarse screenings as a function of openings between bars (ft³/mil. gal × 0.0075 = L/m³; in. × 25.4 = mm).

in the hydraulic design. Grit and other solids impose severe wear on grinders; therefore, they require routine inspection every six months and replacement of bearings and cutter teeth every one to three years.

Screenings Quantities and Characteristics

Both the quantities and characteristics of screenings to be collected and removed require careful consideration before design of screening systems. The quantity of coarse screenings to be removed can vary significantly depending on the bar screen opening, wastewater flow, characteristics of the municipality, type of collection system, and type of screen. These quantities are difficult to estimate without actual operating data, which should be obtained whenever possible.

For separate collection systems, the most important variable affecting screenings quantity is the clear opening between bars. Information, such as that shown in Figure 4.7, provides average and maximum volumes of screenings per unit of wastewater flow for various bar openings.

For bar openings between 30 and 50 mm (1 and 2 in.), the volume of screenings removed per unit of flow is approximately proportional to the clear opening dimension. In this range, for each 13-mm (0.5-in.) reduction of clear opening size, the volume of screenings will approximately double. For screen openings smaller than 30 mm, the volume of screenings removed will increase more rapidly as the clear opening decreases to approximately 6 mm (0.25 in.). Removals by trash racks have little relation to the volume of flow.

Table 4.2 Typical design properties of coarse screenings (U.S. EPA 1979, 1987; WPCF, 1989).

Item	Range[a]	Comment
Quantities		
Separate sewer sytem		
Average	3.5–35 L/1000 m^3	Function of screen opening size and system characteristics
Peaking factor (hourly flows)	1:1–5:1	
Combined sewer system		
Average	3.5–84 L/1000 m^3	
Peaking factor (hourly flows)	2:1–>20:1	
Solids content	10–20%	
Bulk density	640–1100 kg/m^3	
Volatile content of solids	70–95%	
Fuel value	12,600 kJ/kg	

[a] L/1000 m^3 × 0.134 = ft^3/mil. gal: kg/m^3 × 0.0624 = lb/ft^3; kJ/kg × 0.43 = Btu/lb.

The quantity of screenings removed will partly depend on the length and slope of the collection system and whether pumping stations exist. Screenings quantities may be greater with a short, gently sloping collection system with low turbulence than with lengthy, steep interceptor systems or systems with pumping stations. This difference stems from disintegration of solids exposed to long-duration turbulence. The effect of turbulence will be more significant for bar openings smaller than 13 mm, which capture more organic solids.

Composition, as well as volume, affects the disposal of screenings. Coarse screenings consist of rags, sticks, leaves, food particles, bones, plastics, bottle caps, and rocks. With smaller openings near 6 mm, cigarette butts, fecal matter, and other organic matter are included in the screenings. The large organic portion of screenings readily decomposes; plastics and inorganics are not biodegradable. In screenings, some of the material is inherently odorous and other material absorbs odorous compounds from the wastewater. Some of the pathogenic organisms from the wastewater will also likely enter the screenings.

Screenings, containing approximately 10 to 20% dry solids, will have a bulk density ranging from 600 to 1100 kg/m^3 (40 to 70 lb/ft^3). The volatile content of the solids ranges from 70 to 95%. Table 4.2 summarizes information pertaining to screenings quantities and composition.

Design Practice

A bar opening of 19 mm (0.75 in.) or smaller will adequately protect downstream equipment. The smaller the opening, the greater the removal of solids and degree of protection. However, exercise caution when selecting screen openings smaller than 13 mm (0.5 in.) for plants served by gently sloping gravity collection systems because of the potential increased capture of fecal and other organic matter. Excessive capture of such solids should be avoided because they are better handled by downstream primary and secondary treatment processes. In general, selection of screen openings smaller than 6 mm is best applied where steeply sloping sewers or pumping stations provide sufficient turbulence in the collection system to fragment organic matter ahead of the screen.

Trash racks are now seldom used ahead of coarse screens or grinders, except for combined collection systems with significant quantities of stormwater. Trash rack bar openings range from 38 to 80 mm (1.5 to 3 in.). Coarse screen openings range from 6 to 30 mm (0.25 to 1 in.). A 19-mm bar spacing is sometimes too large because twigs, ball-point pens, and other plastics can pass sideways through the bars. Hence, an increasing number of plants are switching to continuous-type screens with smaller openings in the 6- to 13-mm (0.25- to 0.5-in.) range. Alternatively, the primary sludge from plants with larger screen openings passes through grinders to prevent problems with downstream solids-handling processes.

The approach velocity to the screen must not fall below a self-cleaning value or rise enough to dislodge screenings. The lower the velocity, the more material that will be removed from a given waste and the more solids will be deposited in the channel. Ideally, the velocity in the rack chamber of bar screens should be greater than 0.4 m/s (1.3 ft/s) at minimum flows to avoid grit deposition if grit chambers follow bar screens. However, this is not always possible with the typical diurnal fluctuation in wastewater flows. As a reasonable compromise, the channel could be designed so that a velocity of at least 0.8 m/s (2.5 ft/s) for resuspending solids is attained during peak flow periods of the day. Where significant stormwater must be handled, approach velocities of approximately 0.9 m/s (3 ft/s) are needed to avoid grit deposition at the bottom of a mechanically cleaned screen, which might otherwise become inoperative during storms when it is most needed.

The velocity of flow ahead of and through a bar screen affects its operation substantially. Satisfactory designs have provided for velocities of 0.6 to 1.2 m/s (2 to 4 ft/s) through the openings of mechanically cleaned screens and velocities of 0.3 to 0.6 m/s (1 to 2 ft/s) through the openings of manually cleaned screens. Presently, lower velocities are preferred for low flows, and a maximum velocity of 0.9 m/s for peak instantaneous flows is preferred in common design practice.

Accepted practice calls for a minimum head loss allowance through a manually cleaned bar screen of 150 mm (6 in.), assuming frequent screen inspection. The maximum head loss allowance through clogged racks and bar screens is limited to 0.76 m (2.5 ft). A nearly constant head loss for continuously cleaned mechanical screens of any type can be maintained with a constant flow. Curves and tables for head loss through the screening device are available from equipment manufacturers.

The head loss for comminutors ranges from 50 to 300 mm (2 to 12 in.), depending on the flow and opening size; nevertheless, head loss can reach 1 m (3 ft) or more in large units at maximum flow.

The design engineer must take steps to ensure that the head loss through the screening device at maximum flows will not surcharge influent sewers enough to impair upstream services. To prevent flooding of the screening area caused by blinding of the screen because of a power failure or another problem, an overflow weir or gate and a parallel bypass channel allowing overflows to go around the screen should be included in the design.

The head loss through a clean or partially clogged bar screen can be represented by

$$h_L = k(v_B^2 - v^2)/2g \qquad (4.1)$$

where
h_L = head loss, m (ft);
v_B = velocity through bar screen, m/s (ft/s);

Table 4.3 Kirshmer's values of β.

Bar type	β
Sharp-edged rectangular	2.42
Rectangular with semicircular upstream face	1.83
Circular	1.79
Rectangular with semicircular upstream and downstream faces	1.67

v = velocity upstream of bar screen, m/s (ft/s);
g = acceleration due to gravity, 9.81 m/s² (32.2 ft/s²); and
k = friction coefficient (1.43).

The head loss through only a clean bar screen is indicated by Kirschmer's equality

$$h_L = \beta \, (w/b) \, 1.33h \sin \varnothing \qquad (4.2)$$

where
h_L = head loss, m (ft);
β = a bar shape factor (Table 4.3);
w = maximum cross-sectional width of bars facing upstream, m (ft);
b = minimum clear spacing of bars, m (ft);
h = upstream velocity head, m (ft); and
$\varnothing$ = angle of bar screen with horizontal.

Management of Screenings

The method of cleaning screens, manual or mechanical, relates to the method of removing and transporting screenings from the screens to a disposal site. When designing a manually cleaned screen, relatively shallow screening channels (<2 m [<7 ft]) are needed with a drainage plate to allow drainage of the screenings before shoveling. The container used to carry the screenings to a truck or other transport may range from a wheelbarrow to a bin carried by an overhead crane or monorail. Whatever the means of conveyance, operator safety, including nonslip platforms and railings, deserves special attention during design.

Conveyors and pneumatic ejectors are the primary means of mechanically transporting screenings; grinders followed by sludge-handling pumps are occasionally used. Compactors can transport screenings a short distance to a receiving container. Although reliable, conveyors, unless covered, generate odors and attract insects. In the past few years, conveyors that curve around corners and up inclines have been used more frequently. If direction changes are needed, these conveyors eliminate the need for multiple conveyor units. An advantage of conveyors is that the belt is the only portion of the system that contacts the screenings, thus reducing plugging or jamming of mechanical components.

Drainage must be provided for a conveyor system. One method maintains at least a slight incline throughout the conveyor length. Another method uses a perforated conveyor to allow water to drip into a trough beneath the conveyor; the trough drains back to the wastewater flow stream. The holes should be located to prevent dripping on the carriage assembly. The belt material is either rubber or another material that will not rust. A spray wash cleans the belt and prevents the buildup of sticky solids. The best belt will be concave across its width to prevent spillage and will have small ribs across the belt to prevent screenings from sliding back if the conveyor is steeply inclined.

Pneumatic ejectors offer several advantages over conveyors because the ejectors are less odorous and require less space. However, conveying sticks, rags, stones, and other debris in screenings through pipelines can clog and excessively wear the bends. Therefore, the system design needs to include clean-outs at bends and a high-pressure water service for flushing the discharge piping.

Sometimes screenings hauled from the plant site are deposited in municipal solid waste landfills. At other locations, screenings are mixed with municipal refuse and then burned at the municipal incinerator. Although screenings contain more water and have a higher organic content than does municipal solid waste, the relatively small volume of screenings has negligible effect on combustion. Nevertheless, at some locations, washing and drying of screenings is necessary prior to disposal with municipal solid waste.

Adequate ventilation must be provided in the screening room if it is enclosed. Screening areas are the first points of release of hydrogen sulfide, volatile organics, and other odoriferous compounds in the wastewater. In some cases, hazardous conditions may exists if ventilation is not adequate.

Grit Removal

Purpose of Grit Removal

Grit removal, an important part of wastewater treatment, prevents unnecessary abrasion and wear of mechanical equipment, grit deposition in pipelines and channels, and accumulation of grit in anaerobic digesters and biological reactors. Plant operational experience has shown that much grit is present in wastewater conveyed by either separate or combined sewerage systems, with more in the latter. Without grit removal, grit would be removed in primary clarifiers or, if the plant lacks primary treatment, in biological reactors and secondary clarifiers. Provisions for grit removal is now common practice for any treatment plant with mechanical equipment and sludge-handling processes that grit could otherwise impair.

Depending on the type of grit removal process used, removed grit is often further concentrated in a cyclone, classified, and washed to remove lighter organic material captured with the grit. The washed grit is more readily stored and disposed of than unwashed grit.

Wastewater grit materials, characterized as nonputrescent, have a subsidence velocity greater than that of organic putrescent solids: grit particles are discrete rather than flocculent. Grit materials include particles of sand, gravel, other mineral matter, and minimally putrescent organics such as coffee grounds, egg shells, fruit rinds, and seeds.

The earliest use of grit removal systems was for plant influent from combined sewers. Use of these systems has since shown that whether a sewerage system is separate or combined, large or small, operation of the treatment facility can benefit from removal of grit from incoming flow. Grit removal is critical for protection of dewatering centrifuges and high-pressure progressing cavity and diaphragm pumps; all are easily damaged by grit. Grit removal is omitted in natural treatment systems where grit and biosolids are applied to land or allowed to accumulate on the bottoms of lagoons.

Methods of Grit Removal

The quantity and characteristics of grit and its potential adverse effects on downstream processes are important considerations in selecting a grit removal process. Other considerations include head-loss requirements, space requirements, removal efficiency, organic content, and economics. A variety of grit removal devices have been applied successfully over the years. Categories of grit removal processes include

- Aerated grit chambers;
- Vortex-type (paddle or jet-induced vortex);
- Detritus tank (short-term sedimentation tank);
- Horizontal-flow type (velocity-controlled channel); and
- Hydrocyclone (cyclonic inertial separation).

Aerated Grit Chambers

In aerated grit chambers, air introduced along one side near the bottom causes a spiral-roll velocity pattern perpendicular to the flow through the tank. The heavier particles with their correspondingly higher settling velocities drop to the bottom, while the roll suspends the lighter organic particles, which are eventually carried out of the tank. The rolling action induced by the air diffusers is independent of flow through the tank. The non-flow-dependent rolling action allows the aerated grit chamber to operate effectively over a wide range of flows. The heavier particles that settle on the bottom of the tank are moved by the spiral flow of the water across the tank bottom and then into a grit trough or hopper. Chain-and-bucket collectors, screw augers, clamshell buckets, or recessed impeller or air-lift pumps remove collected grit from the trough or hopper. The grit collection trough, depending on the chamber design, requires special design attention. A properly designed aerated grit chamber will produce a relatively clean grit.

When the wastewater flows into the aerated grit chamber, grit particles will settle to the bottom at rates dependent on the size and specific gravity of the particles and on the velocity of roll in the tank. The rate of air diffusion and the tank shape govern the rate of roll, and thereby the size of the particle with a given specific gravity that will be removed. Diffused air serves essentially as a velocity-control method with sufficient flexibility to accommodate varying conditions. Diffused air also requires only minimal head loss through the unit. Proper design depends on an understanding of the variables affecting the air-lift pumping energy and its effect on the roll pattern in the tank.

A typical layout of an aerated grit chamber is shown in Figure 4.8. Typical aerated grit chamber design criteria that will prevent operation problems are summarized in Table 4.4. Air rates range from 5 to 12 L/m·s of tank length (3 to 8 cfm/ft). Rates as low as 2 L/m·s (1 cfm/ft) have been used for shallow, narrow tanks. Rates higher than 8 L/m·s (5 cfm/ft) are often used for deep, wide tanks. Also, providing valves and flow meters is good practice for monitoring and controlling the airflow rate to each bank of diffusers. Aeration—preferably with coarse-bubble, wide-band diffusers—is tapered to allow even removal of grit along the length of the chamber (Finger and Patrick, 1980). The plant's process air supply may provide the air for the grit chamber, but separate dedicated blowers are preferred.

As good practice, a minimum hydraulic detention time of 3 minutes at maximum instantaneous flowrates will reliably capture 95% of the 0.21-mm (0.008-in.) grit (specific gravity of 2.6). Longer detention times improve grit removal and may be necessary to provide additional preaeration or capture smaller grit particles. Detention time,

Figure 4.8 An aerated grit chamber.

however, is a less important criterion than baffle and diffuser location, airflow rate, and system design (Morales and Reinhart, 1984).

A longitudinal baffle is positioned approximately 1 m (3.0 ft) from the wall along the air diffusers to help control the roll pattern. With proper adjustment, the aerated grit chamber, at a level 150 mm (6 in.) below the top of the water, should produce a roll velocity of 0.6 m/s (2.0 ft/s) near the tank entrance and 0.45 m/s (1.5 ft/s) at the tank exit (WEF, 1996).

Table 4.4 Typical design criteria for aerated grit chambers (Albrecht, 1967; Morales and Reinhart, 1984; WPCF, 1989).

Item[a]	Range	Comment
Dimensions		
Depth, m	2–5	Varies widely
Length/width ratio	2.5:1–5:1	
Width/diameter ratio	1:1–5:1	2:1 typical
Minimum detention time (at peak flow), minutes	2–5	3 typical
Air supply		
L/m·s	5–12	
Type of diffuser	Medium to coarse bubble	
Distance from bottom, m	0.6–1.0	
Transverse roll velocity, m/s	0.6–0.75	Provide valves and flow meters to allow proper adjustment

[a] m × 3.281 = ft; L/m·s × 1.549 = scfm/ft; m/s × 3.281 = ft/s.

Vortex Grit Removal

The vortex grit-removal system relies on a mechanically induced vortex to capture grit solids in the center hopper of a circular tank. Incoming flow straightens in the inlet flume to minimize turbulence at the inlet of the chamber. At the end of the inlet flume, a ramp causes grit that may already be on the flume bottom to slide downward along the ramp until reaching the chamber floor where the grit is captured. At the end of the flume, an inlet baffle—positioned so that the flow entering the chamber and the flow inside the chamber impinge on its sloped surface—deflects the flow downward. At the center of the chamber, adjustable rotating paddles maintain the proper circulation within the chamber for all flows. This combination (paddles, inlet baffle, and inlet flow) produces a spiraling, doughnut-shaped flow pattern that tends to lift lighter organic particles and settle the grit. Grit solids, removed from the center hopper by air-lift or recessed-impeller pumps, are further concentrated and washed.

Detritus Tanks

The detritus tank, a constant-level, short-detention settling tank, is one of the earliest grit chambers (square tank degritter). Because these tanks settle heavy organics in addition to grit, they require a grit-washing step as part of the process to remove organic material. Some designs incorporate a grit auger and a rake that removes and classifies grit from the grit sump.

Horizontal-Flow Grit Chambers

The horizontal-flow grit chamber is another early type of grit removal system and operates by controlling the velocity through the use of proportional weirs or rectangular control sections (such as Parshall flumes) to vary the depth of flow and keep the velocity of the flow stream at a constant 0.3 m/s (1 ft/s). Operational experience has shown that this velocity allows heavier grit particles to settle while lighter organic particles remain suspended, or become resuspended, and are carried out of the channel. Chain-and-flight setups are used in larger installations to scrape the grit to a hopper at the chamber's inlet end, where screw augers or chain-and-bucket elevators remove the collected grit. In some small plants, grit is shoveled manually from the channel.

In designing a horizontal-flow grit chamber, the settling velocity of the target grit particle and the flow-control section and depth relationship govern the length of the channel. To determine the actual length of the channel an allowance for inlet and outlet turbulence must be added. The cross-sectional area will be governed by the rate of flow and the number of channels. Allowances for grit storage and removal equipment are included in determining the channel depth. Table 4.5 presents a summary of typical design criteria for horizontal-flow grit chambers.

Hydrocyclones

The hydrocyclone-type system is used for separating grit from organics in grit slurries (see the following grit-washing section), or for removing grit from primary sludge. Hydrocyclone systems are sometimes used to remove grit and suspended solids directly from the wastewater flow by pumping at a head of 3.7 to 9 m (12 to 30 ft). Coarse screening ahead of these units is needed to prevent them from clogging with sticks, rags, and plastics. Centrifugal forces developed in the cyclone cause heavier grit and suspended solids particles to concentrate along the sides and on the bottom, while lighter solids, including scum, are removed from the center through the top of the

Table 4.5 Typical design criteria for horizontal-flow grit chambers.

Item[a]	Range	Comment
Dimensions		
Water depth, m	0.6–1.5	Depends on channel area and flowrate
Length, m	3–25	Function of channel depth and grit-settling velocity
Allowance for inlet and outlet turbulence, %	25–50	Based on theoretical length
Detention time (at peak flow), seconds	15–90	Function of velocity and channel length
Horizontal velocity, m/s	0.15–0.4	Optimal velocity is 0.3 m/s

[a] m × 3.281 = ft; m/s × 3.281 = ft/s.

cyclone. Theoretically, cyclones could remove as much sludge as a primary clarifier, but this would entail the problematic disposal of combined primary sludge and grit. Cyclones operate best at constant flow and pressure. If flows depart from design flows, solids will be lost to the centrate stream.

Grit Quantities and Characteristics

The quantity and characteristics of grit removed from wastewater will vary over a wide range. Variables influencing grit quantity include the type of collection system (combined or separate), characteristics of the drainage area, use of household garbage grinders, condition of sewer system, sewer grades, types of industrial waste, and efficiency of grit removal equipment. If available, actual plant data should be used for design. In lieu of data, grit removal quantities may be estimated based on data in Table 4.6.

Grit solids characteristics will vary, with solids content from 35 to 80% and volatile content from 1 to 55% (U.S. EPA, 1979). A well-washed grit should achieve a solids content of 70 to 80% with a minimum of putrescent solids. The moisture and volatile content will be influenced by the degree of washing the grit receives. The bulk density of dewatered grit will range from 1400 to 1800 kg/m^3 (90 to 110 lb/ft^3). Dewatered grit can contain pathogens unless it is incinerated.

Design Practice

Grit particle sizes, for design purposes, have traditionally included particles larger than 0.21 mm (0.008 in.) (65 mesh) with a specific gravity of 2.65 (U.S. EPA, 1987).

Table 4.6 Estimated grit quantities (U.S. EPA, 1979, 1987; WPCF, 1989).

Type of system	Average grit quantity, m^3/1000 m^3 (typical range)[a]	Ratio of maximum day to average day
Separate	0.004–0.037	1.5 to 3.0:1
Combined	0.004–0.18	3.0 to 15.0:1

[a] m^3/1000 m^3 × 133.7 = ft^3/mil. gal.

Removal of 95% of these particles has traditionally been the target of grit removal equipment design. Modern grit removal designs are now capable of removing up to 75% of 0.15-mm (0.006-in.) (100-mesh) material because of recent recognition that plants often need to remove particles that are small so as to avoid their adverse effects on downstream processes.

Grit removal follows bar screening or comminution to prevent large particles from interfering with grit handling equipment. The grit removal process should precede primary clarification; in secondary treatment plants without primary clarification, grit removal should precede aeration. In some cases, grit is allowed to settle in primary clarifiers and is removed from primary sludge. In such systems, the primary sludge is not allowed to exceed 0.5% solids, so that it can be pumped out using recessed impeller pumps and processed through hydrocyclones for removal of the grit. The degritted sludge must then be thickened before further processing or disposal.

Grit removal ahead of raw wastewater pumping equipment requires deep placement of grit removal processes with the associated high construction and operation and maintenance expenses. As a more economical approach, raw wastewater is pumped with the grit included, despite the increased wear of the pumps. This allows the grit chamber to be located conveniently ahead of other treatment processes.

Commonly, a single grit removal unit with a bypass channel around the unit will suffice for small installations ($<1.5 \times 10^4$ m^3/d [<4 mgd]) or for plants where infrequent flows of wastewater containing grit can be tolerated in downstream processes. For large plants, plants served by combined sewers, or plants with grit-sensitive unit processes such as centrifuges, multiple grit removal units are necessary to allow periodic removal of units from service for cleaning, maintenance, and repair.

Removing Grit from the Chambers

Removal of grit from each of the grit chambers can be accomplished with varying degrees of success by a number of different methods, primarily automatic, although some small plants use manual methods. Manual grit removal (shoveling) requires at least one redundant tank, capable of handling peak flows, to allow isolation and dewatering of the tank to be shoveled.

Large treatment plants or plants that accept wastewater from combined sewerage systems require some type of automatic grit removal equipment. There are four methods that automatically or semiautomatically remove grit from the chamber hoppers: inclined screw or tubular conveyors, chain-and-bucket elevators, clamshell buckets, and pumping.

Inclined Screw or Tubular Conveyors

Inclined screw or tubular conveyors not only lift the grit out of the chamber but also provide, as ancillary benefits, some washing and dewatering of the grit. Such washing and dewatering may suffice for the selected mode of transportation to the disposal site and the disposal method. A specific design concern for inclined conveyors is their inefficient use of plant area because of slope limitations. For the screw conveyor, long screws may require intermediate support and intermediate couplings, both of which are sensitive to wear. When selecting the screw conveyor system, ensure that the motor has sufficient power to drive the selected screw conveyor when it is fully loaded with grit. The motor size selected for a combined sewer system must handle sudden, high peak loads.

Chain-and-Bucket Elevators

Chain-and-bucket elevators are used primarily with horizontal-flow and aerated grit chambers. Design considerations related to peak loading and motor sizing are similar to those for inclined-screw collector mechanisms. Additional concern with the chain-and-bucket system is the wear and potential jamming of submerged sprockets. Because of grit's abrasive nature, this system should always include metallic chains. A disadvantage of the elevator system is that at least one additional system, mechanical or manual, is required to move the grit to the transport vehicle. Also, this system does not provide effective grit washing and dewatering.

When using either screw conveyors or chain-and-bucket mechanisms, chambers must be dewatered for periodic repair or routine maintenance. Therefore, at least one additional or redundant chamber is required beyond those needed to handle peak flows.

Clamshell Buckets

A clamshell bucket arrangement moved by an overhead monorail track offers yet another means of removing settled grit from grit chambers. This method, however, provides inconsistent grit removal, requires discontinuing flow to the chamber during grit removal, lacks effective dewatering and washing, and may generate odors.

Pumping

Pumping grit from hoppers in the form of a slurry offers distinct advantages over other methods of grit removal but also has some disadvantages. Pumping grit from hoppers makes economical use of space because pumps are located in a dry well adjacent to the chambers. This flexible arrangement allows each pump to serve any of the multiple chambers. Grit is highly abrasive, however, and the impellers will not be expected to last very long before requiring replacement.

Grit Washing

After removing collected grit from the grit hopper, grit is washed to ease handling. Grit removal methods have historically attempted to minimize the organic material removed with the grit while efficiently removing grit particles. Facilities have met this dual objective with varying degrees of success. Thus, when choosing a possible washing system, the design engineer should judge the nature of the grit to be removed from the grit chamber. This assessment is easily accomplished for facility renovations, where grit chambers are already in place.

In most cases, a reduction in grit volume by removing the water contained in the grit saves transportation costs and eases transport and handling during disposal. Also, washing the grit (accomplished by some removal techniques) to remove putrescent organic material makes grit handling and disposal more manageable. Removing putrescent organic material prevents odors and nuisance conditions caused by decomposition of the material and the added difficulty of disposing of sloppy material. Perhaps of most importance, removing organic material from the grit will reduce the number of odor complaints that facility operators would otherwise receive from citizens near grit storage facilities, transportation routes, and disposal sites.

Grit Disposal

Grit can be conveyed directly to trucks, Dumpsters, or storage hoppers. Containers need covers to prevent odors during storage and hauling. Conveyors are frequently used for transporting grit from handling facilities to containers. If grit is transported directly to trucks or Dumpsters, the container should have enough capacity to handle daily-peak grit loads because inclement weather or other reasons might prevent replacement trucks from arriving or the loaded truck from leaving. Also, two bays are advisable to ensure continued loading if the loaded container must remain because of a truck breakdown.

Septage

Separate pretreatment of septage may not be required if a limited amount of septage is discharged to an interceptor upstream of a plant or if a relatively small volume is discharged to the headworks of a large existing treatment facility with adequate preliminary processes. However, preliminary treatment facilities often must be designed to accept septage loads from septic tank pumpers.

Septage contains hair, grit, rags, stringy material, and plastics and is highly odorous. Septage is difficult to feed at controlled rates unless a receiving station exists. Septage receiving and storage facilities with separate screening and grit removal constitute the best arrangement, and, as a minimum, 150-mm (6-in.) diameter line sizes should be provided. A covered aerated grit chamber with air scrubbing for preliminary treatment of septage should be provided.

Receiving facility design must account for the anticipated volume of septage, effects of septage on plant processes, and odor control. When discharging septage directly to the head of a treatment plant, equalization facilities are necessary to control the flow of septage proportionately to the wastewater flow. Equalization is not necessary where septage is discharged to an interceptor at a point far enough upstream from the plant to allow complete mixing with wastewater, provided that the total quantity of septage discharged represents less than 1% of the wastewater flow at that time and location. This can often be achieved by avoiding septage discharge during daily low-flow periods.

The *dumping station*, the initial point where septage enters a receiving facility, needs a slightly sloped ramp to tilt the truck for complete drainage and accommodate washed spillage to a central drain. Figure 4.9 shows a basic layout of a dumping station (U.S. EPA, 1984). The dumping station design should prevent tank trucks from releasing septage without any hose connection. Otherwise, spillage and release of odors will inevitably result. Hose bibs or water hydrants must be provided to facilitate washdown and cleanup.

Flow Equalization

Accommodating wide variations in flowrates and organic mass loadings is one of the primary challenges faced in the design of wastewater treatment plants. Because of naturally occurring variations in the generation of wastewater and the related effects of infiltration and inflow, all municipal plants must process unsteady wastewater flows. Efficiency, reliability, and control of unit process operations within the plant can be adversely affected by the cyclic nature of waste generation, resulting in possible violations of effluent standards. Equalization of influent flow can dampen diurnal variations and variations caused by infiltration and inflow to achieve a relatively constant loading of downstream treatment processes. Equalization of effluent discharge is used occasionally before subsurface discharge or other wastewater reuse applications.

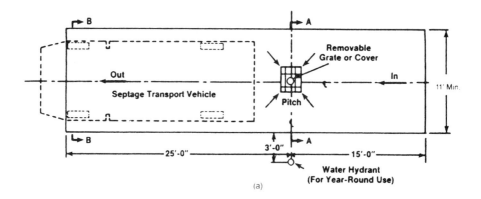

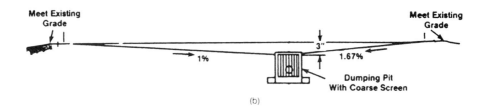

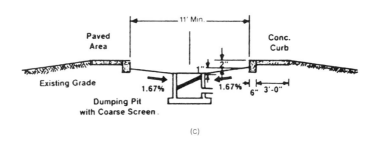

Figure 4.9 A septage dumping station: (a) plan, (b) profile at pavement centerline, (c) section A-A, and (d) section B-B (in. × 25.4 = mm; ft × 0.3048 = m).

Two types of equalization of influent flows may be used: flow equalization and waste-strength equalization. The primary objective of flow equalization tanks for municipal wastewater treatment plants is simply to dampen diurnal flow variation, thus achieving a constant or nearly constant flowrate through downstream treatment processes. An additional benefit is a reduction in the variability of concentration and mass flow of wastewater constituents by blending in the equalization tank. This more uniformly loads downstream processes with organics, nutrients, and other suspended and dissolved constituents. Waste-strength equalization, commonly used in industrial applications, dampens the variability of the strength of the waste by blending wastewater in the equalization tank. For this purpose, the volume of wastewater in the equalization tank remains constant. With this type of equalization, flow remains variable.

Equalization may be used to minimize the effects of loading variations from dry-weather flows, wet-weather flows from separate sanitary sewers, or storm-related flows from combined sewers. In some cases, providing equalization storage for wet-weather flows in sanitary sewers may be feasible, depending on the magnitude of the infiltration and inflow component. For combined sewerage systems, the design engineer should consider temporary storage of peak flows either within the collection system sewers or offline for regulated release back to the plant later.

Flow equalization, a useful upgrading technique, can significantly improve the performance of an existing treatment facility. In the case of new plant design, flow equalization can reduce the required size of downstream facilities. However, economic and operations analyses should determine the feasibility of equalization versus increasing the size of the required unit operations based on the loading variations anticipated. Flow equalization basins are not cost effective for new plants with separate sanitary sewer systems, nonexcessive infiltration and inflow, and small contributions of industrial waste.

Equalization tanks are commonly located in the plant following the grit and screening units. However, based on individual circumstances, equalization storage is sometimes located in the collection system to relieve overloaded trunk lines or pumping station force-main systems. Also, equalization of sludge-handling sidestreams is used frequently to regulate and reduce the effect of sidestream returns on the liquid treatment system. Some design engineers include equalization tanks following primary clarification. This simplifies cleanup of the equalization tanks after use.

In some instances, large interceptor sewers entering the treatment plant can be effectively used as storage tanks to dampen peak diurnal dry-weather flow variations. In such cases, nightly or weekly drawdown of the interceptor system is necessary to increase flow velocity, thereby flushing sludge that has accumulated during the storage period. In some larger cities such as Chicago, equalization tunnels are used to store wet-weather flow and maintain a nearly constant flow to the treatment plants.

Equalization tanks, located at the treatment plant site, may be designed as either in-line or side-line units. For the in-line design shown by Figure 4.10 all of the flow passes through the equalization tanks (U.S. EPA, 1974). This results in significant concentration and mass-flow dampening. The side-line design, also shown in Figure 4.10, diverts only the flow portion exceeding the daily average flow through the equalization tank. This design minimizes pumping requirements but less effectively dampens variations of pollutant concentrations.

The design methodology for equalization involves determining the necessary volume, mixing, and aeration requirements and the control of flows leaving the tank. The first step in estimating the volume necessary for equalization involves determining the diurnal variations of the wastewater. Whenever possible, this step should be based on actual operating data. Diurnal flow patterns will vary from day to day, even seasonally, depending on the nature of the community (e.g., tourist related, winter residences, or agricultural food processing). It is, therefore, important to select a pattern that will ensure a large enough volume for equalization, taking into account conditions such as infiltration and storm-related inflow that influence variability of the influent flow. A flow balance (mass diagram) can be used for estimating equalization volumes required for variable wastewater flows.

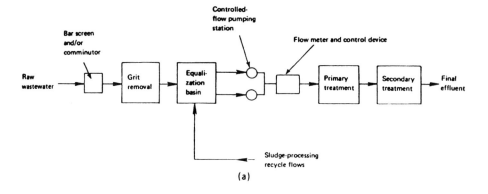

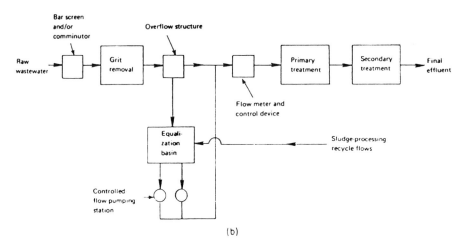

Figure 4.10 Two equalization schemes: (a) in-line equalization and (b) side-line equalization.

Conclusion

When I was a lad (editor Vesilind talking here) my hometown was on the banks of the Ohio River, about 20 miles downstream from Pittsburgh, Pennsylvania. Many of us dreamed of pulling a Huck Finn, or at least venturing out to the Ohio on a summer day on a homemade raft. But all it took to dissuade us of such plans was to go down the steep bank to the river and smell it. It was, by any stretch of the imagination, totally foul. Most towns in those days had only screening for their "wastewater treatment". This was all that was required. The wisdom of the day was to let the river clean itself, and eventually it did. But as the river flowed past my small community, it carried all the untreated effluents of thousands of people all along the Ohio River and its tributaries. We joked about how dirty the water was, talking about what we called "Allegheny white fish". And we did not know then that a river can be anything but dirty. The air was foul, then why not also the river?

Much has changed in the past 50 years. The Ohio River now proudly flows past Beaver, Pennsylvania, in its Sunday best. People use the river for recreation and enjoyment, and the engineers at Allegheny County Sanitary District can unabashedly take pride in what they have done.

Just as screening was only the beginning of wastewater treatment for communities along the Ohio River, so is screening and other preliminary treatment steps just a beginning in the design of any effective wastewater treatment plant. Now on to the rest of the job.

References

Albrecht, A. E. (1967) Aerated Grit Chamber Operation and Design. *Water Sew. Works,* **114** (9), 331.

Finger, R. E.; Patrick, J. (1980) Optimization of Grit Removal at a WWTP. *J.—Water Pollut. Control Fed.,* **52,** 2106.

Morales, L.; Reinhart, D. (1984) Full-Scale Evaluation of Aerated Grit Chambers. *J.—Water Pollut. Control Fed.,* **56,** 337.

U.S. Environmental Protection Agency (1974) *Flow Equalization;* EPA Technology Transfer: Washington, D.C.

U.S. Environmental Protection Agency (1979) *Process Design Manual for Sludge Treatment and Disposal;* EPA-625/1-79-011: Washington, D.C.

U.S. Environmental Protection Agency (1984) *Handbook of Septage Treatment and Disposal;* EPA-625/6-84-009: Washington, D.C.

U.S. Environmental Protection Agency (1987) *Preliminary Treatment Facilities—Design and Operational Considerations;* EPA-430/09-87-007: Washington, D.C.

Water Environment Federation (1996) *Operation of Municipal Wastewater Treatment Plants,* 5th ed.; Manual of Practice No. 11: Alexandria, Virginia.

Water Pollution Control Federation (1989) Technology and Design Deficiencies at Publicly Owned Treatment Works. *Water Environ. Technol.,* **1** (4), 515.

Symbols Used in this Chapter

b = minimum clear spacing of bars, m (ft)
g = acceleration due to gravity, 9.81 m/s^2 (32.2 ft/s^2)
h_L = head loss, m (ft)
h = upstream velocity head, m (ft)
k = friction coefficient
v_B = velocity through bar screen, m/s (ft/s)
v = velocity upstream of bar screen, m/s (ft/s)
w = maximum cross-sectional width of bars facing upstream, m (ft)
β = a bar shape factor
$\emptyset$ = angle of bar screen with horizontal

5

Primary Treatment

Introduction

Primary settling, the principal form of primary treatment, is the oldest and most widely used unit operation in wastewater treatment. Primary treatment is the first "line of defense" in many wastewater treatment plants and reduces suspended solids and biochemical oxygen demand (BOD) loading on downstream treatment processes. Efficient solids and BOD removal in primary clarifiers enhances the effectiveness of downstream biological treatment processes, lowers the oxygen demand, and decreases the rate of energy consumption for oxidation of particulate matter. These effects enhance soluble substrate removal during aeration and reduce the volume of waste activated sludge that is generated. Primary treatment also removes floating material, thereby minimizing operational problems in downstream treatment processes (that is, buildup of scum in secondary treatment processes) and improving the plant's overall aesthetics.

Primary treatment equalizes raw wastewater quality and flow to a limited degree, thereby protecting downstream unit processes from unexpected surges in flow. Money spent on primary treatment often provides the greatest return on investment in terms of dollars per kilogram (pound) of pollutant removed. The most common form of primary treatment is quiescent settling with skimming; collection; and removal of settled primary sludge, floating debris, and grease. Preaeration or mechanical flocculation, often with chemical addition, can be used to enhance primary treatment.

While primary treatment (settling) is used in almost all larger wastewater treatment plants, it can be omitted for smaller plants. The solids that would have been captured in primary treatment then become part of the secondary sludge. Industrial wastewater with low solids concentrations may also not need primary treatment.

Primary Settling

Raw wastewater contains suspended particulates heavier than water; these particles settle by gravity under quiescent conditions. The design of *settling tanks* (often called *clarifiers*) has historically been based on empirically derived relationships or design criteria. These criteria, when coupled with a theoretical understanding of the settling process, may be used for the design of reliable and efficient settling tanks.

Settling Theory

Gravity settling is an effective removal method for raw wastewater suspensions, which range from a low concentration of nearly discrete particles to a high concentration of flocculent solids. In discrete particle settling, the settling velocity of a particle is a function only of the fluid properties of wastewater and the characteristics of solid particles.

Camp (1936, 1953) is credited with the idea of creating an "ideal tank" in which settling occurs. He suggested dividing the settling tank into four zones: the inlet, settling, outlet, and primary-sludge zones (Figure 5.1). In the inlet zone, influent solids and flow are uniformly distributed over the cross-sectional area of the settling tank. In the settling zone, a uniform concentration of particles settles at the terminal-settling velocity to the primary-sludge zone at the bottom of the tank where the settled solids are collected and removed. Effluent discharges over a weir in the outlet zone.

A number of assumptions are necessary to analyze the settling process within this ideal tank

(1) There is uniform flow within the zone. That is, the horizontal flow velocity is the same everywhere within that zone.
(2) The flow is steady; that is, it does not vary with time.

Note on Terminology

Because wastewater treatment technology has evolved over many years, the terminology is often muddled. A prime example is the settling tank. Many engineers and operators will call settling tanks *clarifiers*, and the first settling tank in a wastewater treatment plant becomes a *primary clarifier*, while the secondary settling tank, following the activated-sludge biological reactor, is the *secondary clarifier*. The most general term to describe these unit operations is "settling tank", and settling tanks used in wastewater treatment are often called "clarifiers". Both terms are used in this text.

Some engineers and operators use the term *sedimentation tank* synonymously with settling tanks and clarifiers. This last term is inappropriate and should be discouraged. Sediment is what settles out in river deltas and lakes. In wastewater treatment we settle solids, not sediment.

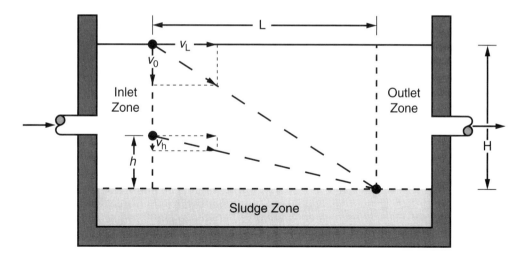

Figure 5.1 Zones of an ideal settling tank.

(3) The settling of particles is laminar. This allows for the use of the Stokes equation for estimating particle settling velocities.
(4) There is no flocculation. Particles settle as discrete entities and do not bump into each other, stick to each other, nor grow in size.
(5) All particles are spheres, with their size defined by their diameter.
(6) Once a particle enters the sludge zone at the bottom, it is removed.
(7) If a particle does not enter the sludge zone and instead travels to the end of the tank and enters the outlet zone, it will be carried over the weir with the effluent.

Consider a single particle, the *critical particle*, that enters the ideal settling zone at the far left and on the water surface, and has a settling velocity, defined by the Stokes equation, of v_O. It is also traveling horizontally with the flow at a horizontal velocity of v_L. As shown in Figure 5.1, its trajectory will carry it just to the intersection of the sludge and outlet zones, so it is just barely removed. All particles that start at the same place and have settling velocities greater than v_O, will enter the sludge zone and be removed. Similarly, all particles that enter the tank at the water surface and have settling velocities less than v_O will escape with the effluent.

If a particle enters that tank anywhere else below the water surface, say at some height h, its settling velocity will have to be v_h if the particle is to be removed (Figure 5.1). By proportion,

$$\frac{v_h}{v_O} = \frac{h}{H} \qquad (5.1)$$

where
v_O = settling velocity of the critical particle, m/s (ft/s);
v_h = settling velocity of any particle entering the ideal tank zone at height h, m/s (ft/s);
h = any height less than H, m (ft); and
H = height of the ideal tank zone, m (ft).

Consider now the triangle formed by the vectors v_O and v_L. If the ideal tank has proportions of height H and length L, then

$$\frac{v_L}{L} = \frac{v_O}{H} \tag{5.2}$$

where
v_O = settling velocity of the critical particle, m/s (ft/s);
v_L = horizontal velocity through the tank, m/s (ft/s);
L = length of ideal tank, m (ft); and
H = height of ideal tank, m (ft).

The hydraulic retention time is

$$\bar{t} = \frac{V}{Q} = \frac{L}{v_L} = \frac{H}{v_O} = \frac{h}{v_h} \tag{5.3}$$

where
V = tank volume, $L \times H \times W$, m³ (ft³);
Q = flowrate, m³/s (ft³/s);
L = ideal tank length, m (ft);
H = ideal tank height, m (ft);
W = ideal tank width, m (ft);
h = height at which a particle with settling velocity v_h enters and is just removed, m (ft);
v_O = settling velocity of critical particle, m/s (ft/s); and
v_L = horizontal velocity through the tank, m/s (ft/s).

Starting with the above equation, through a series of substitutions

$$v_O = \frac{H v_h}{h} = \frac{H}{h} \times \frac{h}{\bar{t}} = \frac{H}{\bar{t}} = H\frac{Q}{V} = \frac{HQ}{L \times H \times W} = \frac{Q}{L \times W} = \frac{Q}{A} \tag{5.4}$$

where $\bar{t}$ = hydraulic retention time, d, and A = surface area of the ideal tank, or $L \times W$, m² (ft²).

This is a curious outcome. The critical particle velocity is defined as the flowrate divided by the surface area, and this also has dimensions of meters per second (feet per second). In the United States, engineers have called this relationship, Q/A, the *overflow rate*, although it has nothing to do with overflow. Instead it is another way of defining the critical velocity.

If a tank is designed with a low critical velocity, then most of the particles that have settling velocities greater than v_O will be removed. If the tank is designed for a high critical velocity, this means that most of the particles will have settling velocities less than this critical settling velocity and there is no time for them to be removed. The overflow rate is therefore another way of defining the critical settling velocity.

Ideal settling tank theory presents a methodology for settling tank design but, in reality, the actual settling performance cannot be adequately predicted because of the

Figure 5.2 Typical rectangular primary settling tank.

unrealistic assumptions regarding discrete particle settling. Departures from ideal discrete particle settling include particle interaction and currents in the settling zone. Suspended solids in wastewater are not discrete particles and vary in size and other characteristics. Under quiescent settling conditions, large and heavy particles settle faster than small and light particles. As these two types of particles pass each other and make contact, they agglomerate and grow in size in a process known as *flocculation*. The flocculation process increases removal efficiency but cannot be adequately represented by equations.

Types of Settling Tanks

Rectangular (Figure 5.2), circular (Figure 5.3), square, and stacked (Figure 5.4) are four types of settling tanks. Rectangular and circular settling tanks are most commonly used for wastewater treatment.

Rectangular settling tanks range from 15 to 90 m (50 to 300 ft) in length and 3 to 24 m (10 to 80 ft) in width. Depths should exceed 2 m (7 ft) (WPCF, 1985). Rectangular tanks with common-wall construction (unit design) are advantageous for sites with space constraints.

Circular settling tanks vary from small, 3 m (10 ft), to more than 90 m (300 ft). Depths range between 2.4 and 4.0 m (8 and 13 ft) (WPCF, 1985). Circular settling tanks can use relatively trouble-free circular primary-sludge removal equipment (drive bearings are not under water). Walls of circular tanks act as tension rings, which permit thinner walls than those for rectangular tanks. As a result of such advantages, circular tanks have a lower capital cost per unit surface area than rectangular tanks (unless the

Figure 5.3 Typical circular primary settling tank.

rectangular tanks use common wall construction). Circular tanks require more yard piping than rectangular tanks.

Square settling tanks have the same primary-sludge removal equipment as circular tanks. Removal of settled solids from corners can cause problems. Square tanks may use common-wall construction but require thicker walls than circular units and are therefore seldom used.

Stacked settling tanks (tray clarifiers), originally proposed by Camp (1946), are used where land for treatment facilities is not available or is extremely expensive. Except in rare circumstances stacked tanks are not used in wastewater treatment plants as primary clarifiers.

Design Considerations

Historically, settling tank design has relied on empirically derived criteria such as tank overflow rate, depth, surface geometry, hydraulic residence time, and weir rate. These criteria are helpful for design but are not accurate enough to permit prediction of actual settling performance.

Although dry-weather flow conditions prevail throughout most of the United States, peak storm or wet-weather flows must also be considered. Wet-weather flows depend on the location, intensity, and duration of rainfall plus characteristics of the sanitary sewer system. Therefore, wet-weather flows are more difficult to predict than dry-weather flows. Substantial infiltration or inflow from sanitary sewers or the existence of combined storm and sanitary sewers might result in wet-weather flows that are several times higher than normal dry-weather flows.

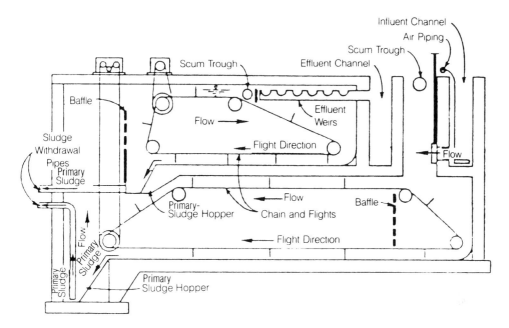

Figure 5.4 Typical stacked settling tank (from Kelly, K. [1988] New Clarifiers Help Save History. *Civ. Eng.*, **58,** 10, with permission from the American Society of Civil Engineers).

Recycle streams, such as waste activated sludge (WAS) or trickling filter recycle (filtrate), may cause flow surges. These surges should be avoided if possible or returned to the plant influent stream during low-flow periods. Such low-flow periods, unfortunately, occur typically during the night whereas the operations producing recycle streams occur during the day.

Overflow rates (surface loading rates) usually govern design. These criteria, guided by the idealized settling theory described above, rely on many favorable conditions that are not attainable in practice. A graphical representation of the relationship between overflow rate and performance on an idealized basis is shown by the curve in Figure 5.5. This figure is basically a scatter-plot, with no obvious line, illustrating that overflow rate, although it is the prime design variable for settling tanks, by itself does not determine the effectiveness of settling. The designer should keep this in mind when considering other options such as inlet and outlet design, which probably have greater effect on tank performance than overflow rate.

Understandably, there is wide disagreement as to the overflow rates to use for design. Table 5.1 shows some of the standards used by various agencies and consulting firms. The design engineer must know what the local and state requirements are for overflow rates and adhere to these, keeping in mind Figure 5.5.

Depth

The opportunity for contact between particles and flocculation increases with depth. Hence, theoretically, removal efficiency should increase with depth. In actual practice, it is uncertain whether better removals can be obtained or higher overflow rates can be applied with deeper settling tanks. Settling tanks must be deep enough to accommodate

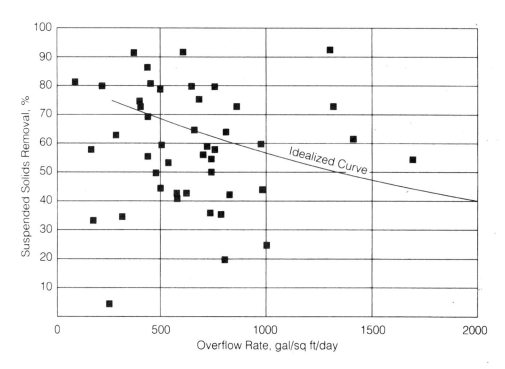

Figure 5.5 Primary settling tank suspended solid removal as estimated by the overflow rate (WPCF, 1985, 1989) (gpd/ft^2 × 0.0407 = m^3/m^2·d).

mechanical primary-sludge removal equipment, store settled solids, prevent scour and resuspension of settled solids, and avoid washout or carryover of solids with the effluent. Shallower depths may be acceptable with continuous primary-sludge removal. Excessive depth is to be avoided if the solids retention time could cause anaerobic conditions.

Hydraulic Residence Time
Sufficient time for contact between solids particles is necessary for flocculation and effective settling. Hydraulic residence times for primary settling tanks are from 1.5 to 2.5 hours. Some states have set limits (high and low) for residence times. Design considerations should include effects of low-flow periods to ensure that longer residence times will not cause septic conditions. Septic conditions increase potential odors, solubilization, and loading to downstream processes.

Weir Rate
Weir rates have little effect on the performance of primary settling tanks, especially with sidewall depths greater than 3.7 m (12 ft) (Graber, 1974). In practice, weir loadings often do not exceed 37 m^3/m·d (3000 gpd/ft) in treatment plants handling 3800 m^3/d (1.0 mgd) or less and 62 m^3/m·d (5000 gpd/ft) in treatment plants handling more than 3800 m^3/d (Merritt, 1983). Launders are often used for rectangular primary clarifiers, but circular clarifiers use only a single, perimeter weir. Some state regulations govern weir loadings.

Table 5.1 Summary of overflow rates and side water depths recommended by various entities for primary settling tanks (WPCF, 1985).

Source	Recommendations and remarks[a]
Metcalf and Eddy, Inc. (1991)	For primary settling followed by secondary treatment: 32 to 48 m³/m²·d at average flow, 80 to 120 m³/m²·d at peak flow. For primary settling with waste activated sludge: 24 to 32 m³/m²·d at average flow, 48 to 70 m³/m²·d at peak flow. Recommended side water depth: 3 to 5 m for rectangular clarifiers, 3.6 m typical, 3 to 5 m for circular clarifies, 4.5 m typical.
Civil Engineering (1979)	49 m³/m²·d at maximum 24-hour flow with all units in service. 81 m³/m²·d at peak flow with all units in service. 163 m³/m²·d at peak flow with one unit out of service. Side water depth = 3 m.
Great Lakes (1997)	For primary clarifiers: 41 m³/m²·d at average design flow, minimum side water depth = 2.1 m. 61 m³/m²·d at peak hourly flow, minimum side water depth = 2.1 m. Area used is the larger of the two areas calculated using the criteria above. For intermediate tanks following fixed-film processes: 61 m³/m²·d at peak hourly flow, minimum side water depth = 2.1 m.
U.S. EPA (1975)	For primary settling followed by secondary treatment: 33 to 49 m³/m²·d at average flow, 81 to 122 m³/m²·d at peak flow, side water depth = 3 to 4 m. For primary settling with waste sludge: 24 to 33 m³/m²·d at average flow, 49 to 61 m³/m²·d at peak flow, side water depth = 4 to 5 m.
U.S. Army (1978)	Allowable overlow rate depends on plant design flow. Varies from 12 m³/m²·d for design flow, not exceeding 38 m³/d to 41 m³/m²·d for design flow higher than 37,850 m³/d. Side water depth depends on clarifier dimensions, between 2.5 and 4.5 m.
Steel (1979)	24 to 60 m³/m²·d side water depth = 1 to 5 m.

[a] m³/m²·d × 24.55 = gpd/ft²; m × 3.281 = ft; m³/d × 0.1835 = gpm.

Flow-Through Velocity

In practice, the linear flow-through velocity (scour velocity) has been limited to 1.2 to 1.5 m/min (4 to 5 ft/min) to avoid resuspension of settled solids (Theroux and Betz, 1959). The critical scour velocity may be calculated from the following equation (Camp, 1946):

$$v_s = \left[\frac{8k(s-1)gd}{f}\right]^{0.5} \tag{5.5}$$

where

v_s = critical scour velocity, m/s (ft/s);
k = constant for type of scoured particles;
s = specific gravity of scoured particles;
g = acceleration due to gravity, 9.8 m/s² (32.2 ft/s²);
d = diameter of scoured particles, m (ft); and
f = Darcy–Weisbach friction factor.

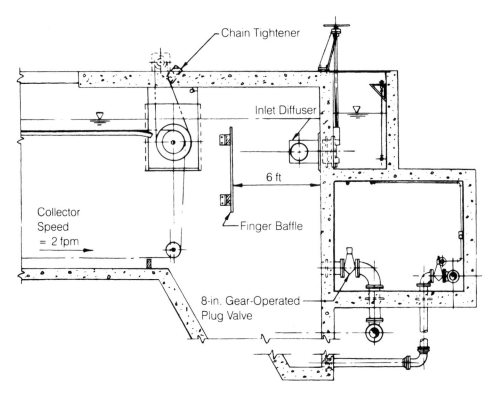

Figure 5.6 Inlet configuration of a rectangular primary settling tank (ft × 0.3048 = m; ft/min × 5.080 = mm/s).

Typical k values are 0.04 for unigranular material and 0.06 for sticky, interlocking material. Values for f range between 0.02 and 0.03 and are a function of the Reynolds number and characteristics of the settled solids surface.

Surface Geometry
Common length/width ratios should be 5:1 or higher, although length/width ratios for existing rectangular tanks range between 1.5:1 and 15:1 (WPCF, 1985). Width is often controlled by the availability of primary-sludge collection and removal equipment.

Inlets
Inlets should be designed to dissipate the inlet port velocity, distribute flow and solids equally across the cross-sectional area of the tank, and prevent short circuiting in the settling tank. Concentration and density differences between influent and the tank contents significantly affect the hydraulic performance of the tank. Inertial currents and wind direction may also affect hydraulic performance. Horizontal velocity variations across the width of rectangular tanks can adversely affect settling efficiency (Hamlin, 1972).

Velocities are dissipated through some type of inlet diffuser or baffle. Baffles are installed 0.6 to 0.9 m (2 to 3 ft) downstream of the inlets and submerged 460 to 610 mm (18 to 24 in.), depending on tank depth. The top of the baffle should be far enough below the water surface to allow scum to pass over the top. Influent flow can be distributed by inlet weirs, submerged ports or orifices with velocities between 3 and 9 m/min (10 and 30 ft/min), and gate valves and perforated baffles. Figure 5.6 shows one inlet configuration using an orifice controlled pipe penetrating the inlet and feeding a horizontal diffuser pipe. This design provides three such diffusers across a tank. The flow from the distribution channel flows into the tank and immediately is split and turned, dissipating its energy. Adjacent diffusers emit flows that impinge against one another. The finger baffle may further dissipate the flow and distribute it uniformly across the tank.

Circular settling tanks (Figure 5.7) used as primary clarifiers use a center feed with peripheral withdrawal. The central feedwells have diameters that are 15 to 20% of the tank diameter. Manufacturers' recommendations for submergence vary significantly. In practice, the feedwell has been extended at least one-half of the tank depth. Feedwells are beneficial for flow distribution and for chemical coagulation, allowing for flocculating prior to entering the settling zone.

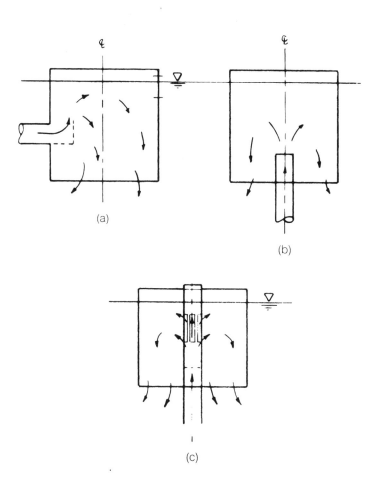

Figure 5.7 Various designs of conventional center-feed inlets: (a) side feed, (b) vertical pipe feed, and (c) slotted, vertical pipe feed.

Chapter 5 Primary Treatment

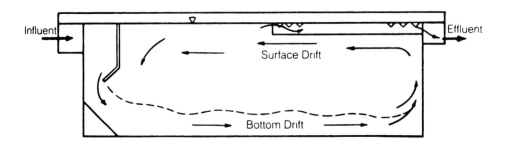

Figure 5.8 Short-circuiting flow in a rectangular basin.

Outlet Conditions

Proper settling tank operation depends on outlet conditions. Effluent should be uniformly withdrawn to prevent localized, high-velocity gradients and short circuiting. Figure 5.8 illustrates prevailing velocity gradients (drift) in a rectangular settling tank. If these velocity gradients reach the scour velocity, settled particles can be swept into the tank effluent. Density currents, rather than a high approach velocity, often causes primary-sludge carryover of the effluent weirs. Therefore, effluent should be withdrawn from the tank in a manner that minimizes these currents. Effluent is withdrawn from a settling tank by an overflow weir into a launder or effluent channel. The overflow weir must be level to control water surface elevation in the settling tank and promote uniform effluent withdrawal. Weirs are almost always V-notched to provide better lateral distribution of outlet flows. Figure 5.9 shows how launders with V-notch weirs can be extended well into the tank to take advantage of the drift.

Submerged launders have also been used for effluent withdrawal (Figure 5.10). Collection pipes or launders with submerged orifices are two types of submerged launders. Orifices should be sized for uniform flow distribution. Compared with overflow weirs, submerged launders offer some advantages. Submerged launders avoid free fall of wastewater, with the consequent release of entrained odorous gases, and allow surface skimming at the end of the tank. A disadvantage of submerged launders is that orifices sized for uniform flow distribution at average flows will not be effective at peak flows. Thus, a separate modulating flow-control device or primary effluent pumps are required with submerged launders. These devices should be located and sized properly for effective scum removal. The submerged orifices will, however, clog more easily and require more effort to clean.

Wave harmonics from wind, earthquakes, or fluid flow may cause launders to oscillate or vibrate, thereby possibly deflecting or deforming the launders and damaging the structural support system. New light materials and long launders aggravate this problem. Launders and weirs should be anchored to resist seismic forces due to wastewater sloshing in the settling tank. The large launders in center-feed, circular settling tanks are particularly vulnerable to this type of damage. In some cases, designing a break-away launder support system that would allow for easy replacement might be more economical and practical than designing a structure to withstand earthquake-induced loadings from sloshing wastewater. Break-away designs must prevent sections from falling to the bottom of the tanks and potentially damaging the primary-sludge collection and removal equipment. This

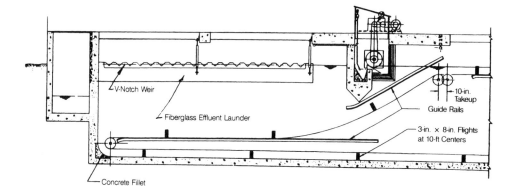

Figure 5.9 Outlet configuration using V-notch weirs of a rectangular primary settling tank (ft × 0.3048 = m; in. × 25.40 = mm).

problem is best solved, however, by not using launders on circular primary clarifiers. A single peripheral weir seems to be just as effective as launders in primary clarification.

The placement of the launder in both rectangular and circular tanks is also important. Moving the launders toward the inlet end will often result in adequate removal at higher overflow rates, but will produce a more dilute sludge.

Weather conditions can affect the performance of settling tanks and must be considered in their design. Rectangular settling tanks should be oriented, if possible, with the length perpendicular to prevailing winds to shelter tanks from the wind. However, some believe that the width should be perpendicular to the predominant wind direction because the water surface would tend to remain level at the outflow weirs. Wind shelter is especially needed for large (>46-m, or 150-ft, diameter) circular settling tanks to avoid nonuniform withdrawal rates and short circuiting caused by wind-created turbulence.

Wind may cause the water surface on the leeward side to be higher than on the windward side. This may lead to unbalanced weir rates, especially for large circular settling tanks. Surface skimmers should be oriented so that prevailing winds will push scum towards the collector. Design considerations for wind mitigation include orientation of tanks, installation of windbreaks or covers, increase of tank freeboard, and reduction of circular tank diameters to 37 m (120 ft) or less.

Cold-weather conditions that also deserve consideration include freeze protection of surface sprays, insulation of piping, installation of underground piping at greater depths (below the freeze line), and provisions for drainage piping that conveys intermittent flow. Because scum collection equipment is especially prone to freezing, it needs adequate protection. Occasionally, these steps are insufficient in areas of severe cold and freezing and settling tanks will require covers to avoid operational problems.

Maintenance Provisions

Primary treatment facilities should include provisions for necessary maintenance. For example, pumps or other equipment located in buildings or vaults require access for maintenance and repair. Access measures include lifting eyes in the overhead, traveling-bridge cranes; access openings for use of a crane outside of the structure; sufficient

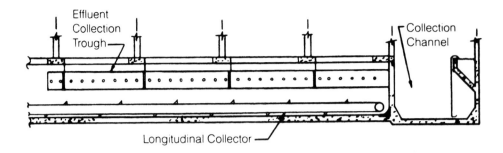

Figure 5.10 Outlet configuration using submerged collection trough of a rectangular primary settling tank.

ventilation; and adequate lighting. Sufficient clear space should be provided around pumps, meters, valves, and other equipment to accommodate maintenance and repair. This is particularly important when replacing the rotor and stator on progressing cavity pumps. Valves should be operable from the floor level.

A good design will allow dewatering of settling tanks for servicing primary-sludge collection equipment or removing an obstruction from an inlet baffle. Dewatering measures can include sloped bottoms for draining, permanent pumps, and piping or connections for temporary equipment. Also necessary are provisions to isolate a tank that is out of service from the remainder of the plant, which must remain in service. Primary-sludge collection and removal mechanisms with their critical components remaining above the water line merit consideration. Ample flushing ports and cleanouts are needed at critical tees, elbows, and ends.

Routine maintenance needs should be taken into account. For example, the plant influent wet well should be pumped down periodically to control the buildup of grease and other floating material. To relieve the heavy load of floating material on flow distribution devices it must be possible to remove scum in inlet distribution boxes or channels with submerged orifices when necessary. Sluice gates should be downward opening wherever possible to avoid the buildup of scum on the water surface and prevent deposition of solids in the track, which impedes full closure of the gate. Eliminate corner pockets and dead ends to minimize the possibility of septic conditions and use corner fillets and channeling where necessary. In practice, the tops of submerged troughs, beams, and other construction features have often been sloped 1.4:1 and their bottoms have been sloped 1:1; this will reduce or prevent the accumulation of solids and scum (Great Lakes, 1978). Provisions should also be made for cleanup after maintenance (for example, frequent use of hose bibs and sump pumps). Hose bibs, spaced no more than 30 m (100 ft) apart should be provided at each tank, scum trough, sump, and pumping station.

Enhanced Settling

Sometimes it is advantageous to have the primary settling tank remove more solids than would occur in normal tanks. Primary settling can be enhanced by preaeration or chemical coagulation, each discussed below.

Preaeration

Preaeration of raw wastewater before settling promotes flocculation of finely divided solids into more readily settleable flocs, thereby increasing suspended solids and 5-day BOD (BOD_5) removal efficiencies. Preaeration also improves scum flotation and removal. Other benefits include scrubbing volatile organic chemicals from raw wastewater, adding of dissolved oxygen, and preventing septicity during primary settling. A grit chamber with preaeration before each settling tank also promotes uniform distribution of flow to settling tanks.

Detention times of 20 to 30 minutes are necessary for floc formation and improved suspended solids and BOD_5 removals. This range is more than the 10 to 15 minutes suggested for odor control. The exact quantity of air required is a function of wastewater characteristics and tank geometry. The minimum air rate provided is 0.82 m^3/m^3 (0.11 cu^3/gal).

Chemical Coagulation

Chemical coagulation of raw wastewater before settling promotes flocculation of finely divided solids into more readily settleable flocs, thereby increasing suspended solids, BOD, and phosphorus removal efficiencies. Settling with coagulation and flocculation may remove 60 to 90% of the suspended solids, 40 to 70% of the BOD_5, 30 to 60% of the chemical oxygen demand (COD), 70 to 90% of the phosphorus, and 80 to 90% of the pathogens attached to the solids (U.S. EPA, 1987). In comparison, settling without coagulation may remove only 40 to 70% of the suspended solids, 25 to 40% of the BOD_5, 5 to 10% of the phosphorus loadings, and 50 to 60% of the pathogens.

Advantages of coagulation include greater removal efficiencies, the ability to use higher overflow rates, and more consistent performance. Disadvantages of coagulation include an increased mass of primary sludge, production of solids that are sometimes more difficult to thicken and dewater, and an increase in operational cost and operator attention. The designer of chemical coagulation facilities should consider the effect of enhanced primary settling on downstream sludge-processing facilities.

Historically, iron salts, aluminum salts, and lime have been the chemical coagulants used for wastewater treatment. Iron salts have been the most common of the coagulants used for primary treatment. However, modern enhanced treatment invariably involves the use of both inorganic coagulants, such as iron and aluminum salts, and organic polyelectrolytes. The inorganic chemicals act as the coagulant (reducing particle charge) and the polymers function as floc builders. While it is possible to have the inorganic coagulants form oxides and hydroxides that help build flocs, the required concentrations are too high, the chemical costs are high, and the process produces extra solids that must be managed in the sludge-processing facilities. The use of polymers as floc builders reduces sludge production and total chemical costs.

Mechanical mixers, in-line blenders, pumps, baffled compartments, baffled pipes, or air mixers can be used to mix the chemicals into the water (Klute, 1985). Mechanical mixers and in-line blenders cost considerably more than other types and might become clogged or entangled with debris. Air mixing eliminates the problem of debris and can offer advantages for primary settling, especially if aerated channels or grit chambers already exist. Pumps, Parshall flumes, flow-distribution structures, baffled compartments, or baffled pipes—methods often used for upgrading existing facilities—offer a lower-cost but less efficient alternative to separate mixers for new construction.

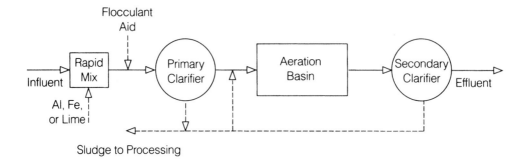

Figure 5.11 Possible scheme for coagulant addition for enhanced primary settling.

Methods listed above are less efficient than separate mixers because, unlike separate mixing, the mixing intensity depends on the flowrate.

Supplementing conventional primary settling with chemical coagulation requires minimal additional construction. The optimal point for coagulant addition is as far upstream as possible from primary settling tanks. Figure 5.11 shows a possible flow diagram for coagulant addition. If possible, several different feed points should be considered for additional flexibility.

Chemically enhanced primary treatment has been shown to produce effluents that are almost equivalent to those from secondary treatment, at a fraction of the cost. This technology seems especially applicable to wastewater treatment plants in developing economies.

In addition to chemical coagulation, primary sludge removal can be increased by physically weighing down the particles using a process called *ballasted settling*. Any number of materials, from kaolin clay to incinerator ash, have been used. The disadvantage of this system is that the total mass of primary sludge produced is significantly increased, and often the process is not economical if the added expense of sludge disposal is included in the economic calculations. Ballasted settling has been most useful in short-term emergency situations.

Sludge Collection and Removal

Sludge Collection

Settled primary sludge is scraped into a hopper where it is removed by gravity or pumping. The hopper for rectangular tanks is located at the inlet end of the tank to minimize the travel time of particles to the hopper. For circular tanks, the hopper is located in the center of the tank. The hopper, up to 3 m (10 ft) deep, has steep sides with a minimum slope of 1.7:1 (Great Lakes, 1978). Hopper wall surfaces should be smooth with rounded corners to avoid any sludge buildup and, in practice, the hopper bottom has a maximum dimension of 0.6 m (2 ft) (Great Lakes, 1978). Settling tanks with steep sides and widths of more than 3 m often need more than one hopper to reduce its depth.

Common withdrawal pipes from two or more hoppers often result in unequal primary-sludge removal from the hoppers. Therefore, multiple tanks and hoppers need separate pipes and pumps or valves on each outlet.

Primary-sludge collection equipment for rectangular tanks have chains and flights or a traveling bridge. Chain and flights (Figures 5.2 and 5.4) consist of two endless loops of chains with cross scrapers (flights) attached at approximately 3-m intervals. Revolving flights push the settled primary sludge to the hopper at the end of the tank. Chain-and-flight collectors are limited in width to approximately 6 m (20 ft); some installations, however, have used side-by-side collectors without common walls in tanks as wide as 24 m (80 ft) or more. Historically, cast iron chains and wood flights were used. Designers now select almost exclusively nonmetallic (plastic) chains and fiberglass flights. A flight speed of 0.9 m/min (3 ft/min) is typical.

Flights travel along the long axis of the tank and, as the upper flights move away from the primary-sludge hopper, they can skim the surface, pushing floating material toward the scum-removal mechanism. At the end of the tank, the flights drop to the floor and drag heavy, settled material to a hopper for removal. Single tanks can have either a single or double hopper.

A traveling bridge (Figure 5.12) consists of a scraper blade mechanism mounted on a bridge or carriage that travels approximately 1.8 m/min (6 ft/min) toward the hopper on tracks or rails mounted on top of the tank. As it travels away from the hopper at approximately 3.7 m/min (12 ft/min), the mechanism, largely out of the water, acts as a skimmer, pushing floating material toward the scum removal mechanism. As it reaches the end of the tank, the mechanism drops to the floor of the tank, reverses direction, and travels toward the hopper end of the tank, pushing settled primary sludge to the hopper.

Traveling-bridge collectors cannot be used if covers are required on primary settling tanks. These collectors may span tank widths up to 30 m (100 ft), and in earthquake zones a restraining mechanism should be considered to resist derailing from seismic ground motion. In cold climates, design of traveling-bridge collectors should provide for control of snow and ice buildup on rails or tracks. Otherwise, this buildup could derail the bridge or reduce traction between the wheel and rail.

Circular primary settling tanks have plow-type primary-sludge collection equipment. The plow type (Figure 5.3) consists of scrapers that drag the tank floor at a top speed of approximately 1.8 to 3.7 m/min (6 to 12 ft/min) (ASCE and AWWA, 1990). Plows are located at an angle to the radial axis to force primary sludge towards the hopper, at the center of the tank, as the device rotates. The center hopper is a vertical-sided sump where the primary sludge is removed by pumping. The rotating element of the device can be driven from either the center or the outside tank wall. Torque must be sufficient to move the densest primary sludge expected.

Primary-Sludge Quantities and Properties

Primary-sludge production can be estimated in SI (metric) units as

$$S_M = \frac{Q \times SS \times E}{1000} \qquad (5.6)$$

and in common U.S. units as

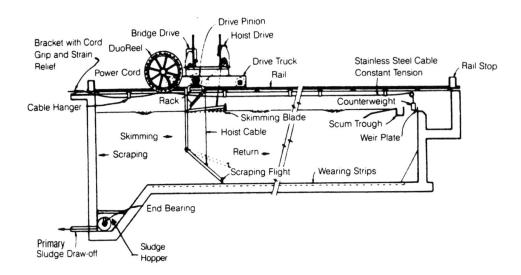

Figure 5.12 Typical traveling-bridge sludge collector in a primary rectangular settling tank.

$$S_M = Q \times SS \times E \times 8.34 \tag{5.7}$$

where

- S_M = mass of dry primary-sludge solids produced, kg/d (lb/d);
- SS = incoming suspended solids concentration (mg/L);
- Q = influent to the settling tank, m³/d (mgd);
- E = efficiency of suspended solids removal (as a decimal); and
- 8.34 = conversion factor.

The above equations yield the mass of the *dry* solids. To obtain the total (water plus solids) mass of primary sludge, S_M in the above equations has to be divided by the solids concentration. Total suspended solids removal efficiencies in primary settling tanks range between 50 and 65%. Many designers assume a removal efficiency of 60% for estimating purposes.

Enhanced primary settling with chemical coagulation can increase primary-sludge mass by 50 to 100%. Chemical sludge quantities can be estimated by the stoichiometric relationship between raw wastewater and coagulants. The stoichiometric quantity should be increased by approximately 35% for aluminum and iron salts to account for increased BOD, COD, and suspended solids removal (Mertsch, 1985).

Composition of primary sludge is variable and depends on the nature and degree of industrial development in the collection area. Chemical sludges can be gelatinous, with a high water content, low suspended solids content, and high resistance to mechanical or gravitational dewatering. Feed solids composition merits careful consideration in design of solids-handling and processing units. More on the composition of primary sludge is discussed in Chapter 12.

Scum Collection and Disposal

Removal of floating materials, or scum, is an important function of primary treatment. Oil, grease, plastics, and other floating materials increase the organic load to downstream treatment processes and might cause various operational problems, including odors and the buildup of scum in downstream treatment processes.

Scum Collection

Scum collection is usually located on the effluent end of the rectangular primary settling tank (Figure 5.2). Some plants, however, have located scum collection on the influent end of the settling tank to decrease the travel distance of scum to the collection point and ensure rapid removal of all flotage.

Automated scum-removal mechanisms may be operated by the primary-sludge collection mechanism or a separate operating device. The scum-removal mechanism should extend the full width of the tank to prevent floating material from reaching the effluent weir. The inlet design should allow scum to freely enter settling tanks without being trapped in inlet channels or behind baffles. Slots should be provided on the circular feed well for this purpose. In some cases (for example, influent high in oil and grease content), a separate skimmer mechanism or water-spray system is necessary for the center feed well. A nearly constant water surface elevation in settling tanks should be maintained for proper operation of scum-collection equipment.

Two types of mechanisms—tilting trough (Figure 5.13) and sloping beach (Figure 5.14)—are used for removing scum from primary settling tanks. A tripping device on the primary sludge collector activates the tilting trough. The trough tilts to allow collected material on the surface to flow into the trough and then to a wet well. Wet well scum is pumped to a scum holding tank for eventual disposal. The sloping beach is a stationary device with a collector trough.

Collectors for rectangular tanks are manually tilted with a lever, rack and pinion, or worm gears, or automatically tilted with a motor-operated device. In circular clarifiers, the scum is skimmed off the top and pushed to either a sloping beach or a tilting-trough skimmer. A spring-loaded section of the rotating arm rides up the beach, wipes material into a trough, and drops back in the water on the far side. The tilting trough extends into the tank just short of the rotating arm. As the arm or surface collector passes, it physically tilts (rotates) the trough, allowing it to skim the surface.

Scum Management

Scum solids range between 0.1 and 19 mL scum/L of wastewater, with a median value of approximately 5 mL/L. Quantity and chemical composition of scum are both highly variable and depend on several factors, including the degree and type of industrial development in the collection area, recycled plant sidestreams, and scum removal efficiency of upstream processes and scum-removal equipment.

Progressing cavity pumps, pneumatic ejectors, and recessed-impeller centrifugal pumps, both with and without cutting bar attachments, have been used to pump scum. The design of scum-removal equipment includes measures to keep the scum tank or hopper contents mixed during pumping to prevent scum from crusting or coning. The bottom of the scum tank should be sloped.

Historically, scum has been landfilled or processed with other wastewater treatment sludges. Adequate digester mixing must be provided when scum is discharged to a

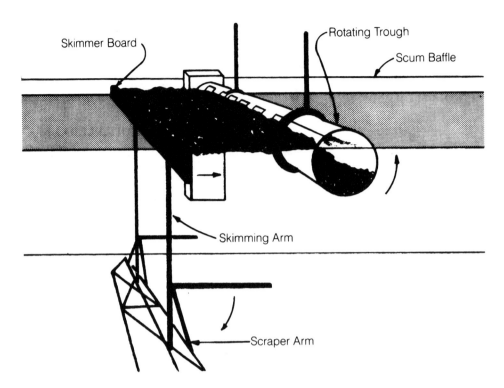

Figure 5.13 Tilting trough scum collector (WPCF, 1985).

digester to ensure complete digestion and minimize scum blanket formation. Scum digestion, often based on convenience instead of value, sometimes defers the scum disposal problem or shifts it to another process.

Thickening with Waste Activated Sludge

Waste biological sludge is sometimes discharged to the influent end of primary settling tanks so that the biological sludge can thicken when it is settled with the primary sludge. Separate waste activated sludge thickening has replaced this practice in most treatment plants because of its detrimental effect on primary settling. Thickening with WAS will decrease the solids concentration to as low as 2 to 4%, instead of the 4 to 8% solids achievable with primary sludge settling and thickening only. Combining sludges and thickening them together using gravitational thickeners seems to be counterproductive. On the other hand, dissolved air flotation thickening has been shown to be successful with some mixed sludges.

Imhoff Tanks

As shown in Figure 5.15, the Imhoff tank consists of a top compartment, which serves as a settling tank, and an unheated lower compartment in which the settled sludge is anaer-

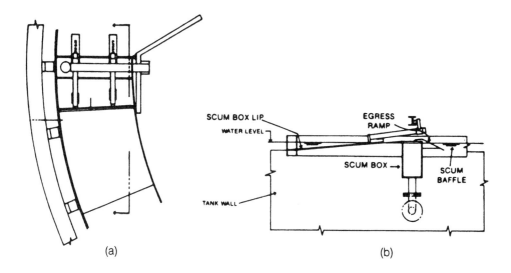

Figure 5.14 Sloping beach scum collector: (a) partial plan and (b) section (WPCF, 1985).

obically stabilized. Effluent flows over the sidewall weir to a channel and gas vents to the atmosphere through openings along the side. Digested sludge is periodically removed by gravity from the tank. Imhoff tanks have no mechanical equipment and have low maintenance requirements; they nonetheless have operational problems, including the periodic production of an odorous foam, excessive accumulation of scum in the gas vents, and production of odorous digested sludge (Linvil, 1980). Heating the lower compartment of the Imhoff tank is not economical because heat will dissipate through the gas vents to the settling compartment. Hence, the tank volume required exceeds that for separate, heated tanks. Imhoff tanks have a long history of successful application for small plants and where maintenance and proper operation is problematic.

Variations of Imhoff tanks are available commercially, under various trade names. These units are usually circular, with a central feed and a deep, central sludge trough in which digestion can take place. The gas escapes inside the circular baffle, thus not interfering with settling in the area outside of the baffle. Such tanks are commonly used in small wastewater treatment plants and in "package plants", small plants that are assembled off-site.

Fine Screens

Fine screens can be used in lieu of settling for primary treatment but will not achieve removal efficiencies of settling. Fine screens can, however, upgrade existing primary settling facilities. Fine screens with openings from 1 to 6 mm (0.04 to 0.25 in.) only achieve removal efficiencies of 15 to 30% for suspended solids, 15 to 25% for BOD, and 10 to 20% for bacteria loadings (Steel, 1979). The most common fine screens used for primary treatment are the inclined, self-cleaning (static) screen (Figure 5.16), and the rotary drum screen (Figure 5.17). Multiple screens are needed to allow some units to be out of service for routine cleaning and maintenance.

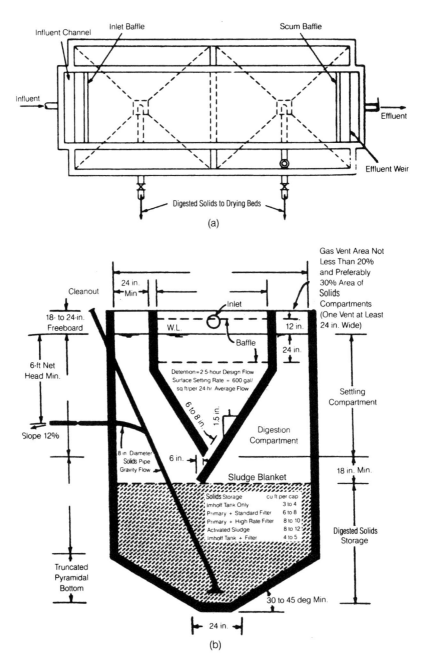

Figure 5.15 Typical Imhoff tanks with design details: (a) plan and (b) section. Use larger sludge storage allowances for garbage solids and communities with less than 5000 population (in. × 25.4 = mm; ft³ × 0.0283 = m³; gal/ft² × 0.04074 = m³/m² (from Metcalf and Eddy, Inc., *Wastewater Engineering: Treatment, Disposal, Reuse,* 3rd ed. Copyright © 1991, McGraw-Hill, Inc.: New York, with permission; from Salvato, J. A. *Environmental Engineering and Sanitation.* Copyright © 1972, Wiley & Sons: New York, with permission).

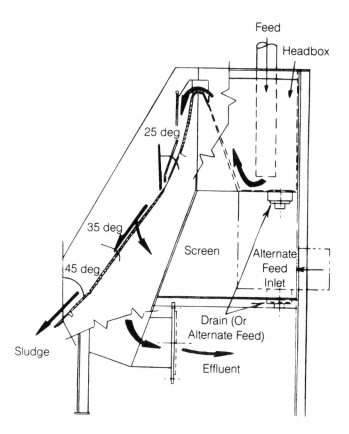

Figure 5.16 Inclined, self-cleaning static screen.

Conclusions

Some years ago, a Princeton musicologist published an influential paper with a title to the effect "It does not matter if nobody listens". His point was that music, at least academic music, was concerned not with writing music people can enjoy, but rather with developing new musical ideas. In his opinion it did not matter if the musical composition was not pleasing to anyone. What mattered was that it was new and unique.

In contrast to the musicologist, Duke Ellington believed that as far as music was concerned, "If it sounds good, it is good." He judged the value of music not by its originality or its complexity, but by how much it was appreciated by the public. He also meant, of course, that many different kinds of music would "sound good", as long they are appreciated by the public.

In the case of wastewater treatment, is *does* matter if it does not work. Design engineers cannot hide behind scholarly nonsense and still earn a living. Their work is out there for everyone to judge, every day. And many things work well, like many different settling tanks. The Duke had the right philosophy: "If it works, it is good." The trick, of course, is designing something that works.

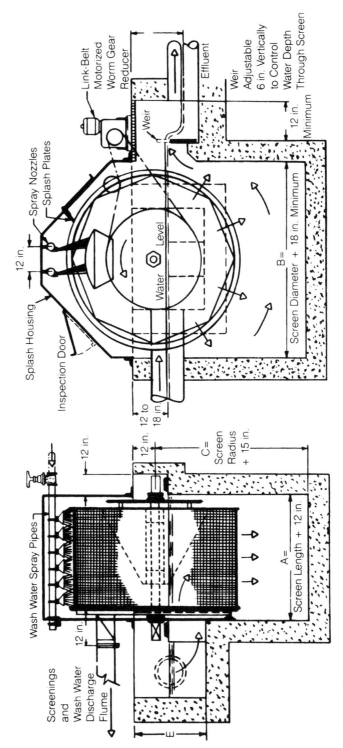

Figure 5.17 Rotary drum screen (in. × 25.4 = mm).

References

American Society of Civil Engineers; American Water Works Association (1990) *Water Treatment Plant Design*; McGraw-Hill: New York.

Camp, T. R. (1936) A Study of the Rational Design of Settling Tanks. *Sew. Works J.*, **8**, 742.

Camp, T. R. (1946) Sedimentation and the Design of Settling Tanks. *Trans. Am. Soc. Civ. Eng.*, **111** (3), 895, (paper no. 2285).

Camp, T. R. (1953) Studies of Settling Basin Design. *Sew. Ind. Wastes*, **25**, 1.

Civil Engineering Pollution Control Systems (1979) U.S. Naval Facilities Design Manual 5.8; Philadelphia, Pennsylvania.

Graber, S. D. (1974) Outlet Weir Loading and Settling Tanks. *J.—Water Pollut. Control Fed.*, **46**, 2355.

Great Lakes—Upper Mississippi River Board of State Sanitary Engineering Health Education Services, Inc. (1978) *Recommended Standards for Sewage Works*; Albany, New York.

Great Lakes—Upper Mississippi River Board of State Sanitary Engineering Health Education Services, Inc. (1997) *Recommended Standards for Sewage Works*; Albany, New York.

Hamlin, M. J. (1972) Preliminary Treatment Sedimentation, Paper 2. In *Advances in Sewage Treatment,* Proceedings of the Conference of the Institute of Civil Engineers: London.

Kelly, K. (1988) New Clarifiers Help Save History. *Civ. Eng.*, **58**, 10.

Klute, R. (1985) Rapid Mixing in Coagulation/Flocculation Processes—Design Criteria. Chem. Water Wastewater Treatment Schr.—Reine Verein Wa Bo Lu 62 G, Fischer, Springer-Verlag: New York.

Linvil, R. G. (1980) *Low-Maintenance Mechanically Simple Wastewater Treatment Systems*; McGraw-Hill: New York.

Merritt, F. S. (1983) *Standard Handbook for Civil Engineers*, 3rd ed.; McGraw-Hill: New York.

Mertsch, V. (1985) Characteristics of Sludge from Wastewater Flocculation/Precipitation. New York, N.Y.

Metcalf and Eddy, Inc. (1991) *Wastewater Engineering: Treatment, Disposal, Reuse,* 3rd ed; McGraw-Hill: New York.

Salvato, J. A. (1972) *Environmental Engineering and Sanitation*; Wiley & Sons: New York.

Steel, W. E. (1979) *Water Supply and Sewerage*; McGraw-Hill: New York.

Theroux, R. J.; Betz, J. M. (1959) Sedimentation and Preaeration Experiments at Los Angeles. *Sew. Ind. Wastes*, **31**, 1259.

U.S. Environmental Protection Agency (1987) *Design Manual for Phosphorus Removal*; EPA-625/1-87-001; Office of Research & Development: Cincinnati, Ohio.

Water Pollution Control Federation (1985) *Clarifier Design*; Manual of Practice No. FD-8: Washington, D.C.

Weber, W. J. (1972) *Physiochemical Processes for Water Quality Control*; Wiley & Sons: New York.

Symbols Used in this Chapter

A = surface area of the ideal tank, or length times width, m² (ft²)
d = diameter of scoured particles, m (ft)
f = Darcy–Weisbach friction factor
g = acceleration due to gravity, 9.8 m/s² (32.2 ft/s²)
h = any height less than H, m (ft)
h = height at which a particle with settling velocity v_h enters and is just removed, m (ft)
H = height of the ideal tank zone, m (ft)

k = constant for type of scoured particles
L = ideal tank length, m (ft)
Q = flowrate, m³/s (ft³/s)
s = specific gravity of scoured particles
$\bar{t}$ = hydraulic retention time, days
V = tank volume, $L \times H \times W$, m³ (ft³)
v_h = settling velocity of any particle entering the ideal tank zone at height h, m/s (ft/s)
v_L = horizontal velocity through the tank, m/s (ft/s)
v_o = settling velocity of critical particle, m/s (ft/s)
v_s = critical scour velocity, m/s (ft/s)
W = ideal tank width, m (ft)

6

Suspended-Growth Biological Treatment

Introduction

Suspended-growth systems are predominantly aerobic processes that achieve high microorganism (biomass) concentrations through the recycle of biological solids. These microorganisms convert biodegradable, organic wastewater constituents and certain inorganic fractions into new cell masses and byproducts, both of which can subsequently be removed from the system by gaseous stripping, settling, and other physical means. Available suspended-growth systems, system modifications, and processes include complete-mix, plug, and series-flow systems; high-rate, extended aeration, contact stabilization, and tapered aeration modifications; and deep shaft, pure oxygen, and sequencing batch reactor processes.

The Activated-Sludge Process

Process Description

Figure 6.1 presents a general schematic of typical, flow-through, activated-sludge systems and its modifications. Wastewater and biological solids are first combined, mixed, and aerated in a reactor (aeration tank). The process operates in a continuous-flow mode, but can also be operated as a batch process. Contents of the

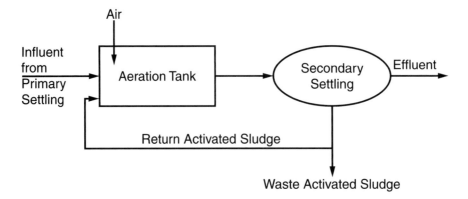

Figure 6.1 The activated-sludge process.

reactor, referred to as *mixed liquor*, consist of wastewater; microorganisms (living as well as dead); and inert, biodegradable, and nonbiodegradable suspended and colloidal matter. The particulate fraction of the mixed liquor is termed *mixed liquor suspended solids* (MLSS).

From the aeration tank the mixed liquor goes to a settling tank (clarifier) to allow gravity separation of the MLSS from the treated wastewater. Settled MLSS are then recycled to the aeration tank to maintain a concentrated microbial population for degradation of influent wastewater constituents. This is known as *return activated sludge*. Because microorganisms are continuously synthesized in this process, a means must be provided for wasting some of the MLSS from the system. This flow is known as *waste activated sludge*, or often as WAS. Wasting is from the return sludge line, although removal from the aeration tank is an alternative. Depending on the design and operation of the process, either maximizing or minimizing production of biological sludge is possible.

A basic suspended-growth system consists of a number of interrelated components, including

- A single aeration tank or multiple tanks designed for completely mixed flow, plug flow, or intermediate patterns and sized to provide adequate solids retention time, organic loading, or other criteria. Results often provide a hydraulic retention time in the range of 1 to 24 hours or more.
- An oxygen source and equipment to disperse atmospheric or pressurized air or oxygen-enriched air to the aeration tank at a rate sufficient to keep the system aerobic.
- A means of mixing the aeration tank contents to keep the MLSS in suspension so as to provide contact between the microorganism, dissolved organic materials, and oxygen.
- A clarifier to separate the MLSS from treated wastewater. (In a sequencing batch reactor, mixing and aeration are stopped for a time interval to permit MLSS settling and treated wastewater decanting, thereby eliminating the need for a separate clarifier.)
- A method of collecting settled solids in the clarifier and recycling them to the aeration tank.
- A means of wasting excess solids from the system.

The basic activated-sludge system, commonly used for carbonaceous biochemical oxygen demand (BOD) removal, can be designed to achieve nitrification as well. The basic activated-sludge system can also be modified to include the removal of both phosphorus and nitrogen.

Effect of Influent Load and Characterisitics

As explained earlier in this text, municipal loads and wastewater characteristics vary substantially with season, day of the week, and hour of the day. Unless these variations are correctly addressed in the design of a plant, process performance can be seriously affected. The activated-sludge system, namely the aeration tank–final clarifier combination, are particularly vulnerable to these large variations. Excessive hydraulic peaks shift aeration tank solids inventory to the clarifiers, which may not be able to contain it. Temperature changes may adversely affect the settling of solids in the clarifier, resulting in a loss of solids to the effluent. Large increases in organic load may lead to deterioration of mixed liquor settling and turbid effluent. Sporadic discharges of toxic compounds may sharply reduce biological activity in the aeration tank or result in residual toxic compounds in plant effluent, thereby causing failure to pass bioassay tests.

For decades, engineers have designed activated-sludge plants by making certain provisions for these peaking or sporadic changes in quality. Such features have included flow equalization, oversizing clarifiers, and providing alternative aeration tank feed patterns and large return activated-sludge capacities compared with those needed for a steady flow. Such modifications increase the final cost of the wastewater treatment plant but can make the difference between a design success and a design failure.

The introduction of dynamic simulation by computer technology adds a new dimension to the design of activated-sludge plants. Various flow patterns, including diurnal curves, can be simulated to quantify the effects of solids inventory shifting and effluent quality effects. Furthermore, higher levels of in-plant sensing and automation facilitate flow diversion strategies that allow process units to be reduced in size from those with less sophistication.

Historical Perspective

Based on experiments conducted at the Lawrence Experiment Station in Lawrence, Massachusetts, in the early 1900s, the activated-sludge process was developed at a Manchester, England, wastewater treatment plant by Ardern and Lockett (1914). (It is rumored that the researchers at the Lawrence Experiment Station did all the right research, but they did not understand the significance of their results. A visitor from England realized what was going on and told Ardern and Lockett, who then performed the work in England.) In the 1920s several installations started operation in the United States; however, because of litigations on patent infringements, widespread use of the process did not begin until the 1940s. Early investigators noted that the amount of biodegradable organics applied to a system affected microorganism metabolic rate. Initial design methods were entirely empirical in nature and aeration tank detention time was one of the first parameters used. Short hydraulic retention times were chosen for what was considered to be weak wastewater and long hydraulic retention times were chosen for strong wastewater.

Various design loading criteria eventually were developed, relating to the mass of 5-day biochemical oxygen demand (BOD_5) applied per day per mass of microbial solids

> **Note on Terminology**
> When a liquid flows into a reactor, its average *hydraulic retention time* is calculated as
>
> $$\bar{t} = \frac{V}{Q} \qquad (6.1)$$
>
> where
> $\bar{t}$ = hydraulic retention time (d);
> V = volume (m³); and
> Q = flowrate (m³/d).
>
> Another way of thinking of hydraulic retention time is the time necessary for any flow Q to fill up a tank of volume V. Still another way to think of retention time is the time it takes for any particle of water to flow from the influent end to the effluent end of the tank (if the tank has uniform flow).
>
> In aerobic treatment, the concept of *solids retention time* is very important and it is necessary to distinguish between these two terms. The solids retention time is defined as
>
> $$\theta_c = \frac{XV}{Q_w X_w} \qquad (6.2)$$
>
> where
> θ_c = solids retention time (d);
> X = concentration of solids in the reactor (g/L);
> V = volume of the reactor (L);
> Q_w = flowrate of sludge wasted (L/d); and
> X_w = concentration of the solids wasted (g/L).
>
> The numerator in this equation is the mass of solids in the reactor (concentration times volume) and the denominator is the mass of solids wasted per day.
>
> Wastewater treatment plant operators use a different term for the solids retention time, namely *sludge age*. Sludge age is calculated exactly like solids retention time, and is expressed in days.
>
> To make matters even more confusing, biochemical engineering researchers (and some environmental engineering researchers) talk about *mean cell residence time*. This is again exactly the same as solids retention time, and expressed in units of time. The three terms are identical and the reader should not be confused by their interchangeable use in some literature. In this book we use *solids retention time*.

present in the aeration tank. Only within the last 40 years have design equations been developed based on the concepts of microbial-growth kinetics and mass balances. The different design approaches offered by Eckenfelder (1966), McKinney (1962), Lawrence and McCarty (1970), and Ramanathan and Gaudy (1971) have been shown to yield similar results (Gaudy and Kincannon, 1977).

The Activated-Sludge Environment

An activated-sludge process uses a suspension of flocculent microorganisms composed of bacteria, fungi, protozoa, and rotifers to treat wastewater. The dry weight of these microorganisms is 95%, or more, organic in composition while the suspension of microorganisms in an activated-sludge process is composed of 70 to 90% organics. Composition of the organic fraction is represented by the empirical formula, $C_5H_7O_2NP_{0.2}$. Other formulas in the literature include $C_5H_9O_{2.5}NP_{0.2}$ and $C_{60}H_{87}O_{23}N_{12}P$. Inorganics include potassium, sodium, magnesium, sulfur, calcium, iron, and other trace elements. The characteristics of the wastewater, environmental conditions, process design, and its mode of operation determine the predominant microorganism group that will develop. Successful plant performance will depend on the development of a microbial community that will assimilate target waste materials and form a flocculent biomass that is readily removed by gravity separation.

The microbial population is dominated by heterotrophic organisms that require biodegradable organic matter for energy and new cell synthesis. These microorganisms include bacteria, fungi, and some protozoa. Autotrophic bacteria, including those that oxidize ammonia to nitrite and nitrate, have the ability to use inorganic materials for energy and cell synthesis. Such autotrophs are present in varying concentrations. Their relative presence depends on many conditions, including the mode of operation of the facility and reactor concentrations of biomass and organic and inorganic materials. Many protozoa and all rotifers in suspended-growth systems are predators, feeding on bacteria and likely enhancing both flocculation and clarification because of their feeding on dispersed organics and organisms. Protozoa and rotifers can constitute as much as 5% of the mass of organisms in a system. Nuisance organisms might occur sometimes and can interfere with successful plant operation. Most nuisance organism problems are associated with poor solids settling and the formation of heavy foam and scum.

Biological Growth and Substrate Oxidation

An empirical relationship depicting the stabilization of biodegradable organic matter in aerobic suspended-growth systems can be expressed by the following equation:

$$\text{Organic matter} + \text{Nutrients} + \text{Microbes} \rightarrow \text{New microbes} + CO_2 + H_2O \quad (6.3)$$

The equation summarizes a complex series of biochemical reactions that can be simplified into three fundamental activities: oxidation, synthesis, and autooxidation. Oxidation is the coupled release of energy through the conversion of organic matter to lower-energy products (carbon dioxide and water). Synthesis is the conversion of a portion of the organic matter, assisted by the energy released during oxidation, into new biomass. Autooxidation is the conversion of some of the cell constituents to low-energy products, with the release of additional energy. Details of these biochemical reactions and their stoichiometry can be found elsewhere (Grady et al., 1999; McCarty, 1972, 1975).

An equation (unbalanced) similar to that above represents the oxidation of ammonia to nitrate (nitrification) by select autotrophs

$$NH_4^+ + O_2 + CO_2 + HCO_3^- + \text{Microbes} \rightarrow \text{New microbes} + H_2O + NO_3^- + H^+ + CO_2 \quad (6.4)$$

This equation represents the net result of both oxidation and synthesis reactions for conversion of ammonium to nitrite and then to nitrate. The stoichiometry of this reaction is better understood than that for the oxidation of organic matter because the substrates are well defined and only a few species of microorganisms can carry out nitri-

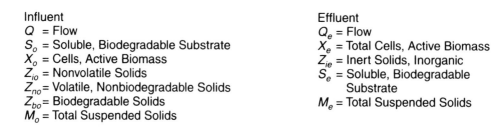

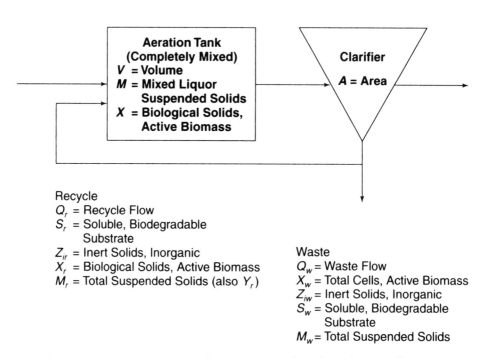

Figure 6.2 Nomenclature for activated-sludge flow sheet (volatile and nonvolatile represent organic and inorganic solids, respectively).

fication. Details of nitrification biochemistry and stoichiometry can be found in the literature (Gaudy and Gaudy, 1980; Grady et al., 1999; McCarty, 1972, 1975; Water Research Commission, 1984).

Process Design for Carbon Oxidation

Figure 6.2 illustrates a typical suspended-growth system flow chart. A reactor with a volume (V) receives an influent or influent flow (Q) plus a recycle flow (Q_r). Influent flow contains a soluble, biodegradable substrate at concentration S_o and suspended solids at concentration M_o. Suspended solids comprise microorganisms (X_o) and other particulates (X_{io}). These "other particulates" include nonvolatile (also called *inorganic*) solids (Z_{io}), nonbiodegradable volatile (also called *organic*) solids (Z_{no}), and biodegradable volatile solids (Z_{bo}). Recycle flow contains soluble biodegradable substrate at concentration S_r, biological solids at concentration X_r, and inert suspended solids at concentration Z_{ir}.

In design, it is important to evaluate all components of incoming wastewater that will influence calculations of solids production and oxygen demand. Many simplistic models do not properly account for incoming suspended solids and, therefore, underestimate sludge generation. Bench-scale, pilot-plant, and full-scale investigations are valuable in providing useful estimates of these factors.

When pilot-plant studies are not possible, wastewater components must be estimated. For domestic wastewater, influent microorganisms (X_o) are assumed to be negligible relative to those in the reactor. However, acknowledging their presence is critical during system start-up. They can also influence oxygen demand patterns in a system's aeration tank (Grady et al., 1999). Nonvolatile suspended solids (Z_{io}) can be calculated as the difference between influent suspended solids and volatile suspended solids. Nonbiodegradable volatile (organic) suspended solids (Z_{no}) have been approximated as 40% of influent organic or volatile suspended solids (Dague, 1983). Biodegradable volatile suspended solids (Z_{bo}) are often presumed to be rapidly adsorbed onto the biomass and subsequently hydrolyzed (solubilized). Therefore, they are often ignored in calculations. However, slowly hydrolyzable biodegradable volatile suspended solids can significantly affect system kinetics and mass balances.

Work of Eckenfelder (1966), Lawrence and McCarty (1970), McKinney (1962), and McKinney and Ooten (1969) led the way in developing a quantitative understanding of the activated-sludge process. The work of Lawrence and McCarty was particularly significant in providing a more unified approach and emphasizing the importance of solids retention time. Basic Lawrence and McCarty design equations used for sizing systems are presented as follows (Lawrence and McCarty, 1970).

The hydraulic retention time is

$$\bar{t} = \frac{V}{Q} \tag{6.5}$$

The solids retention time (sludge age) is

$$\theta_c = \frac{VX}{Q_w X_r} = \frac{VM}{Q_w M_r} \tag{6.6}$$

For completely mixed aeration tanks, the solids retention time is related to the basic growth equation as

$$\frac{1}{\theta_c} = \frac{Y\mu_{max}S_e}{K_s + S_e} - b \tag{6.7}$$

The total biomass (the living organisms) in the aeration tank is calculated as

$$XV = \theta_c QY \frac{S_o - S_e}{1 + b\theta_c} \tag{6.8}$$

where X is the biomass concentration (mass of living organisms per volume of aeration tank). The observed yield of solids in an activated-sludge system can be estimated from

$$Y_{OBS} = \frac{Y}{1 + b\theta_c} \tag{6.9}$$

Note on Terminology

When the original work on growth kinetics as applied to wastewater treatment was published, the data were obtained from *chemostats*, or small continuously mixed laboratory reactors. The substrate was often some type of readily decomposable sugar solution, which resulted in the solids in the chemostats being made up almost totally of microorganisms that could be measured by running the suspended solids test on the reactor contents. There was some minor nonliving matter produced from the death of the microorganisms, but this was also organic. No inorganic or fixed solids were added to these small laboratory reactors.

When the equations derived from such tests were applied to wastewater treatment, the original definition of mixed liquor suspended solids remained, and the X in eq 6.8 referred to the suspended solids in the aeration tank of the activated-sludge system where all of the suspended solids are assumed to be living biomass. But in the real world, the substrate fed to a wastewater treatment plant activated-sludge system is not pure sugar, but contains all manner of organic and inorganic inert matter as well. Hence the definition of solids concentration in the real-world activated-sludge system is defined instead by eq 6.6 as the sum of the mass of living biomass, the dead organic matter, and the inert or fixed solids, all divided by the volume. When it is necessary to calculate the amount of sludge produced in an activated-sludge system (waste activated sludge), eq 6.7 must be used because this takes into account the nonliving solids, which represent a significant fraction of the waste activated sludge.

The total mixed liquor suspended solids in the aeration tank is

$$M = \frac{\theta_c}{t} \left\{ \left[\frac{Y(S_o - S_e)}{1 + b\theta_c} \right] + Z_{io} + Z_{no} \right\} \tag{6.10}$$

The production of excess sludge (waste activated sludge) is calculated as

$$P_x = Q \left\{ \left[\frac{Y(S_o - S_e)}{1 + b\theta_c} \right] + Z_{io} + Z_{no} \right\} \tag{6.11}$$

The recycle ratio, or the ratio of the return sludge flowrate to the influent flowrate, is

$$\alpha = \frac{Q_r}{Q} = \frac{X}{X_r - X} = \frac{M}{M_r - M} \tag{6.12}$$

Note that both the biomass (X) and the suspended solids (M) can be used to calculate the recycle ratio.

The mass of oxygen required per unit time to satisfy the carbonaceous oxidation can be calculated as

$$R_c = \frac{Q(S_o - S_e)(1 + b\theta_c - BY)}{1 + b\theta_c} \tag{6.13}$$

and the oxygen needed for nitrification can be calculated as

$$R_n = 4.57Q (N_o - N) - 2.86Q (N_o - N - NO_3) \tag{6.14}$$

where
- b = endogenous decay coefficient based on biomass in aerated zone (1/d);
- B = oxygen equivalent of cell mass, often calculated as 1.42 kg O_2/kg volatile suspended solids (kg/kg);
- K_s = half-velocity coefficient, or the substrate concentration when half of substrate is used (kg/m^3);
- M = total mixed liquor suspended solids (kg/m^3);
- M_r = total sludge recycle flow suspended solids (kg/m^3);
- N = effluent oxidizable nitrogen (kg/m^3);
- N_o = influent oxidizable nitrogen (kg/m^3);
- NO_3 = effluent nitrate-nitrogen (kg/m^3);
- P_x = mass of total activated-sludge solids generated or wasted per day (kg/d);
- Q = wastewater inflow (m^3/d);
- Q_r = sludge recycle flow (m^3/d);
- Q_w = sludge waste flow (m^3/d);
- R_c = mass of oxygen required per unit time to satisfy carbonaceous oxidation (kg/d);
- R_n = mass of oxygen required per unit time to satisfy nitrification oxygen demand (kg/d);
- S_e = effluent soluble substrate (kg/m^3);
- S_o = influent soluble biodegradable substrate (kg/m^3);
- $\bar{t}$ = hydraulic retention time (d);
- V = aeration tank volume (m^3);
- X = reactor biological solids (kg/m^3);
- X_r = sludge recycle flow biological solids (kg/m^3);
- Y = true cell yield (kg/kg);
- Y_{OBS} = observed cell yield (kg/kg);
- Z_{io} = influent nonvolatile suspended solids (kg/m^3); and
- Z_{no} = influent volatile nonbiodegradable solids (kg/m^3).
- α = sludge recycle ratio;
- μ_{max} = maximum rate of substrate use per unit weight of biomass (1/d); and
- θ_c = solids retention time (d).

In the above definitions, the units for each variable are expressed in terms of metric units. Common U.S. units are seldom used in such calculations. Substrate can be expressed as either chemical oxygen demand (COD) or BOD as long as the coefficients selected are dimensionally correct.

Design of a system requires determination of the following:

- Volume of the aeration tanks (V);
- Quantity of sludge wasted (P_x);
- Total oxygen demand ($R_c + R_n$);
- Sludge recycle requirements; and
- Size of clarifiers.

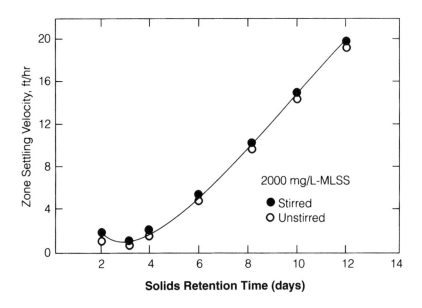

Figure 6.3 Effect of solids retention time, θ_c, on the settling properties of activated sludge.

The mathematical development assumes that the loss of suspended solids in the clarifier effluent and seed in the influent are negligible. Because up to 20% of the suspended solids that should be removed from a system to maintain equilibrium can be in the clarifier effluent, the designer should add an additional term, $Q_e X_e$ (where Q_e = effluent flowrate and X_e = effluent suspended solids), to the denominator of eq 6.6, resulting in

$$\theta_c = \frac{VX}{Q_w X_r + Q_e X_e} \qquad (6.15)$$

With the use of this revised equation for solids retention time, θ_c, subsequent equations must also be modified. Note also that a constant ratio of volatile suspended solids to suspended solids in the mixed liquor, recycle flow, and effluent is assumed. As a further caution, such constancy does not necessarily occur.

Finally, the development assumes that all of the active solids (biomass) are in the aeration tank. This is not correct, of course, because a substantial amount of activated sludge is in the final clarifiers. In the pure-oxygen system, for example, fully 50% of the solids may be in clarifiers, and this should be taken into account when estimating the total mass of solids in the activated-sludge system.

Volume of Aeration Tanks

Sizing of aeration tanks is based on two important factors. The first is providing sufficient time to remove soluble BOD (and oxidize ammonia-nitrogen, if required) and to allow biomass activity, as estimated by oxygen uptake rate measurements, to return to an endogenous level. The second is maintenance of flocculent, well-settling MLSS that can be effectively removed by gravity settling.

For municipal systems the process design should be based on the assumption that $S_e = 0$. Experience has shown its exact value to be somewhat unpredictable. It is inadvisable to reduce aeration tank volume by assuming a higher S_e value because discharge permits are based on total BOD or COD, which includes the contribution of effluent organic solids.

These equations are not well supported by data derived from full-scale activated-sludge systems. For this reason the relationship between S_e and solids retention time, θ_c, in eq 6.7 is not used for design. When it is used, the value of θ_c is predicted to be unrealistically low and is scaled upward by a large so-called safety factor so that the resulting value of the solids retention time is in agreement with operating experience (Dague, 1983; Grady et al., 1999; Lawrence and McCarty, 1970; Water Research Commission, 1984). Alternatively, information from the literature or pilot-plant studies may be useful for estimating θ_c and other kinetic parameters.

The selection of values of θ_c for carbonaceous BOD removal systems is not based on kinetics considerations but rather on experience. The design is based on providing a high enough value of θ_c for the system to yield a well-flocculated sludge that settles well. Figure 6.3, which represents a nonfilamentous sludge grown on a soluble waste (glucose plus yeast extract), shows that a minimum θ_c value of approximately 3 days is required. In practice, a value of θ_c from 1 to 5 days is used during warm weather and up to 15 days during cold weather. Nitrification may well occur in these ranges and should be taken into account during design. Values of θ_c outside of this range are selected in situations where environmental conditions warrant lower or higher values. In warm climates where nitrification is not desirable, θ_c values of 1 to 2 days are used. Also, long θ_c values are often used in extended aeration systems where secondary goals require minimization and stabilization of the excess sludge solids generated.

Once a design value of θ_c has been selected, eq 6.8 can be used to calculate the required aeration tank volume, V. This calculation does not, however, include any influent inerts or nonbiodegradable volatile suspended solids. Calculation of V requires an estimation of the mixed-liquor biomass concentration (X) and the stoichiometric coefficients (Y and b). Selection of X is not a trivial exercise. It may be determined by trial and error in the design process, optimizing the aeration tank and clarifiers design based on the θ_c required for wastewater treatment, oxygen transfer limitations, solids-settling characteristics, and the allowable solids loading rate to the secondary clarifiers. Conventional air-activated-sludge system MLSS concentrations ranging from 1500 to 3000 mg/L are often used. Values of 2000 mg/L mixed liquor volatile suspended solid (MLVSS) and 2500 mg/L MLSS are most common. It is possible, however, for these systems to accommodate higher concentrations.

Solids settling and thickening properties often dictate final selection of the MLSS concentration. For air-activated-sludge systems, design for concentrations more than approximately 5000 mg/L is seldom economical. Figure 6.4 shows suggested values as functions of sludge volume index and temperature. (The sludge volume index (SVI) is a measure of how well the sludge settles. A low SVI (<100) is desirable.) In pure-oxygen systems, the upper boundary of the figures may be higher because better-settling sludges can be generated; however, most operating pure-oxygen plants in the United States do not exceed 5000 mg/L MLSS.

Selection of stoichiometric coefficients (Y and b) can be based on designer experience, pilot-plant studies, or values found in the literature. The decay constant, b, decreases with increasing solids retention time because the active biomass fraction of the

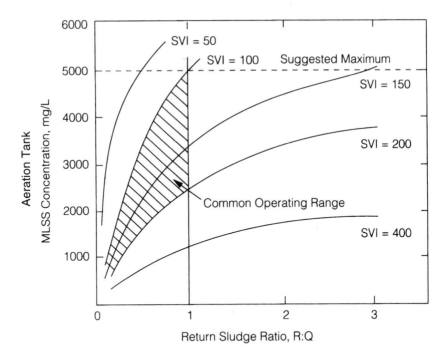

Figure 6.4 Typical operating range for activated-sludge systems.

MLSS decreases and inert organic and inorganic fractions increase, especially for plants with little or no primary treatment. Typical values of the kinetic coefficients are listed in Table 6.1.

Design of completely mixed aerated lagoons with sludge recycle can be based on the design equations presented previously because they function as a system operated in the extended aeration mode. Aerated lagoons without recycle can also be designed using the above equations, with the solids retention time (θ_c) set equal to the hydraulic retention time. Another approach to designing aerated lagoons without recycle is to assume that the observed BOD removal (either total or soluble BOD) can be described by first-order kinetics. For a single, completely mixed lagoon, the first-order equation is

$$\frac{S_e}{S_o} = \frac{1}{1 + k_1 \bar{t}} \quad (6.16)$$

where k_1 = observed BOD$_5$ removal rate constant (base e, d^{-1}) and $\bar{t}$ = hydraulic retention time (d). Reported values of k_1 range from 0.25 to 1.0 d^{-1} for overall BOD removal.

Sludge Generated and Wasted

The amount of sludge generated can be estimated using eq 6.11. Note that the total mass includes nonvolatile, biodegradable volatile, and nonbiodegradable volatile suspended solids. Any precipitates that form from the addition of iron or aluminum salts for phosphorus removal or other purposes should also be included in this calculation. Figure 6.5 illustrates net secondary treatment system sludge production (to be

Table 6.1 Typical kinetic constants for suspended-growth reactors (Dague, 1983; Metcalf and Eddy, Inc., 1991; Water Resource Commission, 1984).

Coefficient	Basis	Range	Typical
μ_{max}	d^{-1}	0.3–3.0	1.0
K_{sn}	NH_4-N, mg/L	0.2–5.0	1.4
Y_n	NH_4-N, mg VSS/mg NH_3-N	0.04–0.29	0.15
b_n	d^{-1}	0.03–0.06	0.05

removed as waste activated sludge and secondary effluent suspended solids) for the stated waste characteristics, both with and without primary sedimentation. This figure assumes typical values for coefficients Y and b and assumes primary sedimentation removes 60% of the influent suspended solids.

Oxygen Demand

Oxygen demand in an aeration tank can be calculated using eqs 6.13 and 6.14. Additional oxygen demand can also result from the presence of certain readily oxidizable components in wastewater such as the sulfide ion. Oxygen demand varies both spatially and temporally in a suspended-growth system. Temporal variations can be estimated from statistical analyses of data collected for influent loadings (carbonaceous BOD and nitrogenous oxygen demand) of the process. Spatial variations depend on kinetic relationships between growth rates of the biomass and substrate removal rates and dissolved oxygen concentrations; they also depend on the flow regime and hydraulic retention time of the process. Total oxygen demand for design should be based on peak loadings anticipated during the design year. As a minimum, the design requirement for a conventional system should be based on the 24-hour demand of the average day of the peak month.

Return Activated-Sludge Capacity Requirements

Requirements for return activated-sludge pumping capacity can be estimated from eq 6.12. The recycle ratio (α) deserves careful consideration because it affects the size of final clarifiers without influencing the size of aeration tanks. No generalizations can be offered as to the best ratio because that will depend on sludge settling and thickening characteristics. As a rough guide, the design value of α should range from 20 to 100% of the facility design flow for conventional systems and as high as 150% for some extended aeration systems.

Clarifier Sizing

The sizing of the secondary clarifier is an important function, integral with the design of other components of a suspended-growth system. Details of clarifier sizing are found later in this chapter.

Process Design for Nitrification

Nitrogen contained in municipal raw wastewater occurs predominantly in the organic and ammonium nitrogen forms. Values reported in the literature for total nitrogen range from 20 to 85 mg/L (as N), with a median value of 40 mg/L. Approximately 40% of the total occurs in the organic form and 60% in the ammonium form: less than 1% is present as nitrate or nitrite unless influenced by industrial waste contributions.

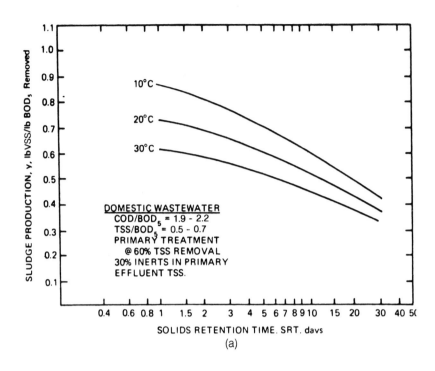

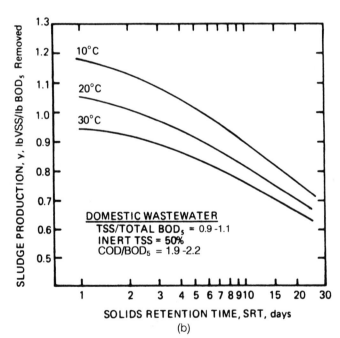

Figure 6.5 Sludge production in activated-sludge systems, with and without primary settling.

The growth of new cells will remove some of the influent nitrogen. This nitrogen removal will approximate 11 to 12% of the volatile suspended solids mass (on a dry-weight basis) of the net biomass accumulation, although some ammonia is released again through endogenous respiration. For given conditions of solids retention time and carbonaceous BOD removal, nitrogen removal by assimilation can be estimated from eq 6.7 and will range from 8 to 20% of the total influent nitrogen. This range indicates the interdependence of solids retention time and endogenous decay on net biomass retention of nitrogen. Nitrogen assimilation depends on the ratio of BOD to nitrogen in the influent, and thus can be significant in systems treating wastewater with high concentrations of BOD.

Ammonia nitrogen is oxidized to nitrate by the staged activities of the autotrophic species *Nitrosomonas* and *Nitrobacter*. Each gram of ammonia oxidized to nitrate (both expressed as N) will result in 4.57 g of oxygen consumed, 7.1 g of alkalinity (as calcium carbonate) destroyed, and 0.15 g of new cells (nitrifiers) produced.

The degree of biological nitrification will depend on the mass of nitrifying organisms allowed to remain in the system. Their presence depends on the relative growth rates of the autotrophic and heterotrophic populations involved, system solids retention time, and other conditions such as temperature, and ammonium ion, organic substrate, and dissolved oxygen concentrations. Thus, mechanisms exist to either limit or promote nitrification.

Biological oxidation of ammonia to nitrate can be achieved in combined carbonaceous BOD removal–nitrification (single-stage) systems or in separate nitrification (two-stage) systems. The degree of nitrification in a combined, single-stage process depends on the solids retention time of the system. The degree of nitrification is governed to a large extent by design parameters ($\bar{t}$ and θ_c for a nitrification system). Two-stage systems allow some separation of carbonaceous and nitrogenous oxidation processes. In the first stage (aeration tank with clarification and sludge recycle), most of the carbonaceous BOD_5 removal occurs and nitrification is limited. The second stage (separate aeration tank and clarifier with sludge recycle) can, therefore, maintain more favorable conditions for nitrification of the wastewater. Two-stage systems have been found to be more costly and are rarely, if ever, designed and built in the United States now.

Other Environmental Effects

Temperature, dissolved oxygen, nutrients, toxic and inhibitory wastes, pH, and the inherent variability of wastewater flows and characteristics affect the performance, thus the design, of activated-sludge systems. Each of these effects is described in this section.

Temperature

Temperature will affect reaction rate, stoichiometric constants, and oxygen-transfer rates. Most temperature corrections used in biological treatment designs follow the modified van't Hoff–Arrhenius equation

$$K_{T1} = K_{T2} \beta^{(T2-T1)} \tag{6.17}$$

where

K_{T1} = a specific kinetic, stoichiometric, or mass-transfer coefficient at temperature T_1;
K_{T2} = a specific kinetic, stoichiometric, or mass-transfer coefficient at temperature T_2; and

β = a constant, 1.00 to 1.04 for carbonaceous BOD removal systems except for aerated lagoons or 1.06 to 1.12 for aerated lagoons.

Novak (1974) has shown that both k, the maximum rate of substrate unitization, and K_S, the substrate concentration when one half of the substrate has been used (the half-velocity constant), are affected by temperature, as

$$k_T = k_0 e^{C_1 T} \tag{6.18}$$

and

$$K_{ST} = k_{S0} e^{C_2 T} \tag{6.19}$$

where
 k_T = maximum substrate utilization constant at temperature T,
 k_0 = maximum substrate utilization constant at some reference temperature,
 C_1 = a constant equal to the slope of the log k versus temperature line,
 K_{ST} = half-velocity constant at some temperature T,
 K_{S0} = half-velocity constant at some reference temperature, and
 C_2 = a constant equal to the slope of the log K_S versus temperature curve.

For aerobic systems, both k and K_S increase with temperature, but for anaerobic systems, k increases but K_S decreases with increasing temperature. These variations with temperature are important because they can significantly affect treatment efficiency. Substituting the temperature relationships from eqs 6.18 and 6.19 into the equation for the mean solids retention time, eq. 6.6, and ignoring the decay constant,

$$\frac{1}{\theta_c} = \frac{Y(k_0 e^{C_1 T})S}{K_{S0} e^{C_2 T} + S} \tag{6.20}$$

clearly shows that the efficiency of treatment (reduction in substrate) is affected by both the solids retention time and the temperature. Using typical values of the constants and common temperatures, the efficiency of treatment can be expected to be greatly reduced at low temperatures when the solids retention time is low, as shown in Figure 6.6. This figure shows that if the solids retention time, θ_C, is low, the effect of low temperature can be severe. Operators of activated-sludge systems in colder climates should therefore maintain high solids retention times to attain effective treatment.

Dissolved Oxygen
Dissolved oxygen concentration is another important control parameter. In systems designed for carbonaceous BOD removal, a minimum acceptable average tank dissolved oxygen concentration of 0.5 mg/L is under peak loading conditions and 2.0 mg/L under average conditions. Using low values increases oxygen-transfer efficiency but can lead to filament formation and poor settleability. In nitrifying systems, a minimum average tank dissolved oxygen of 2.0 mg/L is required under all conditions.

Nutrients
An adequate nutrient balance is necessary to ensure an active biomass that settles well. Nutrient requirements are based on a ratio of BOD_5/nitrogen/phosphorus ratio of

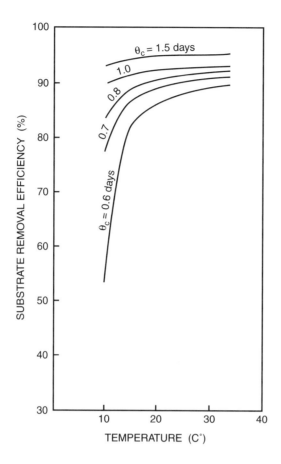

Figure 6.6 Predicted relationship between process efficiency and temperature at several values of θ_c under aerobic conditions (Novak, 1974).

approximately 100:5:1. Systems with higher values of θ_c require fewer nutrients because of their recycle resulting from biomass autooxidation. Because nutrient requirements depend on θ_c, some engineers calculate nutrient requirements based on waste volatile suspended solids. The minimum nitrogen requirement should be 12% of the waste volatile suspended solids and the phosphorus requirement should be 2% of the waste volatile suspended solids. Normal domestic wastewater contains ample nutrients. Wastes with substantial industrial compounds may require nutrient addition.

Toxic and Inhibitory Wastes
The presence of certain inorganic and organic constituents can inhibit or destroy suspended-growth system microorganisms. An excellent listing of many of these is presented in Grady et al., 1999. Nitrification processes are particularly sensitive to toxic inhibition.

pH
The pH of mixed liquor should range from 6.5 to 7.5 for optimum cell growth in both carbonaceous BOD removal and nitrification. Pure-oxygen systems often depress pH more

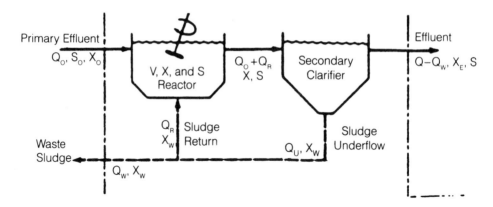

Figure 6.7 Completely mixed reactor type activated-sludge system.

than air systems because the former lacks nitrogen gas flow to help strip dissolved carbon dioxide (formed in respiration) from the mixed liquor. Unless stripped out by downstream channel aeration or similar processes, at least some of the carbon dioxide is recirculated through the clarifiers and back to the reactors. To avoid pH reduction, residual alkalinity of at least 60 mg/L (as calcium carbonate) for either pure-oxygen or classical aeration systems should be provided, as well as a means for supplemental alkalinity addition if required. Operating at even the 50-mg/L level is risky. A higher range of 80 to 100 mg/L gives more protection for operating over daily and diurnal load fluctuations.

Process Configuration Alternatives

Activated-sludge aeration tanks have been designed in a large number of different configurations. They can be categorized by tank shape, loading rates, feeding and aeration patterns, type of aeration, and other such criteria. For smaller plants, oxidation ditches are more popular and for larger ones, plug flow (some with configuration flexibility) is favored. Completely mixed activated sludge seems to be less vulnerable to toxic slug loads, but these systems apparently promote the formation of filamentous microorganisms. The plug-flow configuration has the added flexibility to be converted for bionutrient removal by creating anaerobic and anoxic zones. Compartmentalization (using baffles or walls within a tank) can include flexibility for intermediate zones to be used for more than one reactor purpose.

Tank Shape
Categorization of tanks by shape leads to the definitions of complete-mix, plug-flow, oxidation ditch, and aerated lagoon. Each of these, in turn, has subcategories.

Complete Mix
By definition, a completely mixed reactor has uniform characteristics throughout the contents of the entire reactor. In this configuration, shown in Figure 6.7, the influent waste is rapidly distributed throughout the tank and operating characteristics of MLSS, respiration rate, and soluble BOD are uniform throughout. Because the total body of the tank liquid has the same quality as the effluent, only a low level of food is available

at any time for the large mass of microorganisms present. This characteristic is cited as the primary reason why the completely mixed process can handle surges in organic loading and toxic shocks to a limited extent without producing a change in effluent quality. In recent years, some engineers have rejected completely mixed systems because low food concentrations and low or variable DO levels can stimulate the growth of filamentous bacteria that settle poorly. Nevertheless, there are many such plants that are properly operated and produce excellent results.

The completely mixed tank is conventionally square, round, or rectangular. Tank dimensions may be controlled by the size and mixing pattern of the aeration equipment and local site considerations. Surface turbine units provide aeration for the tanks, although diffused air is becoming increasingly common. Some submerged turbines are being used, but few if any remain. Factors that influence mechanical aeration mixing effectiveness include biological reactor length/width ratio, mixing power introduced, and wastewater velocity through the tank. Achieving complete mixing in a tank is difficult but can be approached. From a treatment point of view, any square or circular tank with a reasonable detention time and level of mixing intensity can be considered a completely mixed reactor, regardless of the type of aeration system used. The length/width ratio of a tank should be maintained at less than 3:1 to remain primarily completely mixed using mechanical aeration and no baffles. Multiple mechanical aeration units in long, narrow tanks (for example, length/width ratio greater than 5:1) create a mixing pattern that tends to approach mixed tanks in series and start to resemble plug flow. If diffused air is used, full tank width influent feed and effluent removal weir structures are provided as good practice to approach complete mix results. Multiple feed points and withdrawal weirs along opposite sides of an aeration tank could also be used for this purpose.

Plug Flow

A plug-flow reactor can be viewed as an infinite number of small, completely mixed tanks in series. Plug flow, the oldest tank geometry in use, originally was used to meet the mixing and oxygen-transfer requirements of diffused aeration systems. Such tanks are 5 to 9 m (15 to 30 ft) wide and up to 120 m (400 ft) long (length/width ratio of more than 10:1). Long tanks may be constructed side by side or in a folded arrangement. Figure 6.8 is a schematic of a plug-flow reactor.

As one of its characteristics, the plug-flow configuration has a high organic loading at the influent end of the tank. Loading is reduced over the length of the tank as organic material in wastewater is assimilated. At the downstream end of the tank, oxygen consumption primarily results from endogenous respiration. The high organic loading at the head end of this process discourages most types of filamentous bacteria growth and often improves sludge settling beyond that realized from a complete-mix reactor if sufficient dissolved oxygen concentrations are maintained throughout the tank. Keeping a low dissolved oxygen level too low may encourage filamentous growth. As discussed elsewhere in this chapter, polysaccharide formation may result from high loading and low dissolved oxygen concentrations at the inlet end.

Whereas a complete-mix reactor is noted for its ability to handle small shock loads, plug-flow configurations have a superior ability to avoid the passage of untreated substrate during peak flows. Plug-flow reactors also have an advantage where high effluent dissolved oxygen concentrations are desirable. In a complete-mix configuration, the entire tank contents would have to be maintained at the elevated dissolved oxygen

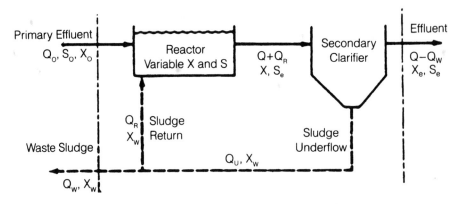

Figure 6.8 Plug-flow reactor type activated-sludge system.

level to achieve that objective.

Oxidation Ditch

In a classical oxidation ditch system, wastewater and mixed liquor are pumped around an oval pathway (racetrack) by brushes, rotors, or other mechanical aeration devices and pumping equipment located at one or more points along the flow circuit. Figure 6.9 shows an oxidation ditch with a rotor to maintain motion and aerate the ditch contents. As mixed liquor passes the aerator, the dissolved oxygen concentration is sharply raised but then declines as the flow traverses the circuit. Oxidation ditches operate in an extended aeration mode with long hydraulic retention times (24 hours) and solids retention times (20 to 30 days). Depending on the relative location of wastewater input and removal, sludge return, and aeration equipment, oxidation ditches can also achieve nitrification and denitrification. For BOD removal or nitrification, the influent enters the reactor near the aerator and the effluent exits the tank upstream of the entrance.

Oxidation ditches may be viewed as a complete-mix reactor even though they have some plug-flow characteristics as flow traverses the loop. The reason some consider it to be complete mix is that the influent concentration of substrate is immediately diluted by the large mixed liquor flow to a value nearly equal to that of the biological reactor effluent.

Oxidation ditches have depths ranging from approximately 0.9 to 5.5 m (3 to 18 ft) and channel velocities from 0.24 to 0.37 m/s (0.8 to 1.2 ft/s). Ensuring that ditch geometry will be compatible with the aeration device and mixing equipment calls for

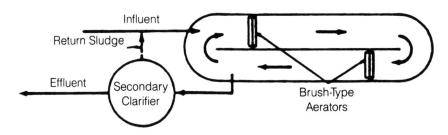

Figure 6.9 Oxidation ditch reactor activated-sludge system.

Table 6.2 Operational characteristics of activated-sludge processes (from Metcalf & Eddy, Inc., *Wastewater Engineering: Treatment, Disposal, Reuse*, 3rd ed. Copyright © 1991, McGraw-Hill, Inc.: New York, with permission).

Process modification	Flow model	Aeration system	BOD removal efficiency, %	Remarks
Conventional	Plug flow	Diffused-air, mechanical aerators	85–95	Use for low-strength domestic wastes; process is susceptible to shock loads
Complete-mix	Continuous-flow stirred-tank reactor	Diffused-air, mechanical aerators	85–95	Use for general application; process is resistant to shock loads but is susceptible to filamentous growths
Step feed	Plug flow	Diffused air	85–95	Use for general application for a wide range of wastes
Modified aeration	Plug flow	Diffused air	60–75	Use for intermediate degree of treatment where cell tissue in the effluent is not objectionable
Contact stabilization	Plug flow	Diffused-air, mechanical aerators	80–90	Use for expansion of existing systems and package plants
Extended aeration	Plug flow	Diffused-air, mechanical aerators	75–95	Use for small communities, package plants, and where nitrified element is required; process is flexible
High-rate aeration	Continuous-flow stirred-tank reactor	Mechanical aerators	75–90	Use for general applications with turbine aerators to transfer oxygen and control floc size

consultation with individual equipment suppliers. Mechanical brushes, surface turbines, and jet devices are used to aerate and move the liquid.

Aerated Lagoon Reactors
Aerated lagoon reactors with clarifiers and sludge return have been used on occasion to create activated-sludge processes. Various types of aeration devices have been installed to provide mixing and aeration. Reactor shapes have been square or rectangular. These have often been mixing limited and led to the development of oxidation ditches because the horizontal flow in a ditch was able to keep particles in suspension at lower energy input than mechanical or diffused aeration in lagoons.

Loading Rates
Activated-sludge processes can also be classified by loading rate or organic feed rate. The terms are conventional, low-rate, and high-rate. Table 6.2 provides a summary of operational characteristics for various processes. Table 6.3 specifies typical ranges for relevant design parameters.

Conventional
The term conventional activated sludge applies to a plug-flow or completely mixed system with a food/microorganism (F/M) loading of approximately 0.2 to 0.5 kg BOD_5/kg MLVSS·d (0.20 to 0.5 lb BOD_5/d/lb MLVSS). This system can obtain BOD_5 removal efficiencies in the range of 85 to 95%. Conventional system MLSS design concentrations often range from 1500 to 3000 mg/L. Design MLSS concentration has

Table 6.3 Design parameters for activated sludge processes (from Metcalf & Eddy, Inc., *Wastewater Engineering*, 4th ed. Copyright © 2003, McGraw-Hill, Inc.: New York, with permission).

Process name	Type of reactor	SRT, d	E/M kg BOD/kg MLVSS·d	Volumetric loading		MLSS, mg/L	Total τ, h	RAS % of influent[b]
				lb BOD/ 1000 ft³·d	kg BOD/ m³·d			
High-rate aeration	Plug flow	0.5–2	1.5–2.0	75–150	1.2–2.4	200–1000	1.5–3	100–150
Contact stabilization	Plug flow	5–10	0.2–0.6	60–75	1.0–1.3	1000–3000[c] 6000–10,000[d]	0.5–1[c] 2–4[d]	50–150
High-purity oxygen	Plug flow	1–4	0.5–1.0	80–200	1.3–3.2	2000–5000	1–3	25–50
Conventional plug flow	Plug flow	3–15	0.2–0.4	20–40	0.3–0.7	1000–3000	4–8	25–75[e]
Step feed	Plug flow	3–15	0.2–0.4	40–60	0.7–1.0	1500–4000	3–5	25–75
Complete mix	CMAS[f]	3–15	0.2–0.6	20–100	0.3–1.6	1500–4000	3–5	25–100[e]
Extended aeration	Plug flow	20–40	0.04–0.10	5–15	0.1–0.3	2000–5000	20–30	50–150
Oxidation ditch	Plug flow	15–30	0.04–0.10	5–15	0.1–0.3	3000–5000	15–30	75–150
Batch decant	Batch	12–25	0.04–0.10	5–15	0.1–0.3	2000–5000[g]	20–40	NA
Sequencing batch reactor	Batch	10–30	0.04–0.10	5–15	0.1–0.3	2000–5000[g]	15–40	NA
Counter-current aeration system	Plug flow	10–30	0.04–0.10	5–10	0.1–0.3	2000–4000	15–40	25–75[e]

[a]Adapted from WEF (1998); Crites and Tchobanoglous (1998).
[b]Based on average flow.
[c]MLSS and detention time in contact basin.
[d]MLSS and detention time in stabilization basin.
[e]For nitrification, rates may be increased by 25 to 50%.
[f]CMAS = complete-mix activated sludge.
[g]Also used at intermediate SRTs.
NA = not applicable.

increased steadily over the history of the process because of improvements in the oxygen-transfer capability of aeration devices, clarifier performance, and understanding of system concepts.

As an important consideration in the design of conventional systems, nitrification might occur, even when not desired. This often happens with low loading conditions during summer months or high solids retention time because of wasting inadequacies. When nitrification occurs, denitrification might begin in final clarifiers, possibly causing rising sludge problems. As an easy remedy for limiting nitrification or denitrification, the MLSS concentration can be reduced. This increases the F/M ratio and lowers the solids retention time.

Low Rate

Low-rate (also called extended aeration) plants are characterized by the introduction of pretreated (for example, screened and degritted) wastewater directly to an aeration tank with a long aeration time, high MLSS concentration, high return activated-sludge pumping rate, and low sludge wastage. This system, initially used in the United States for flows of approximately 4000 m³/d (~1 mgd) or less, often incorporated complete-mix reactors. During the past three decades, low-rate systems have been applied to larger sizes in oxidation ditches and similar processes.

A particular advantage of using long hydraulic retention times (16 to 36 hours) is that they allow the plant to operate effectively over widely varying flow and waste loadings and provide lower overall solids production. The more stable solids are often advantageous for subsequent solids-handling processes. Secondary clarifiers must be designed to handle variations in hydraulic loadings and high MLSS concentrations associated with this process.

One of the process goals is to maintain the biomass in the endogenous respiration phase. Because microorganisms are essentially undergoing aerobic digestion in the biological reactor, more oxygen is required than for other single-stage systems. Many low-rate plants experience a dissolved oxygen deficiency during the hours when the waste load is high. However, the long solids retention time and excess dissolved oxygen at night allow some nitrification, causing a daily, but noncoincidental, nitrification–denitrification cycle.

High Rate

High rate is the term applied to an activated-sludge system characterized by a short hydraulic retention time, a high-sludge recycle ratio, and a high organic loading rate. Mixed liquor suspended solids concentration may vary over a wide range from 800 to as high as 10,000 mg/L, and F/M ratios are higher (>0.5 g BOD_5/kg MLVSS·d) than those used in conventional systems. Process integrity depends on maintaining the biomass in the growth phase. Although high-rate systems can produce an effluent quality approaching that of a conventional system, they encourage a higher ratio of dispersed organisms, often resulting in a more turbid clarifier effluent. Therefore, high-rate systems must be operated with special care. For example, inadequate return activated sludge flowrates, insufficient wasting, and high sludge flux rates make the clarifiers of these systems more sensitive to washout. A high-rate, single-stage system can be used to partially remove carbonaceous BOD as the first stage of a two-stage nitrification system.

Feeding and Aeration Patterns

Changing the number and location of feed points of an activated-sludge biological reactor can appreciably alter acceptable loading rates and quality of clarifier effluent. Using this concept and the pattern of aeration, one can further classify activated-sludge process options.

Conventional

Conventional activated-sludge design would introduce influent to the head end of a rectangular tank. Return activated sludge could be mixed with the influent before the tank or be added to the inlet end separately. Keeping return activated sludge separate facilitates subsequent conversion to other feed patterns (for example, step feed). If return activated sludge is blended with the influent ahead of multiple biological reactors, care must be taken to see that the influent is well mixed before flow splitting occurs.

Contact Stabilization

Contact stabilization is a modification of the activated-sludge process in which the feed point is moved downstream in the biological reactor (or into a separate tank), thereby

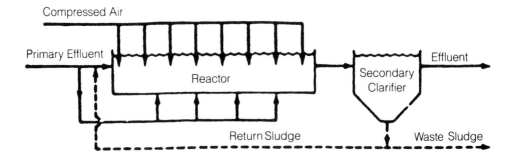

Figure 6.10 Step-feed process activated-sludge system.

providing a relatively short detention time (~20 minutes). Return activated sludge is added to the tank inlet separately and aerated for a period of time before being blended with the mainstream influent. Because the upstream end of the biological reactor contains liquid at the return activated sludge concentration, instead of the MLSS concentration, a given volume of the biological reactor would contain a larger mass of mixed liquor solids and, therefore, longer solids retention time. The reduced contact time provides less opportunity to oxidize ammonia, organic nitrogen, and soluble organics. The process was originally developed to remove carbonaceous BOD, which was largely in the suspended rather than the soluble form. This is a preferable waste characteristic for the process because suspended organics are readily adsorbed by MLSS bacteria and subsequently stabilized in the reaeration (upstream) part of the biological reactor. If the type of wastewater is amenable, a given biological reactor's capacity can, therefore, be increased by conversion to contact stabilization provided that there is flexibility in relocating the feed point and returning the return activated sludge in a separate conduit.

Step Feed

The step-feed or step-aeration process, a modification of a plug-flow reactor, allows entry of influent wastewater at two or more points along the length of the aeration tank. With this arrangement, the oxygen uptake rate becomes more uniform throughout the tank. Other operating parameters are similar to those of the conventional process. Return activated sludge would be added to the biological reactor in a separate conduit at the inlet end of the tank. Step aeration configurations include diffused-aeration equipment. An existing plug-flow reactor can be modified for step feed by simply dividing the tank into compartments and redirecting the flow so that each compartment receives wastewater input. This design results in storage of the biomass in the front end of the biological reactor and affords some protection to the clarifiers from large solids loading because the clarifiers only have contact with mixed liquor from the far end of the tank, which is at a much lower solids concentration. A step-feed tank configuration is shown in Figure 6.10.

Tapered Aeration

An aeration system design that modulates the oxygen supply along the length of a plug-flow reactor is called tapered aeration (Figure 6.11). This approach is commonly associated with diffused-air systems. Plug-flow reactor designs should incorporate tapered aeration for operational control. Mixing requirements govern the minimum air supply

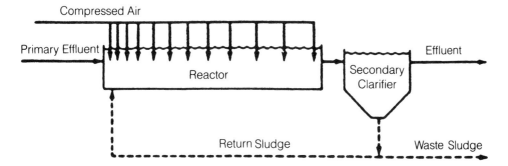

Figure 6.11 Tapered aeration process activated-sludge system.

rate. Design parameters for the tapered-aeration process are consistent with those for conventional treatment. Adding more air at the influent end of the biological reactor than at the effluent end (for example, by increasing density of diffusers) produces a number of beneficial results, including reduced blower capacity and operating costs, greater operational control, and possible inhibition of nitrification (when desired) by reducing dissolved oxygen concentration in downstream segments of the biological reactor.

Other Variations

Pure Oxygen

Use of pure oxygen (also referred to as high-purity oxygen) rather than air for aeration in the activated-sludge process was evaluated in the 1950s by Budd and Lambeth (1957). The process achieved commercial status in 1970. Primary advantages claimed by manufacturers of the process include reduced power for dissolving oxygen into the mixed liquor, improved biokinetics, the ability to treat high-strength soluble wastewater, and a reduction in bulking problems from dissolved oxygen-deficit stress.

Pure-oxygen systems in the past were characterized by high MLSS concentrations (3000 to 8000 mg/L) and short hydraulic retention times (1 to 3 hours). Recent practice indicates that a more appropriate MLSS concentration will likely range from 1000 to 3000 mg/L for municipal wastewater. *Nocardia* accumulations in the reactor often cause foaming at high MLSS concentrations.

Pure oxygen is fed concurrently with wastewater flow in reactors that are covered (Figure 6.12). Oxygen feed is controlled by maintaining a constant gas pressure within the tanks. A dissolved oxygen concentration of 4 to 10 mg/L is maintained in the mixed liquor. Less than 10% of the inlet oxygen vents from the last stage of the system.

Pure-oxygen systems reduce pH values because of high partial pressure of carbon dioxide, which is not stripped out by the nitrogen gas of air as in conventional aeration tanks and potentially by loss of alkalinity because of nitrification. When the pH is depressed below 6.5, nitrification will attenuate and the system can require a longer solids retention time, greater aeration tank volume, and perhaps additional settling tank area if the mixed liquor suspended solids concentration is increased to compensate for decreased efficiency. These effects have led to consideration of separating the carbonaceous BOD removal and nitrification stages when using oxygen. Some design engineers

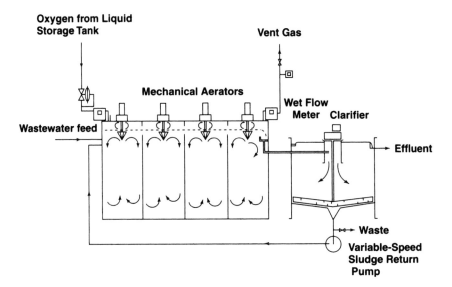

Figure 6.12 Pure-oxygen activated-sludge system.

suggest that the first stage of a two-stage system receives oxygen and the second stage receives air. Another option is to open up the last cell of the train to air rather than pure oxygen.

Covered tanks using oxygen must have provisions for warning of potential explosions that could result from the presence of combustible, volatile hydrocarbons in influent wastewater. A detector system is used to automatically purge the tanks with air if the volatile hydrocarbon level becomes excessive. Covered tanks also capture volatile organic chemical emissions and reduce the offgas volume compared with an air system.

An atmosphere of high-purity oxygen and carbon dioxide in the reactor tank requires careful selection of construction materials. Compared with air, this atmosphere is more corrosive and reactive with organic compounds such as oils and greases. Some plants have even experienced corrosion of materials used in the construction of secondary clarifiers. Suppliers of high-purity oxygen systems have evaluated materials suitable for safe and reliable construction.

Mechanical surface turbines are used to mix the reactors. For deep tanks, submerged turbines or surface turbines with extended shafts to provide additional mixing blades closer to the bottom are used.

Sequencing Batch Reactors

The sequencing batch reactor process involves a fill-and-draw, complete-mix reactor in which both aeration and clarification occur. Settling is permitted when aeration is turned off and a decanter device is used to withdraw supernatant. Discrete cycles are used during prescribed, programmable time intervals and MLSS remain in the reactor during all cycles. A treatment cycle consists of the following steps:

(1) Fill (raw or settled wastewater fed to the reactor),
(2) React (aeration and mixing of the reactor contents),

(3) Settle (quiescent settling and separation of MLSS from the treated wastewater),
(4) Draw (withdrawal of treated wastewater from the reactor), and
(5) Idle (removal of waste sludge from the reactor bottom).

Advantages of sequencing batch reactors include elimination of a secondary clarifier and return activated sludge pumping, high tolerance for peak flows and shock loadings, avoidance of MLSS washout during peak-flow events, clarification under ideal quiescent conditions, and process flexibility to control filamentous bulking.

Activated Carbon Addition
The addition of powdered activated carbon to the aeration tank of an activated-sludge treatment plant has demonstrated some beneficial properties, including

- Improved solids-settling characteristics;
- Increased dewaterability of waste sludge;
- Ability of powdered activated carbon to adsorb biorefractory materials and toxic compounds (especially in nitrifying systems) to improve effluent quality and lessen shock-loading effects;
- Reduction in odor, foaming, and bulking problems; and
- Improved color and BOD removal.

A key disadvantage of carbon addition is the need to regenerate the carbon for reuse or to purchase virgin powdered activated carbon if the plant lacks regeneration facilities. Another disadvantage is that this process requires tertiary filtration.

Hybrid Systems
To enhance performance and increase capacity of aeration tanks, the concept of adding inert support media to the aeration tank was developed decades ago. This allows fixed-film biomass to grow on the media and augment the microbial population of the mixed liquor. The first widely accepted application of this concept, also known as combined fixed-film–suspended-growth systems, had flat asbestos sheets suspended in a biological reactor (termed contact aeration) (Wilford and Conlon, 1957). More recent applications include the use of synthetic trickling filter media; polyurethane foam pads; loops of fiber bundles; or small, plastic elements as the inert material in the biological reactor.

Oxygen Transfer
The supply of oxygen to suspended biomass represents the largest single energy consumer in an activated-sludge facility. Over the years, oxygen-transfer equipment has evolved enough to give an engineer a wide selection of devices to meet the specific needs of a facility. Transfer rates for diffused-air systems are reported as oxygen-transfer efficiency, expressed as a percentage; oxygen transfer rate, expressed in units of mass per time; or aeration efficiency, expressed in units of mass per time per unit of power.

As a secondary function, aeration devices furnish sufficient energy for mixing. Ideally, mixing energy should be sufficient to thoroughly disperse dissolved substrate and oxygen throughout a given segment of an aeration tank and keep MLSS suspended. This does not necessarily mean that both soluble and suspended material should be uniformly mixed throughout the entire aeration tank. For example, plug-flow tanks and reactors with point-source oxygen addition do not rely on uniformity for proper operation, promoting mixing vertically and transversely, but not longitudinally.

Note that the power required to satisfy oxygen demand depends on substrate and biomass concentrations, flowrate, and reactor volume; the power for mixing depends on aeration tank volume and, to a lesser degree, MLSS concentration. For certain combinations of biomass and substrate concentrations and other variables listed above, power requirements for mixing may exceed those for oxygen transfer. In systems with high biomass concentrations, oxygen demand will control power requirements; in plug-flow systems, mixing may control near the effluent end of the biological reactor; and power requirements for aerated lagoons are often dictated by mixing.

Diffused Aeration

Diffused aeration has been used since the turn of the last century. Early applications introduced air through open tubes or perforated pipes located at the bottom of biological reactors. The desire for greater efficiency led to the development of porous plate diffusers that produce small bubbles. These diffusers, used as early as 1916, became the most popular method of aeration by the 1930s. Unfortunately, serious fouling problems occurred at a number of installations, which gradually discouraged their use. Systems requiring lower maintenance gained dominance during the period of relatively inexpensive energy (before 1972). These low-maintenance devices used fixed orifices (6 mm [0.25 in.] or more in diameter) to produce large bubbles. Rapid escalation of power costs commencing in the early 1970s rekindled interest in porous media devices and triggered vigorous efforts to increase the oxygen-transfer efficiency of all types of aeration systems.

Diffused aeration, defined as the injection of a gas (air or oxygen) below a liquid surface, covers all equipment described in this section. However, some systems combine gas injection with mechanical pumping or mixing equipment and are arbitrarily classified as diffused aeration equipment. These devices include jet aerators and U-tube aerators. Another device, the combination turbine and sparger aerator, is arbitrarily classified as a mechanical device. These are discussed later.

The wastewater treatment industry has witnessed the introduction of a wide variety of air diffusion equipment. In the past, the various devices commonly were classified as either fine bubble or coarse bubble, designations that supposedly reflected oxygen-transfer efficiency. Unfortunately, the demarcation between coarse and fine bubbles is difficult to define. Also, applying this classification to specific equipment generated confusion and controversy. For these reasons, the industry now prefers to categorize air diffusion systems by the physical characteristics of the equipment. In the following discussion, various devices are divided into three categories: porous diffusers, nonporous diffusers, and others. These classifications relate more to organization than performance and should not be used for estimating performance.

Porous Diffuser Systems

Use of porous diffusers has gained renewed popularity because of their high oxygen transfer efficiency. The oldest and most common rigid porous diffuser is produced from ceramic media, including aluminum oxide, aluminum silicates, and silica. Media consist of rounded or irregular-shaped mineral particles bonded together to produce a network of interconnected passageways through which compressed air flows. As air emerges from the diffuser surface, pore size, surface tension, and airflow rate interact to produce a characteristic bubble size. Currently, the most common rigid porous diffusers are manufactured from aluminum oxide. Porous plastics are made from a number of thermosetting polymers. The two most common are high-density polyethylene and styrene–acrylonitrile. Porous plastic materials are lighter in weight; inert in composition; and, depending on the actual material, may have greater resistance to breakage. Disadvantages include the brittleness of some plastics and lack of quality control on others.

Membrane diffusers differ from rigid diffusers because the former does not contain a network of interconnected passageways. Instead, mechanical methods create preselected patterns of small, individual orifices (perforations) in the membrane to allow gas passage. Over the past few years, perforated membranes have continued to change in composition, shape, and perforation pattern. Two types of membrane materials are being used: thermoplastics and elastomers. Currently, membrane diffusers on the market are dominated by elastomers, mostly ethylenepropylene dimers (EPDM). Ethylenepropylene dimer formulations all contain EPDM, carbon black, silica, clay, talc, oils, and various curing and processing agents. Oils make up a significant proportion of the mix and give the membrane its flexibility. Each manufacturer has their own formulations, and each can be distinctively different in characteristics including tensile strength, hardness, elongation at failure, modulus of elasticity, tear resistance, creep, compression set, and resistance to chemical attack. Thus, selection of the membrane for a particular application requires a thorough analysis of the material to ensure that it will perform effectively over the expected design life of the material. Some of the many types of porous diffusers are shown in Figure 6.13.

Nonporous Diffusers

Nonporous diffusers, available in a wide variety of shapes and materials, have larger orifices than porous devices. Fixed orifices vary from simple holes drilled in piping to specially designed openings in metal or plastic fabrications. Perforated piping, spargers, and slotted tubes are typical nonporous diffuser designs.

Valved orifice diffusers include a check valve in an attempt to prevent backflow when the air is shut off. There are other types of diffusers that also allow adjustment of airflow rate by changing either the number or size of the orifices through which air passes.

System layouts for fixed and valved nonporous diffusers closely parallel those for porous diffuser systems. The most prevalent configurations are the single- and dual-roll spiral patterns using either narrow- or wide-band diffuser placement. Mechanical lift-out headers with either swing joints or removable diffusers, which allow removal of the diffusers for cleaning without process interruption, are common. Cross-roll and full-floor coverage patterns may also be used. Fixed and valved orifice diffusers are commonly used where fouling may be a problem for porous diffusers. These applications include aerated grit chambers, channel aeration, sludge and septage storage tank aeration, flocculation tank mixing, aerobic digestion, and industrial waste applications.

The static tube, another type of nonporous diffuser, resembles an air-lift pump, except that the tube has interference baffles placed within the riser. These baffles are intended to mix the liquid and air, shear coarse bubbles, and increase contact time. With this type of system, the tubes, approximately 1.0 m (3 ft) in length, are anchored to the tank floor in a full-floor coverage pattern.

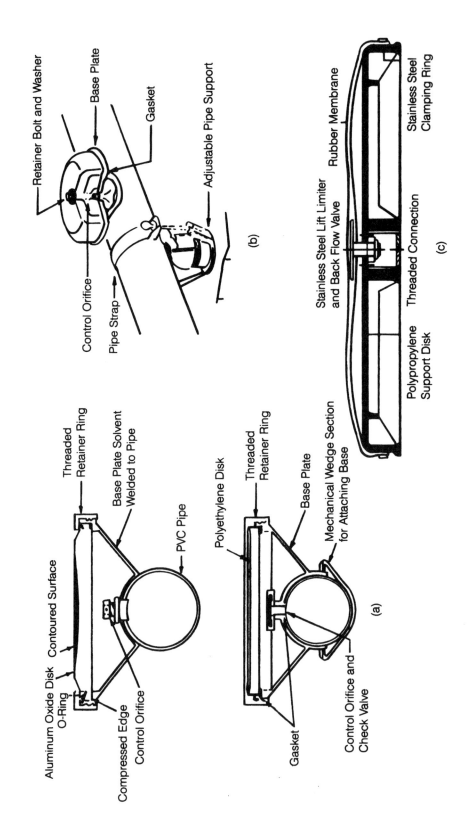

Figure 6.13 Some types of porous diffusers: (a) disks, (b) dome diffusers, and (c) perforated membrane diffusers.

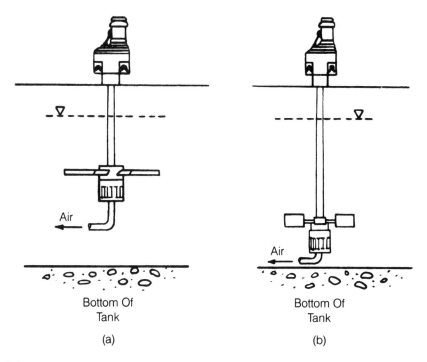

Figure 6.14 Submerged turbine aerators: (a) axial flow and (b) radial flow.

Other Diffused Aeration Systems

Jet systems, which are placed on the floor of aeration tanks, combine liquid pumping with air diffusion. The pumping system circulates mixed liquor in the aeration tank, ejecting it through a nozzle assembly. Air, supplied from a blower, is introduced to the mixed liquor before its discharge through the system's nozzles. Figure 6.14 shows two such turbine jet systems.

A *U-tube system* consists of a 9- to 150-m (30- to 500-ft) deep shaft that is divided into an inner and outer zone. Air is added to the influent mixed liquor in the downcomer zone. The mixture travels to the bottom of the tube and then back to the surface through the return zone where the effluent is removed. The great depths of mixed liquor result in higher partial pressures of oxygen that may enhance oxygen transfer.

Countercurrent aeration involves using a circular biological reactor with a center-pivoted traveling bridge supporting air diffusers. Rotating aerators continually resuspend mixed liquor while leaving a veil of fine bubbles providing the aeration. Another set of fixed-bubble aerators can also be provided. Its rising bubbles would be swept along with the rotating liquid current induced by traveling diffusers. The rotating velocity of the liquid causes bubbles from both sources to lead or trail away from their point of release, thereby eliminating the vertical lifting action of air lift common to conventional, stationary diffusers. It also increases bubble–liquid contact time (bubble residence time) and reduces bubble coalescence, resulting in smaller bubbles, slower rise velocities, and better transfer efficiencies. The energy input required to rotate the diffuser mechanism serves to offset this advantage.

Chapter 6 Suspended-Growth Biological Treatment

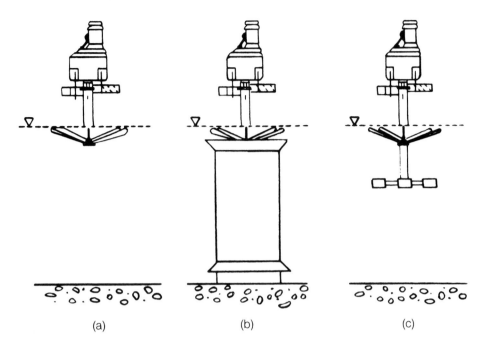

Figure 6.15 Some types of surface aerators: (a) standard surface aerator, (b) surface aerator with draft tube, and (c) surface aerator with impeller.

Mechanical Surface Aeration

Surface aerators can be grouped into four general categories: radial flow, low speed; axial flow, high speed; aspirating devices; and horizontal rotors. Each is used widely and has distinct advantages and disadvantages, depending on the application.

Surface aerators are float-, bridge-, or platform-mounted. Platform and bridge designs reduce torque and vibration problems. Bridges should be designed for at least four times the maximum moment (torque and impeller side load) anticipated. The aerator manufacturer can provide the magnitude of this moment. The efficiency and power draw of platform- and bridge-mounted aerators are sensitive to changes in the depth of impeller submergence. An increase in submergence results in increased fluid pumpage at an increase in power draw and can decrease gearbox life expectancy. High-speed (radial and axial) surface aerators are most often float-mounted, providing portability and low cost.

Some surface aerators are equipped with submerged draft tubes that tend to mix by bringing liquid from the bottom of the tank up through the tube and to the impeller (Figure 6.15). Draft tubes are used where tanks are deeper than 4.6 m (15 ft) and where mechanical aerators alone might not provide enough mixing throughout the entire tank. This feature adds to the cost of the system and will require additional power.

All the mixed liquor pumped through a draft tube is dispersed into the air. Without draft tubes, a portion of the pumped fluid flows beneath the liquid surface and is not aerated. Mixed liquor pumped in either way creates liquid momentum that tends to circulate the mixed liquor directly around the aerator. Mechanical aerators provide point-source oxygen input and the pumped mixed liquor flows radially outward from the aerator with decreasing velocity. Dissolved oxygen reaches its maximum near the impeller blade where surface turbulence is greatest and decreases as fluid flows back below the surface of the aeration tank toward the aerator.

As an alternative to the draft tube, an auxiliary submerged mixing impeller can be provided. This submerged impeller will increase the amount of liquid pumped from the bottom of the tank. The impeller is of an axial-flow design to maximize pumping efficiency, but submerged impellers and draft tubes increase system power requirements. The optimum location of the impeller depends on its configuration. Radial-flow impellers are located 0.5 to 0.7 times the impeller diameter above the tank bottom; axial-flow impellers are located at 60 to 65% of the tank depth, measured from the water surface. Water depths using unsupported shafts (no bottom bearings) range up to 9 m (30 ft). With this unsupported length, shafts can transmit high side loads that the gearbox must be designed to withstand.

The action of surface aeration devices, particularly splashing from high-speed units, can generate mists with attendant health concerns and nuisance odors. Odors can result from insufficient oxygen supplied by the aerator or influent wastewater containing sulfides or other volatiles. Mists can freeze in cold climates, coating equipment with ice. Such freezing can cause hazards to facility staff and equipment. Splashing effects can be minimized with proper geometric design of the aeration tank and use of deflector plates. Another cold-weather problem is tank heat loss induced by surface aerators.

Secondary Settling

The separation of solids from the liquid stream is vital to the operation and performance of suspended-growth treatment systems. This is achieved by gravity settling tanks (or *secondary clarifiers*). These tanks not only separate the MLSS but also thicken the sludge before it is returned to the aeration process or wasted.

In secondary clarification, mixed liquor is fed to the clarifier and the MLSS settle, forming a sludge blanket with an overlying clear water zone. Within the clear water zone, discrete floc particles settle, resulting in a clarified effluent. In the lower zone, the blanket of MLSS thickens before withdrawal as clarifier underflow. Secondary clarifier process failure can involve either clarification or thickening stages; thus, proper design must address both.

Suspended-growth system variables include process feeding configuration, solids retention time, hydraulic retention time, dissolved oxygen concentration, influent wastewater characteristics, and level of turbulence (causing floc shear) in the biological reactors or conduit. Clarification variables include hydraulic and solids loading rates, turbulence, degree of flocculation, hydraulics, baffles, and tank geometry. Variables, including solids loading rate, MLSS characteristics, clarifier and blanket depths, detention time, clarifier underflow rate, and type of sludge collection mechanism, have been shown to influence the thickening performance of secondary clarifiers.

General Design Considerations

The design of secondary clarifiers must be considered integrally with that of the biological reactors, transfer conduits, and return sludge pumping facilities. Area and depth requirements directly depend on settleability of the mixed liquor suspended solids, average and peak wastewater flowrates, and return sludge pumping rates.

As its primary function, the clarifier provides an overflow with low suspended solids and low BOD. Typical effluent standards require <30 mg/L of suspended solids or less on a monthly average basis for suspended-growth systems and <30 mg/L BOD_5. Because effluent suspended solids contribute to the BOD, low BOD effluents cannot be achieved without simultaneously reducing the effluent suspended solids.

To achieve the second function of the clarifier, namely sludge thickening or compaction, adequate settling time and depth are needed. The ability of the mixed liquor suspended solids to settle and compact in the final clarifier is commonly measured using the sludge volume index (SVI). This test is run by placing a sample of the mixed liquor in a liter cylinder and observing the level of the settled sludge at the end of 30 minutes. Knowing the suspended solids concentration of the mixed liquor, the SVI is calculated as

$$SVI = \frac{(\text{Volume of sludge in mL at 30 min}) \times 1000}{\text{MLSS concentration}}$$

Because the volume of the sludge is divided by the suspended solids concentration, SVI is intended to be independent of the solids concentration. That is, small changes in the MLSS concentration will not affect the final SVI value. This is not true for higher concentrations, and the SVI is not a useful test for solids concentrations higher than approximately 4000 mg/L (Dick and Vesilind, 1969).

The advantages for the plant operator of using SVI are that the test is simple, inexpensive, fast, and gives the operator excellent information about trends in the settling characteristics of the MLSS. An SVI of less than 100 is considered to be a well-settling sludge, while an SVI of over 250 is often a sign that the sludge is not settling well in the final clarifiers. The causes for such poor settling may be many, including a growth of filamentous microorganisms.

Steady-state operating conditions seldom occur in final clarifiers. Fluctuations in wastewater flowrate, return activated sludge, and MLSS concentrations, and settleability occur seasonally, daily, and even diurnally. Some conservatism in design, therefore, is essential to maintain performance with these uncontrollable and unpredictable fluctuations.

Process Design

The secondary settling tank (clarifier) is an integral part of the activated-sludge system and its design is just as important as the design of the biological reactor. The important designing steps include determining the solids loading rate, overflow rate, depth, influent structure, internal baffles, scum removal, weirs, and sludge collection and withdrawal.

Solids Loading Rate

Because the secondary clarifier acts not only as a settling tank but also a thickener, its ability to handle solids must be a part of the design. Establishing the maximum allowable solids loading rate is of primary importance to ensure that the clarifier will function adequately. One way to quickly determine limiting solids loading rates is shown in Figure 6.16, which shows the limiting rate as a function of SVI for single-point and multipoint sludge-removal mechanisms. This figure seems to be useful if the SVI values are with the normal range of 100 to 250.

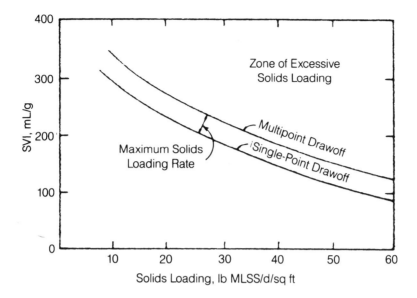

Figure 6.16 Design solids loading on a secondary clarifier (lb MLSS/d/ft² × 4.883 = kg MLSS/m²·d).

Most design engineers prefer to keep the maximum solids loading rate (including full return activated sludge capacity) in the range of 100 to 150 kg/m²·d (20 to 30 lb/d/ft²). These values, according to Figure 6.16, are quite conservative. Rates of 240 kg/m²·d (50 lb/d/ft²) or more have been observed in some well-operating plants with low SVI, deep tanks, and effective solids removal.

Overflow Rate
Overflow rates for secondary clarifiers vary from 0.5 to 2 m³/m²·h (300 to 1000 gpd/ft²). Some plants are known to operate without difficulty at the upper end of this range and produce a high-quality effluent. In many other documented cases, diurnal or maximum pumping peak rates of 2.7 to 3.1 m³/m²·h (1600 to 1800 gpd/ft²) do not exceed a secondary clarifier's capacity. In as many or even more cases, however, clarification efficiency suffers with lower average and peak overflow rates. The area necessary for achieving a required overflow rate is calculated, and this is compared with the area necessary for thickening derived from the solids loading calculation. The larger of these two surface areas governs design.

To drive home the point about the overflow rate not necessarily being the proper design variable, consider the data in Figure 6.17. The data for surface overflow rate is plotted against the effluent suspended solids, and shows an impossible amount of scatter, with no discernable trend. One would expect that if the surface overflow rate was doubled from 1 m/h to 2 m/h, that some deterioration of performance would result, but none seems to be evident. The conclusion is that the overflow rate must be but one variable, and not even an important one, that determines the performance of secondary clarifiers.

Depth
Increasing depth has been observed to improve performance in terms of suspended solids removal and return activated sludge concentration. Too much depth, however,

Chapter 6 Suspended-Growth Biological Treatment

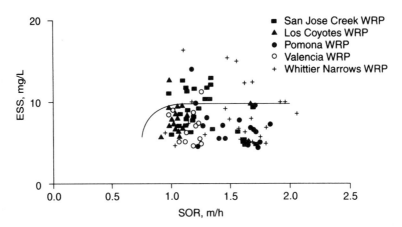

Figure 6.17 Surface overflow rate and effluent suspended solids recovery for five cities in California (Stahl and Chen, 1996).

increases construction cost. Most tanks are less than 4.5 or 5 m (15 or 16 ft) deep at the wall.

Influent Structure

Das et al. (1993) demonstrated that velocities in excess of 0.6 m/s (2 ft/s) will cause deflocculation of biological flocs. Thus, the maximum day velocity in the influent pipe and center column should be limited to less than 0.6 m/s (2 ft/s). The velocity through outlet ports to the clarifier feed well area should be limited to approximately 0.5 m/s (1.6 ft/s) to minimize floc breakup.

Inlets must dissipate influent mixed liquor energy, distribute flow evenly in vertical or horizontal directions, reduce density current effects, minimize blanket disturbances, and promote flocculation. Poor distribution or jetting of incoming flow results in short circuiting as evidenced by concentration–time distributions in dye studies (Crosby, 1984). Density currents from downflowing mixed liquor are commonly observed and are considered detrimental to clarifier performance. Some inlets are designed to reduce density current effects by discharging the mixed liquor to the lower portions of a clarifier or distributing the mixed liquor over a large area. Small deflectors to direct inflow in an upward direction have not been successful (Crosby, 1983). Figure 6.18 shows one alternative influent structure for a circular clarifier. The flow enters the tank through a series of baffled gates, producing a circular motion to further dissipate energy.

Scum Removal

Scum baffles and removal devices have been common for decades. In most designs, scum is handled by skimmers that move up a beach and drop it into a scum trough. During the past decade or so, the ducking skimmer has been applied to an increasing number of circular tanks.

Outlet Structure

Numerous state regulations limit maximum allowable weir loadings to 120 m³/m·d (10,000 gpd/ft) for small treatment plants (less than 4000 m³/d [1 mgd]) and 190

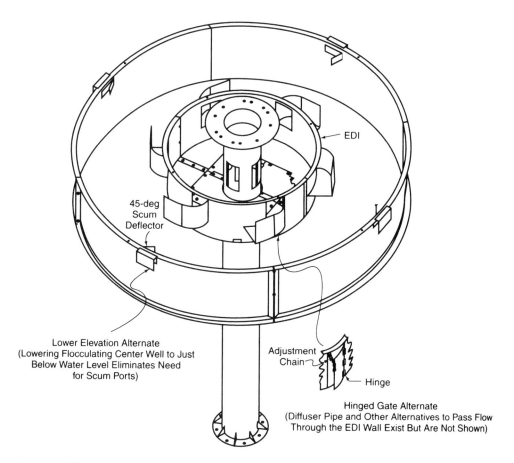

Figure 6.18 Center-column, energy-dissipating inlet and hydraulic flocculating feed well in a circular clarifier (EDI = energy dissipating inlet).

m³/m·d (15,000 gpd/ft) for larger plants. Experiences of many operators and design engineers has led to a general consensus that substantially higher weir loading rates do not impair performance.

For radial-flow (circular or square) clarifiers, a single peripheral weir is considered adequate, especially if some baffling is provided to prevent an updraft wall effect that results in suspended solids approaching the weir. Other engineers prefer to handle this problem by locating double inboard launders at a distance of approximately 30% of the tank radius from the outer wall. The double launder concept increases construction cost but, as demonstrated by Anderson (1945) more than 40 years ago, improves performance over that of simple peripheral weirs without baffling.

For rectangular tanks, launders that extend 25 to 30% of the tank length from the effluent end and are spaced at approximately 3-m (10-ft) intervals have worked well. Some engineers continue to believe that a simple full-width weir at the effluent end is sufficient. Regardless, providing extensive launder structures to meet arbitrary criteria of 120 to 190 m³/m·d (10,000 to 15,000 gpd/ft) seems unwarranted unless necessary to meet certain state criteria. Providing adequate baffling is less costly than installing

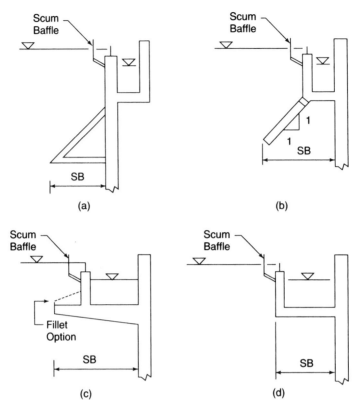

Figure 6.19 Alternative peripheral baffle arrangements: (a) Stamford, (b) unnamed, (c) Lincoln, and (d) interior trough (SB = sludge baffle).

complex launders on cantilevers. Effluent troughs on the inside of the tank wall, extending to the bottom of the trough inwardly to obtain a horizontal deflection baffle, is a cost effective way of achieving excellent settling performance.

In rectangular tanks, density and thermal currents can be reduced by installing baffles at the effluent end of the settling tank, as shown in Figure 6.19. These baffles force the flow away from the weir and back into the tank where the floc can settle.

Sludge Withdrawal

Mixed liquor solids that settle out in suspended-growth secondary clarifiers are removed by plows that transport the sludge to hoppers, or by hydraulic suction (vacuum) devices that pick the sludge off the floor. Most rectangular tanks use chain-and-flight, traveling bridge, or floating pontoon mechanisms. Circular clarifiers use rotating mechanisms. Likewise, square clarifiers use rotating devices that may or may not have corner sweep plows. In rectangular tanks, the most common collector mechanism is the chain-and-flight design. Hopper shapes for rectangular clarifiers are often inverted pyramids with a rectangular top opening. A single rectangular tank may have two or more sludge withdrawal hoppers, each equipped with a separate withdrawal pipe. If these pipes are arranged in a manifold, one hopper may start to pull in diluted sludge while the other allows sludge to accumulate. To prevent this, separate controls must be provided for each withdrawal pipe.

Another concept is the use of siphoned suction piping for rectangular clarifiers. One advantage of a hydraulic suction traveling device is that its pattern can be programmed. That is, it can spend a higher percentage of its travel time at the front end of the tank where larger quantities of solids are deposited.

Sludge removal in circular tanks is accomplished by rotating scrapers or hydraulic suction. There are basically two types of scraper mechanisms. The first and most common in the United States uses two sets of staggered plows that progressively move the sludge toward a central hopper. Figure 6.20 illustrates the revolving scraper mechanism typically used in circular clarifiers. Some units use curved scraper blades (called *spirals*), shaped to keep a constant angle between the blade tangent and radial line to that point on the spiral.

Control Strategies

In addition to the design features discussed above, the success of an activated-sludge clarifier also depends in part on the control strategy of the operator. Many operators keep the return activated sludge flowrate constant, whereas others make adjustments either manually or by assistance from automation. The flow versus time pattern of waste activated sludge is not as significant as the return activated sludge, but the waste activated sludge capacity must be large enough to adequately control the solids retention time of the system and handle inventory shifts that may be necessary because of changes in operating the downstream solids-processing systems.

Return Activated Sludge Flowrates

Most conventional activated-sludge plants are designed with a return activated sludge flowrate that can be lowered to 20% of the average daily dry-weather flow to as much as 100% of that flow. For extended-air activated-sludge systems, the range may be further extended to 150% of average dry-weather flow. It is common practice to return sludge continuously. Adequate standby pumping and power should be provided to ensure that flow is not interrupted.

Return Activated Sludge Flow Patterns

Many operators choose to operate the return activated sludge flow at a constant rate, adjusted upward or downward depending on sludge settleability. If sludge settles well and SVI values are low, the return activated sludge flowrate may be reduced. If bulking occurs, conversely the flowrate may be increased. If conditions become severe, operators may even choose to turn on standby pumping capacity for short periods of time.

Another operational strategy is to return sludge at a flow proportional to plant influent. Thus, at peak wet-weather conditions, the return activated sludge capacity is increased and, at diurnal minima, it is lowered. A disadvantage to this is that the hydraulic flow to the clarifier is then compounded with the higher hydraulic flow of return activated sludge. The turbulence generated may offset the advantages of increased solids-removal capacity. Another strategy is to maintain a constant blanket level in the clarifier. Some plants use blanket-level sensing devices, which are used to increase or decrease the rate of return activated sludge pumping.

The blanket thickness and rate of return activated sludge pumping may also be adjusted to reduce the amount of dissolved oxygen in the return activated sludge. This is

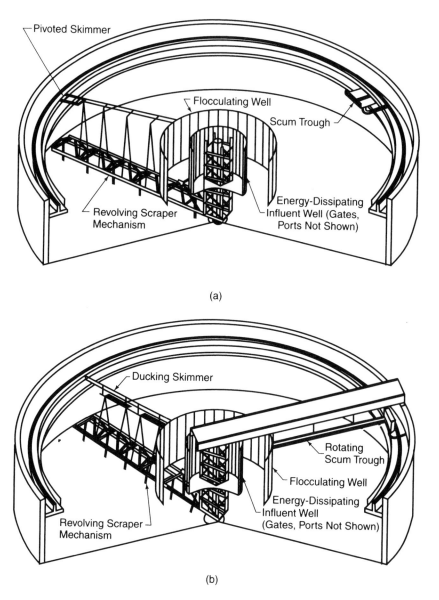

Figure 6.20 Alternative skimming devices and sludge collection mechanisms for circular clarifiers: (a) revolving skimmer and fixed scum trough and (b) rotary ducking skimmer.

particularly important for biological phosphorus removal. If the sludge is held too long in the clarifier, however, denitrification and resulting rising sludge may develop.

Waste Activated Sludge Rate Control

The primary operating tool in activated-sludge systems is the amount of sludge wasted. The mass of solids wasted per day determines the sludge solids retention time (or sludge age, a term more familiar to operators), recalling that solids retention time age is calculated as the mass of sludge solids under aeration divided by the mass per day of solids wasted. By increasing the amount of solids wasted per day, the operator can reduce the solids retention time, which affects the type of biomass produced in the biological reactor. Foaming problems, for example, are often the result of excessively high solids retention times, and often such problems can be solved by wasting more sludge.

Many plants, especially large ones, operate with a constant rate of waste activated sludge flow. For smaller plants, waste activated sludge may be withdrawn from biological reactors rather than from the clarifier or return activated sludge lines. Wasting mixed liquor simplifies solids retention time calculations but results in a dilute waste activated sludge and may hydraulically limit belt presses or centrifuges for dewatering.

Types of Pumps and Systems

Most return activated sludge stations use centrifugal pumps for return and waste activated sludge. For intermediate-sized plants using circular clarifiers, these pumps are often connected directly to clarifier withdrawal pipes. For some larger plants, however, a wet well for these pumps to draw suction from is provided. Wet wells introduce a source of potential odor, but may have the advantage of reducing the number of pumps required.

Conclusions

In H. G. Wells' classic story *The War of the Worlds*, written in 1898, we humans are about to become domesticated farm animals to feed the Martians, new masters of the earth. In a wonderful exchange of views by two humans hiding out in the sewers, a debate ensues on whether being taken care of like a farm animal and periodically harvested would not be too bad with our hero holding out for human dignity. Before they both become hamburger meat for the Martians, however, the world is saved by an unlikely ally. The Martians had not counted on microorganisms, and lacking immunity against these pathogens, succumb to the attack by our little friends.

At the time of the publication of *The War of the Worlds*, the "germ theory" was still being debated. The microbes that attacked the Martians were not benign, but were deadly to life, in this case Martian life. Most people did not understand how microorganisms could also be beneficial. Of course, beer had been made for many centuries but the beer-making process was thought to be a chemical reaction. Similarly the pollution of rivers was thought of as a chemical problem and therefore treatment systems all relied on chemical principles. The recognition that minute creatures actually were doing the work of cleansing our aquatic environment was slow in coming and even slower in being applied to wastewater treatment.

In 1885, when William Dibdin applied biological treatment to London's wastewater and opened up a new chapter in environmental engineering and public health, the newspapers were critical of the engineers who were so slow in understanding what then was very obvious (once they figured it out!) that the germs can be our friends as well as enemies.

References

Anderson, N. E. (1945) Design of Final Settling Tanks for Activated Sludge. *Sew. Works J.*, **17**, 50.

Ardern, E.; Lockett, W. T. (1914) Experiments on the Oxidation of Sewage Without the Aid of Filters. *J. Soc. Chem. Ind.*, **33**, 523.

Bisogni, J. J.; Lawrence, A. W. (1971) Relationships Between Biological Solids Retention Time and Settling Characteristics. *Water Res.*, **5**, 753.

Budd, W. E.; Lambeth, G. F. (1957) High Purity Oxygen in Biological Sewage Treatment. *Sew. Works J.*, **29**, 237.

Crosby, R. M. (1983) *Clarifier Newsletter;* Crosby, Young, and Associates: Plano, Texas.

Crosby, R. M. (1984) *Evaluation of the Hydraulic Characteristics of Activated Sludge Secondary Clarifiers;* U.S. Environmental Protection Agency, Office of Research and Development; Washington, D.C.

Dague, R. R. (1983) A Modified Approach to Activated Sludge Design. Paper presented at the 56th Annual Water Pollution Control Federation Conference, Atlanta, Georgia, October.

Das, D.; Keinath, T. M.; Parker, D. C.; Wahlberg, E. J. (1993) Floc Breakup in Activated Sludge Plants. *Water Environ. Res.*, **65**, 138.

Dick, R. I.; Vesilind, P. A.(1969) The Sludge Volume Index—What is It? *J. Water Pollut. Control Fed.*, **41**, 1285.

Eckenfelder, W. W., Jr. (1966) *Industrial Water Pollution Control;* McGraw-Hill: New York.

Gaudy, A. F.; Gaudy, E. T. (1980) *Microbiology for Environmental Scientists and Engineers;* McGraw-Hill: New York.

Gaudy, A. F.; Kincannon, D. F. (1977) Comparing Design Models for Activated Sludge. *Water Sew. Works*, **123**, 66.

Grady, C. P. L.; Daigger, G.; Lim, H. C. (1999) *Biological Wastewater Treatment Theory and Applications;* Marcel Dekker: New York.

Lawrence, A. W.; McCarty, P. L. (1970) Unified Basis for Biological Treatment Design and Operation. *J. Sanit. Eng. Div., Proceedings of the American Society of Civil Engineers*, **96**, 757.

McCarty, P. L. (1972) Energetics of Organic Matter Degradation. In *Water Pollution Microbiology;* Mitchell, R., Ed.; Wiley & Sons: New York.

McCarty, P. L. (1975) Stoichiometry of Biological Reactions. *Prog. Water Technol.*, **7**, 157.

McKinney, R.E. (1962) Mathematics of Complete Mixing Activated Sludge. *J. Sanit. Eng. Div., Proceedings of the American Society of Civil Engineers.*, **88**, SA3.

McKinney, R. E.; Ooten, R. J. (1969) Concepts of Complete Mixing Activated Sludge. Transcripts of 19th Annual Conference of Sanitary Engineering and Architectural Bulletin No. 60; University of Kansas, Lawrence.

Metcalf and Eddy, Inc. (1991) *Wastewater Engineering: Treatment, Disposal, Reuse*, 3rd ed.; Tchobanoglous, G., Ed.; McGraw-Hill: New York.

Metcalf & Eddy, Inc. (2003) *Wastewater Engineering*, 4th ed.; McGraw-Hill: New York.

Novak, J. T. (1974) Temperature-Substrate Interactions in Biological Treatment. *J. Water Pollut. Control Fed.*, **46**, 1984.

Ramanathan, M.; Gaudy, A. F., Jr. (1971) Steady State Model for Activated Sludge with Constant Recycle Sludge Concentration. *Biotechnol. Bioeng.*, **13**, 125.

Stahl, J. F.; Chen, C. L. (1996) Review of Chapter 8. Rectangular and Vertical Secondary Settling Tanks. Presented at the Secondary Clarifiers Assessment Workshop at the 69th Annual Water Environment Federation Technical Exposition and Conference, Dallas, Texas.

Water Pollution Control Federation (1988) *Aeration;* Manual of Practice No. FD-13; Washington, D.C.

Water Research Commission (1984) *Theory: Design and Operation of Nutrient Removal Activated Sludge Processes;* Pretoria, South Africa.

Wilford, J.; Conlon, T. P. (1957) Contact Aeration Sewage Treatment Plants in New Jersey. *Sew. Ind. Wastes*, **29**, 845.

Symbols used in this Chapter

B = oxygen equivalent of cell mass, O_2/kg volatile suspended solids
b = endogenous decay coefficient based on biomass in aerated zone (1/d)
K_{T1} = a specific kinetic, stoichiometric, or mass-transfer coefficient at temperature T_1
K_{T2} = a specific kinetic, stoichiometric, or mass-transfer coefficient at temperature T_2
k_1 = observed BOD_5 removal rate constant (base e, d^{-1})
K_s = half-velocity coefficient (kg/m^3)
k = maximum rate of substrate utilization per unit weight of biomass (1/d)
M = total mixed liquor suspended solids (kg/m^3)
M_r = total sludge recycle flow suspended solids (kg/m^3)
N_o = influent oxidizable nitrogen (kg/m^3)
N = effluent oxidizable nitrogen (kg/m^3)
NO_3 = effluent nitrate-nitrogen (kg/m^3)
P_x = mass of total activated-sludge solids generated or wasted per day (kg/d)
Q = wastewater inflow (m^3/d)
Q_r = sludge recycle flow (m^3/d)
Q_w = sludge waste flow (m^3/d)
R_c = mass of oxygen required per unit time to satisfy carbonaceous oxidation (kg/d)
R_n = mass of oxygen required per unit time to satisfy nitrification oxygen demand (kg/d)
S_o = influent soluble biodegradable substrate (kg/m^3)
S_e = effluent soluble substrate (kg/m^3)
$\bar{t}$ = hydraulic retention time (d)
V = aeration tank volume (m^3)
X = reactor biological solids (kg/m^3)
X_r = sludge recycle flow biological solids (kg/m^3)
Y_{OBS} = observed cell yield (kg/kg)
Y = true cell yield (kg/kg)
Z_{io} = influent nonvolatile suspended solids (kg/m^3)
Z_{no} = influent volatile nonbiodegradable solids (kg/m^3)
α = sludge recycle ratio
β = constant
θ_c = solids retention time (d^{-1})

7

Attached-Growth Biological Treatment

Introduction

The fact that slime-producing organisms attached to rocks and other surfaces accomplishes the self-cleansing of water in streams has been known for many years. The trickling filter and related treatment processes simply accelerate this naturally occurring phenomenon. Attached-growth biological treatment processes provide a surface (medium) on which the microbial layer can grow and expose this surface repeatedly to wastewater for adsorption of organic material and to the atmosphere for oxygen.

Early installations of fixed-film biological treatment used rectangular beds of rock media where presettled wastewater was sprayed from fixed nozzles onto the media surface. Automatic siphons were used with dosing tanks to provide rest periods. Circular beds with rotary distributors were later introduced to allow for continuous feeding and these became the *trickling filters* still in use today.

The original filter medium for trickling filters was rocks, but plastic packing has become standard practice, providing more surface area per unit volume and improving ventilation, thereby reducing clogging and odor problems. Synthetic media have been fabricated as assemblies of corrugated sheets that can be stacked, or as small ring structures that are placed in a loose (random) packing configuration.

The *rotating biological contactor* uses a fixed-film biomass on rotating media for biological treatment. The rotating medium, made from sheets of high-density plastic,

Figure 7.1 A typical trickling filter.

provides a surface on which organisms grow and make contact with organic wastewater constituents and oxygen from the air. The rotating reactor carries a film of wastewater into the air. The wastewater trickles down the surfaces of the contactor and absorbs oxygen from the air.

Dual or coupled biological treatment systems have a fixed-film reactor and a suspended-growth process reactor in series. This combination results in a two-stage unit process that has unique design parameters; its treatment efficiency capabilities often exceed the individual performance of both parent systems. The activated-sludge (suspended-growth) unit provides a variety of functions, including flocculation to improve clarification, reduction of biochemical oxygen demand (BOD), nitrification, denitrification, and phosphorus removal to meet advanced wastewater treatment requirements.

Trickling Filters

Processes that use a static medium such as rocks for growing a film of biomass and then trickling the wastewater over this medium are known as *trickling filters* (Figure 7.1). Trickling filters are not filters at all because they do not filter anything. They are in fact compact and efficient streams, with the same bio-processing going on in a controlled and effective manner that occur naturally in flowing watercourses.

Trickling filters are used either as *roughing filters* early in wastewater treatment where they remove some of the high BOD loadings (particularly shock loads from industrial effluents), or as the secondary treatment step in a wastewater treatment plant, processing effluent from the primary settling tanks and producing plant effluents in the range of 30 to 45 mg/L 5-day BOD (BOD_5) and suspended solids, respectively. With proper design, however, trickling filters can produce effluent qualities of about 10 mg/L BOD_5 and suspended solids, comparable to effluent from activated-sludge treatment. Trickling filters can also be designed to achieve nitrification.

Theory of Operation

In principle, the trickling filter process has the biomass attached to a fixed medium. Thus, recycling of settled biomass is not required. On the other hand, the activated-sludge process involves aeration of wastewater and maintenance of a suspended microbiological culture by recycling settled solids. The two processes are similar because they both depend on biochemical oxidation of a portion of the organic matter in wastewater to carbon dioxide and water. Remaining organic matter is incorporated or transformed to biomass; the energy produced is released as heat into the medium.

Settled primary treated or screened wastewater is applied to the filter medium over and around which the flow percolates. The surface of the medium quickly becomes coated with a viscous, jelly-like, slimy substance containing bacteria and other biota. The biota remove organics by adsorption and assimilation of soluble and suspended constituents. For aerobic metabolism, oxygen is supplied from the natural or forced circulation of air through interstices in the filter medium. Oxygen transfer may be direct or by diffusion through the liquid films.

In the trickling filter, the liquid retention time is only a few (8 to 20) minutes and the biomass present must be adequate to remove the substrate by sorption or oxidation in that time period. Efficient flow distribution and contact with biomass on the media is necessary for optimum performance.

After an initial start-up period, the microbial buildup may create an anaerobic interface with some of the filter media. This furthers the growth of facultative and possibly anaerobic organisms, especially if the accumulation of biomass is excessive. However, aerobic organisms at the upper microbial surfaces provide the basic mechanism for organic removal and conversion. True anaerobic functions of hydrolysis and gasification are minimal or absent in a properly operating aerobic fixed-film reactor.

The quantity of biomass produced is controlled by available food. The amount of biomass on the medium surface increases as organic load and strength increase until a maximum effective thickness is reached. This maximum thickness is controlled by several factors, including hydraulic dosage rate, type of media, type of organic matter, amount of essential nutrients present, temperature, and the nature of the biological growth. During filter operation, a portion of the biological slime periodically sloughs off. Excess biomass that cannot retain an aerobic condition impairs performance. Uniform sloughing, as measured by trickling filter effluent suspended solids, provides an indication of a properly operating trickling filter.

Filter flies and snails are often encountered in trickling filter plants. Filter flies may be prevented or controlled by designing trickling filters to allow flooding of the medium—a simple way for operators to control nuisance organisms. Reduced or uneven dosing also helps control nuisance organisms and odors.

Significant populations of snails sometime develop, accumulating in pumping stations, causing plugging and excessive wear of pumps and creating problems with other equipment in both liquid and solids processes downstream from the filters. One control procedure is to provide a depressed, low-velocity channel between the trickling filter and the final settling tank, with bypass facilities (or mechanized removal) to allow cleaning of collected snails.

Primary settling is necessary ahead of stone filters to minimize problems with clogging but may not be required in plants where comminution and fine screening of coarser raw wastewater solids occur and the medium consists of corrugated plastic or material with large void spaces. If primary clarifiers are not used, screening units should remove sludge particles larger than 3 mm (0.12 in.). Adequate final settling must be provided downstream of trickling filters to collect biomass sloughed from the medium.

Recirculation of trickling filter effluent is a convenient operational tool that has been used to improve filter efficiency in numerous applications. Increasing the total hydraulic flowrate provides improved liquid distribution, enhances oxygen transfer, and reduces the likelihood of dry or partially wetted surface areas within the filter, thereby maintaining maximum treatment capability. In addition, a higher flowrate maintains shear forces sufficient to slough excess growth, thereby reducing clogging problems associated with excessive solids accumulation. Finally, organic matter that may have missed exposure to the slime on its first passage through the filter may be treated on the second. However, this latter aspect can be insignificant if the proper dosing procedure is used and media are fully wetted.

Loading Terminology

Trickling filters are classified according to applied hydraulic and organic loadings. Hydraulic loading is the total volume of liquid, including recirculation, per unit of time per unit of filter cross-sectional area. Although hydraulic loading has been expressed in various units, the current practice is to use $L/m^2 \cdot s$ or $m^3/m^2 \cdot h$ (gpm/ft^2), but $m^3/ha \cdot d$ (mgd/ac) is sometimes used. Organic loading is currently expressed as kg $BOD_5/m^3 \cdot d$ ($lb/d/1000\ ft^3$) of filter media, and $kg/m^2 \cdot d$ (BOD_5 and chemical oxygen demand) and $kg/m^3 \cdot d$ (nitrification) ($lb\ NH_4\text{-}N/d/1000\ ft^2$) for process loadings and performance on a unit surface area basis. Metric units consistent with current practice will be used for design equations. However, referenced equations will retain their original derived format.

When wastewater is recycled through the filter, organic loading calculations become more complex. General practice is to ignore the recycled organic load, but recycle flow can improve treatment efficiency, an important consideration when evaluating subsequent design criteria.

Roughing filters are high-rate filters receiving high hydraulic or organic loadings. Although these filters may provide a high degree of organic load removal per unit volume, their settled effluent still contains substantial soluble BOD and solids-related BOD. Roughing filters are used to provide intermediate treatment or as the first step of multistage biological treatment.

Development of plastic trickling filter media has led to the development of *superrate* filters. These filters, used either as roughing units for strong wastewater or as complete secondary treatment units, do not represent a significant departure from the roughing filters discussed above. Synthetic media filters are often referred to as *oxidation towers*, *biotowers*, or *biofilters*.

The term BOD as used throughout this section is intended to represent uninhibited biochemical oxygen demand, whether referring to total BOD or the fraction of BOD resulting from suspended solids plus the fraction resulting from soluble materials.

$$BOD = BOD_{soluble} + BOD_{solids} \qquad (7.1)$$

or

$$BOD = BOD_{soluble} + f(\text{suspended solids})$$

where

$f = BOD_{solids}/SS$
= the oxidation or respiration of suspended solids (SS) expressed as BOD.

The value of f varies from 0.2 to 0.9 mg/mg, depending on the stage in the treatment process, loading, and temperature.

Factors Affecting Performance and Design

Numerous variables affect trickling filter performance, including wastewater composition, wastewater treatability, pretreatment and primary treatment, media type, depth, recirculation, hydraulic and organic loadings, ventilation, and temperature. All of these process factors of the system are interrelated and must be considered jointly for effective design. Finally, the important step of solids–liquid separation (secondary clarification) is critical to the overall performance of the plant.

Wastewater Composition

Biochemical oxygen demand and suspended solids determinations are the principal measurements used in BOD removal systems to assess both the strength of the applied wastewater and the quality of the settled effluent. Both the volume and strength of plant influent may exhibit significant hourly, daily, and seasonal variations.

Wastewater Treatability

The treatability of a particular wastewater depends, in part, on the ratio of suspended–colloidal concentration to soluble organic concentration. The trickling filter process readily removes suspended and colloidal organics by the combined processes of biological flocculation, adsorption, and enzyme-complexing—not just through biological oxidation and synthesis. Soluble, small-molecule organics, such as simple sugars, organic acids, alcohols, and so on, are readily removed from the waste stream even in the short residence times provided by the trickling filter. Therefore, the process efficiently treats industrial waste having a high percentage of soluble, small-molecule organic material.

Pretreatment and Primary Treatment

The degree of pretreatment affects the performance and design of trickling filters. Grit and screening removal, chemical treatment, clarification, equalization, neutralization, prechlorination, and preaeration are pretreatment processes that can improve the performance of trickling filters. Consideration of both capital expenditures and operating costs is necessary as part of the evaluation of improved treatability. For domestic wastewater, primary clarification without equalization has provided more than adequate pretreatment.

Trickling Filter Media Type

The introduction of synthetic media for trickling filters has extended the ranges of hydraulic and organic loadings well beyond those of stone media. Unlike rocks or wood media, the increase in slime thickness on plastic media reduces the aerobic biological

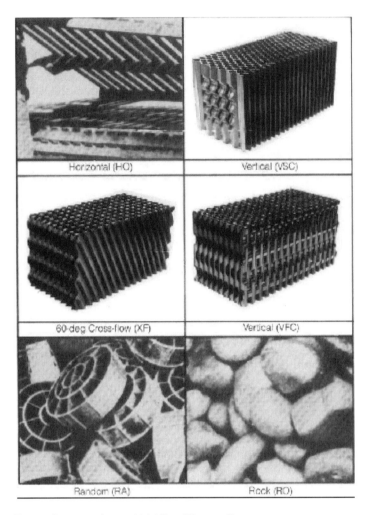

Figure 7.2 Types of commonly used trickling filter media.

surface area (Albertson and Eckenfelder, 1984). Thus, as the specific (clean) surface area of the media increases, quantity of biomass increases and the net effect is that the aerobic surface area does not increase and may decrease because of biofouling. Figure 7.2 illustrates typical media used in trickling filters.

Trickling Filter Depth
In the United States, rock filters are 1 to 2 m (3 to 6 ft) deep and occasionally 2.4 m (8 ft) deep. This depth limitation is associated with lack of adequate ventilation produced by natural draft and an increased tendency to pond. Plastic media trickling filters, the so-called *biofilters*, are most commonly constructed between 5 and 8 m (16 and 26 ft) deep, although units up to 12.8 m (42 ft) deep exist. The limiting depth is associated more with tower height aesthetics, serviceability, pumping requirements, and structural design of media than with biological treatment efficiency. Increasing the depth of the filter is worthwhile to reduce the minimum flow required for high wetting

efficiency, and therefore, to reduce recirculation. In taller filters that have high loadings, oxygen deficiency may occur in the uppermost layers. However, adequate ventilation and routine hydraulic flushing can prevent odor problems from developing.

Trickling Filter Hydraulics and Loadings

Recirculation, recirculation arrangement, recirculation ratio, effect of recirculation, distributor operation, organic loading, and hydraulic and organic loading models are interrelated factors that affect trickling filter design and performance.

As an often important element, a portion of the trickling filter effluent is recycled through the filter. This practice is known as *recirculation* and the ratio of returned flow to incoming flow is called the *recirculation ratio*. Recirculation is an important element in stone filter design, because of apparent increases in BOD removal efficiency, and is important in synthetic media filter design because it can ensure that the filter is adequately wetted. Figure 7.3 shows a number of recirculation arrangements. Clarification of trickling filter effluent before recycle to the trickling filter or passing on to the second stage of trickling filtration is known to enhance the process performance. However, if a recycle flow overloads the primary, intermediate, or secondary clarifier, then most, and often all, of the benefits can be lost. In some cases, clarifier performance is reduced because of hydraulic overloading by the recirculated flow, especially during diurnal flow peaks and wet-weather flow. Only when clarifiers have been sized to handle the recycle flow should this flow pass through the primary clarifiers. Conventional practice is direct recirculation of the trickling filter effluent back ahead of the trickling filter as shown in Figure 7.3b.

Decisions on whether to use recirculation and what magnitude to use depend on comparisons of annual costs of various designs providing equal treatment. Where recirculation is used, recirculation ratios range from 0.5 to 4.0 but occasionally reach 10 or higher for strong industrial wastewater. However, a recirculation ratio greater than 4 does not materially increase the efficiency of rock filters and is also uneconomical (Galler and Gotaas, 1964).

If the noted benefits of recirculation are mostly a result of increased wetting efficiency, the use of recycle may be unnecessary or can be reduced if distributor speed is significantly reduced below conventional, hydraulically driven rates. Typical distributor operation in the United States over the past 30 to 40 years used a rotational speed of 0.5 to 2 min/rev. With two or four arms, the filter is then dosed every 10 to 60 seconds. Subsequent research (Albertson, 1995) provide further support for periodically slowing distributors to a fraction of the conventional rotational speed and advocate use of the instantaneous dosing. The uneven dosing seems to result in better performance, a noticeable reduction of odors, and reduced or an absence of high solids-sloughing cycles. These results also suggest that adoption of this mode of operation enhances performance of all types of media and significantly improves the general background database for trickling filters.

Trickling filters are designed for a wide range of organic loadings that sometimes, but not always, correlate with the hydraulic rate. Biochemical oxygen demand loading rates may range from 0.08 to 8.0 kg BOD_5 /$m^3 \cdot d$ (5 to 500 lb/d/1000 ft^3), depending on the objective. The wastewater applied concentration is only a design criterion when BOD_5 concentrations exceed 400 mg/L. In such cases, dilution by recirculation is necessary. The minimum recycle rate should result in a raw applied BOD_5 less than or equal to 400 mg/L.

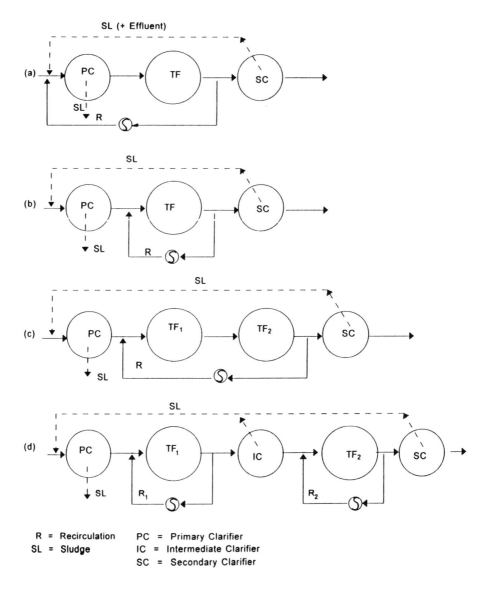

Figure 7.3 Flow diagrams for single- and two-stage trickling filter plants (R = recirculation, SL = sludge, PC = primary clarifier, IC = intermediate filter, and SC = secondary clarifier).

Ventilation

Ventilation of filters is essential to maintain aerobic conditions necessary for effective performance. If enough passageways are provided, the difference in air and wastewater temperature and humidity differences between ambient air and trickling filter air provide a natural draft that often provides sufficient aeration. In many cases, however, these differences are minimal, resulting in inadequate airflow. These periods occur in the morning and evening, especially in the spring and fall. Thus, odors often occur during these periods and in warmer climates where temperature differences are smaller.

Good design practice requires provisions of adequate sizing of underdrains and effluent channels to permit free flow of air. Passive devices for ventilation include vent stacks on the filter periphery, extensions of underdrains through filter side walls, ventilating manholes, louvers on the sidewall of the tower near the underdrain, and discharge of filter effluent to the subsequent settling tank in an open channel or partially filled pipes. However, these methods are inadequate if high performance is required or natural draft forces are low. For municipal wastes, plastic media manufacturers have recommended 0.1 m² (1 ft²) of ventilating area for each 3 to 4.6 m (10 to 15 ft) of trickling filter tower periphery and 1 to 2 m² of ventilation area in the underdrain area per 1000 m³ of media.

The effect of temperature on filter performance has been expressed by (Onda et al., 1968)

$$E_t = E_{20} \beta^{(T-20)} \tag{7.2}$$

where

β = constant assumed equal to 1.035, ranging from 1.015 to 1.045;
E_t = filter efficiency at temperature T;
E_{20} = filter efficiency at 20 °C; and
T = wastewater temperature (°C).

The range of β varied from 1.015 to 1.045 (Howland, 1953) for carbonaceous BOD removal; the use of 1.035 produces a conservative design.

Odor, Vector, and Macroorganism Control

Odor problems have been a serious liability for trickling filters, and yet trickling filters have been used to remove odors from the air. Odors are generated either at the influent where volatile components are stripped by upflowing air, or off the media. Generation of odors within media is caused by lack of adequate ventilation or excess biomass buildup in the media. When excess biomass is present in the media, odors will be generated from the septic mass even when ventilation is adequate.

Preventing odors from being generated is accomplished by designing adequate hydraulics or routinely flushing the media and assuring adequate and continuous ventilation. The controlled flow of water and air should be provided with electrical and mechanical equipment. Also, to minimize stripping of volatile components from influent wastewater, airflow should be countercurrent to the liquid, allowing the biology in media an opportunity to cleanse the air.

Flies, the larvae of macroorganisms, and snails represent serious problems for some trickling filter facilities. Flies are an aesthetic nuisance and larvae can reduce the treatment efficiency. Controlled uneven dosing seems to control flies at the larval stage in rock filters.

Design Formulas

Numerous investigators have attempted to delineate the fundamentals of the trickling filter process by developing relationships among variables that affect operation. Analyses of operating data have been made to establish equations or curves to fit available data. Results of these data analyses have led to the development of various empirical formulas.

National Research Council Formula

The National Research Council (NRC) formula (1946) resulted from an extensive analysis of operational records from stone-media trickling filter plants serving military installations. The NRC data analysis is for stone-media based on two principles: the amount of contact between filter media and that organic matter removed depends on filter dimensions and the number of passes (the greater the effective contact, the greater the performance efficiency). However, the greater the applied load, the lower the efficiency. Therefore, the primary determinant of efficiency is the combination of effective contact and applied load.

The empirical model based on the NRC study uses the principle of applied load/effective contact area (W/VF)

First stage or single stage

$$E_1 = 100/[1 + 0.0085(W_1/VF)^{0.5}] \tag{7.3}$$

Second stage

$$E_2 = \frac{100}{1 + \frac{0.0085}{1 - E_1/100}\sqrt{\frac{W_2}{VF}}} \tag{7.4}$$

where
- E_1 = BOD removal efficiency through the first-stage filter and settling tank (%);
- W_1 = BOD loading to the first- or single-stage filter, not including recycle, kg/d (lb/d);
- V = volume of the particular filter stage (surface area times depth of media), m³ (ac·ft); and
- F = number of passes of the organic material, equal to $(1 + R/Q)/[1 + (1 - P)R/Q]^2$;
- R/Q = recirculation ratio (recirculated flow-to-plant influent flow);
- P = a treatability factor that, for military trickling filter plants, was found to be approximately 0.9;
- E_2 = BOD removal efficiency through the second-stage filter and settling tank (%); and
- W_2 = BOD load to the second-stage filter, not including recycle, kg/d (lb/d).

Figure 7.4 compares trickling filter operational data for recirculation ratios of 0 to 2 with predicted values using the first- or single-stage NRC formula with a similar range of recirculation ratios (NRC, 1946). Clearly the NRC formula is not a good predictor of BOD removal. It is, however, of historical importance and seems to have a rational, if empirical, basis. It is still used by some states in establishing design standards for trickling filters.

Velz Formula

In 1948, Velz proposed a model based on the assumption that BOD removal is a first-order reaction with regard to depth. He suggested that

$$L_D/L_O = 10^{-KD} \tag{7.5}$$

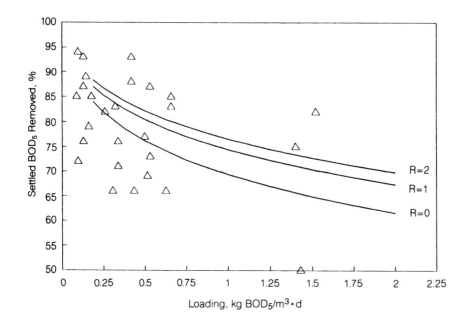

Figure 7.4 Comparison of trickling filter operating data with performance predicted by the National Research Council formula (R = recirculation ratio) (kg/m³·d × 0.0624 = lb/d/ft³) (NRC, 1946).

where
- L_D = BOD removed at depth D (mg/L);
- L_O = influent BOD (mg/L);
- K = first-order rate constant (d⁻¹); and
- D = filter depth, m (ft).

This formula implies that K is constant for all hydraulic rates; however, K seems to vary with the hydraulic rate (Albertson and Davies, 1984; Dow Chemical Company, 1964). For high-rate filters, the value of rate constant K_{20} has been determined to be 0.15 d⁻¹. Low-rate trickling filters yielded an approximate K_{20} value of 0.18 d⁻¹.

Kinematic Wave—Velz Model

The Velz model and subsequent later refinements of the Velz model by Eckenfelder (1963) requires a determination of a first-order rate constant and uses a geometric parameter such as depth and specific surface area to simulate holdup time. The combined rate coefficient required pilot testing to determine a value. Kinematic wave theory allows determination of the holdup time independent of the rate constant. This effectively allows the equation to reflect biological rate and reaction time as two different components in the Velz equation. The reaction rate constant is relatively stable for normal domestic waste at influent concentrations of BOD_5 soluble of 100 mg/L. Pilot testing would still be needed for high strength wastes.

Chapter 7 Attached-Growth Biological Treatment

$$S_e = \frac{S_o(1 - e^{-k_T \cdot t})}{1 + (R/Q)(1 - e^{-k_T \cdot t})} \tag{7.6}$$

where

k_T = reaction rate coefficient (1/min);

t = holdup time, minutes = $H_n \left[\dfrac{Q}{V} \left(\dfrac{R}{Q} + 1 \right) \right]^{-0.44}$;

S_e = effluent soluble BOD$_5$ (kg/d);
S_o = influent soluble BOD$_5$ (kg/d);
Q = influent flow (m³/d);
V = reactor volume (m³);
R/Q = recycle ratio;
R = recycle flow (m³/d);

H_n = hydraulic number = $\dfrac{D A_s^{1/3}}{\sin \phi \, C^{2/3}}$;

D = depth (m);
A_S = specific surface area (m²/m³);
ϕ = media angle from horizontal; and
C = Chezy coefficient = 3 to 10, (Reynolds number <1000, $C = 2 \cdot N_R^{0.5}$).

The reaction rate coefficient k is approximately 1.0 for 20 °C and for a domestic wastewater feed of 100 mg/L soluble BOD$_5$. It varies with temperature as $k_T = k_{20} \beta^{T-20}$ (eq 7.2).

Kincannon and Stover Model

Kincannon and Stover (1982) developed a mathematical model based on a relationship between the specific substrate utilization rate and total organic loading, which followed a Monod plot. The relationship follows:

$$A_s = [(8.34 Q S_o / \mu_{max} S_o)/(S_o - S_e)] - K_b \tag{7.7}$$

where

A = surface area, 1000 ft²;
Q = influent flowrate, mgd;
S_o = influent BOD$_{5\text{-soluble}}$ (mg/L);
S_e = effluent BOD$_{5\text{-soluble}}$ (mg/L);
μ_{max} = maximum specific substrate utilization rate of A, lb BOD$_5$/d/1000 ft²; and
K_b = proportionality constant of A, lb/d/1000 ft².

The biokinetic constants μ_{max} and K_b must be determined by pilot-plant tests, full-scale results, or a summary of prior experiences. They may be determined graphically by plotting BOD loading versus the inverse of BOD removed. The y-intercept is μ_{max}^{-1}, and the slope is K_b. These investigators noted that the variability in correlated data is normal. Biochemical oxygen demand removal is controlled by volumetric loading and treatability, and BOD$_5$ removal is not influenced by media depth.

Logan Model

The Logan model is an example of the new generation of trickling filter models presently being evaluated. The computer model (Logan et al., 1987a, 1987b, 1990) was developed to predict soluble BOD removal in plastic media trickling filters as a function of media geometry. The Logan model is based on characterizing the plastic module as a series of inclined plates covered with a thick (undefined) biofilm. Dissolved wastewater organics that exert soluble BOD are equally distributed into a five-component, molecular-size distribution. The rate of soluble BOD removal is determined using a numerical model to solve transport equations that describe the rate of mass transfer of soluble BOD components through the thin liquid film into the biofilm.

Although the model was calibrated with a single data set for one type of plastic media, it was shown to predict soluble BOD removal in a variety of laboratory, pilot-plant, and full-scale trickling filter studies. Equations on which the Logan model is based (Logan et al., 1987a) cannot be solved in closed form, but the model is available as a computer program.

Nitrification

The nitrification process is sensitive to the availability of oxygen and reportedly to high and low temperatures, the level of soluble organics, ammonium-nitrogen concentration, and perhaps media configuration and hydraulics of the trickling filter. In recent years, trickling filters have been used for nitrification, that is, oxidation of ammonium-nitrogen. This has been achieved for many years by standard-rate rock filters and tertiary plastic media filters. Since 1980, use of plastic media to simultaneously remove BOD and total Kjeldahl nitrogen has expanded. These are single-stage units, even though the design may provide for two separate filters directly coupled in series used to treat smaller flows.

Nitrification rates of suspended-growth or fixed-film processes are highly dependent on the available oxygen level. The rate of nitrification also depends on the concentration of ammonia-nitrogen. At higher levels of ammonia-nitrogen the removal rate is zero-order with as much as 100% ammonia-nitrogen removed. At loadings above approximately 0.0015 kg NH_4-N /$m^2 \cdot$d not all of the ammonium-nitrogen may be removed. With even higher loadings, it appears that the maximum removal rate, regardless of loading rate, is 0.003 kg NH_4-N/$m^2 \cdot$d (Okey and Albertson, 1989). Apparently, oxygen availability becomes a dominant control variable for removal rates higher than an ammonium-nitrogen loading of approximately 0.0015 kg NH_4-N /$m^2 \cdot$d.

The effect of temperature on nitrification is similar to that reported for carbonaceous filters (Onda et al., 1968). The rate of nitrification can be expected to increase by a factor of 1.5 when the temperature increases from 10 to 20 °C (Parker et al., 1990).

Rotating Biological Contactors

During the late 1960s, interest in plastic media led to the development and commercialization of rotating biological contactors (RBCs), which provided many of the advantages of old rock-media trickling filters without some of their disadvantages.

The RBC consists of circular plastic disks mounted on a horizontal shaft in a tank. The shafts are rotated (1 to 2 revolutions per minute, or rpm) by either a mechanical or compressed air drive. For typical aerobic systems, approximately 40% of the media are immersed in the wastewater. One RBC equipment manufacturer has developed an alter-

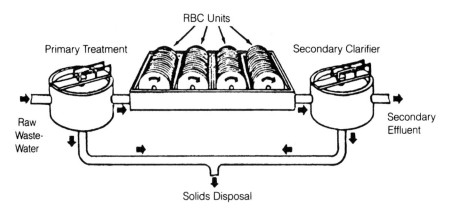

Figure 7.5 Rotating biological contactor.

native design that uses deeper submergence (approximately 85%) with claimed advantages of less stress on media, media supports and shafts because of biomass weight, and fewer RBC units required because of a larger media volume provided on units offered in sizes up to 5.5 m diameter.

The wastewater being treated flows through the contactor by simple displacement and gravity. The rotation of disks alternately exposes the biofilm to the organic material in the wastewater and atmospheric air. Bacteria and other microorganisms that are naturally present in wastewater adhere to and grow on the surface of rotating media. Within 1 to 2 weeks of startup, bacteria will form a fixed biological film, covering the entire media surface. Depending on organic loading conditions, each stage will show varying slime thickness and color. The first stage shows a characteristic brownish-gray color when operated within the reasonable loading range; the last (nitrifying) stages show a reddish-bronze color. Because of shearing forces, the biological film tends to slough off whenever the biomass growth on the media surface becomes too heavy. Growth and sloughing is continuously repeated. The sloughed biofilm and other suspended solids are carried away in wastewater and are removed in the secondary clarifier.

Figure 7.5 shows an RBC system being used for secondary treatment. The wastewater flows from a primary clarifier through a series of RBC units and then to a final clarifier where sloughed biomass is settled and removed. In addition, screens and grit removal equipment should be used ahead of the primary clarifier. With the right wastewater and environmental conditions and appropriate design, RBCs can achieve secondary treatment, nitrification, and denitrification.

Organic overloading, insufficient rotational speed, or other conditions may cause excessive biomass buildup and result in structural damage to shafts or media, the inability to maintain rotational speeds, and other process problems. Process performance below design expectations, structural problems with shafts and media, excessive biomass buildup on media, uneven shaft rotation ("loping") for air-driven units, and other process problems have been serious concerns at many installations. These problems at first resulted in the process falling out of favor with designers but many of these issues have been resolved and rotating biological contactors appear to be much more reliable using contemporary design standards.

Process Concepts and Principles

The RBC process relies on providing an adequate biomass, effective contact between biomass and substrate, necessary nutrients and oxygen, and sufficient retention time in the process to affect treatment. An adequate biomass is maintained in the process by provision of necessary RBC media surface area or volume and by imparting rotational or scour energy between the media and bulk fluid sufficient enough to shear excess biomass from the media. This promotes a relatively thin and active layer of biomass. Effective contact between biomass and substrate is maintained by attention to mixing conditions and proper baffling between stages to optimize reactor performance. Nutrient concentrations are suitable for the process in most domestic wastewater situations but may be adjusted if necessary. Necessary oxygen levels (for aerobic systems) are provided in the process by maintaining adequate rotational speeds and aeration levels (for systems with diffused or supplemental air). Adequate retention time is provided in the process when normal media surface area or volume determinations are made and standard tankage configurations are used.

Mass transfer–diffusion of substrate and oxygen dominates organic removal in most aerobic RBC systems. Mass transfer–diffusion resistance associated with both the liquid phase and the biofilm results in significant concentration gradients from the bulk liquid to biological reaction sites within the biofilm. Oxygen transfer becomes limiting and controls the overall reaction rate in heavily loaded systems. Oxygen transfer from the air directly to the attached wastewater and biofilm represents the primary oxygen source for organisms. In addition, oxygen also enters the bulk liquid as a result of turbulence generated by the rotation of media and return to the bulk liquid of wastewater that is lifted into the atmosphere and then flows freely back across the media.

The substrate removal mechanism within an attached-growth process is complex. The biological requirements of RBC processes are essentially the same as for fixed-media systems. Phenomena that occur when a fixed film is brought into contact with a wastewater containing substrate and oxygen are discussed at the beginning of this chapter. Maximum oxygen-transfer and substrate removal rates are important for RBC performance; however, because of complex transfer–diffusion phenomena, substrate removal from bulk liquid is not expected to follow any simple mathematical model.

Factors Affecting Performance

The primary factors controlling operation and performance of RBC systems are organic and hydraulic loading rates, influent wastewater characteristics, wastewater temperature, biofilm control, dissolved oxygen levels, and operational flexibility.

Organic and Hydraulic Loading

A limiting factor in the design of aerobic RBC systems is organic loading to the first stages, which should be compatible with the oxygen-transfer capability of the system. With excessive organic loadings, the biofilm thickness increases, oxygen becomes limiting, and a white–gray biomass (*Beggiatoa*) frequently forms. Problems caused by overloading are septic odors, process deterioration, structural overloading, and possible equipment failure. Experience has shown that whenever the first-stage loading limit exceeded 0.031 kg $BOD_5/m^2 \cdot d$ (6.4 lb BOD/d/1000 ft^2), sulfur-oxidizing organisms proliferated, causing a heavier-than-normal biofilm thickness, nuisance organism growth, dissolved oxygen depletion, and overall deterioration of process performance.

Influent Wastewater Characteristics

Municipal wastewaters are exceedingly diverse with respect to the number of biodegradable components and the range of particle sizes. Wastewater characteristics and their effects on biodegradability are important considerations in the design of an RBC system. High influent hydrogen sulfide (H_2S) concentrations diminish RBC performance because of the acceleration of nuisance organism growth. When high influent hydrogen sulfide concentrations to aerobic systems are anticipated (e.g., summertime low flows), appropriate provisions in the design (such as preaeration, supplemental aeration, and reduced loadings to the first RBC stage) should be considered. The effect of sidestreams, particularly from solids processing, and their effect on the RBC treatment system also deserve consideration during the design.

Wastewater Temperature

Studies have indicated that organic removal efficiency deteriorates below 13 °C (55 °F) (Antonie, 1976).

Biofilm Control

Biofilm thickness is important to the RBC process; thus, a distinction must be made between total film thickness and active film thickness. Measurements of biofilm thickness in several studies suggested that total film thickness varies from 0.07 to 4.0 mm, depending on hydrodynamic conditions (Atkinson and Fowler, 1974). Substrate removal rates within the thicker biofilm are not likely to be greater than those in thin films because of diffusional resistance within the biofilm. The active portion of the biofilm that contributes to substrate removal has been called the active biofilm thickness.

Dissolved Oxygen Levels

The importance of dissolved oxygen in aerobic wastewater treatment is well-known as inadequate oxygen may be a primary cause of process failure. In the RBC process, the level of available dissolved oxygen, among several operational variables, affects process performance. A minimum dissolved oxygen level of 2 mg/L is commonly accepted as a requirement for proper performance of aerobic systems.

Beggiatoa, whitish autotrophic sulfur bacteria, use hydrogen sulfide and sulfur as energy sources in the presence of oxygen. These sulfide-using organisms compete with heterotrophic organisms for available oxygen and space on the RBC media surface. The growth of *Beggiatoa* often suggests oxygen-limiting conditions. Low dissolved oxygen levels at high loadings promote sulfide production within a thick biofilm, which enhances the growth of sulfide-oxidizing organisms such as *Beggiatoa*. Sulfide present in influent wastewater can also stimulate *Beggiatoa* growth. *Beggiatoa*, a filamentous organism, structurally reinforces biofilm, thus complicating biomass control. This causes excess biomass and weight and possible shaft or media failures.

Process Design

Rotating biological contactors may be used for organic carbon removal, combined organic carbon removal and nitrification, or separate-stage nitrification or denitrification. Numerous established design approaches can achieve the above objectives, including the use of pilot-plant studies, mathematical models, and empirical procedures. Pilot studies are suggested where economic considerations and feasibility warrant such efforts or where atypical wastewater characteristics are anticipated. Whenever possible, use of a suitable mathematical model, combined with an RBC pilot program to obtain values for model coefficients, is recommended. When pilot-plant evaluations are not possible, the designer must use empirical design approaches, exercising technical judgment regarding their applicability and adaptability to specific situations and conditions.

Rotating biological contactor design information can best be obtained by a comprehensive, on-site, full-scale pilot-plant evaluation of the wastewater. The use of full-scale-diameter media for pilot studies is recommended to avoid scale-up problems because studies have indicated higher removal capabilities per unit surface area when less than full-diameter units are used (Murphy and Wilson, 1980; Reh et al., 1977). Using a 0.5-m diameter RBC overestimates the effectiveness of a full size (3.5-m unit) by almost 25% (Wilson et al., 1980).

Although project-specific pilot testing and system modeling is preferred for design, most RBC systems designed for BOD_5 removal have historically been designed using empirical approaches developed by manufacturers of RBC process equipment or others. An empirical design equation was developed as part of a study for the U.S. Environmental Protection Agency (Benjes, 1977) to predict the performance of RBC systems. The equation was based on empirical relationships developed for the performance of trickling filters and packed biological towers. The equation predicts effluent BOD_5 based on influent BOD_5, media volume, and hydraulic loading

$$\frac{S_e}{S_i} = e^{-K(0.000125 V/q)^{0.5}} \tag{7.8}$$

where
S_e = secondary clarifier effluent total BOD_5 (mg/L);
S_i = rotating biological contactor influent total BOD_5 (mg/L);
e = 2.7183;
V = media volume (m^3);
q = hydraulic loading (L/s); and
K = reaction constant (0.30).

The effect of cold-weather operation must also be considered in the design of aerobic RBC systems for BOD removal. Manufacturers of RBC equipment have empirical design relationships to account for reduced performance during low-temperature operation. Benjes (1977) reported a reduction in K rate from a value of 0.3 at approximately 13 °C to a value of 0.2 at approximately 5 °C.

Significant variables influencing nitrification in an RBC system include influent organic and nitrogen concentrations, dissolved oxygen, wastewater temperature, pH and alkalinity, and influent flow and load variability. Nitrification designs for RBCs are often based on pilot-plant studies (Brenner et al., 1984). However, some equipment manufacturers have developed empirical procedures for the design of RBC nitrification systems.

Staged or plug-flow configurations promote development of nitrifying organisms. Therefore, appropriate staging is essential for nitrification to occur in RBC systems. The growth of nitrifiers in any stage depends primarily on the soluble organic concentration present in that stage's wastewater. Heterotrophic bacteria will dominate when the organic concentration is high. Field data indicate that nitrification is observed when soluble BOD declines to approximately 15 mg/L, and maximum nitrification occurs when soluble BOD_5 drops to 10 mg/L or lower (Brenner et al., 1984).

When ammonium-nitrogen exceeds 5 mg/L, removal is claimed to proceed at a zero-order rate of approximately 0.0015 kg NH_4-N/m²·d (0.3 lb NH_4-N/d/1000 ft²) at 13 °C (55 °F). Because of the scatter of data from full-scale plants, Brenner et al. (1984) recommends a conservative approach to nitrification design.

Denitrification

Rotating biological contactor systems for denitrification have not been widely used, and designs should be based on careful evaluation of pilot-plant data and results for representative full-scale systems. Designs incorporating a denitrifying RBC process downstream of organic carbon removal and nitrification RBC processes have been proposed as well as the use of anoxic RBC stages at the beginning of organic carbon removal processes, which incorporate nitrate recycle from the nitrifying stages of the overall RBC configuration. Murphy et al. (1977), based on pilot-plant study results, reported a zero-order nitrate removal rate of 0.0041 kg NO_3-N/m²·d (0.85 lb/d/1000 ft²) at 13 °C.

Combined Biological Treatment

Most combined processes use a fixed-film reactor in series with a suspended-growth reactor. In combined processes, the fixed-film reactor consists of a biological tower, and the suspended-growth reactor is a biological reactor or small contact channel.

Other types of combined processes have been used where RBCs, lagoons, or other treatment processes are in series with biological towers. Combined processes have been described by various names such as two-stage, series, coupled, or dual processes. However, in this text, the term *combined process* is used to denote the in-series coupling of two different reactors of which at least one is a fixed-film reactor.

Combined processes have been widely used, especially where designers have attempted to compensate for weaknesses of systems by combining several processes. For example, fixed-film processes resist shock loads, are energy efficient, and require little maintenance. By combining a shock-resistant fixed-film reactor with a suspended-growth process known for producing high-quality effluent and capable of operating under various treatment modes, designers have found the combination can often yield an overall biological treatment method that highlights the advantages of the two processes.

Trickling Filter–Solids Contact

The trickling filter–solids contact process includes a fixed-film reactor that has low to moderate organic loads followed by a small contact channel. The contact channel is 10 to 15% the size that would typically be required in an aeration tank for activated sludge alone. By combining the fixed-film reactor with the contact channel, the fixed-film reactor size is reduced by 10 to 30% of that typically required if treatment had been accomplished with a trickling filter (Krumsick et al., 1984).

Roughing Filter–Activated Sludge

A common method of upgrading existing activated-sludge plants is to install a roughing filter ahead of the activated-sludge process. As part of the roughing filter–activated sludge process, the roughing filter is 15 to 30% of the size required if treatment had been accomplished through the use of the trickling filter process alone. Hydraulic retention time in the biological reactor is 35 to 50% that required with the use of the activated-sludge process alone

Biofilter–Activated Sludge

The biofilter–activated sludge process is similar to that of the roughing filter–activated sludge process except that, with the biofilter system, the return activated sludge is recycled over the fixed-film reactor. Incorporating the return activated sludge recycle over the fixed-film reactor has sometimes reduced bulking from filamentous bacteria, especially with food-processing wastes, which are difficult to treat. There is no evidence that sludge recycle improves the oxygen-transfer capability of the biological filter (Harrison et al., 1984).

Trickling Filter–Activated Sludge

The trickling filter–activated sludge process is designed for high organic loads. A unique feature of this process is that an intermediate clarifier is provided between the fixed-film and suspended-growth reactors. The intermediate clarifier removes sloughed solids from the fixed-film reactor before the underflow enters the suspended-growth reactor. A benefit of using this mode of combined process is that solids generated from BOD removal can be separated from second-stage treatment. This is often a preferred mode where ammonia removal is required and the second stage of the process is designed to be dominated by nitrifying microorganisms. Another advantage in using intermediate clarification is reduced effects from sloughing of the fixed film on the suspended-growth portion of the plant. However, designers generally do not believe there is evidence of significant reduced oxygen demand or improved solids settleability from use of intermediate clarification.

Developing Fixed-Film Processes

There are perhaps hundreds of developing ideas on how to use fixed-film treatment more effectively. Described below are two examples to give a sense of the thinking by process-design engineers.

Subsided Fixed-Bed Reactors

The biological aerated filter is similar to the design of high-rate sand filters except that air is continually sparged into the lower regions of the filter and a relatively coarse medium is used. The wastewater flows either down (Rogalla et al., 1990) or up (Pujol et al., 1992) and the granular medium retains influent suspended solids and provides a surface area for biofilm development. The media size in the reaction zone is tailored to the specific application.

Floating Bed Aerated Filters

These processes use a floating bed of media to provide the biological surface area and simultaneously act as a filtration system. Owing to the media size, the clean surface area is high. When coarse-bubble aeration diffusers are used, the media encourage excellent contact of air, water, and biomass.

Secondary Clarification

Selection of proper design criteria will result in removal of nearly all of the settleable suspended solids in the effluent from fixed-film reactors. Trickling filter plants on average produce an effluent with suspended solids of 10 mg/L. The use of sand filtration reduced the effluent concentrations of suspended solids to 5 mg/L. Secondary clarifiers for trickling filter plants have low solids loading so the overflow rate governs design. Secondary clarifiers for trickling filter plants should have average overflow rates of approximately 0.09 m/h, and should not exceed 0.28 m/h. If the tanks are very deep (>3 m), experience shows that these overflow rates may be doubled (IAWQ, 1992). For plants employing both fixed-film and suspended-growth processes, the clarifier design criteria must be the same as for activated-sludge plants.

Conclusions

The Roman goddess of the hearth, Vesta, was worshiped in a temple that contained a perpetual fire tended by six virgins. The Vestal Virgins were highly respected and took on demigoddess status within the community. Mortals were not allowed to even touch them. Occasionally, one of the Vestal Virgins fell from grace and had a liaison. When this violation was discovered the offending Virgin was led to a deep underground cell, given some bread and water, and left to die. This punishment seemed to be a compromise between two strict rules: do not touch the Vestal Virgins and the Vestal Virgins must remain chaste or be put to death. By leaving her in the underground cell to die she was not touched by anyone and yet she paid for her transgression.

The lore of the Vestal Virgins illustrates that there are two ways to kill anything, be it a human or any other living thing: by harming it directly, or killing it indirectly by removing the sustaining environment. The moral responsibility of environmental engineers is to sustain the environment for present and future generations. Unfortunately, the engineering codes of ethics are inadequate in addressing the problem of environmental quality. Engineers must in all situations consider the protection of our environment for future generations, in addition to the promise to promote the health, safety, and welfare of the present public. Designing and constructing wastewater treatment plants that function so that the watercourses are not permanently damaged is a laudable and worthy goal, and those who undertake this task should be conscious of their responsibilities. They should take heart that they are certainly part of the solution, and not a part of the problem.

References

Albertson, O. E. (1989) Slow-Motion Trickling Filters Gain Momentum. *Oper. Forum,* **6** (8), 28.

Albertson, O. E. (1995) Excess Biofilm Control by Distributor Speed Modulation. *J. Environ. Eng.,* **121** (4), 330.

Albertson, O. E.; Davies, G. (1984) Analysis of Process Factors Controlling Performance of Plastic Bio-Media. Paper presented at 57th Annual Water Pollution Control Federation Conference, New Orleans, Louisiana, October.

Albertson, O. E.; Eckenfelder, W. W. (1984) Analysis of Process Factors Affecting Plastic Media Trickling Filter Performance. *Proceedings of the Second International Conference on Fixed-Film Biological Processes,* Washington, D.C.

Antonie, R. L. (1976) *Fixed Biological Surfaces—Wastewater Treatment: The Rotating Biological Contactor;* CRC Press: Cleveland, Ohio.

Atkinson, B.; Fowler, H. W. (1974) The Significance of Microbial Film in Fermenters. *Adv. Biochem. Eng.,* **3,** 221

Benjes, H. H., Sr. (1977) *Small Community Wastewater Treatment Facilities—Biological Treatment Systems.* U.S. Environmental Protection Agency, Technology Transfer National Seminar, Chicago, Illinois.

Bowker, R. P. G.; Stensel, H. D. (1987) *Design Manual for Phosphorus Removal;* EPA-625/1-87-001; U.S. Environmental Protection Agency: Washington, D.C.

Brenner, R. C.; et al. (1984) *Design Information on Rotating Biological Contactors;* EPA-600/2-84-106; U.S. Environmental Protection Agency: Cincinnati, Ohio.

Dow Chemical Company (1964) Surfpac Process Evaluations Report; Report No. AA-1925; Midland, Michigan.

Eckenfelder, W. W. (1963) Trickling Filter Design and Performance. *Trans. Am. Soc. Civ. Eng.,* **128** (Part III), 371.

Galler, W. S.; Gotaas, H. G. (1964) Analysis of Biological Filter Variables. *J. Sanit. Eng. Div., Proc. Am. Soc. Civ. Eng.,* **90** (6), 59.

Harrison, J. R.; Daigger, G. T.; Filbert, J. W. (1984) A Survey of Combined Trickling Filter and Activated Sludge Processes. *J. Water Pollut. Control Fed.,* **56,** 1073.

Howland, W. E. (1953) Effect of Temperature on Sewage Treatment Processes. *Sew. Ind. Wastes,* **25,** 161.

International Association on Water Quality (1992) *Design and Retrofit of Wastewater Treatment Plants for Biological Nutrient Removal;* Randall, C. W., Barnard, J. L., Stensel, H. D., Eds.; Technomic Publishing Company: Lancaster, Pennsylvania.

Kincannon, D. F.; Stover, E. L. (1982) Design Methodology for Fixed Film Reactors, RBCs and Trickling Filters. *Civ. Eng. Practicing Design Eng.,* **2,** 107.

Krumsick, T. A.; et al. (1984) Trickling Filter Solids Contact Process Demonstration, Salt Lake City, Utah. Paper presented at the Annual Conference of the Utah Water Pollution Control Association, Salt Lake City.

Logan, B. E.; Hermanowicz, S. W.; Parker, D. S. (1987a) A Fundamental Model for Trickling Filter Process Design. *J. Water Pollut. Control Fed.,* **59,** 1029.

Logan, B. E.; Hermanowicz, S. W.; Parker, D. S. (1987b) Engineering Implications of a New Trickling Filter Model. *J. Water Pollut. Control Fed.,* **59,** 1017.

Logan, B. E.; Parker, D. S.; Arnold, R. G. (1990) O_2 Limitations in CH_4 and NH_4 Utilizing Biofilms. *Proceedings of the American Society of Civil Engineers Conference on Environmental Engineering,* Washington, D.C., April 8–11; pp 31–38.

Murphy, K. L.; et al. (1977) Nitrogen Control: Design Considerations for Supported Growth Systems. *J. Water Pollut. Control Fed.,* **49,** 549.

Murphy, K. L.; Wilson, R. W. (1980) Pilot Plant Studies of Rotating Biological Contactors Treating Municipal Wastewater; Report SCAT-2; Environment Canada: Ottawa, Ontario.

National Research Council (1946) Sewage Treatment at Military Installations. *Sew. Works J.,* **18,** 787.

Okey, R. W.; Albertson, O. E. (1989) Evidence for Oxygen-Limiting Conditions During Tertiary Fixed-Film Nitrification. *J. Water Pollut. Control Fed.,* **61,** 510.

Onda, K.; Takeuchi, H.; Okumoto, Y. (1968) Mass Transfer Coefficients between Gas and Liquid Phases in Packed Columns. *J. Chem. Eng. Jpn.,* **1** (1), 56.

Parker, D. S.; Lutz, M. P.; Pratt, A. M. (1990) New Trickling Filter Applications in the USA. *Water Sci. Technol.,* **22,** 215.

Pujol, R.; Canler, J. D.; Iwema, A. (1992) Biological Aerated Filters: An Attractive and Alternative Biological Process. *Water Sci. Technol.* **26** (3), 373.

Reh, C. W.; et al. (1977) An Approach to Design of RBCs for Treatment of Municipal Wastewater. Paper presented at the American Society of Civil Engineers National Environmental Engineering Conference, Nashville, Tennessee.

Rogalla, F.; Payraudeau, M.; Bacquet, G.; Bourbigot, M.-M.; Sibony, J.; Gilles, P. (1990) Nitrification and Phosphorus Precipitation with Biological Aerated Filters. *J. Water Pollut. Control Fed.*, **62**, 169.

Velz, C. J. (1948) A Basic Law for the Performance of Biological Filters. *Sew. Works J.*, **20**, 607.

Wilson, R. W.; et al. (1980) Scale-Up in Rotating Biological Contactor Design. *J. Water Pollut. Control Fed.*, **52**, 610.

Symbols Used in this Chapter

A_S = specific surface area (m²/m³)
A = total media surface area, 1000 ft²
C = Chezy coefficient = 3 to 10, (Reynolds Number <1000, $C = 2 \cdot N_R^{0.5}$)
D = depth, m (ft)
E_t = filter efficiency at temperature T
E_{20} = filter efficiency at 20 °C
E_1 = BOD removal efficiency through the first-stage filter and settling tank (%)
E_2 = BOD removal efficiency through the second-stage filter and settling tank (%)
F = number of passes of the organic material, equal to $(1 + R/Q)/[1 + (1 - P)R/Q]^2$
H_n = hydraulic number
K = first-order rate constant (d⁻¹)
k_T = reaction rate coefficient
K = reaction constant
K_b = proportionality constant of A_s, g/m²·d (lb/d/1000 ft²)
L_O = influent BOD (mg/L)
L_D = BOD removed at depth D (mg/L)
P = a treatability factor that, for military trickling filter plants, was found to be approximately 0.9
Q = influent flow, m³/d (mgd)
q = hydraulic loading (L/s)
R = recycle flow (m³/d)
S_e = secondary clarifier effluent total BOD₅ (mg/L)
S_i = rotating biological contactor influent total BOD₅ (mg/L)
S_e = effluent soluble BOD₅, kg/d (or mg/L)
S_o = influent soluble BOD₅, kg/d (or mg/L)
t = holdup time (minutes)
T = wastewater temperature (°C)
V = reactor volume (m³)
V = volume of the particular filter stage (surface area times depth of media), m³ (ac-ft)
W_2 = BOD load to the second-stage filter, not including recycle, kg/d (lb/d)
β = constant
ϕ = media angle from horizontal
μ_{max} = maximum specific substrate utilization rate of A_s, g BOD₅/m²·d (lb BOD₅/d/1000 ft²)

Biological Nutrient Removal

Introduction

Many wastewater treatment plants face increasingly stringent effluent standards, often because of the sensitivity of surface and groundwater to degradation by nitrogen and phosphorus. These limits can often be met with integrated processes for nutrient control. An integrated biological process for nutrient removal is a wastewater treatment method that combines biological and chemical or physical unit operations and processes to reduce concentrations of nitrogen and phosphorus in plant effluent below levels that would be attainable solely by a typical secondary treatment facility.

Activated-sludge systems form the heart of most integrated systems for nutrient removal; however, these biological systems and the less numerous attached-growth systems have inherent limitations stemming from many variables, including raw wastewater characteristics, methods of solids handling, and population dynamics of mixed cultures within a given treatment facility. Because of these limitations, the biological system often needs either a physical process, such as final filtration to remove additional phosphorus- and nitrogen-bearing solids, or chemical addition to precipitate additional phosphorus or provide an external energy source for denitrification. In general, integrated systems for nutrient control require supplemental chemical addition or additional physical unit processes to produce effluent concentrations of less than 2 mg/L total nitrogen and less than 1 mg/L total phosphorus.

In their biological forms, integrated processes simply serve as an extension of the activated-sludge process. Integrated processes are more complicated than conventional activated-sludge processes because the former involves sequences or series of stages, phases, or zones with anaerobic, anoxic, and aerobic conditions in mixed liquors, often with selected recycling of mixed liquor between stages. The exact sequencing and sizing of stages, zones, or phases and the location and magnitude of internal recycle streams depends on the design effluent total nitrogen and phosphorus concentrations with the known behavior of wastewater microorganisms.

Similar to other biological treatment systems, a mixed culture of bacteria and other microorganisms accomplishes the actual treatment in integrated processes. Specific types of bacteria indigenous to municipal wastewater can function differently, depending on the type of substrate and ultimate electron acceptors available in the mixed liquor. As these bacteria grow and reproduce within the confines of the treatment system, they rely on carbonaceous or nitrogenous compounds in raw wastewater for their energy. Organisms obtain their energy by cascading electrons through complex biochemical cycles from the raw substrate to an ultimate electron acceptor. The designer or operator of biological waste treatment systems can use the principles of microbial biochemistry to tailor the microbial environment and, thereby, encourage the microorganism behavior necessary to remove nitrogen and phosphorus biologically.

Many flow schemes have been proposed and developed for the integration of carbon oxidation, nitrification, denitrification, and enhanced biological phosphorus removal. The selection of a particular process flow sheet will depend on the specific application, especially effluent criteria and influent characteristics. A fair comparison of alternative processes to meet specific effluent criteria must use equivalent design assumptions for influent wastewater flowrate and strength, effluent criteria versus process abilities, reliability, solids production, aesthetics, and capital and annual operating costs.

For example, consider a comparison between process A that can meet a 0.5-mg/L phosphorus limit and process B that can meet a 2.0-mg/L phosphorus limit. An equivalent comparison must include sufficient chemical costs for process B to also meet an 0.5-mg/L limit. Likewise, if process B were to create additional solids and higher phosphorus concentrations in plant recycle streams, process B costs must include incremental costs for the treatment and disposal of additional solids and sidestream treatment of the plant recycle streams.

A significant consideration often overlooked is the degree of conservatism, or safety factor, to be applied in the sizing of each process. A fair comparison must apply an equivalent safety factor for each alternative process. Ultimately, the selection of any particular process is a largely subjective decision that must weigh present worth costs versus process reliability and performance potential.

Phosphorus Removal Processes

Basic Theory
Conventional secondary biological treatment systems take up phosphorus from solution for biomass synthesis during biochemical oxygen demand (BOD) oxidation.
Phosphorus, required in intracellular energy transfer, becomes an essential cell component. For this reason, phosphorus is taken up in an amount related to the stoichiometric

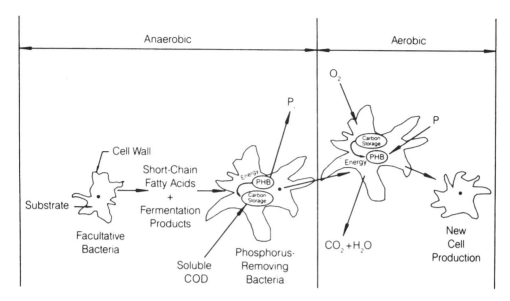

Figure 8.1 Removal mechanisms for excess biological phosphorus (COD = chemical oxygen demand, PHB = poly-β-hydroxybutyrate) (Bowker and Stensel, 1987).

requirement for biosynthesis. A typical phosphorus content of microbial solids is 1.5 to 2% on a dry-weight basis.

A sequence of an anaerobic zone followed by an aerobic zone results in the selection of a population rich in organisms capable of taking up phosphorus at levels beyond stoichiometric requirements for growth. With this environment, the biomass accumulates phosphorus to levels of 4 to 12% of microbial solids. Wastage of these solids results in approximately 2.5 to 4 times more phosphorus removal from the system than that from conventional treatment. The organism associated with enhanced biological phosphorus removal was thought at first to be the genus *Acinetobacter* (Lotter et al., 1986), but this appears not to be the case. Research is continuing on determining what microorganisms are actually responsible for this process.

Rational design and operation require an understanding of the mechanism by which enhanced biological phosphorus uptake occurs. The currently accepted mechanism of enhanced biological phosphorus removal is summarized in Figure 8.1. Acetate and other short-chain fatty acids (fermentation products), produced by fermentation reactions in the anaerobic zone, are taken up and stored intracellularly, most commonly as poly-β-hydroxybutyrate (PHB). In performing the anaerobic uptake of soluble organics and forming intracellular storage products, microorganisms must expend energy. They obtain this energy anaerobically through the cleavage of high-energy phosphate bonds in stored long-chain inorganic polyphosphates. This process produces orthophosphate that is released from the cell into solution. Thus, removal of soluble BOD with concomitant release of phosphorus occurs in the anaerobic zone.

In the aerobic zone (stage), a rapid uptake of soluble orthophosphate provides for the resynthesis of the intracellular polyphosphates. Accompanying this uptake, previously stored PHB is aerobically oxidized to carbon dioxide, water, and new cells. The aerobic metabolism of residual soluble BOD will also occur in this zone. Figure 8.2 shows typical profiles of phosphorus and BOD concentration through anaerobic and

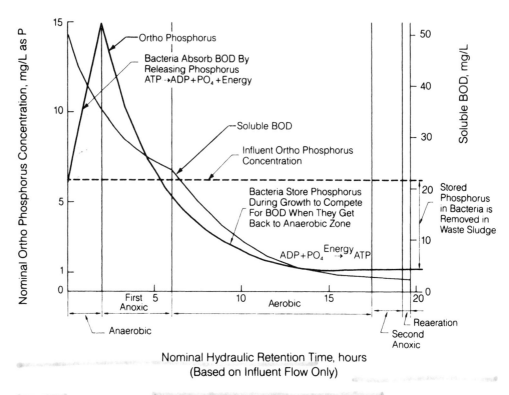

Figure 8.2 Typical profile of soluble phosphorus concentrations in a biological nutrient removal process (ATP = adenosine triphosphate, ADP = adenosine diphosphate).

aerobic zones. In this figure, ATP is adenosine triphosphate, the chemical used to store phosphorus in the microbes. The ATP is reduced to adenosine diphosphate, ADP, as phosphates are released along with energy. In the aerobic zone, the phosphorus is then again stored as ATP and this requires energy.

Unlike many other biologically mediated reactions of significance in wastewater treatment, the stoichiometry and kinetics of phosphorus release and uptake are not fully understood. Thus, the design engineer must rely on empirical observations to obtain information for process design and modifications.

The rate and extent of phosphate release in the anaerobic zone are related to the type and quantity of soluble substrate available for uptake and storage as PHB. In this regard, lower molecular weight fatty acids are preferred substrates. Research has indicated that approximately 1 mg phosphorus/L will be released for every 2 mg/L acetate as chemical oxygen demand (COD) is removed anaerobically (Ekama et al., 1984). The actual rate of uptake of readily biodegradable COD (expressed here as BOD) and the rate of release of phosphorus in the anaerobic zone with municipal wastewater are first-order reactions with respect to the BOD (Wentzel et al., 1988). This implies that the division of the anaerobic zone into two to four compartments will enhance biological release and subsequent uptake of phosphorus (Grady et al., 1998; Levenspiel, 1972). The behavior described above depends on the volatile acid fraction of BOD and is controlled by the rate of conversion of BOD to volatile fatty acids. Acetate addition to the anaerobic zone increases the rate of release of phosphorus and it

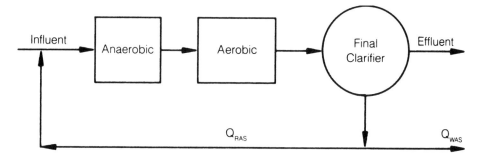

Figure 8.3 The A/O process (RAS = return activated sludge, WAS = waste activated sludge).

becomes a zero-order reaction with respect to acetates (Wentzel et al., 1988). Phosphorus removal does not appear to be greatly affected at lower temperatures (Ekama et al., 1983; Sell et al., 1981).

The aerobic zone performance for enhanced biological phosphorus removal depends on the amount of phosphorus release achieved and the amount of organic matter present for growth. If anaerobic detention time is sufficient for complete excess phosphorus release and a favorable incoming ratio of organic matter to phosphorus exists, rapid soluble phosphorus uptake can be expected in the aerobic zone. Phosphorus removal of approximately 2.0 to 2.5 mg PO_4-P/100 mg influent COD (3 to 4 mg/100 mg/L 5-day BOD [BOD_5]) is typical.

Design Options
Various options for implementing the theory of biological phosphorus removal are available, some of which are proprietary.

Two-Stage Process
This process has two stages—an anaerobic stage preceding an aerobic stage—for biological phosphorus removal (Figure 8.3) (Irvine et al., 1982). The two-stage process, developed primarily for carbon oxidation and phosphorus removal, is specifically designed to optimize biological phosphorus removal. The anaerobic and aerobic stages are each divided into a number of equally sized, completely mixed compartments. The clarifier underflow returns to the first compartment of the anaerobic zone.

This process will likely function most effectively when nitrification does not occur because the process lacks any provision for the removal of nitrates returned to the anaerobic zone with return sludge. The process can be adapted for nitrification by allowing the necessary detention time in the aerobic stage. A significant degree of denitrification is not attainable because the process lacks an anoxic zone for nitrate reduction. Typical effluent total phosphorus concentrations from such processes are about 1 mg/L. The two-stage process was originally marketed as the A/O process.

Combined Chemical and Biological Process
This process combines biological and chemical phosphorus removal (Figure 8.4). It diverts phosphorus-rich biomass in a sidestream of the return activated sludge (RAS) to an anaerobic stripper where phosphorus is released to solution. The phosphorus-rich stripper supernatant is then precipitated with lime while the biomass, stripped of phosphorus, returns to the aeration tank.

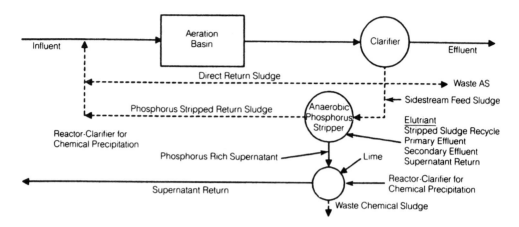

Figure 8.4 PhoStrip process.

This process removes phosphorus from the wastewater by two methods. Phosphorus is chemically precipitated from the stripper supernatant, and it is removed from the process with the waste sludge. Similar to other biological phosphorus removal processes, exposure of microorganisms to a sequence of anaerobic and aerobic conditions enhances the concentration of phosphorus in waste sludge higher than normal levels. The phosphorus removal performance of this sidestream process has been enhanced by the addition of a source of readily biodegradable organics, such as raw wastewater or primary effluent, to the stripper tank. It has been reported that the stripper underflow will contain a high concentration of soluble BOD. This process, originally marketed as the PhoStrip process, has consistently achieved levels of effluent total phosphorus below 0.5 mg/L without filtration; however, filtration is a good practice when designing for such low concentrations.

Sidestream Fermentation Processes

The anaerobic tank provides for two basic functions: the detention time required for phosphorus release and the formation of low molecular weight volatile fatty acids, principally acetic and propionic. Both of these functions need not occur in a single tank. Separate unit processes, used successfully for the sidestream production of volatile fatty acids, include primary sedimentation tanks, gravity thickeners for primary sludge fermentation, or "activated" primaries.

Originally marketed as the OWASA Nutrification process and developed at the Orange Water and Sewer Authority plant in Carrboro, North Carolina, it evolved through modifications and operational changes at an existing biological treatment facility using primary clarifiers and roughing filters followed by six-stage, jet-aerated activated sludge. Figure 8.5 shows the process schematic. This process uses anaerobic fermentation of primary sludge, after which the supernatant, rich in volatile fatty acids, enters an anaerobic contact tank to be mixed with the return activated sludge. In this plant, trickling filter effluent is introduced to the aeration tank. As its advantage, this process is independent of the BOD/phosphorus ratio and fluctuations in wastewater strength. The effluent phosphorus concentration from this process is typically between 0.6 and 1.2 mg/L.

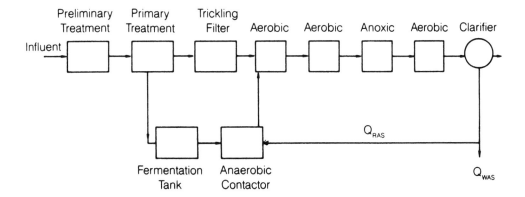

Figure 8.5 OWASA process (RAS = return activated sludge, WAS = waste activated sludge).

The sequence of anoxic and anaerobic zones, in addition to achieving biological phosphorus removal, is instrumental in the nitrification and denitrification process, as discussed later in this chapter.

Sequencing Batch Reactors

Biological phosphorus removal can be achieved in sequencing batch reactors by creating a sequence of anaerobic conditions followed by aerobic conditions if readily biodegradable organics are present during the anaerobic phase (Manning and Irvine, 1985) (Figure 8.6). Biological phosphorus removal will occur in sequencing batch reactors when the operating cycle begins with an anoxic period to eliminate nitrates followed by an anaerobic period to induce phosphorus release. Effluent phosphorus concentrations of less than 1 mg/L total phosphorus, without supplemental chemical addition, have been achieved with sequencing batch reactors.

Chemical Polishing

Enhanced biological phosphorus removal processes have inherent limitations as to minimum effluent concentrations of phosphorus attainable. Providing standby chemical storage and feeding equipment for these processes is considered good practice because such equipment will ensure more reliable phosphorus removal at a capital cost of only a small fraction of the overall facility cost. Biological phosphorus removal can be chemically supplemented by adding a source of readily biodegradable organics, such as acetate or other low molecular weight carboxylic acids, to the anaerobic tank or by adding a metal salt of iron, aluminum, or calcium. In most situations, the most reliable and economical option will be use of metal salts. Using biological phosphorus removal and sand filters, effluent levels of 0.5 mg/L of phosphorus can be met with confidence. At the present time, lower levels can only be attained with chemical addition.

Nitrogen Removal Processes

Basic Theory

From an overall viewpoint, biological denitrification is a two-step process that requires nitrification in an aerobic environment followed by denitrification in an anoxic environ-

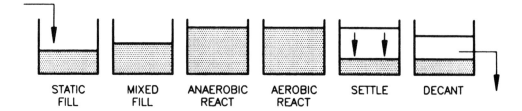

Figure 8.6 Sequencing batch reactor for removal of phosphorus.

ment. As with all biological activity, these reactions are affected by the specific environmental conditions in the reactor, including pH, wastewater temperature, dissolved oxygen concentration, substrate type and concentration, and the presence or absence of any toxic substances. The design engineer is primarily interested in the stoichiometric equations and the kinetic or rate expressions for microbial growth. The stoichiometry informs the designer of reactions that will occur and to what extent; the kinetic expression describes how fast the reaction will occur (Characklis and Gujer, 1979). Using this information, the designer can determine the size and type of reactor needed, environmental conditions to be maintained in the reactor, and quantities of external reactants such as oxygen or a carbon source such as methanol that must be supplied.

Nitrification is the sequential oxidation of ammonium-nitrogen to nitrite-nitrogen and then to nitrate-nitrogen. Nitrifying organisms are obligate aerobic lithotrophic organisms relying on an inorganic substrate such as ammonia for an energy source, but using oxygen as the electron receptor. Common organisms that perform nitrification in aeration plants include *Nitrobacter* and *Nitrospira*.

Biological denitrification reduces nitrate-nitrogen to nitrogen gas (N_2). Denitrifying organisms are primarily facultative heterotrophs that oxidize organics anaerobically by reducing nitrite or nitrate in the absence of molecular oxygen or other oxygen sources. Nitrifying organisms are obligate aerobic organisms relying on an inorganic substrate such as ammonia for an energy source, but using oxygen as the electron acceptor. A relatively broad range of heterotrophic bacteria and some autotrophic bacteria can denitrify, among them *Nitrosomonas*.

As the heterotrophic mechanism predominates for wastewater treatment, this discussion is restricted to pathways using organic compounds as an energy source. Both assimilatory and dissimilatory enzyme systems are involved in nitrate reduction. Assimilatory nitrate reduction converts nitrate-nitrogen to ammonium-nitrogen, which can subsequently be used for biosynthesis. This reaction occurs only when a more reduced nitrogen form is unavailable. Dissimilatory nitrate reduction transforms nitrate-nitrogen to the sparingly soluble nitrogen gas, which may then be liberated from solution. Dissimilatory nitrate reduction can result in a decrease of total system nitrogen rather than just a transformation in state. The dissimilatory pathway is of primary importance in wastewater denitrification; therefore, further discussion focuses on this reaction.

Biological mechanisms and stoichiometry for biological denitrification are relatively well established. Basic reactions for the reduction of nitrate to nitrogen gas are

$$NO_3^- \rightarrow NO_2^- \rightarrow NO \rightarrow N_2O \rightarrow N_2 \qquad (8.1)$$

The overall reduction is

$$NO_3^- + \frac{5}{6} CH_3OH \rightarrow \frac{1}{2} N_2 + \frac{5}{6} CO_2 + \frac{7}{6} N_2O + OH^- \qquad (8.2)$$

The reaction including synthesis CH_3OH, carbon source, and nitrate-nitrogen source is

$$NO_3^- + 1.08\ CH_3OH + 0.24\ H_2CO_3 \rightarrow 0.056\ C_5H_7O_2N + 1.68\ H_2O + HCO_3^- \qquad (8.3)$$

If the ammonia nitrogen source is included

$$NO_3^- + 2.5\ CH_3OH + 0.5\ NH_4 + 0.5\ H_2CO_3 \rightarrow 0.5\ C_5H_7O_2N \\ + 0.5\ N_2 + 4.5\ H_2O + 0.5\ HCO_3^- \qquad (8.4)$$

The overall reaction, including synthesis and municipal wastewater as the carbon source and ammonia-nitrogen source, is then

$$NO_3^- + 0.315\ C_{10}H_{19}O_3N + H^+ + 0.267\ HCO_3^- + 0.267\ NH_4^+ \rightarrow \\ 0.655\ CO_2 + 0.5\ H_2 + 0.612\ C_5H_7O_2N + 2.30\ H_2O \qquad (8.5)$$

The conclusions can be summarized as follows:

- Nitrate is converted to nitrogen gas;
- Oxygen recovery is 2.86 mg O_2/mg NO_3-N reduced;
- Alkalinity recovery is 3.6 mg $CaCO_3$/mg NO_3-N; and
- Heterotrophic biomass production is approximately 0.4 mg volatile suspended solids (VSS)/mg COD removed.

Nitrogen removal requires not only the oxidation of ammonia-nitrogen to nitrate-nitrogen, but denitrification (the conversion of nitrate-nitrogen to nitrogen gas). This denitrification step is somewhat more difficult to achieve than nitrification only because the former requires the presence of both a degradable carbon source and nitrate. Denitrification can be achieved in three general ways

(1) Supplying an exogenous carbon source such as methanol or acetate to the denitrification zone or reactor;
(2) Using carbonaceous BOD in the wastewater as a degradable carbon source by either (a) recycling a large amount of nitrified effluent back to an anoxic reactor at the head of the flow scheme or (b) diverting a portion of the raw influent or primary effluent flow to a zone containing nitrate; or
(3) Using endogenous carbon present in cell mass as the degradable carbon source.

The important variables that significantly affect biological denitrification kinetics include

- Carbon substrate type and concentration,
- Dissolved oxygen concentration,
- Alkalinity and pH, and
- Temperature.

Table 8.1 Monod kinetic coefficients (Baillod, 1988; Henze et al., 1986).

Coefficient	Symbol	Typical range	Suggested
Maximum specific growth rate of heterotrophs	$\mu_{max,H}$	3–13	4.0–6.0
Heterotrophic biomass yield	Y_H	0.46–0.69	0.67
Half-saturation coefficient organic substrate	K_{COD}	10–180	10–20
Half-saturation coefficient nitrate-nitrite	K_{NO}	0.06–0.5	0.2–05
Correction factor for μ_H under anoxic conditions	η_g	0.5–1.0	0.8
Half-saturation coefficient for dissolved oxygen for heterotrophic biomass	$K_{O,H}$	0.10–0.28	0.1–0.2
Mass nitrogen per mass of COD in biomass	$C_{X,B}$	0.06–0.12	0.06–0.086
Decay coefficient for heterotrophic biomass	b_H		0.05

The most critical variable is the type and concentration of carbonaceous substrate available in the mixed liquor. Two primary substrate conditions, identified by Grau (1982) for suspended-growth denitrification, are denitrification under non-carbon-limiting conditions and denitrification under carbon-limiting conditions. The first condition occurs in anoxic tanks (first anoxic or preanoxic tanks) and the second occurs in postaeration anoxic tanks (second anoxic tanks).

The performance of nitrification can be simulated using Monod growth kinetics. One such equation (based on 20 °C temperature) was developed by Grau (1982)

$$r_{V,NO} = \left(\frac{1-Y_H}{2.86 Y_H}\right) \mu_{max,H} \left(\frac{S_{BOD}}{K_{COD} + S_{COD}}\right) \left(\frac{S_{NO}}{K_{NO} + S_{NO}}\right) \eta_g X_{b,h} \qquad (8.6)$$

where

$r_{V,NO}$ = reaction rate per unit volume nitrate- and nitrite-nitrogen,
Y_H = biomass yield coefficient,
$\mu_{max,H}$ = maximum specific growth rate of heterotrophs,
S_{COD} = soluble material concentration organic substrate,
K_{COD} = half-saturation coefficient organic substrate,
S_{NO} = soluble material concentration nitrate- and nitrite-nitrogen,
K_{NO} = half-saturation coefficient nitrate-nitrite,
η_g = correction factor for μ_H under anoxic conditions, and
$X_{b,h}$ = particulate material concentrations.

Table 8.1 summarizes typical values for the Monod kinetic coefficients.

Another approach in modeling denitrification is to use a zero-order expression

$$r_{V,NO} = kX \qquad (8.7)$$

where k is the reaction-rate coefficient. If external substrate is available, then

$$r_{X,NO} = 0.03 \, (F/M) + 0.029 \qquad (8.8)$$

where M = total mass of the mixed liquor suspended solids (kg) and F = applied BOD_5 (kg/d).

If no external substrate is available, then

$$r_{X,NO} = 0.12\, \theta_c^{-0.706} \tag{8.9}$$

where θ_c is the solids retention time. Alternatively, the reaction can be expressed as (Batchelor, 1982)

$$r_{X,NO} = \frac{\eta_g}{2.86} \frac{A_n}{Y_n} \frac{1}{\theta_c} \tag{8.10}$$

where A_n = net or observed oxygen use coefficient and Y_n = net or observed biomass yield coefficient.

The correction factor (η_g) applied to the specific growth rate of heterotrophs was proposed to account for observed reductions in the growth of heterotrophs under anoxic conditions (Batchelor, 1982). This reduction occurs either because a part of the heterotrophic biomass cannot use nitrate as an electron acceptor or because microorganisms cannot grow as rapidly with nitrate as with oxygen.

Based on research studies (Parkins et al., 1978; Stern and Marais, 1974; Wilson and Marais, 1976), it appears that denitrification with the influent wastewater BOD as the carbon source exhibits two distinct kinetic regions. The first region, an initial fast primary phase of short duration (1 to 10 minutes), is followed by a slower secondary phase of approximately 15% the rate of the primary phase. This second region precedes a final phase with even lower rates resulting from exhaustion of external carbon sources. Data obtained in these experiments also showed that the reduction of nitrate in the primary phase was approximately proportional to the concentration of BOD_5 in the influent (S_{bsi}). The total nitrate reduction equation for an anoxic reactor that precedes the aeration tank can be expressed as follows:

$$\Delta S_N = \alpha S_{bsi} + K_2 X_1 \bar{t}\, (1.08)^{T-20} \tag{8.11}$$

where
ΔS_N = reduction in the nominal or system nitrate concentration (mg NO_3-N/L)
   = $(R_{ir} + R_{RAS} + 1)\Delta S_{Na}$, where R_{ir} = internal recycle ratio, R_{RAS} = return activated sludge recycle ratio, ΔS_{Na} = reduction in the actual concentration of nitrate (mg NO_3-N/L);
α = empirical first-phase denitrification constant (mg N/mg BOD_5);
K_2 = denitrification rate constant for secondary phase of denitrification (t^1);
X = biomass concentration (mg/L);
$\bar{t}$ = nominal hydraulic retention time (V/Q);
V = volume;
Q = liquid volumetric flowrate;
S_{bsi} = influent BOD_5 (mg/L); and
T = temperature (°C).

For raw or settled wastewater, the mean value of α is approximately 0.028 mg N/mg BOD_5 and is independent of the temperature investigated.

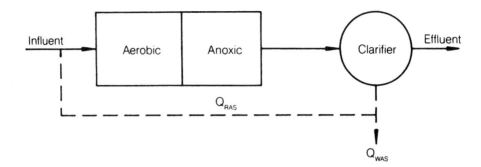

Figure 8.7 Wuhrmann process for nitrogen removal (RAS = return activated sludge, WAS = waste activated sludge).

The overall system nitrogen removal for a reactor configuration with anoxic zones both preceding and following the aeration tank may thus be written as

$$\Delta S_N = \alpha S_{bsi} + K_2 X_1 \bar{t} (1.08)^{T-20} + K_3 X \bar{t} (1.03)^{T-20} \tag{8.12}$$

where K_3 is the reaction-rate coefficient for the third phase. This equation is valid only if the nitrate is not reduced to zero in either anoxic zone.

The optimal pH for denitrification ranges from 6.5 to 8.5. The pH effects on specific growth rate can be estimated as

$$r_{X,NO} = r_{X,NO,max} \left(1/[1 + 10^{5.5 - pH} + 10^{pH - 9}]\right) \tag{8.13}$$

Denitrification rates are strongly influenced by temperature; therefore, temperature correction factors must be selected carefully. The most commonly used expression is

$$\ln \frac{r_{x,1}}{r_{x,2}} = \beta^{(T_1 - T_2)} \tag{8.14}$$

where
- β = is the empirical temperature coefficient, 1.06 to 1.15;
- $r_{x,1}$ = denitrification reaction rate at temperature T_1; and
- $r_{x,2}$ = denitrification reaction rate at temperature T_2.

Suspended-Growth Processes for Nitrogen Removal

The various suspended-growth processes for nitrogen removal can be grouped into three categories—single sludge, dual sludge, and triple sludge—each described below.

Single-Sludge Processes

Wuhrmann (1954) proposed the single-sludge configuration for nitrogen removal shown in Figure 8.7. The *Wuhrmann approach* is sometimes called *postdenitrification*.

Without the addition of an exogenous electron donor, the design relies on residual organic matter passing through the first stage or on the endogenous respiration of biomass to provide the energy sources for denitrification. If complete nitrification (thus,

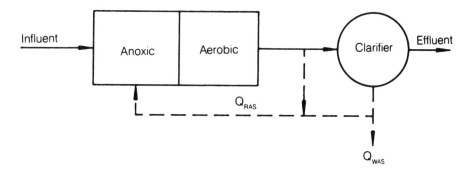

Figure 8.8 Ludzack–Ettinger process for nitrogen removal (WAS = waste activated sludge).

complete carbon oxidation) is achieved, endogenous respiration would provide the principal energy source. Nitrogen removals of 29 to 89% have been achieved in bench- and pilot-scale studies (Christensen and Harremöes, 1972).

The *Ludzack–Ettinger design* shown in Figure 8.8 reverses the sequence of anoxic and aerobic stages in the Wuhrmann design (Ludzack and Ettinger, 1962). The advantage of this design is the provision of influent BOD to the anoxic stage as an exogenous electron donor. Total nitrogen removal efficiency in this process is a function of return activated sludge flowrate. An 88% reduction in total nitrogen from an influent of 130 mg/L has been reported (Sutton and Bridle, 1980).

A modification of the Ludzack–Ettinger design was proposed by Barnard (1973). This process, shown in Figure 8.9, incorporates an internal recycle (Q_{ir}) of mixed liquor from the aeration stage to the anoxic stage. This modification increased both the denitrification rate and overall nitrogen removal efficiency. The modified Ludzack–Ettinger process provides for control over the fraction of nitrate removed through variation of the internal recycle ratio. In addition, higher denitrification rates are attained because the anoxic reactor receives a source of BOD. This allows smaller

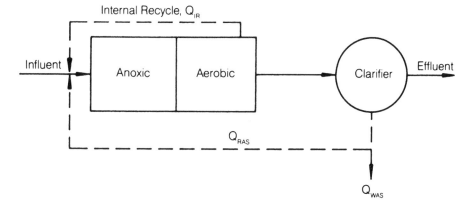

Figure 8.9 Modified Ludzack–Ettinger process for nitrogen removal (WAS = waste activated sludge).

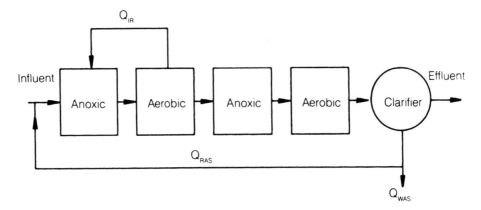

Figure 8.10 Four-stage Bardenpho process for nitrogen removal (WAS = waste activated sludge).

anoxic volumes for a given nitrate removal requirement compared with the Wuhrmann and Ludzack–Ettinger processes. This process can be used when only nitrification is required. Because the process effluent will contain 5 to 7 mg/L or more of nitrate-nitrogen, it is often used when partial denitrification is required before effluent disposal to groundwater via high-rate application, as in rapid infiltration tanks.

The process using dual anoxic and aerobic zones is known as the *Bardenpho process*. In this process, the mixed liquor is recycled from the first aerobic zone to the first anoxic zone at a rate as high as four to six times the influent flowrate. This process (see configuration shown in Figure 8.10) is intended to achieve more complete nitrogen removal than is possible with a two- or three-stage process. Complete denitrification cannot be attained with preaeration anoxic zones because part of the aerobic-stage effluent is not recycled through the anoxic zone. The second anoxic zone provides for additional denitrification using nitrate produced in the aerobic stage as the electron acceptor and endogenous organic carbon as the electron donor.

The second (postaeration) anoxic zone is capable of almost completely removing the nitrate in the aeration tank effluent. The final aeration stage strips residual gaseous nitrogen from solution and minimizes phosphorus release in the final clarifier by increasing the oxygen concentration. The Bardenpho process can achieve an effluent concentration of total nitrogen as low as 2 to 4 mg/L (Ekama et al., 1984).

Biological nitrogen removal can also be accomplished in *sequencing batch reactors* by creating, in one reactor, the proper combinations of aerobic and anoxic conditions in time sequence. Control strategies for biological nutrient removal take into account reaction time, tank water level, and mixed liquor dissolved oxygen concentrations. Sequencing batch reactors appear well suited for relatively small systems with highly variable wastewater flow and strength.

Oxidation ditch processes are readily adaptable for carbon oxidation, nitrification, and denitrification as illustrated in Figure 8.11. Horizontal rotors, slow-speed mechanical aerators or rotating disks, or draft-tube aerators provide aeration at one or more locations in the ditch. The dissolved oxygen concentration will be highest at the points of aeration and will subsequently decrease because of oxygen uptake by the biomass as the mixed liquor moves around the looped reactor. After sufficient travel time, anoxic zones will form upstream from aeration devices. The location and

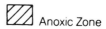
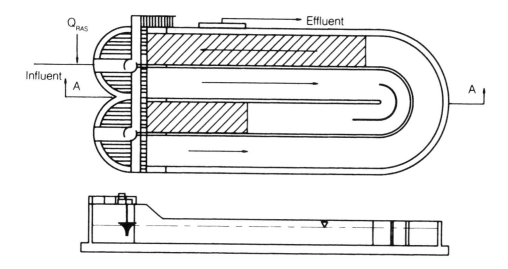

Figure 8.11 Oxidation ditch process for nitrogen removal.

size of these anoxic zones will vary with time because oxygen uptake and transfer rates will vary with wastewater quality and flow. Therefore, reliance on this mechanism for denitrification requires a comprehensive control system to monitor dissolved oxygen throughout the tank and control the amount of oxygen transferred to the tank. Also, the energy input for mixing and aeration must be carefully controlled to maintain the mixed liquor in suspension. This system must afford sufficient operational flexibility with adjustable weirs or two-speed aerators for varying the oxygen input to match diurnal and seasonal changes in oxygen demand. Otherwise, during periods of low loading, necessary anoxic zones will not develop. Nitrogen removals greater than 90% have been reported with oxidation ditch processes (Rittmann and Langeland, 1985).

Dual-Sludge Processes

Multiple-sludge systems, by definition, house the various process stages in physically separate tanks, each with their own clarifier and return sludge systems. Three dual-sludge configurations have been proposed and are shown in Figure 8.12. In Figure 8.12a, the aerobic system first performs carbon oxidation and nitrification. Then, the nitrate-laden stream is supplemented with an external carbon source before contacting the denitrifying biomass in the anoxic system.

The system in Figure 8.12b uses the same configuration, except that feeding a portion of influent wastewater to the second stage supplies the organic carbon to the anoxic system. Although this system eliminates the need for a carbon supplement, some ammonia and organic nitrogen (total Kjeldahl nitrogen, or TKN) will pass through because ammonia present in the feed to the anoxic system will not be oxidized.

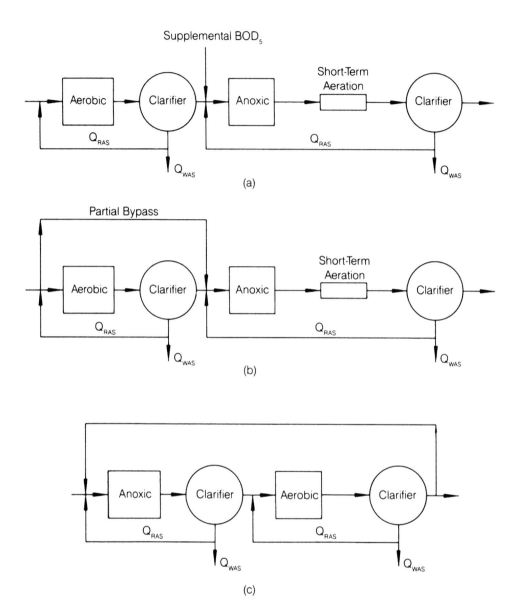

Figure 8.12 Dual-sludge process for nitrogen removal (WAS = waste activated sludge) (Grady et al., 1998).

The third configuration, shown in Figure 8.12c, also eliminates the need for supplemental carbon. In this configuration, the anoxic system precedes the aerobic system, thus providing sufficient BOD for denitrification. An additional recycle stream supplies nitrate to the anoxic system. This flow scheme offers the possibility of reducing the aeration requirement because a substantial portion of the BOD can be oxidized in the anoxic system. Although some oxidized nitrogen will be discharged, its magnitude will be related to the recycle flow used.

Attached-Growth Processes for Nitrogen Removal

Denitrification Filters

Granular media denitrification (or deep-bed filter denitrification) has been used successfully in the United States since the early 1970s. In current practice, denitrification filters follow nitrification and serve the dual functions of filtration and denitrification. Denitrification filters rely on heterotrophic microorganisms that grow on and, to a certain degree, between the coarse media in deep-bed filters. Any carbon source suitable for denitrification with any other denitrification process is also suitable for denitrification filters. Methanol, at a dosage of 3 mg/L per mg/L of NO_3-N, is used. Raw wastewater or primary effluent are unsuitable carbon sources because of their high ammonium and suspended solids concentrations, but these sources could be used if the resulting ammonium levels and shorter filter runs were acceptable.

The carbon source can be controlled by using automatic analyzers (Chen, 1980), but manual sampling and adjustment often is adequate and more reliable (Pickard et al., 1985). As an interesting aspect of denitrification filters, if carbon (e.g., methanol) is overdosed, this will not necessarily result in a proportional increase in effluent BOD_5 if the wastewater contains sufficient sulfate to serve as an electron acceptor. With sufficient sulfate, however, hydrogen sulfide odors may be present (English et al., 1974).

Operation of denitrification filters is similar to that for any filter except for necessary nitrogen-release cycles (referred to as bumps). This need results from nitrogen gas accumulating in the denitrification filter and causing an increase in head loss across the filter. This head loss is relieved by periodically backwashing the filter (with water) for approximately 1 to 5 minutes to release this accumulated gas. Most or all of this backwash is not captured; that is, this temporary reversal of flow is not used to clean the filters. Denitrification filters require bumps 4 to 5 times per day (Chen, 1980), but they can be required as often as 14 to 16 times per day (Pickard et al., 1985).

Process design of denitrification filters must consider the kinetics of denitrification and the practical limitation on the number of times per day a filter can be bumped. Kinetics dictate an approximately 20-minute empty bed detention time for warm waters (higher than 20 °C) and up to approximately 60 minutes in cold (approximately 10 °C) waters. This is achieved by using 1.8-m (6-ft) deep beds loaded at annual average design conditions of 1.7 to 5 m/h (0.7 to 2 gpm/ft^2).

Fluidized Bed Denitrification

Upflow fluidized bed reactors use another attached-growth process for denitrification. In contrast to the typical downflow denitrification filter, the wastewater feed to an upflow reactor enters the bottom of the reactor with sufficient velocity to expand the media bed. The biomass in these reactors grows on the reactor media. The media can be plastic shapes such as used in air pollution scrubbers. Soluble carbon is added to provide substrate for denitrification. As a principal advantage, upflow reactors can achieve biomass concentrations in the range of 25,000 to 30,000 mg/L (Shieh, 1981). This high concentration allows smaller reactor volumes for equal loadings compared with suspended-growth processes.

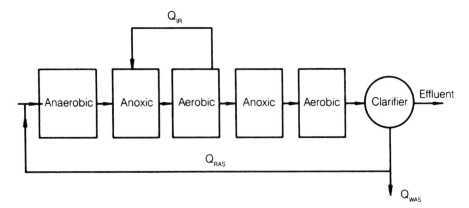

Figure 8.13 Modified Bardenpho process for phosphorus and nitrogen removal (WAS = waste activated sludge).

Simultaneous Phosphorus and Nitrogen Removal Processes

The similarities between the process flow diagrams for phosphorus removal and for nitrogen removal would suggest that it ought to be possible to achieve the removal of both nutrients simultaneously. Below are some processes that have been applied to achieve this end.

Modified Bardenpho

The modified Bardenpho process (Figure 8.13), sometimes referred to as the Phoredox process in South African literature, provides anaerobic, anoxic, and aerobic stages for removal of phosphorus, nitrogen, and carbon. The difference between the Bardenpho and modified Bardenpho process is an anaerobic stage at the beginning of the five-stage modified process for biological phosphorus removal. This process will reduce phosphorus to less than 1 mg/L and total nitrogen to less than 10 mg/L. The nitrogen removal, as would be expected, is strongly a function of temperature.

A^2/O Process

Figure 8.14 shows the flow schematic of a typical three-stage A^2/O process. Each stage is divided into equally sized, completely mixed compartments. Mixed liquor is recycled from the end of the nitrification (aerobic) stage to the anoxic stage for denitrification at an internal recycle rate ranging from 100 to 300% of the influent flow. Clarifier underflow returns to the first stage of the anaerobic reactor. The performance of this process is similar to that of the modified Bardenpho.

University of Cape Town Process

The University of Cape Town (UCT) process is shown in Figure 8.15. In the UCT process, both the return activated sludge and aeration tank contents are recycled to the anoxic zone, and the contents of the anoxic zone are then recycled to the anaerobic zone. This recycle sequence decreases the chance of introducing residual nitrate to the anaerobic zone. The internal recycle can be controlled to maintain zero nitrates in effluent from the anoxic reactor, thereby ensuring that no nitrates will be returned to the anaerobic reactor.

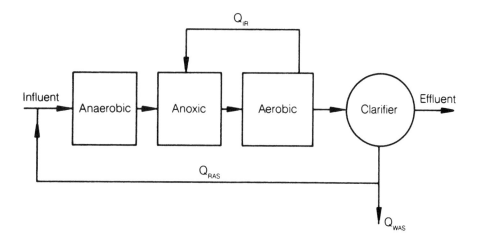

Figure 8.14 A²/O process for phosphorus removal (WAS = waste activated sludge).

PhoStrip II Process

This process can achieve combined phosphorus and nitrogen removal to levels less than 1.0 mg/L of total phosphorus and 10 mg/L of total nitrogen by including an anoxic zone in the process. The PhoStrip II process is shown in Figure 8.16. Nitrogen removal will require additional reaction volume to provide for nitrification and denitrification.

Denitrification in a PhoStrip II process is accomplished by adding a prestripper tank ahead of the phosphorus stripper, increasing the detention time in the stripper, and providing series reactors for phosphorus release. High concentrations of nitrates in return sludge will require increased anaerobic retention time, thus a larger stripper. The prestripper tank accepts underflow from the secondary clarifier containing the nitrate produced in nitrification. The stripper underflow has been found to have high concentrations of soluble BOD and provides the carbon source for denitrification. The prestripper tank hydraulic detention time approximates 2 hours. Up to 70% denitrification has been observed in operating systems (Kang et al., 1988; Matsch and Drnevich, 1987).

Fixed-Growth Reactor–Suspended-Growth Reactor

The fixed-growth reactor–suspended-growth reactor (FGR–SGR) process combines one or more trickling filters with activated-sludge basins for biological removal of

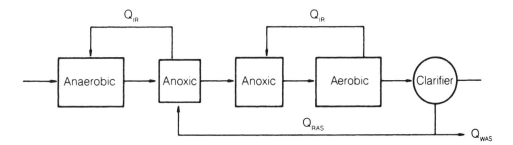

Figure 8.15 Modified University of Cape Town process for phosphorus and nitrogen removal (WAS = waste activated sludge).

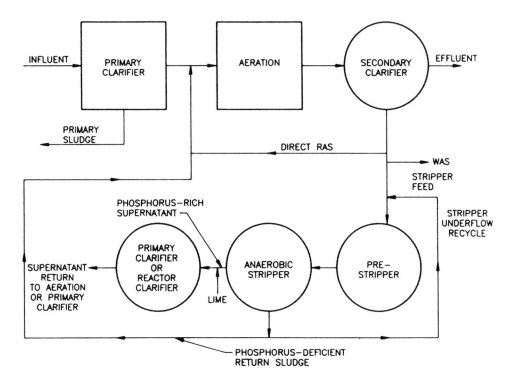

Figure 8.16 PhoStrip II process for phosphorus and nitrogen removal (WAS = waste activated sludge).

phosphorus and nitrification–denitrification. The fixed-growth component is included to enhance process stability, improve the settling qualities of the biological solids, and reduce aerated hydraulic retention times (Gibb et al., 1998; Kelly et al., 1995). A schematic of the FGR–SGR process configured for excess biological phosphorus removal and nitrification–denitrification is shown in Figure 8.17.

The anaerobic–anoxic sequence, which is modeled on the UCT activated-sludge process, is designed to prevent nitrates in the return biological solids from reaching the anaerobic tank. Settled biological solids from the final clarifier are returned to the anoxic tank, where they meet the mixed liquor leaving the anaerobic tank. Denitrified mixed liquor from the anoxic tank is returned to the anaerobic tank, where it meets the primary clarifier overflow. Biological uptake and storage of phosphorus, nitrification, and removal of residual BOD are achieved by irrigating the mixed liquor from the anoxic tank over one or more trickling filters. Oxygen for phosphorus uptake, nitrification and BOD removal by suspended- or fixed-growth organisms is provided by cascade aeration. A time flushing cycle is used to prevent clogging of the trickling filter media. Mixed liquor leaving the trickling filters flows to the solids contact tank, where flocculation of sloughed trickling filters solids and removal of residual phosphorus, ammonia, and BOD by suspended bacteria occurs. Mixed liquor from the solids contact tank flows to the final clarifier, where the clarified effluent is discharged, and settled biological solids are returned to the anoxic tank.

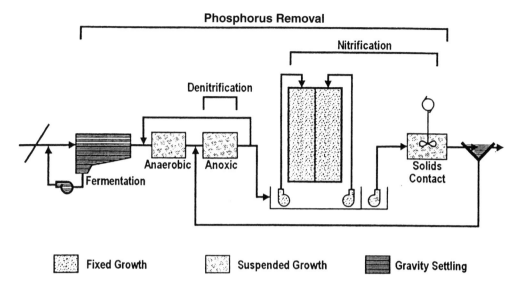

Figure 8.17 Schematic of the fixed-growth reactor–suspended-growth reactor (FGR–SGR) process.

Phased Isolation Ditches

Oxidation ditch technology can readily be modified to remove biological phosphorus by incorporating anaerobic and anoxic stages and zones to the process. Such phase-isolation ditch processes create sequential aerobic, anoxic, and anaerobic conditions within one tank or a series of tanks.

Design Considerations

Design Procedures for Phosphorus Removal

Unlike nitrogen, phosphorus does not exist in a gaseous form under the conditions found in an activated-sludge process; thus, any phosphorus entering the plant must leave the plant either in the effluent or waste solids. The design of processes for phosphorus removal is based on the concept of converting soluble orthophosphorus to a solid form and then removing this solid phosphorus in either a chemical or biological form with the waste solids. Regardless of the design procedure used, the key to successful phosphorus removal is the presence of sufficient BOD in the anaerobic zone.

Despite advances in understanding of the basic biochemistry of biological excess phosphorus removal, current practice for sizing anaerobic tanks depends on empirical methods. Hydraulic retention times range from 0.5 to 2 hours, depending on the expected difficulty in establishing anaerobic conditions.

The anaerobic zone detention time should be increased as required to ensure that the biomass receives adequate anaerobic conditioning. Sizing the anaerobic zone is based on selecting an appropriate fraction of the activated-sludge mixed liquor suspended solids (MLSS) to be contained at any time in the anaerobic zone. This method is founded on the hypothesis that the ultimate level of phosphorus uptake depends on readily biodegradable organics in the anaerobic reactor, the fractional mass of solids passing through the reactor, and the actual retention time in the reactor (Marais et al., 1983). A practical approach will provide an anaerobic mass fraction of at least 0.10 that can be increased to 0.20 provided that the total unaerated (anaerobic plus anoxic) mass fraction does not exceed 0.60. With a large anaerobic biomass, however, a phenomenon called *secondary release* occurs, releasing phosphorus into solution.

South African researchers have developed a semiempirical, parametric model (Ekama et al., 1983) and a theoretical, mechanistic model (Wentzel et al., 1988, 1989a, 1989b, 1990) for phosphorus removal. The parametric model, which was developed from extensive testing, predicts phosphorus removal using the following equation (Martin and Marais, 1975):

$$P_s = S_{ti} \{[(1 - f_{us} - f_{up})Y_h/(1 + b_{hT}\theta_c)](\gamma + f_p f b_{hT} \theta_c) + f_p (f_{up}/f_{cv})\} \qquad (8.15)$$

where
- P_s = phosphorus removal by incorporation in the solids (mg/L);
- S_{ti} = total influent COD (mg/L);
- f_{us} = unbiodegradable, soluble COD fraction in the influent (mg/mg COD [0.05]);
- f_{up} = unbiodegradable particulate COD fraction in the influent (mg/mg COD [0.13]);
- f_{cv} = ratio of COD to volatile suspended solids = 1.48 mg COD/mg VSS;
- b_{hT} = decay coefficient for heterotrophic bacteria (d [0.24]);
- θ_c = solids retention time (d);
- Y_h = yield coefficient for heterotrophic organisms, mg VSS/mg COD (0.45);
- γ = coefficient of excess removal (phosphorus content of volatile solids) mg phosphorus/mg VSS = $0.35 - 0.29 \exp(0.242 P_f)$;
- P_f = excess phosphorus removal propensity factor = $(S_{bsa} - 25)f_{xa}$;
- S_{bsa} = readily biodegradable COD concentration in the anaerobic reactor (mg/L);
- f_{xa} = anaerobic mass fraction of MLSS;
- f_p = phosphorus content of the nonbiodegradable volatile solids, mg phosphorus/mg VSS (0.015); and
- f = unbiodegradable fraction of active mass, mg/mg VSS (0.20).

This parametric model predicts that biological phosphorus removal will improve as the concentration of readily biodegradable COD in the influent increases and that biological phosphorus removal will increase as the fraction of the MLSS contained in the anaerobic zone increases. This equation also predicts that phosphorus removal will increase as the solids retention time decreases. A minimum concentration of 25 mg/L of readily biodegradable COD is needed to induce excess biological phosphorus removal.

Design Procedures for Nitrogen Removal

Integration of anoxic zones to a system for nitrification and denitrification requires the development of a nitrogen mass balance to determine total system nitrogen reduction and the selection of an appropriate method for determining denitrification rates.

The quantity of nitrogen to be removed will be governed by influent nitrogen concentrations, the effluent criterion, and nitrogen mass balances around the entire plant and the specific denitrification unit process. When developing nitrogen balances and evaluating effluent limitations, it is helpful to consider the various nitrogen forms that occur in an activated-sludge process. These forms include ammonium, particulate and soluble organic nitrogen, nitrite, and nitrate. Most of the influent nitrogen is in the form of ammonium and organic nitrogen. Because only a small fraction of the organic nitrogen, perhaps 5 to 7%, is not biodegradable, most of the influent nitrogen is available for nitrification or building new cells via synthesis. The following is a typical overall plant nitrogen balance:

- Influent ammonia-nitrogen = 30.0 mg/L N;
- Plus influent organic nitrogen = 10.0 mg/L N;
- Total influent nitrogen = 40.0 mg/L N;
- Less nitrogen in primary sludge (40 to 70% removal of particulate TKN) = 2.7 mg/L N;
- Less nitrogen in biological sludge (7 to 10% of VSS in waste sludge) = 11.2 mg/L N;
- Less effluent ammonia (design parameter; set by permit) = 0.5 mg/L N;
- Less effluent particulate organic nitrogen (7 to 10% of effluent VSS) = 1.5 mg/L N;
- Less effluent soluble organic nitrogen (0.5 to 1.5 mg/L N) = 0.5 mg/L N; and
- Total nitrogen oxidizable to nitrate-nitrogen (N_{ox}) = 23.6 mg/L N.

Nominal concentrations are based on the mass of nitrogen divided by the plant influent flowrate. The nitrogen concentration of plant influent is diluted by the return sludge and internal recycle but, thereafter, the concentration remains unchanged as the flow enters the aeration tank, with the exception of any uptake for synthesis in the anoxic tank. For example, the actual nitrogen concentration in the nitrification zone, assuming complete nitrate reduction, can be calculated as follows:

$$NO_3\text{-}N, \text{mg/L} = N_{ox}/[1 + (Q_{ir}/Q) + (Q_{RAS}/Q)] \qquad (8.16)$$

where

Q = influent flowrate;
Q_{ir} = internal recycle flowrate from the aeration tank to the first anoxic tank;
Q_{RAS} = return activated sludge flowrate; and
N_{ox} = nominal concentration of nitrate, based on organic and ammonia nitrogen in the influent oxidized to nitrate, mg/L.

Table 8.2 is a summary of the design parameters for biological nutrient (phosphorus plus nitrogen) removal.

Many researchers and practicing engineers have adopted the use of the empirical Monod equation for mathematically describing or modeling overall biological reactions occurring in integrated systems for nutrient control. This equation appears to predict the biological oxidation of organics and ammonium and the biological reduction of nitrate, all of which have been reported as both zero order and first order, depending on the substrate concentration. While single Monod-type equations reasonably approximate solids retention times and reactor volumes for a single complete-mix suspended-growth reactor system, such equations alone are not sufficient to predict the performance of multiple-stage processes, particularly more complex biological nutrient removal systems with alternating anoxic and aerobic zones and internal recycle streams. Such

Table 8.2 Design parameters for biological nutrient removal processes.

Parameter[a]	Orbal Bionutre	A/O	A²/O	Modified Bardenpho	Bardenpho	UCT	PhoStrip	SBR	Dual-sludge methanol	Triple-sludge
F:M, kg TBOD/kg MLVSS·d	0.1–0.5	0.2–0.7	0.15–0.25	0.1–0.2	0.1–0.2	0.1–0.2	0.1–0.5	0.1	0.1–0.2	1–2
SRT, days	5–40	2–5	4–27	10–40	10–40	10–30	0.3–0.8		6–12	10–30
									5–20	5–20
MLSS, mg/L	2000–6000	2000–6000	2000–4000	2000–5000	2000–5000	2000–5000	2000–4000	600–5000	2000–3000	2500–3000
HRT, hours										
Anaerobic	—	0.5–1.5	0.5–1.5	1–2	—	1–2	8–12	0.0–3.0[b]	—	—
Anoxic	6–12	—	0.5–1.0	2–4	2–5	2–4	—	0.0–1.6	—	—
Aerobic	6–12	1.0–3.0	3.5–6.0	4–12	4–12	4–12	4–10	0.5–1.0	6	1.5–3.0
Anoxic	—	—	—	2–4	2–5	2–4	—	0.0–0.3	2	2.5
Aerobic	—	—	—	0.5–1.0	0.5–1.0	—	—	0.0–0.3	—	—
Settle/decant	—	—	—	—	—	—	—	1.5–2.0	—	4.0
Total	12–24	1.5–4.5	4.5–8.5	9.5–23.0	8.5–23.0	9–22.0	12–22.0	4.0–9.0	8	8.0–9.5
RAS, percent of influent	50–100	25–40	20–50	80–100	80–100	80–100	20–50	—	—	40–50
Internal recycle, percent of influent	100–400	—	100–300	400–600	400–600	100–600	—	—	—	—
Anaerobic recycle flow, percent of influent						50–100	—			
Stripper feed, percent of influent							20–30			
Elutriation flow, percent of influent							13–30			
Stripper underflow, percent of influent							10–20			

[a]TBOD = total biochemical oxygen demand; MLVSS = mixed liquor volatile suspended solids; MLSS = mixed liquor suspended solids; HRT = hydraulic retention time; and RAS = return activated sludge.
[b]Batch cycle times.

multiple-stage processes are better modeled by somewhat more complex systems of equations that more adequately account for the many process variables involved. Henze et al. (1986) prepared the basis for a more complex mathematical model to simulate the basic processes occurring in biological activated-sludge systems for carbon oxidation, nitrification, and denitrification. This model can be applied for design (Baillod, 1988; Bidstrup and Grady, 1987, 1988; Dold and Marais, 1986; Dold et al., 1980; Henze et al., 1986; and University of Cape Town, 1988).

Conclusions

One of the most interesting watercourses in the world is the Oslo fjord in Norway. The fjord, carved out by a glacier, is a large body of water, up to 300 m deep, that narrows down to a neck barely 10 m deep and about 100 m wide. In Viking days this was the perfect harbor, making surprise attacks difficult. Even during the Second World War the Norwegians inflicted damage on German ships attacking through Oslo harbor.

After the war, people began to notice a distinct deterioration of the water quality in the fjord and a project was begun to find out what was going on. After some scientific sleuthing, workers found that the water in the fjord was essentially stagnant, with almost no flushing from the Skagarak, the sea between Norway and Denmark, and that nutrients from human and agricultural wastes were building up to levels that were causing accelerated eutrophication. Because the water in the fjord was flushed only every 5 to 10 years there was plenty of time for the algae to grow.

At that time most Oslo homes were not connected to adequate wastewater treatment plants and the government set to correct this situation. In the late 1960s they decided to build a large secondary treatment plant on the western side of the fjord. But this plant as planned did not have nutrient removal. The engineers at the Norwegian Institute for Water Research knew full well that this was the wrong kind of plant. What they needed was a plant to remove nutrients, not carbon. When the chief of the institute was questioned by a young and brash American engineer as to why they were building a plant that would not solve the problem, the answer was, "This may not solve all of the problems, but we are doing something good."

Engineers did not know at that time how to design a plant that cost-effectively reduces phosphorus and converts organic and ammonia nitrogen to nitrate, or better yet, to nitrogen gas. Here was an immediate and clear problem, and the engineers did not have the science and technology to respond to it. So they did what they could, knowing it was good.

A great deal has happened in the last 30 or so years in improving our abilities to remove nitrogen and phosphorus from wastewater, and the research conducted by the many people who are listed in the references to this chapter contributed to this knowledge. This is a true success story for environmental engineering. Now even Hagar the Horrible would be pleased with the water quality in the Oslo fjord.

References

Baillod, C. R. (1988) *Oxygen Utilization in Activated Sludge Plants: Simulation and Model Calibration*; EPA-600/2-86-065; U.S. Environmental Protection Agency: Washington, D.C.

Barnard, J. L. (1973) Biological Denitrification. *J. Int. Water Pollut. Control Fed.*, **72**, 6.

Batchelor, B. (1982) Kinetic Analysis of Alternative Configurations for Single-Sludge Nitrification/Denitrification. *J. Water Pollut. Control Fed.*, **54**, 1493.

Bidstrup, S. M.; Grady, C. P. L., Jr. (1987) *A User's Manual for SSSP, Simulation of Single-Sludge Processes for Carbon Oxidation, Nitrification and Denitrification*; Clemson University, South Carolina.

Bidstrup, S. M.; Grady, C. P. L., Jr. (1988) SSSPC Simulation of Single Sludge Processes. *J. Water Pollut. Control Fed.*, **60**, 351.

Bowker, R. P. G.; Stensel, H. D. (1987) *Design Manual for Phosphorus Removal*; EPA-625/1-87-001; U.S. Environmental Protection Agency: Washington, D.C.

Characklis, W. G.; Gujer, W. (1979) Temperature Dependency of Microbial Reactors. *Prog. Water Technol.*, **Supplement 1**, 111.

Chen, J. J. (1980) Plant Scale Operation of Biological Denitrification System. Paper presented at the American Society of Civil Engineers Convention and Exposition, Florida.

Christensen, M. H.; Harremöes, P. (1972) *Biological Denitrification in Water Treatment: A Literature Study*; Report No. 2-72; Department of Sanitation Engineering: Technical University of Denmark, Lyngby.

Dold, P. L.; Marais, G. v. R. (1986) Evaluation of the General Activated Sludge Model Proposed by the IAWPRC Task Group. *Water Sci. Technol.*, **18** (6), 63.

Dold, P. L.; et al. (1980) A General Model for the Activated Sludge Process. *Prog. Water Technol.*, **12**, 44.

Ekama, G. A.; et al. (1983) Considerations in the Process Design of Nutrient Removal Activated Sludge Processes. *Water Sci. Technol.*, **15**, 283.

Ekama, G. A.; et al. (1984) *Theory, Design, and Operation of Nutrient Removal Activated Sludge Processes*; Water Research Committee: Pretoria, South Africa.

English, J. N.; et al. (1974) Denitrification in Granular Carbon and Sand Columns. *J. Water Pollut. Control Fed.*, **46**, 28.

Gibb, A. J.; Kelly, H. G.; Berzins, A.; Frese, H.; Robinson, L.; Koch, F. A. (1998) The Importance of Operational Controls for Biological Nutrient Removal in a Trickling Filter Activated Sludge Process. In *19th Biennial Conference Proceedings*; International Association on Water Quality; June 21–26; Vancouver, Canada; pp 1–8.

Grady, C. P. L., Jr.; Daigger, G.; Lim, H. C. (1998) *Biological Wastewater Treatment Theory and Applications*; Marcel Dekker: New York.

Grau, P. (1982) Recommended Notation for Use in the Description of Biological Wastewater Treatment Processes. *Water Res.*, **16**, 1501.

Henze, M.; et al. (1986) IAWPRC Task Group on Mathematical Modeling for Design and Operation of Biological Wastewater Treatment; Activated Sludge Model No. 1. J.W.; Arrowsmith Ltd.: Great Britain.

Hong, S.; et al. (1981) A Biological Wastewater Treatment System for Nutrient Removal. Presented at the 54th Annual Conference of the Water Pollution Control Federation, Detroit, Michigan.

Irvine, R. L.; et al. (1982) Summary Report Workshop on Biological Phosphorus Removal in Municipal Wastewater Treatment. Annapolis, Maryland.

Kang, S. J.; et al. (1988) Biological Nitrification and Denitrification in the PhoStrip Process. Presented at the 61st Annual Conference of the Water Pollution Control Federation, Dallas, Texas.

Kelly, H. G.; Berzins, A.; Gibb, A. J. (1995) Advanced Biological Wastewater Treatment Using Trickling Filters and Thermophilic Sludge Digestion. *Proceedings of the Water Environment Federation Conference on New and Emerging Technologies and Products*, Toronto, Ontario, Canada, June; pp 3-49.

Levenspiel, O. (1972) *Chemical Reaction Engineering*, 2nd ed.; Wiley & Sons: New York.

Lotter, L.T.; et al. (1986) A Study of Selected Characteristics of *Acinetobacter* spp. Isolated from Activated Sludge in Anaerobic/Anoxic/Aerobic and Aerobic Systems. *Water SA*, **12**, 203.

Ludzack, F. J.; Ettinger, M. B. (1962) Controlling Operation to Minimize Activated Sludge Effluent Nitrogen. *J. Water Pollut. Control Fed.,* **34**, 920.

Manning, J. F.; Irvine, R. L. (1985) The Biological Removal of Phosphorus in Sequencing Batch Reactors. *J. Water Pollut. Control Fed.,* **57**, 87.

Marais, G. v. R.; Lowenthal, R. E.; Siebritz, I. P. (1983) Observations Supporting Phosphate Removal by Biological Excess Uptake—A Review. *Water Sci. Technol.,* **15**, 15.

Martin, L. A. C.; Marais, G. v. R. (1975) *Kinetics of Enhanced Phosphorus Removal in the Activated Sludge Process.* Research Report W14; Department of Civil Engineering: University of Cape Town, South Africa.

Matsch, L. C.; Drnevich, R. F. (1987) Biological Nutrient Removal. In *Advances in Water and Wastewater Treatment;* Ann Arbor Science: Chelsea, Michigan.

Parkins, A. R.; et al. (1978) Kinetics of Denitrification in the Activated Sludge Process. *Prog. Water Technol.,* **10**, 225.

Pickard, D. W.; et al. (1985) Six Years of Successful Nitrogen Removal at Tampa, Florida. Presented at the 58th Annual Conference of the Water Pollution Control Federation, Kansas City, Missouri.

Rittmann, B. E.; Langeland, W. E. (1985) Simultaneous Denitrification with Nitrification in Single-Channel Oxidation Ditches. *J. Water Pollut. Control Fed.,* **57**, 300.

Sell, R. L.; et al. (1981) Low-Temperature Biological Phosphorus Removal. Presented at the 54th Annual Conference of the Water Pollution Control Federation, Detroit, Michigan.

Shieh, W. K. (1981) Predicting Reactor Biomass Concentration in a Fluidized-Bed System. *J. Water Pollut. Control Fed.,* **53**, 1574.

Stern, L. B.; Marais, G. v. R. (1974) *Sewage as the Electron Donor in Biological Denitrification.* Research Report W7; Department of Civil Engineering: University of Cape Town, South Africa.

Sutton, P. M.; Bridle, T. R. (1980) Biological Nitrogen Control of Industrial Wastewater. WATER-1980; AIChE Symposium Series, 177.

University of Cape Town and Ninham Shand, Inc. (1988) NutremCA Program for the Design of Nutrient Removal Activated Sludge Processes Under Steady-State Conditions; Rev. 3.0, Water Research Committee.

Wentzel, M.C.; et al. (1988) Enhanced Polyphosphate Organism Cultures in Activated Sludge Systems—Part 1: Enhanced Culture Development. *Water SA,* **14**, 81.

Wentzel, M. C.; et al. (1989a) Enhanced Polyphosphate Organism Cultures in Activated Sludge Systems—Part II: Experimental Behavior. *Water SA,* **15**, 71.

Wentzel, M. C.; et al. (1989b) Enhanced Polyphosphate Cultures in Activated Sludge Systems—Part III: Kinetic Model. *Water SA,* **15**, 89.

Wentzel, M. C.; et al. (1990) Biological Excess Phosphorus Removal—Steady-State Process Design. *Water SA,* **16**, 29.

Wilson, D. E.; Marais, G. v. R. (1976) *Adsorption Phase in Biological Denitrification.* Research Report W11; Department of Civil Engineering: University of Cape Town, South Africa.

Wuhrmann, K. (1954) High Rate Activation Sludge Treatment and its Relation to Stream Sanitation. *Sew. Ind. Wastes,* **26**, 1.

Symbols Used in the Chapter

b_{hT} = decay coefficient for heterotrophic bacteria (d)
f_{us} = unbiodegradable, soluble COD fraction in the influent (mg/mg COD)
f_{up} = unbiodegradable particulate COD fraction in the influent (mg/mg COD)
f_{cv} = ratio of COD to volatile suspended solids
f_{xa} = anaerobic mass fraction of MLSS

f_p = phosphorus content of the nonbiodegradable volatile solids (mg phosphorus/mg VSS)
f = unbiodegradable fraction of active mass (mg/mg VSS)
K_{BOD} = half-saturation coefficient organic substrate
K_{NO} = half-saturation coefficient nitrate-nitrite
N_{ox} = nominal concentration of nitrate, based on TKN in the influent oxidized to nitrate
P_f = excess phosphorus removal propensity factor = $(S_{bsa} - 25)f_{xa}$
P_s = phosphorus removal by incorporation in the solids (mg/L)
Q = influent flowrate
Q_{ir} = internal recycle flowrate from the aeration tank to the first anoxic tank
Q_{RAS} = RAS flowrate
$r_{V, NO}$ = reaction rate per unit volume nitrate and nitrite-nitrogen
S_{bsa} = readily biodegradable COD concentration in the anaerobic reactor (mg/L)
S_{ti} = total influent COD (mg/L)
S_{NO} = soluble material concentration nitrate and nitrite-nitrogen
S_{BOD} = soluble material concentration organic substrate
$X_{b,h}$ = particulate material concentrations
Y_h = yield coefficient for heterotrophic organisms (mg VSS/mg COD)
Y_H = biomass yield coefficient
γ = coefficient of excess removal (phosphorus content of volatile solids) (mg phosphorus/mg)
η_g = correction factor for μ_H under anoxic conditions
θ_c = solids retention time (d)
$\mu_{max, H}$ = maximum specific growth rate of heterotrophs

9

Alternative Biological Treatment

Introduction

Natural systems for wastewater treatment include soil absorption, lagoons (ponds), land treatment, floating aquatic plants, and constructed wetlands. Where sufficient land of suitable character is available, these natural systems can often be the most cost-effective option for both construction and operation. They are better suited for small communities and rural areas because of the need for and availability of suitable land area. Natural systems range in size from the typical, small 1 m^3/d (264 gpd) individual home soil absorption systems to the large 190,000 m^3/d (50 mgd) rapid infiltration land treatment system in Orlando, Florida.

The common element in the use of natural systems for wastewater treatment is the significant contribution made by the "natural" environmental components that provide the desired treatment. In general, these responses by the vegetation, soil, terrestrial and aquatic microorganisms, and, to a limited extent, higher animal life proceed at their "natural" rates.

Soil Absorption Systems

These systems are limited to wastewater flows of approximately 190 m^3/d (0.05 mgd) or less. The typical absorption system is a series of gravel-filled trenches preceded by a septic tank. Effluent from the septic tank flows by gravity into the trenches, often referred to as *leach lines* or *drain lines*. The lines together are often referred to as a *tile field*. The basis for the system design is the ability of applied wastewater to infiltrate and then percolate through the soil profile.

The most widely used natural system for wastewater management is the subsurface soil absorption system, consisting of a septic tank and a tile field or other absorption device. Although most soil absorption systems serve individual homes, in recent years some small communities have adopted the technology.

Lagoon Systems

The second most prevalent natural system is the wastewater treatment lagoon. Wastewater treatment lagoons can be classified based on their depth and the biological reactions that occur in the lagoon. Using this classification, the four primary types of lagoons are aerobic, facultative, aerated, and anaerobic.

Aerobic lagoons are relatively shallow with typical depths ranging from 0.3 to 0.6 m (1 to 2 ft). Oxygen is provided by algae during photosynthesis and wind-aided surface aeration. These lagoons are often mixed by recirculation to maintain dissolved oxygen throughout their entire depth. Aerobic lagoons are limited to warm, sunny climates and are used mostly in the southern portion of the United States and in similar climates (Reed et al., 1995; U.S. EPA, 1983).

Facultative lagoons, the most prevalent lagoon type, are also referred to as *oxidation lagoons*. These lagoons are 1.5 to 2.5 m (5 to 8 ft) deep, with detention times ranging from 25 days to more than 180 days. Depths are kept at 1.5 m or more to avoid the growth of emergent plants. Surface layers of the lagoons are aerobic with an anaerobic layer near the bottom. Oxygen is supplied by the natural diffusion of atmospheric oxygen into the liquid surface aeration and photosynthetic algae. Facultative lagoons are designed in series with a minimum of three cells to reduce short circuiting. The primary problem with facultative lagoons is the algae that remains in the effluent, sometimes causing effluent suspended solids to exceed discharge requirements.

Aerated lagoons or ponds can be either partially mixed or completely mixed. Oxygen is supplied by mechanical floating aerators or diffused aeration. Aerated lagoons are 3 to 6 m (10 to 20 ft) deep, with hydraulic retention times ranging from 5 to 30 days. Aerated lagoons accept higher biochemical oxygen demand (BOD) loadings than facultative lagoons, are less susceptible to odors, and require less land. Aerated lagoons are followed by a facultative lagoon or a settling lagoon (2-day retention or less) to reduce suspended solids before discharge.

Anaerobic lagoons are heavily loaded with organics and do not have an aerobic zone. They are 2.5 to 6 m (8 to 20 ft) deep and have detention times of 20 to 50 days. Biological activity is slow when compared with that of a mixed anaerobic digester. Anaerobic lagoons have been used as pretreatment to facultative or aerobic lagoons for strong industrial wastewater and for rural communities with a significant organic load from industries such as food processing. They are not widely used for municipal wastewater treatment in the United States, but are common in Africa and South America.

Another method of classifying lagoons is based on the duration and frequency of their effluent discharges. This classification includes total containment lagoons, controlled discharge lagoons, hydrograph controlled release lagoons, and continuous discharge lagoons.

The *total containment lagoon* or *evaporation lagoon* applies only to climates in which the evaporation exceeds the precipitation on an annual basis. The *controlled discharge* concept is to discharge only once or twice per year when stream conditions are satisfactory. The *hydrograph controlled release lagoon*, a variation of the controlled discharge concept, is designed with a discharge rate correlated to the stream flowrate. As with the controlled discharge lagoons, the hydrograph controlled lagoon only discharges when stream flow is higher than some acceptable minimum value. Most controlled discharge and hydrograph controlled lagoons are facultative. In the *continuous discharge lagoon*, the effluent is discharged at the same rate (less evaporation and seepage loss) as influent wastewater flow.

Facultative Lagoons

Facultative lagoons receive no more pretreatment than screening; therefore, they store heavy solids and grit in the first or primary lagoons. Typical practice is to operate three or more lagoons in series, although flexibility to discharge raw wastewater to different lagoons and recycle treated effluent back to the primary lagoon deserves consideration.

Treatment Performance

Facultative lagoons are designed to reduce 5-day BOD (BOD_5) to approximately 30 mg/L; however, typical performance ranges from 30 to 40 mg/L. Influent suspended solids are removed by sedimentation; however, algae contribute to the 40 to 100 mg/L of effluent suspended solids that are found during periods of maximum algal growth.

Facultative lagoons can remove significant amounts of nitrogen and pathogens as a result of long detention times. Nitrogen removal often ranges from 40 to 95% (Reed et al., 1995). Phosphorus removal is low, less than 40%. Bacteria and viruses are removed as a result of sedimentation, predation, natural die-off, and adsorption.

Toxic organic compound removals in facultative and aerated lagoons vary, depending on the volatility of the compound. For example, a facultative lagoon removed between 77 and 96% (average 86%) of volatile organics, whereas an aerated lagoon removed between 61 and 80% (average 68%) of the same compounds. Corresponding removals of semivolatile compounds by the two systems were found to be from 25 to 80% and from 22 to 77%, respectively. Only the activated-sludge process removed more of the organic compounds (Hannah et al., 1986).

Design Procedures

At least five methods have been used to design facultative lagoons. The results of a design example using the five common methods are presented in Table 9.1. For each method, this table presents the calculated detention time, surface area, depth, and BOD_5 loading rate for 1900-m^3/d (0.5-mgd) flow, a BOD_5 of 200 mg/L, and a temperature of 0.6 °C (33 °F) (critical period water temperature). The areal loading rate, the most conservative design method, can be adapted to specific standards. Recommended BOD_5 loading rates are based on average winter air temperatures (see Table 9.2).

To calculate the area needed for the facultative lagoon, divide the BOD_5 load (kilograms per day) by the appropriate loading rate from Table 9.2 or from specific state standards. The first cell in a series of cells should not be loaded at more than 100 kg BOD_5/ha·d (90 lb/d-ac) for warm climates with average winter air temperatures higher than 15 °C (50 °F) and 40-kg BOD_5/ha·d (36 lb/d-ac) for cold climates with average air temperatures lower than 0 °C (32 °F). When ice cover forms for extended

Table 9.1 Design criteria for facultative ponds (U.S. EPA, 1981).

Method	Detention time, day[a]	Surface area, ha[b]	Effective depth, m	BOD_5 loading rate, kg/ha·d[b]
Areal loading rate	180	22.3	1.4	17
Gloyna	140	26.5	1.0	14
Marais and Shaw	74	5.6	2.4	68
Plug flow	180	22.3	1.4	25
Wehner and Wilhelm	80–132	10.8–17.9	1.4	30–48

[a]Total system using four ponds in series.
[b]Flow = 1893 m³/d (0.5 mgd); secondary treatment objective; influent BOD_5 = 200 mg/L; temperature = 5 °C; light intensity adequate; influent total suspended solids = 250 mg/L.

periods of time, facultative lagoons perform essentially as cold anaerobic ponds that remove particulate BOD by settling with little biological activity.

The water surface area for a facultative lagoon can be calculated as

$$A = \frac{(BOD)(q)}{(1000)(L_R)} \quad (9.1)$$

where

A = area for facultative lagoon (ha);
BOD_5 = BOD_5 in the wastewater (mg/L);
q = wastewater flow (m³/d);
1000 = conversion from grams to kilograms; and
L_R = BOD_5 loading rate selected from Table 9.2 for appropriate average winter air temperature (kg/ha·d).

Lagoons can be designed to avoid hydraulic short circuiting by providing manifolds or diffusers on inlets and outlets, keeping the inlet and outlet as far apart as possible, using in-lagoon baffles, and providing multiple inlets and outlets. Multiple lagoons in series, three or four, will reduce short circuiting. Recirculation from the last lagoon to the first helps distribution and reduces short circuiting.

Controlled Discharge Lagoons

These are facultative lagoons with detention times of 180 days or more. Lagoons with seasonal discharge have been operated in the northcentral United States with overall

Table 9.2 Facultative pond biochemical oxygen demand loading rates (U.S. EPA, 1975).

Average winter air temperature, °C[a]	Depth, m	BOD_5 loading rate, kg/ha·d
<0 °C	1.5–2.1	11–22
0–15 °C	1.2–1.8	22–45
>15 °C	1.1	45–90

[a]°C × 1.800 + 32 = °F.

organic loading of 22 to 28 kg BOD$_5$/ha·d (20 to 25 lb/ac-d), and water depth of 2 m (6.5 ft) or less in the first cell and 2.5 m (8 ft) or less in subsequent two to three cells. In Alberta, Canada, and other cold climate areas, two cells are used with the working cell of 1.5 m depth and loaded at 35 to 45 kg BOD$_5$/ha·d; the second is a storage cell designed for twice-a-year discharge (spring and fall) so that the inlet structure is covered with wastewater before freeze-up.

Hydrograph Controlled Release Lagoons
A variation of the controlled discharge lagoon concept has been developed to optimize the dilution of lagoon effluent in receiving water. Effluent is discharged based on stream flow and varies, above some minimum flow, with the actual stream flow (Zirschky, 1986).

Partially Mixed Aerated Lagoons
The basic equation for design of partially mixed aerated lagoons is given below (WPCF, 1989).

$$\frac{C_n}{C_o} = \frac{1}{[1 + (k\bar{t}/n)]^n} \qquad (9.2)$$

where
C_n = effluent BOD$_5$ in cell n (mg/L);
C_o = influent BOD$_5$ (mg/L);
k = first-order reaction rate constant (0.14 to 0.3) (d^{-1});
$\bar{t}$ = hydraulic retention time (d); and
n = number of cells (ponds) in series.

This equation is based on three equal-size ponds at the same temperature. The detention time is for each single pond.

In the partially mixed aerated lagoon system, no attempt is made to completely mix the lagoon contents. The aeration demand is calculated to provide an adequate oxygen supply, by providing 1.5 to 2 kg of oxygen/kg of BOD$_5$ loading or more. A portion of the suspended solids and, therefore, some of the particulate BOD settle to the lagoon bottom and degrade anaerobically. Because this bottom reaction resembles that of a facultative lagoon, the partially mixed lagoon is often referred to as a facultative aerated lagoon.

The reaction rate constant k depends on the lagoon temperature (see Chapter 6 for temperature coefficients). At 20 °C (68 °F) the typical k value used for domestic wastewater is 0.276; at 1 °C (34 °F), the k value should be reduced to 0.14 (Reed et al., 1995; U.S. EPA, 1983).

Hydraulic retention times range from 5 to 30 days for most municipal systems and water depths range from 3 to 6 m (10 to 20 ft). Biochemical oxygen demand loading rates of 100 to 400 kg BOD$_5$/ha·d (90 to 356 lb/d-ac) are typical. A settling lagoon of 1- to 2-day retention often follows the last partially aerated lagoon in series. The number of cells ranges from three to five or more. Treatment performance is good, with effluent BOD$_5$ values ranging from 20 to 40 mg/L and effluent suspended solids values ranging from 20 to 60 mg/L. To calculate the average water surface area needed for a partially mixed aerated lagoon, use the following equation:

$$A = \frac{q\bar{t}}{10,000(d)} \qquad (9.3)$$

where
- A = average area of lagoon (ha);
- q = wastewater flow (m³/d);
- $\bar{t}$ = hydraulic retention time (d);
- 10,000 = conversion from square meters to hectares; and
- d = average depth, m (3 to 6 m).

The area calculated should be increased to account for the side slope of the lagoon berm. In partially mixed aerated lagoons, the aeration requirements are based on the oxygen demand of the wastewater, not on the mixing requirements for lagoons, which are approximately 3 W/m³ (15 hp/mil. gal).

Most existing treatment lagoons are not lined but have soil conditions that minimize percolation. The trend in design of wastewater lagoons is toward including liners to minimize percolation. Typical materials used for liners are native clay, bentonite-amended soil, geosynthetic clay liners, and geomembranes. Many states reference the *Recommended Standards for Sewage Works* (Great Lakes, 1990) requirement that the permeability of the liner shall not exceed

$$k = 3.0 \times 10^{-9} (z) \qquad (9.4)$$

where k = permeability (cm/s) and z = thickness of the seal (cm).

For water balance calculations, note that permeability and seepage rates per unit area differ. The seepage rate can be calculated from

$$Q = \frac{kAh}{z} \qquad (9.5)$$

where
- k = permeability (cm/s);
- z = thickness of the seal (cm);
- Q = flow through the liner (cm³/s);
- A = liner area (cm²); and
- h = hydraulic head over the liner (cm).

Installation of a lagoon liner is shown in Figure 9.1. To prevent erosion and desiccation of clay or bentonite liners, interior slopes of the lagoon should have soil cover and riprap. Suggested minima are 0.5 m (1.6 ft) above and below the lagoon water level or twice the impinging wave height calculated for twice the maximum wind velocity anticipated. With surface aerators, consider the need for protecting the basin directly beneath aerators from the vortex or other scouring action. Concrete pads, anti-erosion plates on the aerators, or 150 mm (6 in.) of crushed rock provide adequate protection. Riprap for erosion control on the slopes of a facultative lagoon is shown in Figure 9.2. Embankment tops of a minimum width of 2.5 m (8 ft) permit access of maintenance vehicles. Embankment outer slopes no steeper than 3:1 allow grass growth and tractor mowing. A minimum freeboard of 0.9 m (3 ft) represents typical practice.

Figure 9.1 Installation of a lagoon liner.

Dual-Power Multicellular Aerated Lagoons

The concept of dual-power multicellular aerated lagoons is to combine a series of partially mixed lagoons. The first cell has a depth of 3 m (10 ft) and is aerated and mixed with a surface aerator at 6 W/m^3 (30 hp/mil. gal). The subsequent three cells are aerated at a rate of 1 W/m^3 (5 hp/mil. gal). Hydraulic retention time in the first cell is 1.5 to 2 days, and the overall retention time of all four cells is 4.5 to 5 days (Rich, 1996).

Complete mixing in aerated lagoons requires between 6 and 12 W/m^3 (30 and 60 hp/mil. gal), typically 10 W/m^3 (50 hp/mil. gal) of aeration. The combination of two levels of aeration meets the oxygen requirement for biological conversion while minimizing algae production by the turbulence of mixing (Rich, 1996).

Advanced Integrated Lagoon Systems

The advanced integrated lagoon system concept combines multiple lagoons with recycle (Oswald, 1991). The system consists of a deep, primary facultative lagoon followed by a shallow aerobic lagoon. The primary lagoon has fermentation pits for anaerobic treatment of settled solids. The fermentation pits must be unaerated and unmixed and will then serve as upflow anaerobic digesters.

Land Treatment Systems

Land treatment is the controlled application of wastewater to land at rates compatible with the natural physical, chemical, and biological processes that occur on and in the

Figure 9.2 Riprap for erosion control on the sideslopes of a facultative lagoon.

soil. The three types of land treatment systems are slow rate, overland flow, and rapid infiltration. The features of the three types of land treatment are presented in Table 9.3.

In slow-rate and rapid infiltration systems, wastewater is treated as it percolates through the soil. In overland flow, the treatment occurs in a thin film on the grassy slopes constructed on slowly permeable soil.

Table 9.3 Features of land treatment systems.

Feature	Slow rate	Overland flow	Rapid infiltration
Treatment goals	Secondary or advanced wastewater treatment, zero discharge	Secondary, nitrogen removal	Secondary, advanced wastewater treatment, groundwater recharge, zero discharge
Vegetation	Yes, various crops	Yes, water-tolerant grasses	Only for soil stabilization
Climate restrictions	Storage needed for cold weather and heavy precipitation	Storage needed for cold weather	No storage needed when properly designed and operated
Hydraulic loading, m/y	0.5–6	3–20	6–100
Area needed, ha[a]	23–280	7–46	1.4–23

[a] For design flow of 3785 m^3/d (1 mgd).

Of the three types of land treatment, slow-rate systems achieve the highest level of treatment. Surface runoff is contained on-site, although rainfall-induced runoff is allowed to leave the site. With typical loading rates of 1 to 2 m/y (3 to 7 ft/yr) much of the applied wastewater can be lost to evapotranspiration, particularly in arid climates. The application technology is similar to that for crop irrigation, varying from sprinkler to surface application. Crops range from pasture to forest to row and field crops.

Overland flow systems, similar to other fixed-film biological treatment systems, remove significant amounts of BOD, suspended solids, and nitrogen. Phosphorus, trace elements, and pathogens are less removed. Hydraulic loading rates range from 3 to 20 m/y (10 to 70 ft/yr). Overland flow is best suited to slowly permeable solids that can be graded to mild slopes (2 to 8%) and planted with water-tolerant grasses. Overland flow produces an effluent of better than secondary quality, depending on the application rate. This technology emerged in the 1970s and is now relatively well developed, although it is less frequently used outside the southeastern or southwestern regions of the United States.

Rapid infiltration, considered to be an established treatment technology, consists of shallow spreading basins in permeable soils to which wastewater is intermittently applied. Each basin is dosed for a period of 1 to 7 days and then rested for 6 to 20 days. Treatment is accomplished by natural means as wastewater infiltrates through the soil surface and percolates through the soil. Rapid infiltration systems discharge to groundwater or can be underdrained.

Preapplication Treatment

Historically, land treatment systems have provided a method of treating and dispersing treated water to the environment. As the performance of these systems was documented and became better understood, the need for preapplication treatment was reexamined. Slow-rate systems require the highest level of preapplication treatment, especially when site access is not controlled or when private farmer contracts are used. Under these circumstances, secondary treatment and disinfection commonly precede slow-rate application. For overland flow, the use of a partially aerated lagoon with a 1-day detention time has been successful. The lagoon serves to remove larger solids and adds dissolved oxygen to the wastewater. Short-term detention minimizes the growth of algae that would otherwise not be removed efficiently. Rapid infiltration systems can operate year-round using effluent from secondary treatment plants as an alternative to tertiary treatment. Aerobic or facultative lagoons are not recommended before rapid infiltration unless effluent suspended solids concentrations (algae) are controlled or have been allowed for in design (larger land areas are needed). Algae will clog infiltrative surfaces.

Site Requirements

Requirements for suitable sites for land treatment systems are presented in Table 9.4. Suitable site characteristics for slow-rate and overland flow (slope and soil permeability) overlap somewhat as do those for slow-rate and rapid infiltration (soil depth and permeability). Site investigations are important in the selection of the appropriate land treatment process and the best available site. For slow-rate systems, investigations concentrate on topography mapping, soil type, groundwater depth, and surface drainage features. If a detailed soil survey is available, the preliminary soil investigation is confined to field verification of the survey. If a detailed survey is unavailable, field evaluation of the soils is advisable. Evaluation includes backhoe pits with field analysis

Table 9.4 Site requirements for land treatment processes.

Characteristic	Slow rate	Overland flow	Rapid infiltration
Soil depth, m	>0.6	>0.3	>1.5
Soil permeability class range	Slow to moderately rapid	Very slow to moderately slow	Rapid
Soil permeability, mm/h	1.5–500	<5.0	>50
Depth to groundwater, m	0.6–1	Not critical[a]	1 during flood cycle[b]; 1.5–3 during drying cycle
Slope, %	<20 on cultivated land; <40 on noncultivated land	0–15; finished slopes 2 to 8[c]	<10; excessive slopes require much earthwork

[a]Effect on groundwater should be considered for more permeable soils.
[b]Underdrains can be used to maintain this level at sites with high groundwater table.
[c]Slope as low as 1% and as high as 10% may be considered.

of the soil profile by a soil scientist or experienced land treatment specialist. Where the percolation rate for the slow-rate site is considered significant, a field analysis for design is recommended.

The same site features listed for slow-rate systems are important for overland flow systems. The soil depth and slope are more important than permeability for overland flow because site grading to uniform slopes is necessary and deep percolation is not desired.

For rapid infiltration systems, the most important site features are the soil depth, soil permeability, and depth to groundwater. Site investigations should be performed to determine the soil depth, depth to groundwater, and most importantly, the infiltration rate of the limiting soil layer in the soil profile. The field investigation may have to extend beyond the proposed application site to ensure that the percolate will flow away from the application point and will not emerge as surface seepage at undesirable locations.

Slow-Rate Systems

Slow-rate systems are effective in the treatment of wastewater and are especially useful where high levels of nutrient removal are required.

Treatment Performance

Biochemical oxygen demand removal is accomplished by filtration, soil absorption, and bacterial oxidation. Slow-rate systems effectively remove BOD at loading rates of 500 kg BOD_5/ha·d (450 lb/d-ac) and more (Jewell et al., 1978). Effective BOD_5 removal (more than 90%) can be expected for slow-rate systems loaded up to 500 kg BOD_5/ha·d (Reed et al., 1995). Careful management should be practiced with organic loadings of more than 300 kg BOD_5/ha·d (270 lb/d-ac) to avoid odor production. In practice, municipal slow-rate systems will rarely be loaded beyond 100 kg BOD_5/ha·d (90 lb/d-ac).

In slow-rate systems, a combination of plant uptake, denitrification, and soil storage removes nitrogen. Percolates from slow-rate systems often achieve phosphorus levels less than 0.1 mg/L and total nitrogen levels of less than 10 mg/L. Soil adsorption and chemical precipitation remove phosphorus. Metals are removed by adsorption, chemical precipitation, ion exchange, and complexation. Pathogens are removed by soil filtration, adsorption, desiccation, radiation, predation, and exposure to other adverse environmental conditions. Trace organics are removed in slow-rate systems by photodecomposition, volatilization, sorption, and biological degradation.

Design Objectives

Slow-rate systems, classified according to the design objective, are either type 1 (slow infiltration) or type 2 (crop irrigation). The objective of type 1 systems is wastewater treatment. Design of type 1 systems is based on the limiting design factor, which is either the soil permeability or the allowable loading rate for a particular wastewater constituent, such as nitrogen. Type 1 systems are found in humid areas of the United States and are managed by municipal wastewater agencies.

The objective of type 2 systems is water and nutrient reuse. Crop production is a primary objective in type 2 systems, while wastewater treatment is a secondary objective. Design of type 2 systems is based on applying sufficient water to meet crop irrigation requirements for water and nutrients. Type 2 systems are found in the arid areas of the United States and are managed by municipal wastewater agencies through contract or leases with farmers, or directly by the private farmers themselves.

Selection of the crop is an important early step in the design of slow-rate systems because the crop selected affects the level of preapplication treatment, type of distribution system, and hydraulic loading rate. For type 1 systems, compatible crops have high nitrogen uptake capacity, high evapotranspiration rates, and tolerance to moisture and wastewater constituents. Type 1 system crops include perennial forage grasses, turf grasses, some tree species, and some field crops.

For type 2 systems, a broad variety of crops can be considered (U.S. EPA, 1981). Double cropping increases revenue potential. In warm climates, short-season summer crops (such as corn or sorghum) can be combined with winter grains (such as barley, oats, or wheat).

The choice of distribution system depends on the crop, topography, and soil. Sprinkler systems are commonly used in wastewater application because of their adaptability to different soil and topographic conditions. Variations in sprinkler systems include fixed impact-type sprinklers, continuous move (center pivots and linears) systems, and move-stop systems (wheel line and traveling gun) (Reed and Crites, 1984).

For type 1 systems, the hydraulic loading rate can be calculated from the water balance equation

$$L_w = E - P + W_p \tag{9.6}$$

where
L_w = wastewater hydraulic loading rate based on soil permeability (m/y);
E = design evapotranspiration rate (m/y);
P = design precipitation rate (m/y); and
W_p = design percolation rate (m/y).

The design evapotranspiration rate is estimated as the average rate for the selected crop. The design precipitation rate is the total for the wettest year in a 10-year period. The design percolation rate should be measured in the field using a cylinder infiltrometer, sprinkler infiltrometer, or basin flooding technique.

> **Note on Terminology**
> Agricultural engineers often calculate environmental effects such as precipitation and evaporation on the basis of inches (or meters) per year. The notation for this is "per annum", abbreviated by *a*, as in 20 m/a. This can be confusing to civil engineers who use the letter *y* to indicate "per year", as in 20 m/y. In this text we use the engineering notation (yr in English units), but the reader should be aware that wastewater disposal on land is often the purview of the agricultural engineer and the notation can be different.

For type 2 systems, the hydraulic loading rate equation for a specific crop use is

$$L_w = (E - P)(1 + G/100)(100/E_u) \qquad (9.7)$$

where
- L_w = annual wastewater loading (m/y);
- E = crop evapotranspiration rate (m/y);
- P = precipitation (m/y);
- G = leaching requirement, 15 to 25%; and
- E_u = irrigation efficiency, 65 to 85%.

The specific crop and its sensitivity to wastewater total dissolved solids determines the leaching requirement, which may range from 10 to 40% but is between 15 and 25%. The irrigation efficiency for sprinklers ranges from 70 to 80%; surface irrigation efficiencies range from 65 to 85%. The total percolation is a combination of the leaching fraction and the irrigation inefficiency fraction $(1 - E_u/100)$.

If the slow-rate system percolate enters a potable groundwater aquifer, the percolate nitrogen concentration is often limited to 10 mg/L or less (as nitrate-nitrogen). The nitrogen balance is

$$L_N = \frac{U + 0.01\,C_p P_w}{1 - f} \qquad (9.8)$$

where
- L_N = nitrogen loading rate (kg/ha·y);
- U = crop uptake of nitrogen (kg/ha·y);
- f = fraction of applied nitrogen lost to denitrification, volatilization, and soil storage;
- C_p = percolate nitrate-nitrogen concentration (mg/L); and
- P_w = percolate flow (m/y).

The value of f depends on the BOD_5/nitrogen ratio in wastewater and the air temperature during the application season. High-strength wastewater (BOD_5/N) has the highest f value, as shown in Table 9.5. Lower f values apply to cold climates.

By combining the water balance and nitrogen balance equations, the hydraulic loading rate based on nitrogen limits can be calculated as

$$L_{wn} = \frac{C_p(P - E) + 0.1\,U}{(1 - f)\,C_n - C_p} \qquad (9.9)$$

Table 9.5 Ranges of f values (fraction of applied nitrogen lost) for municipal wastewaters.

Wastewater type	f value
High strength	0.5–0.8
Primary effluent	0.25–0.5
Secondary effluent	0.15–0.25
Advanced treatment effluent	0.10–0.15

where L_{wn} = hydraulic loading rate based on nitrogen limits (m/y) and C_n = wastewater nitrogen concentration (mg/L).

The design limiting loading rate is the lowest of the two calculated values, L_w or L_{wn}, for type 1 systems.

The land area needed for a slow-rate site includes the field application area plus the areas for roads, buffer zones, and any required storage. The field area can be expressed as

$$A = \frac{365\,q + V_s}{10{,}000\,L_l} \tag{9.10}$$

where

A = field area (ha);
q = wastewater flow (m³/d);
V_s = net loss or gain in stored wastewater volume because of evaporation, seepage, or precipitation on the storage lagoon (m³/y); and
L_l = limiting hydraulic loading rate (m/y).

Most slow-rate systems need to store wastewater during cold- or wet-weather periods. In addition, the application rate will vary during the year while the wastewater supply remains relatively constant. A storage lagoon can store the excess wastewater whenever the allowable application rate is lower than average. Storage needed based on climatic data can be estimated from maps (U.S. EPA, 1981) or by using computer programs from the National Oceanic and Atmospheric Administration (Whiting, 1976). A detailed water balance is necessary for final design to determine the storage volume. Design details of storage facilities may be found elsewhere (Reed and Crites, 1984).

Overland Flow Systems

Overland flow systems can be designed to achieve secondary treatment, advanced treatment, or nitrogen removal. Phosphorus removal requires either pre- or post-application treatment.

Treatment Performance

Treated runoff concentrations of BOD and suspended solids differ little among raw, primary, and secondary effluent applications. Algae in lagoon effluent are not effectively removed in most overland flow systems (Witherow and Bledsoe, 1983).

Effluent BOD_5 is typically in the 10 mg/L range, while nitrogen is reduced to less than 10 mg/L. Land treatment requires thorough soil water contact to provide effective phosphorus, metals, and pathogen removal. As a result of the limited soil contact, overland flow removes approximately 40 to 60% of applied phosphorus, 60 to 90% of trace metals, and 99% of bacteria and viruses (U.S. EPA, 1981). Trace organics are adequately removed in overland flow by the same mechanisms as in slow-rate systems.

Design Factors
Overland flow design factors include application rate, slope length, slope grade, and application period. The application rate, expressed in cubic meters per meter per hour, applied to the top of the slope or terrace ranges from 0.03 to 0.37 m³/m·h (0.04 to 0.5 gpm/ft). The length of the slope or terrace is 30 to 60 m (100 to 200 ft). Slope grades are kept between 1 and 12%, with a preferred range of 2 to 8%.

Design Procedures
The following equation presents the relationship between BOD removal and design variables (U.S. EPA, 1981):

$$\frac{C_z - c}{C_o} = B\exp(-KZ/L^n) \tag{9.11}$$

where
- B = constant;
- C_z = effluent BOD_5 at point z (mg/L);
- c = residual BOD_5 at end of slope (mg/L);
- C_o = applied BOD_5 (mg/L);
- Z = slope length (m);
- L = hydraulic application rate (m³/m·h); and
- K, n = empirical constants.

As good practice, the application rate is reduced by dividing by a safety factor of 1.5 before calculating the field area.

Suspended Solids Loadings
Except for algae, wastewater solids will not be limiting in overland flow system designs. Suspended solids are effectively removed on overland flow slopes because of the low velocity and shallow depth of flow. For high-strength and high-solids content wastewater, sprinkler applications will best distribute suspended solids uniformly over the upper 65% of the slope.

Removal of algae by overland flow varies with the application rate, type of algae, and algae concentration. Removal rates range from 45 to 83% (Witherow and Bledsoe, 1983). Algae that are buoyant or motile resist removal by sedimentation or filtration (WPCF, 1989). Where facultative lagoons are used as pretreatment, the hydraulic loading rate on the overland flow slopes should not exceed 0.09 m³/m·h (0.12 gpm/ft). If algae concentrations cause suspended solids values to exceed 100 mg/L, the overland flow system will not be able to reduce the suspended solids below 30 mg/L. In these cases, operating the overland flow system in a nondischarge mode might be possible by using repeated short application periods (15 to 30 minutes) followed by resting for 1 to 2 hours.

Biochemical Oxygen Demand Loadings
Biochemical oxygen demand loadings of up to 100 kg BOD_5/ha·d (90 lb/d-ac) have been used successfully in overland flow systems. When the BOD_5 exceeds 800 mg/L, the

oxygen-transfer capacity of the system becomes limiting and preapplication treatment or effluent recycling may be necessary to obtain successful treatment (Reed and Crites, 1984).

Land Requirements
The following equation is used to calculate the field area needed for overland flow:

$$A = \frac{qZ}{MP\,(10{,}000)} \qquad (9.12)$$

where
- A = field area (ha);
- q = wastewater flowrate (m³/d);
- Z = slope length (m);
- M = application rate (m³/m·h); and
- P = period of application (h/d).

Note that M is cubic meters of wastewater applied per hour per square meter of field area.

Vegetation Selection
Water-tolerant grasses are used in overland flow systems to provide a support medium for microorganisms, minimize erosion, and remove nitrogen. The crop is cut periodically and either removed as hay or green chop or left on the slope. Sod-forming grasses such as reed canary grass are selected. Other cool-season grasses include tall fescue, perennial ryegrass, and redtop. Warm-season grasses include common and coastal Bermuda grass and bahia grass.

Distribution System
Municipal wastewater can be surface-applied to overland flow systems using gated pipe; however, industrial wastewater should be sprinkler-applied. Sprinkler systems for municipal wastewater should be located 30% of the distance down the slope. Typical distances from the edge of the sprinkler-wetted diameter to the runoff collection ditch range from 15 to 20 m (50 to 65 ft). Top-of-the-slope distribution methods, in addition to gated pipe, include low-pressure sprays, bubbling orifices, and perforated pipe (Reed and Crites, 1984).

Monitoring Wells
Because overland flow systems have the potential of contaminating groundwater, monitoring wells have to be driven upstream and downstream (of the groundwater flow) and the water periodically sampled to make sure the groundwater is not being contaminated.

Rapid Infiltration Systems
Rapid infiltration systems require deep, permeable soils for wastewater treatment. This section describes expected treatment performance, design procedures, hydraulic loading rates, organic loading rates, and land requirements.

Treatment Performance
Rapid infiltration systems effectively remove BOD and suspended solids through filtration, adsorption, and bacterial decomposition. Biochemical oxygen demand removals

for rapid infiltration are near 95% and suspended solids are removed to low levels, approaching 1 mg/L.

Nitrogen removal for rapid infiltration systems varies from 40 to 90% as a result of biological denitrification. Important design criteria are the BOD_5/nitrogen ratio, hydraulic loading rate, and ratio of flooding period to drying period. The design objective is to manage these factors to obtain nitrification–denitrification, allowing escape of nitrogen as a gas. The BOD_5/nitrogen ratio should be greater than 3:1 for effective denitrification. The hydraulic loading rate, if kept within the range of 15 to 30 m/y (50 to 100 ft/yr), should provide adequate detention time within the soil profile for effective nitrogen removal (Crites, 1985). The soil profile should be 3 m (10 ft) or deeper to ensure adequate hydraulic retention time at a 30-m/y (100-ft/yr) loading. The wetting and drying pattern is also critical for nitrogen removal (Crites, 1985).

Phosphorus removal is accomplished by absorption and chemical precipitation. Detention time, critical for chemical precipitation, is a function of the percolation rate through the soil and the aquifer and the flow distance to the point of monitoring. Phosphorus removal in rapid infiltration systems is in the range of 90% to as high as 99%. Although phosphorus removal declines with time, the removal rate might, nonetheless, remain high for many years. Rapid infiltration systems are also effective in removing metals, pathogens, and trace organics (U.S. EPA, 1981).

Design Procedures

The design hydraulic loading rate is based on the soil infiltration rate, subsurface flowrate, or loading of BOD or nitrogen. Each of these loading rates must be calculated and the lowest value selected for design. The procedure for calculating the hydraulic loading based on infiltration rate includes converting the hourly infiltration rate to an annual rate (multiply by 8760 h/y) and multiplying the result by a factor to account for the wetting and drying cycle, variability of the soils, and type of infiltration rate field test. The field-measured infiltration rate used in design is the steady-state rate measured during 1 hour or more at the end of a test. The equation for the annual design loading rate is

$$L = al \qquad (9.13)$$

where
  L = design annual wastewater loading rate (m/y);
  a = design factor ranging from 0.02 to 0.15; and
  l = measured steady-state infiltration rate (m/y).

The design factor should be 0.02 to 0.04 for small-scale tests and can be increased to 0.07 to 0.15 for larger installations, depending on the variability of the soils, number of test results, and degree of conservatism used. The design factor must not exceed the fraction of the loading cycle during which the basins are flooded. For example, if the application period is 1 day and the drying period is 9 days (a total cycle period of 10 days), the design factor must be less than 0.10.

For municipal rapid infiltration systems, the BOD loading rate will range from 10 to 200 kg BOD_5/ha·d (9 to 180 lb/d-ac). The suggested maximum rate is 670 kg BOD_5/ha·d (598 lb/d-ac) (Reed and Crites, 1984).

The basin bottom area, basin berms, roads, buffer area, and area for expansion can be estimated by

$$A = \frac{q}{10{,}000 L_l} \qquad (9.14)$$

where
    A = net field area (ha);
    q = wastewater design average flow (m³/d); and
    L_l = limiting loading rate (m/d).

Floating Aquatic Systems

Floating aquatic plants have been used for wastewater treatment in a variety of processes, including the upgrading of facultative lagoon effluent, thereby achieving various degrees of advanced wastewater treatment depending on loading and management. The floating plants that have been studied and used the most are water hyacinths and duckweed.

The concept of using floating aquatic plants, such as water hyacinths and duckweed, in wastewater treatment partially arose from an attempt to control suspended solids concentrations in aerobic and facultative lagoon discharges. The floating plants shield the water from sunlight and reduce the growth of algae. Floating plant systems have also been shown to reduce BOD, nitrogen, metals, and trace organics.

Water hyacinth systems are an emerging technology being developed in large-scale pilot systems. Cold weather restricts the growth of water hyacinths, limiting their suitability to warm climates. The primary characteristics of water hyacinths that make them an attractive biological support medium for bacteria are their extensive root systems and rapid growth rate. The primary characteristic that limits their widespread use is their temperature sensitivity (that is, they are rapidly killed by frost conditions). It may be necessary to purchase water hyacinths through a certified vendor if they are not available locally.

Duckweed (family Lemnaceae) *systems* have been studied alone and together with water hyacinths in polyculture systems. The primary advantage of duckweed is its lower sensitivity to cold climates; its primary disadvantages are its shallow root systems and sensitivity to movement by winds. Experience has shown that loading rates ranging from 140 to as high as 700 m³/ha·d (0.09 to 0.46 mil. gal/ac-d) can reduce BOD by as much as 80%. Both water hyacinths and duckweed need to be harvested. Regular harvesting of duckweed is practiced in existing systems and duckweed is land-applied or composted. Treatment plants in northern climates cannot rely on this system for nutrient removal because duckweed dies during cold winters. Experience in Europe with the construction of so-called "green wastewater treatment plants" has been poor, with the plants using more energy than typical biological nutrient removal systems.

Constructed Wetlands

Constructed wetlands, an emerging technology for treatment of municipal and industrial wastewater, are designed to treat wastewater using emergent plants such as cattails, reeds, and rushes. Applications for constructed wetlands include the treatment of stormwater, acid mine waste, landfill leachate, agricultural runoff, and food-processing wastewater (Crites, 1996).

Figure 9.3 Typical constructed wetland illustrating open water and emergent vegetation.

The three primary categories of constructed wetlands are *free-water surface, subsurface flow,* and *vertical flow*. For free-water surface wetlands, the flow path of the applied wastewater is above the soil surface. For subsurface wetlands, the flow path is lateral through the root zone and the medium (which ranges from sand to coarse gravel to rocks). For vertical flow wetlands, the application is either by spray or surface flooding, and the flow path is down through the medium and out through the underdrains.

A typical free-water surface wetland system is illustrated in Figure 9.3 and is a swampy area with free water. A subsurface flow system consists of beds or channels filled with gravel, sand, or other permeable medium planted with emergent vegetation. Wastewater is treated as it flows horizontally through the medium plant filter.

Performance of Constructed Wetlands

The performance of wetland systems varies widely and depends to a great degree on the strength of the influent. Effluent BODs are normally less then 10 mg/L. Nitrogen removal can be effective in both types of constructed wetlands depending on preapplication treatment, detention times, and loading rates. In cold weather, the ability to nitrify decreases when water temperatures fall below 5 °C (41 °F) (Reed et al., 1995). When plants go into senescence, nutrients are released into the water column. Phosphorus removal in wetlands is achieved through vegetative takeup and precipitation into the bottom muck. This phosphorus can be removed from the wetland if the vegetation is harvested or the bottom muck is dredged.

Land Requirements

The required land area for free-water surface wetlands can be estimated as

$$A_{fw} = \frac{q\,\bar{t}}{10{,}000\,d} \qquad (9.15)$$

where
- A_{fw} = surface area of free-water surface wetland (ha);
- q = wastewater flow (m³/d);
- $\bar{t}$ = hydraulic retention time (days); and
- d = depth (m).

and the required area for a subsurface flow wetland can be estimated as

$$A_{sf} = \frac{q\,(\ln C_o - \ln C_e)}{k\,d\,n\,(10{,}000)} \qquad (9.16)$$

where
- A_{sf} = subsurface flow wetland area (ha);
- q = flow (m³/d);
- C_o = influent BOD_5 concentration (mg/L);
- C_e = effluent BOD_5 concentration (mg/L);
- k_t = first-order rate coefficient, d⁻¹, at temperature T (°C);
- d = depth of media (m);
- n = drainable voids, fraction; and
- 10,000 = conversion from square meters to hectares.

For a bed of medium to coarse gravel, a typical value of k_{20} would be 1.1 d⁻¹ and a typical value of n would be 0.38 (Reed et al., 1995). Typical bed depths for subsurface flow wetlands range from 0.5 to 0.75 m (20 to 30 in.).

Free-water surface wetlands can also provide significant wildlife habitat. Alternating shallow (less than 0.6 m) and deep (greater than 1 m) water areas can provide supplemental oxygen for aerobic treatment, provide open water for waterfowl, and reduce the need for planting and harvesting of emergent plants. Distribution to a free-water surface wetland should be designed as a manifold, as shown in Figure 9.4 or an equivalent method. Outlets can be by manifolds or weirs to allow variation in the water depth.

Conclusions

Tzar Peter the Great built his capital, now again known as St. Petersburg, in the swamps of the Neva river. This was his window to the west and he set about making sure that it was impressive. Constructing the foundations for the buildings proved to be an immensely difficult problem, requiring the driving of piles and bringing in huge stone blocks. Peter spent millions of rubles (which Russia at that time could ill afford) and thousands of lives to turn his swamp into a grand city.

On a much smaller scale, an eminent civil engineer some years ago looked down on the city of Pittsburgh from a restaurant far atop Mt. Washington and marveled: "Imagine. Not too long ago this was just a worthless swamp."

Figure 9.4 Inlet manifold to the freewater surface wetland.

On an even smaller scale, a developer in New Hampshire dumps rock and dirt on a lakeshore to fill in a swampy area, and then years later, when the area is already overgrown and no trace of his crime remains, he builds a waterfront house at a handsome profit.

People do not like muck and squishy surfaces, and we go about diligently paving the entire earth. What many people do not realize is that these mucky and squishy places are absolutely necessary for the global ecosystem. Only when the destruction of wetlands hits us in our pocketbook (or our stomach) will most people acknowledge that they are important. The bumper sticker "No Wetlands, No Seafood" neatly summarizes our environmental ethic. It does not matter until it affects us directly.

Constructing wetlands as a means of wastewater disposal is a drop in the bucket as far as the global wetlands requirement goes, but it is a start. Design engineers should seriously consider the alternative of wetlands or other "natural" means of treatment, and not do a knee-jerk design of concrete boxes and high-energy-demand systems.

References

Crites, R. W. (1985) Nitrogen Removal in Rapid Infiltration Systems. *J. Environ. Eng.*, **111**, 865.

Crites, R. W. (1996) Constructed Wetlands for Wastewater Treatment and Reuse. Presented at the Engineering Foundation Conference, Environmental and Engineering Food Processing Industries XXVI, Santa Fe, New Mexico.

Great Lakes–Upper Mississippi River Board of State Sanitary Engineering Health Education Services, Inc. (1990) *Recommended Standards for Sewage Works*; Albany, New York.

Hannah, S. A.; Auster, B. M.; Eralp, A. E.; Wise, R. H. (1986) Comparative Removal of Toxic Pollutants by Six Wastewater Treatment Processes. *J. Water Pollut. Control Fed.*, **58**, 27.

Jewell, W. J.; et al. (1978) *Limitations of Land Treatment of Wastes in the Vegetable Processing Industries*; Cornell University Press: Ithaca, New York.

Oswald, W. J. (1991) Introduction to Advanced Integrated Wastewater Ponding Systems. *Water Sci. Technol.*, **24** (5) 1.

Reed, S. C.; Crites, R. W. (1984) *Handbook of Land Treatment Systems for Industrial and Municipal Wastes*; Noyes Publications: Park Ridge, New Jersey.

Reed, S. C.; Crites, R. W.; Middlebrooks, E. J. (1995) *Natural Systems for Waste Management and Treatment*, 2nd ed.; McGraw-Hill: New York.

Rich, L. G. (1996) *High Performance Aerated Lagoon Systems*; American Academy of Environmental Engineers: Annapolis, Maryland.

U.S. Environmental Protection Agency (1975) *Wastewater Treatment Lagoons*; EPA-430/9-74-001, MCD-14, U.S. EPA, Washington, D.C.

U.S. Environmental Protection Agency (1981) *Process Design Manual Land Treatment of Municipal Wastewater*; EPA-625/1-81-013; Center for Environmental Research Information: Cincinnati, Ohio.

U.S. Environmental Protection Agency (1983) *Design Manual on Municipal Wastewater Stabilization Lagoons*; EPA-625/1-83-015; Center for Environmental Research Information: Cincinnati, Ohio.

Water Pollution Control Federation (1989) *Natural Systems for Wastewater Treatment*; Manual of Practice No. FD-16: Alexandria, Virginia.

Whiting, D. M. (1976) *Use of Climatic Data in Estimating Storage Days for Soil Treatment Systems*; EPA-600/2-76-250; U.S. Environmental Protection Agency, Office of Research and Development: Cincinnati, Ohio.

Witherow, J. L.; Bledsoe, B. E. (1983) Algae Removal by the Overland Flow Process. *J. Water Pollut. Control Fed.*, **55**, 1256.

Zirschky, J. (1986) Hydrograph Controlled Release Lagoons. *Proceedings of Field Evaluation and I/A Technology*; Technology Transfer Seminar; U.S. Environmental Protection Agency: Washington, D.C.

Symbols Used in this Chapter

A = area (ha)
a = constant
B = constant
C_n = effluent BOD (mg/L)
C_n = nitrogen concentration (mg/L)
C_p = percolation rate of nitrogen (kg/ha·y)
C_0 = influent BOD (mg/L)
C_z = effluent BOD at point z
d = depth (m)
E = evaporation rate (m/y)
E_u = irrigation efficiency (%)
f = fraction
G = leaching requirement (%)
h = hydraulic head (cm)
k = constant

k = permeability coefficient (cm/s)
L_l = limiting hydraulic loading rate (m/y)
L = hydraulic loading rate (m/ha)
L_R = BOD_5 loading rate (kg/ha·d)
L_N = nitrogen loading rate (kg/ha·y)
L_w = wastewater hydraulic loading rate on basis of soil permeability (m/y)
L_{wn} = hydraulic loading rate based on nitrogen limits (m/y)
l = steady-state infiltration rate (m/y)
M = application rate (m³/m²·h)
n = number of cells
n = constant
P = precipitation rate (m/y)
P_w = percolation flow (m/y)
q = flowrate (m³/d)
Q = flow through liner (cm³/s)
$\bar{t}$ = hydraulic retention time (d)
U = crop uptake of nitrogen (kg/ha·y)
V_s = net stored wastewater volume (m³/y)
W_p = percolation rate (m/y)
z = thickness of membrane (cm)
Z = slope length (m)

10 Physical–Chemical Processes

Introduction

Engineers are often required to design wastewater treatment plants for pollutant removals that exceed the usual definition of secondary treatment. For example, removal of plant nutrients such as nitrogen or phosphorus may be required to prevent eutrophication of the receiving water body. Lower effluent concentrations of 5-day biochemical oxygen demand (BOD_5) or suspended solids may be required to achieve local water quality standards. Sometimes, trace pollutants such as heavy metals or refractory organics may require reduction because of their toxicity to aquatic life or interference with downstream potable water supplies.

Effluent toxicity is an issue of ever-increasing importance. New discharge permits or renewals of current permits will, in many cases, include toxicity-based limits. Traditional secondary treatment methods have achieved limited success in removal of toxic substances.

These advanced wastewater treatment needs may be addressed as an integral part of a new treatment flow sheet or as an add-on to the existing secondary treatment train. This chapter presents design information for unit processes that provide effluent polishing, nutrient removal, or removal of toxic constituents. The six unit processes discussed are filtration, adsorption, chemical treatment, membrane processes, air stripping, and breakpoint chlorination.

Process Selection

Each unit process added to a conventional secondary treatment process flow sheet to achieve enhanced degrees of treatment has certain design objectives and capabilities that, in light of treated effluent limits, provide the basis for process selection. Further, within each unit process category are many variations of systems and equipment. In general, as a minimum, the following factors should be considered: effluent goals, process capabilities, process compatibility with overall treatment flow sheet, operational factors, process control, sidestreams and recycles, solids production, air emissions, energy requirements, space requirements, worker health and safety, reliability, and cost.

Granular Media Filtration

Process Description

The objective of wastewater filtration is to remove suspended solids. Granular media filtration has been applied for treating municipal wastewater in a variety of processing sequences. Filtration is used where the effluent limit is equal to or less than 10 mg/L suspended solids. It may be applied following secondary biological treatment to remove particulate carbonaceous BOD and residual insolubilized phosphorus. In addition, combinations of filtration with biooxidation or reduction systems have some applications in tertiary treatment. The degree of suspended solids removal when filtering secondary effluent without the use of chemical coagulation depends on the degree of bioflocculation achieved during secondary treatment. The presence of significant amounts of algae impedes filtration of lagoon effluent. Pretreatment with a primary coagulant is considered to be a good practice for such cases.

Most wastewater filters in operation in the United States are downflow units, but some proprietary systems pass flow upward through the media while others, called *biflow systems*, combine downflow and upflow. Alternate configurations to downflow were developed to accommodate higher solids loads and achieve filtration throughout the bed depth. In single-medium filters, stratification occurs after backwash and the concentration of the fine-grained portion of the medium in the upper portion of the bed prevents penetration and full use of the bed depth. Upflow is one way to achieve fuller use of the bed. Another approach involves use of dual-media or multimedia beds.

The driving force for filtration may be either gravity or applied pressure through pumping. Gravity filters are most commonly used in larger treatment plants; pressure filters are often used in smaller plants and in industrial waste applications. Pressure filters can accommodate higher hydraulic loading rates and higher terminal head losses. Theoretically, this results in longer filter runs and lessens backwash requirements, although these advantages may be offset by increased energy requirements and mechanical complexity.

Design Objectives

An effective design of a filtering unit fulfills three objectives. The primary goal is to consistently obtain desired filtrate quality when treating an influent with a variable suspended solids concentration exhibiting a wide spectrum of particle sizes and composition. This performance measure is called *filtration efficiency*, expressed best in terms of residual suspended solids or turbidity rather than as percent removal.

A second goal is to maintain continuous service under a variety of loading conditions. A measure of this is the filter's capacity for solids retention, best expressed in terms of the mass of suspended solids removed per area of medium per developed head loss. This can be defined as the filter's efficiency. A third goal is to clean the filter bed successfully. The effectiveness of this third step can be measured by the restoration of original head loss and solids storage conditions.

Filtration Theory

A variety of equations relating these process variables and removal mechanisms can be found in the literature. To date, however, no mathematical model accurately predicts optimum operational conditions for filtration of heterogeneous, flocculated suspensions found in wastewater. Nevertheless, because these equations provide some degree of approximation, conclusions derived from them influence filtration design.

Equations characterizing granular medium filtration account for an initial transport stage, where the particles are brought into contact with the surface of the filter grains, and a physical–chemical stage, where the particles become attached and retained within the body of the filter bed. Transport theory predicts that filtration will improve with lower flowrates, larger suspended solids particle sizes, and finer media. This theory rests on the hypothesis that the quantity of suspended solids removed within a given layer of a filter bed is proportional to the concentration of suspended solids in the liquid applied to that layer. As visualized, filtration efficiency would decline as deposited solids increase because of increased velocity in restricted pores with resultant scouring. Scoured solids would then move through the bed and into the final effluent, sharply increasing the suspended solids at virtually the same instant that maximum allowable head loss develops.

This analysis leads to the deduction that increasing the overall efficiency of the filter would require a greater proportion of the solids to reach the lower layers of the bed. This can be accomplished by increasing the depth of the filter bed, providing coarser grains at the influent end and progressively finer media in the direction of flow, and increasing the solids particle size (flocculation) in the direction of flow. Dual-media filters using anthracite coal and sand accomplish this objective. Multimedia filters use both anthracite and garnet sand, which has a high density. After backwashing, the filter stratifies to anthracite on top, then sand, and then garnet sand on the bottom.

Application of Granular Media Filtration to Wastewater Treatment

In advanced wastewater treatment, filtration is used to remove residual biological floc in settled effluent from secondary treatment (tertiary filtration) and deposits produced by alum, iron, or lime precipitation of phosphates in secondary effluent. (The suspended solids to be filtered can behave substantially differently from those in nonchemically treated secondary effluent.)

Filters may be used as the final process of wastewater treatment (polishing secondary or tertiary effluent) or as an intermediate process to prepare wastewater for further treatment (for example, before downflow carbon adsorption columns or clinoptilolite ammonia exchange columns). Developing a design to achieve the necessary filtrate quality may require a pilot study to evaluate flow and solids characteristics of the wastewater to be filtered. In the absence of a pilot study, the design must be based on experience with similar filter influent water at other installations.

Avoiding the potential problem of short filter cycles during peak load periods calls for consideration of flow equalization before filtration as part of the plant flow scheme.

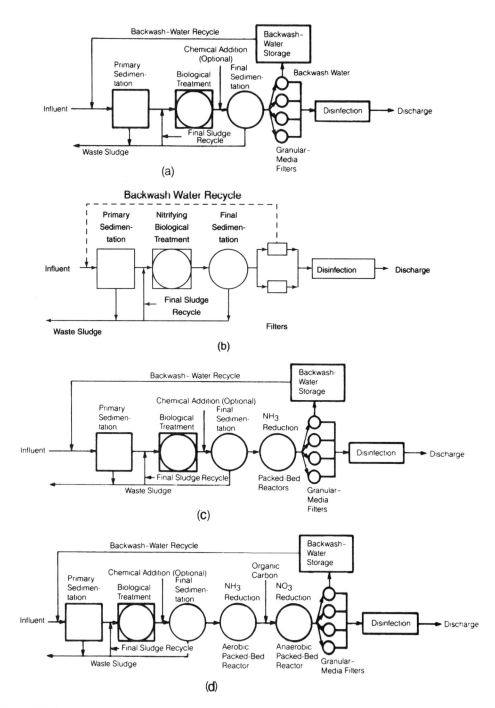

Figure 10.1 Granular-media filters following (a) biological secondary treatment, (b) nitrifying biological treatment plus tertiary filtration for biochemical oxygen demand and suspended solids control, (c) biological secondary treatment plus tertiary treatment for ammonia reduction, and (d) biological secondary and tertiary treatment for ammonia and nitrate reduction.

Storage capacity equal to 15 to 20% of mean daily flow will permit constant-rate flow to the filters for a 24-hour period. Storage capacity equal to 100% of mean daily flow will permit constant flow and nearly constant solids load to the filters. Neither flow equalization practice is widespread and the costs do not justify the filtration advantage alone.

Figure 10.1 shows four typical process flow sheets incorporating granular media filtration for effluent polishing. The first train is a nitrifying activated-sludge process followed by conventional gravity filters. This sequence of processes includes a backwash water surge basin, and backwash water supply that is stored in the chlorine contact basin.

The second sequence consists of a nitrifying activated-sludge process followed by low-head or automatic backwash filters. Because those filters are washed one cell at a time, the backwash water surge basin can be eliminated and the need for a backwash water supply basin is eliminated.

Treatment sequences three and four present various combinations of biological and advanced treatment processes followed by filtration. In each case, filtration is the final solids removal step following various combinations of secondary biological treatment and nitrogen removal. The only process recycle from the granular filtration is the filter backwash water, which is routed back to the primary clarification stage. In systems using granular activated carbon to remove organics, filtration might be necessary, ahead of the carbon columns to prevent clogging.

Process Design

Pretreatment to Enhance Filterability
Each wastewater process stream filters differently. A pilot-plant study of a process stream can provide a good approximation of the stream's filterability; however, such studies do not completely duplicate full-scale stress conditions. Regardless of documented filter performance, good design practice provides for the capability to add inorganic or organic coagulants both upstream of the settling tank that precedes the filter and to the filter influent.

Typical dosages of organic polyelectrolytes added to the settling tank influent are 0.5 to 1.5 mg/L and 0.05 to 0.15 mg/L to the filter influent. Nonetheless, jar testing is advisable to determine the optimum coagulant dosage for a particular wastewater. Overdosing may impair operations as much as or more than adding no coagulant because mudballs may form in the filter bed.

Use of proper coagulant dosages can be expected to produce effluent suspended solids of 5 mg/L or less with proper media sizing and depths. If a daily average influent suspended solids concentration of less than 40 mg/L to the filter cannot be realistically anticipated (an influent average of 25 mg/L or less would be best), the design engineer should consider upstream pretreatment facilities consisting of chemical coagulation, flocculation, and sedimentation or flotation. In addition, during some periods of the day and operating year the suspended solids concentration in the filter influent might significantly exceed the design average concentration. Therefore, the design engineer should assess filter performance under stress-loading conditions and include it among the filter selection criteria.

Filter Type and Loading Rates
As an initial step in the design process, the engineer determines the type of filter to be used. This decision is often based on variables related to the existing secondary treat-

ment plant such as space available for tertiary filters, hydraulic profile of secondary plant, construction time available, regulatory considerations, capital and operating costs, and operation and maintenance factors.

Either gravity or pressure filters may be used. Gravity filters operate with lower filtration rates than pressure filters (5 to 15 m/h [2 to 6 gpm/ft^2]) and lower terminal head losses (2.4 to 3 m [8 to 10 ft]). Pressure filters can be operated with filtration rates of 20 m/h (10 gpm/ft^2) and terminal head losses up to 9 m (30 ft) without solids breakthrough. At large plants with adequate capital resources, multiple gravity filters with fairly low filtration rates and low terminal head losses may be preferable. At a smaller plant with severe space limitations, pressure filters may be a better choice. However, these filters must be selected with caution because internal inspection, maintenance, and media replacement are complicated by size and space limitations and concurrent problems of lighting and ventilation. In practice, the use of pressure filters for municipal application is not common; however, their use is encountered more frequently in industrial wastewater treatment applications.

Multiple filter units are used to allow continuous treatment during backwash or maintenance of a unit. The number of units must be minimized to reduce piping and construction costs, but it should be sufficient to avoid excessive backwash flows and ensure that units in service can properly accommodate flow. Typical length/width ratios for gravity filters vary from 1:1 to 4:1. The size and configuration of pressure filters depend on systems available from manufacturers or local fabricators.

Media Selection and Characteristics

The depth of solids penetration to a filter depends principally on the size of the filter medium. Duration of filter run, hydraulic load, medium shape, and medium size distribution are all of secondary importance. If the medium is too large, a poor filtrate will result. If the medium is too small, removed solids will accumulate near the surface and result in short filter runs.

Medium shape affects both filtration and backwash. Sharp, angular grains may interlock and require increased backwash pressure. A nonuniform medium increases the tendency for channelization during backwash. Rounded grains break up and fluidize more readily; they also tend to rotate during backwash, scouring adjacent grains and freeing adhered solids. Angular grains with flat, plain surfaces tend to stick together, resist rotation, and may float out of the system when backwashing with air. Filter sand, when compared with a perfect sphere, should have a sphericity ratio of approximately 0.9; filter anthracite should have a sphericity ratio of approximately 0.7.

Anthracite, a largely organic medium, has a surface that preferentially adsorbs other organic substances such as fats and oils. This results in an oily film that may resist removal by conventional cleaning techniques and, in turn, accelerate the formation of agglomerates that reduce system effectiveness if they are not flushed from the filter during backwash. Because anthracite is relatively soft, it tends to deteriorate from abrasion, especially during backwash. This deterioration may lead to production of smaller particles that, if not lost during backwash, will diminish solids penetration in the bed. Oily films and deposits can be removed from silica sand using less scouring intensity than for anthracite.

Specific gravities of typical materials used as filtering media are 4.2 for garnet sand, 2.6 for silica sand, and 1.6 for anthracite. Lighter media require greater freeboard to be retained in the filter tanks. However, greater freeboard also increases the tendency for heavy process solids to be retained in the filter.

Silica sand has a Moh scale hardness of 7; anthracite exhibits a Moh scale hardness of 3.0 to 3.75. High-intensity scouring, needed for effective filter cleaning to remove tenacious solids found in wastewater effluent, tends to erode media grains. This attrition, if excessive, shortens filter runs, increases backwash losses, and results in a continuously widening size gradation of media. As a result, the substance with the most hardness is best for wastewater filtration.

The effective size and required depth of filtering media are interrelated. As a good practice, the minimum depth of the finest medium is at least 150 mm (6 in.) and the minimum medium particle diameter will be at least 0.35 mm. Presently, no specific guides relating media size and depth to suspended solids removal performance exist.

Media specifications include the effective size and the uniformity coefficient. Effective size is defined as the size at 60% of the sand passing a stack of sieves. The uniformity coefficient is the 60% passing size divided by the 10% passing size. The closer the uniformity coefficient is to 1.0, the more all the sand grains are of equal size (greater uniformity). Filter sand used in water treatment has an effective size of about 0.5 mm (0.02 in.) and a uniformity coefficient of 1.5. The uniformity coefficient should not exceed 1.7. Uniform media or media with a uniformity coefficient less than 1.3 are unnecessary, except for deep-bed, single-medium filters, especially if the bed is cleaned with an air scour.

Filter beds consist of a single medium in both stratified and unstratified forms, dual media, and multimedia. Single-medium stratified beds, used in the past, are no longer designed for municipal wastewater applications because they tend to clog. Single-medium unstratified beds, now in use, have bed depths up to 2.0 m (6.5 ft). The random pore size allows filtration throughout the bed depth, resulting in longer filter runs before head loss buildup requires backwash. The unstratified form for the single medium can be achieved by using a single-sized medium or a variable-sized medium with a combined air–water backwash. The combined air–water backwash scours accumulated material from the filter bed without fully fluidizing the medium. This avoids stratification of the nonuniform medium.

Automatic Backwashing Filters

The automatic backwashing filter, which is often referred to as a *traveling-bridge filter*, a *continuous backwash filter*, or a *low head filter*, is constructed in modules that are 4.9 m (16 ft) wide. The length of the unit and the number of units are then adjusted to provide the required filter surface area. The automatic backwashing filter varies from conventional granular media filters in two important ways. The filter uses a relatively shallow depth of media, less than 300 mm (12 in.), and the filter is backwashed continuously. The filter underdrain system of an automatic backwashing filter is divided into compartments, or cells, and each cell is individually backwashed as the traveling bridge and backwash hood are positioned over the cell to be cleaned. Water enters through the influent channel, flows into the filter box, and is collected in the filtrate channel.

Moving Bed Filters

Moving bed filters provide a continuous supply of filtered water without the interruption of backwash cleaning cycles. In a typical downflow configuration, influent enters the top of the filter and flows downward through layers of increasingly finer sand. Filtered water collects in a central filtrate chamber before exiting the filter. Solids captured in the filter bed are drawn downward with the sand into the suction of an air-lift pump. The turbulent, upward flow in the air lift provides scrubbing action that effectively separates sand and solids before discharging to the filter washbox. The washbox is a baffled chamber that allows gravity separation of cleaned sand from concentrated waste solids. From the washbox, regenerated sand returns to the top of the filter bed and the solids are piped to a suitable disposal. Filter media size is tailored to individual application requirements, with effective size ranging from 0.6 to 1.0 mm (0.02 to 0.04 in.).

In a typical upflow moving bed filter, influent entering the bottom of the filter flows upward through a series of riser tubes. Wastewater spreads evenly into the sand bed through the open bottom of an inlet distribution hood and flows upward through the downward moving sand bed as solids are removed. Clean filtrate exits the sand bed and overflows a weir as it leaves the filter. Simultaneously, the sand bed, with the accumulated solids, is drawn downward into the suction of an air-lift pipe positioned in the center of the filter. A small volume of compressed air is introduced to the bottom of the air lift. Sand, dirt, and water surge upward through the pipe at a rate of approximately 500 m/h (200 gpm/ft^2). This violently turbulent upward flow scours impurities from the sand. On reaching the top of the air lift, dirty slurry spills over into a central reject compartment. Sand returns to the sand bed through the gravity washer separator that allows fast-settling sand, but not the dirty liquid, to penetrate. The sand bed is continuously cleaned while both a continuous filtrate and reject stream are produced.

Pulsed Bed Filters

A typical pulsed bed filter has quasi-continuous operation with a 300-mm (10-in.) bed depth and a media having a 1.5 uniformity coefficient. Media selection depends on effluent standards and feed solids characteristics. This system incorporates an air-mix–pulse-mix feature that operates once the bed becomes clogged. Once the water level builds up to the air-mix probe, low-pressure air is supplied to the diffuser just above the bed surface. Fine air bubbles leaving the diffuser create a gentle rolling motion above the bed that captures large particles on the bed surface and holds them in suspension to allow continuous filtration. After the process continues for a time, some solids may deposit on the bed; this causes the operating water level to rise to the pulse-mix probe. Once energized, the effluent valve is closed and the backwash pump turns on to force the trapped air in the underdrain through small orifices in individual subcompartments and, ultimately, through clogged media. After expelled solids are trapped in the gentle rolling admixture above the bed surface, the backwash pump is shut down. This again relieves the bed and operation continues. When the pulse-mix system accumulator ultimately reaches a number of preset pulses, 6 to 10, the system is deactivated. The admixture level then rises and activates the true backwash.

Operation

Design of a filtration system requires an understanding of a system's operation, operational requirements, and commonly encountered problems. Constant pressure, constant rate, and variable declining rate are three basic methods of filter operation.

Constant-pressure filtration results in true declining-rate filtration. Filtration begins at a high rate when resistance is low, and the rate declines as head loss develops. Such a system requires a large upstream storage capability and is seldom used with gravity filters.

In constant-rate filtration, the filtration rate or water level is held constant by the action of an effluent flow-control valve. The flow-control valve, nearly closed at the beginning of a filter run, opens slowly as the filter becomes clogged. A run terminates when the valve is fully open. Disadvantages of this system include initial and operating costs of the rate control system, which may require frequent maintenance. True constant-rate filtration is seldom used for wastewater treatment without upstream storage.

As an alternative to filtration rate control with an effluent flow-control valve, constant-rate filtration may be achieved by an influent weir box located above the design terminal head loss. This weir box splits flows nearly equally to all operating filter units and smoothly adjusts to flow changes and head loss indications by changing water levels. Variable, declining-rate filtration has advantages similar to those of constant-rate filtration with influent flow splitting. An effluent control weir at an elevation above the surface of the filtering media may prevent accidental bed dewatering and the possibility of a negative head in the filter from air binding caused by gases escaping solution. The principal difference between this arrangement and the constant-rate method is the former method's location and type of influent arrangement, which maximize working head loss.

Filter influent travels through a common conduit shared with other filters and enters the filter below the wash trough level. By suitable vertical baffling, constant-rate filtration of influent flow occurs until the water level in the filter rises above the level of the wash trough and baffle; thereafter, the unit provides variable declining-rate filtration. The water level will remain the same in all operating filters at all times, but filtration rates will vary among filters and over time, depending on the head loss buildup in each. Providing a flow-restricting valve or orifice in the effluent pipe is advisable to prevent excessive filtration rates when the filter is clean. Variable declining-rate operation sometimes produces better filter effluent quality than constant-rate filtration if the filter effluent quality tends to deteriorate toward the end of a filter run. Further, variable declining-rate configuration requires less available head loss because the filtration rate declines toward the end of a filter run.

Generally, declining-rate filters provide the best mode of gravity filter operation unless the design terminal head loss exceeds 3 m (10 ft). In such a situation, constant-level control of pressure filters may be a more economical choice. This prospective economy, however, must be weighed against a potential disadvantage of the tendency to achieve terminal head loss conditions simultaneously in all filters. This could cause a system failure under a declining-rate filtration mode unless backwash is consistently initiated before terminal conditions, an upstream reservoir is provided, or both.

Common problems identified with the operations of automatically backwashed filters include the following:

- High effluent turbidity caused by high hydraulic and organic loadings;
- Short filter runs caused by high hydraulic and suspended solids loadings;
- Mudball formation caused by long filter run time, inadequate backwashing, or both;
- Loss of media during backwashing;
- Seasonal variation of filter operation caused by secondary effluent variations (not a critical problem);

- Operational difficulties caused by surging water flows; and
- Insect problems.

Activated Carbon Adsorption

Powdered activated carbon has not been widely used in wastewater treatment. New techniques to recover and regenerate powdered carbon and new methods to apply powdered carbon might result in its increased use in wastewater facilities.

Process Description

Activated carbon, when contacted with water containing organic material, will remove these compounds selectively by a combination of adsorption of the less polar molecules, filtration of larger particles, and partial deposition of colloidal material on the exterior surface of the activated carbon. The extent of removal of soluble organics by adsorption depends on the diffusion of the particle to the external surface of the carbon and diffusion within the porous adsorbent. For colloidal particles, internal diffusion is relatively unimportant because of particle size. Organic substances refractory to adsorption—that is, those dissolved molecules passing through the column—consist of strongly hydrophilic organic molecules such as carbohydrates and other highly oxygenated organic compounds.

Adsorption is partially the result of forces of attraction at the surface of a particle that cause soluble organic materials to adhere to the particle surface, and is partially attributable to the limited water solubility of many organic substances. Activated carbon has a large and highly active surface area that results from the activation process; this produces numerous pores within the carbon particle and creates active sites on the surface of pores.

Adsorption is said to occur in three basic steps: film diffusion, pore diffusion, and adhesion of solute molecules to carbon surfaces. Film diffusion is the penetration of the solute molecule, the adsorbate, through the carbon particle's "surface film" (actually, the adsorbate molecule overcomes the resistance of the carbon particle to mass transfer). Pore diffusion involves the migration of solute molecules through carbon pores to an adsorption site; adhesion then occurs when the solute molecule adheres to the carbon pore surface.

Theory suggests that adsorption is a dynamic rather than static process; that is, adsorption–desorption occurs continuously as different organic molecules or solutes approach adsorption sites on particle surfaces. More simplistically, adsorption is a selective process because different organic molecules, or structures, bond to carbon in varying degrees. A loosely bonded organic structure can be displaced at an adsorption site by a molecule with a functional group that more tightly adheres to the carbon. Thus, the displaced solute "desorbs", or is released by the carbon, during adsorption of more preferred solute species.

Two types of adsorption have been hypothesized. Physical adsorption occurs when solute molecules are held loosely to carbon surfaces by van der Waals forces. Theory suggests that molecules are mobile and migrate on the carbon surface. Chemisorption, or chemical adsorption, occurs because molecular functional groups of the adsorbate and the carbon interact to form a stable carbon bond. Desorption is more applicable to adsorbates that are physically adsorbed than to those that are chemically adsorbed.

Activated carbon consists of rigid clusters of microcrystallites, with each microcrystallite made up of a stack of graphitic planes or plates. Each carbon atom within a particular plane is bonded to four adjacent carbon atoms, with carbon atoms at the edges of graphitic planes having highly reactive (active) radical sites. At these sites, which consist of a heterogeneous mix of basal planes and microcrystallite edges, adsorption takes place. The adsorbent capacity of carbon is reached when active sites have been filled. As these sites fill, sorption equilibrium is approached, and effluent quality deteriorates to an unacceptable level. The carbon is then considered spent and disposed of or removed for regeneration to a reactivation furnace.

The carbon transport and regeneration system provides for the movement of spent carbon to and from the carbon regeneration furnace, regeneration of the carbon, and the introduction and transport of makeup carbon within the system. Methods to regenerate granular carbon include passing low-pressure steam through the carbon bed to evaporate and remove the adsorbed solvent, extraction of the adsorbate with a solvent, regeneration by thermal means, and exposure of the carbon to oxidizing gases.

Carbon regeneration is accomplished primarily by thermal means. The two most widely used reactivation methods use rotary kiln and multiple-hearth furnaces. In rotary kilns, carbon moves countercurrent to a mixture of combustion gases and superheated steam, with carbon recovery reportedly more than 90 to 95%.

The multiple-hearth furnace is heated to a temperature sufficient to burn off carbon monoxide and hydrogen produced by the regeneration reaction. A shaft with rabble blades moves the carbon to continuously bring fresh granules to the surface and transports the carbon toward the hearth outlet opening. Thus, carbon can be transferred from hearth to hearth.

System recoveries of 90 to 95% are reported, with adsorptive capacity of the regenerated carbon similar to that of new carbon. Closely controlled heating in a multiple-hearth furnace is the most successful procedure for removal of adsorbed organics from activated carbon. The attrition of activated carbon during the regeneration process is a significant design concern.

Application of Activated Carbon Adsorption to Wastewater Treatment

The most common use of granular activated carbon in wastewater treatment is as a tertiary process following conventional secondary treatment (Figure 10.2). Activated carbon treatment is one of many processes that may be used for advanced wastewater treatment. Processes upstream of the activated carbon are designed to remove essentially all of the soluble biodegradable organics associated with the suspended solids present in the secondary effluent. Chemical clarification precedes the carbon adsorption step.

In tertiary treatment, the role of activated carbon is to remove the relatively small quantities of refractory organics and inorganic compounds such as nitrogen, sulfides, and heavy metals remaining in an otherwise well-treated wastewater. Effluent inorganic concentrations can possibly be higher than those in influent. This phenomenon can be overcome by the addition of air or, in the case of sulfides, chlorine and by the addition of a supplemental carbon source (for denitrification).

Design

The usefulness and efficiency of carbon adsorption for municipal wastewater treatment depend on the quality and quantity of the delivered wastewater. To be fully effective, feedwater to the carbon unit should be of uniform quality, without surges in flow.

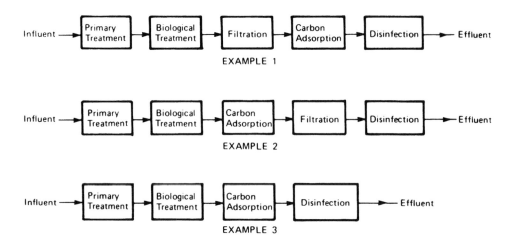

Figure 10.2 Typical flow diagrams for tertiary treatment with carbon adsorption.

Wastewater constituents that can cause potential problems are suspended solids, BOD, organics (e.g., methylene blue active substance or phenol), and dissolved oxygen. Environmental parameters of importance include pH and temperature.

The effect of suspended solids on the efficiency and life of carbon is not precisely known; nonetheless, channeling and short circuiting can occur and reduce bed life and increase carbon losses because of higher localized velocities. However, restriction of pore openings or buildup of ash (from thermal regeneration of the carbon) and other materials within the pore structure, caused by the presence of colloidal materials, reduces the adsorptive capacity of carbon. Such restrictions or buildups interfere with diffusion processes or reduce effective adsorption sites. To avoid or minimize such impairment, pretreated wastewater of the highest level of clarity is fed to activated carbon columns.

In cases where control of wastewater influent is not manageable, carbon unit performance is affected. With high influent suspended solids concentrations (that is, 20 mg/L and higher in secondary effluent), solids can deposit on carbon granules as a floc, resulting in pressure loss and flow channeling or blockages. Also, if a high level of soluble organic removal in secondary treatment is not maintained, more frequent carbon regeneration may be required. Similarly, lack of consistency in pH, temperature, and flowrate may adversely affect carbon adsorption. For these reasons, it is good practice to precede activated carbon treatment with granular media filtration.

Carbon Characteristics

The amount of substance that can be removed from a wastewater by carbon adsorption depends on conditions that provide optimum adsorption. A useful expression relating the amount of impurity in solution to that adsorbed is the empirically derived Freundlich equation

$$x/m = kc^{1/n} \qquad (10.1)$$

where
  x = impurity adsorbed (kg);
  m = unit mass of adsorbing material (carbon) (kg);

k = constant (the log x/m intercept in the graph of log x/m versus log c);
c = unadsorbed concentration of impurity left in solution (the "equilibrium" concentration, kg/L); and
n = constant (where $1/n$ is the slope of the curve on the log x/m versus log c graph).

To use this equation, the quantity x/m is measured for a number of influent concentrations. The log x/m is plotted versus log c for each of the influent concentrations, and the constants k and n are determined. This produces an adsorption isotherm that allows determination of the degree of removal achieved by the adsorption processes and the adsorptive capacity of the carbon.

However, isotherm data are developed by achieving equilibrium conditions, while field adsorption systems operate in a dynamic environment that is not necessarily in equilibrium. Because differences in adsorptive capacities between equilibrium and dynamic conditions exist, isotherm data overestimate the capability of operating systems.

During the development of adsorption isotherms, the type of activated carbon available should be considered. Activated carbons produced from different base materials and by different activation processes will have varying adsorptive capacities (Mattson and Kennedy, 1971).

When an adsorbable solute molecule (adsorbate) contacts an unoccupied adsorption site on the carbon surface, the molecule adheres almost instantly. As the solution passes over a bed of granular carbon, the adsorption rate is initially rapid and subsequently becomes slower as sites become occupied. Granular carbons require more time for their adsorptive potential to be exhausted than pulverized or powdered carbons. For larger grain active carbon, the time increment for total adsorption capacity can be several hours; a powdered carbon, if mixing is adequate, will achieve its adsorptive potential (rate) within 1 hour.

Types of Carbon Adsorption Units

Several types of activated carbon contactor systems are used in plant designs. Either pressure- or gravity-type carbon adsorption columns can be used as an upflow. In addition, countercurrent-type columns can be operated with packed or expanded carbon beds or as an upflow or downflow fixed bed unit having two or three columns in series.

Upflow columns are arranged so that the liquid moves vertically upward, with the wastewater inlet at the bottom of the column and the liquid outlet at the top. As carbon adsorbs organics, the apparent density of carbon particles increases, encouraging migration of heavier or more spent carbon to the bottom of the column. Upflow columns may have more carbon fines in the effluent than downflow columns because the upflow mode tends to expand, not compress, the carbon. Bed expansion creates fines because carbon particles collide, causing particle attrition, and allows these fines to escape through passageways created by the expanded bed.

Downflow carbon columns consist of two or three columns operated in series. Columns are piped and valved to allow operation of contactors in a countercurrent mode. The operational advantage of this design is that two processes—adsorption of organics and filtration of suspended solids—are accomplished in a single step. Claims that capital costs are lower for downflow columns are questionable as they require extensive interconnecting valves and piping to permit interchanging the relative positions of individual contactors in the adsorption system. Also, because a filter composed solely of carbon is a surface-type filter, increased head loss requiring more frequent backwashing often results. Finally, physical plugging of carbon pores with suspended materials may require premature removal of the carbon for regeneration and increase the probability of ash buildup, thus decreasing the useful life of the carbon.

Process water flows through steel or concrete contactor beds in which carbon granules remain fixed in the downflow mode or in which carbon separates to form an expanded bed in the upflow mode. Fixed beds remove particulates (if present in wastewater influent), thus requiring backwashing to dispose of accumulated particulate material. Fixed beds use downward flow to lessen the chance of accumulating particulate material at the bottom of the bed where it would be difficult to remove by backwashing. Sand and gravel resting on a filter block form the supporting media for downflow contactors.

The upflow column leads to the development of the moving, or pulse, bed, in which wastewater flows upward through a descending fixed bed of carbon. When the adsorptive capability of the carbon at the bottom section of the column is exhausted, that portion of the carbon is removed, and an equivalent quantity of regenerated or virgin carbon is added to the top of the column. Expanded beds provide a degree of organic removal similar to that achieved by fixed beds, but they require less pumping pressure and less downtime. Aeration of carbon surfaces also may be provided more readily.

When impurities that are difficult to adsorb are present in large concentrations, removal efficiency requirements for a single-stage unit may dictate the use of more carbon than is practical. However, because carbon in equilibrium with a dilute solution of partially treated wastewater may remove additional contaminants from a more concentrated wastewater, use of countercurrent techniques may reduce the amount of carbon used.

The operating principle of this process involves using two separate beds of carbon (infrequently, more than two are used). A quantity of wastewater is partially treated by a once-used carbon. This twice-used carbon is discarded, and the once-treated wastewater receives a second treatment with sufficient virgin or regenerated carbon to produce the required effluent quality. The process then is repeated for additional wastewater. By exposing the carbon to different concentrations of wastewater constituents, the amount of material held by a unit weight of carbon is increased, thus decreasing the amount of carbon required.

Unit Sizing

The sizing of carbon contactors is based on four factors: contact time, hydraulic loading rate, carbon depth, and number of contactors. The carbon contact time, calculated on the basis of the volume of the column occupied by the activated carbon, ranges from 15 to 35 minutes depending on the application, wastewater constituents, and desired effluent quality. For tertiary treatment applications, carbon contact times of 15 to 20 minutes are used where the effluent quality limits require a chemical oxygen demand (COD) of 10 to 20 mg/L, and 30- to 35-minute contact times are used where the effluent COD requirement is 5 to 15 mg/L. For chemical–physical treatment plants, carbon contact times of 20 to 35 minutes are used, with a typical contact time of 30 minutes.

A standard column (Figure 10.3) is a vessel having a flat, conical, or dished head and a carbon-retaining screen and supporting grid installed in the bottom. The minimum height/diameter ratio of a column is 2:1.

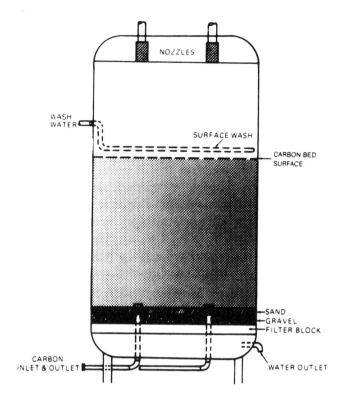

Figure 10.3 Typical downflow activated carbon contactor (U.S. EPA, 1973).

Hydraulic loading rates of 10 to 20 m/h (4 to 10 gpm/ft²) of cross section of the bed are used for upflow carbon columns. For downflow carbon columns, hydraulic loading rates of 7 to 12 m/h (3 to 5 gpm/ft²) are used. Actual operating pressure is seldom more than 7 kPa (1 psi) for each 0.3 m (1 ft) of bed depth.

Bed depth varies, within a range of 3 to 12 m (10 to 40 ft), depending primarily on carbon contact time. A minimum carbon depth of 3 m is advisable. Typical total carbon depths range from 4.5 to 6 m (15 to 20 ft). Freeboard must be added to the carbon depth to allow a bed expansion of 10 to 50% during backwash or expanded bed operation. Carbon particle size and water temperature determine the required quantity of backwash water to attain the desired level of bed expansion.

A minimum of two parallel carbon contactor units is necessary for a plant of any size. Two units in series are also advisable to permit removal of a spent carbon unit while the plant remains in operation. The number of contactors should be sufficient to ensure enough carbon contact time to maintain effluent quality while one column is offline during removal of spent carbon for regeneration or maintenance.

Backwashing

Backwashing consists of exposing the column media to a fluid flow sufficient to remove solids that either have accumulated in the bed or have been created by abrasive action. The rate and frequency of backwash depend on hydraulic loading, the nature and concentration of suspended solids in the wastewater, carbon particle size, and adsorber type (that is, expanded or fixed bed). Backwash frequency may be specified arbitrarily (each day at a specified time) or determined based on operating criteria (head loss or turbidity). Backwash duration is 10 to 15 minutes. The equipment used for backwashing and control is similar to that used for the granular-media filtration system.

One of the alternatives for carbon contacting systems is the downflow of wastewater through the carbon bed. The principal reason for selecting a downflow contactor is dual-purpose use of carbon for adsorption of organics and filtration of suspended materials. A disadvantage associated with downflow systems is that provisions are necessary for backwashing beds periodically to relieve the pressure drop resulting from the accumulation of suspended solids. Otherwise, continuous operation for several days without thorough backwashing eventually compacts or fouls beds. Normal quantities of backwash water are less than 5% of the product water for a filter 0.8 m (2.5 ft) deep and 10 to 20% for a filter 4.5 m (15 ft) deep. Typical backwash flowrates for granular carbons 29 to 50 m/h (12 to 20 gpm/ft^2).

Control of Biological Activity

A close relationship between adsorption phenomena and some aspects of biological behavior has been observed because both biomass and activated carbon can adsorb and remove certain dissolved substances from wastewater. Therefore, removal of 50 to 100% more organics in full-scale columns than that predicted by laboratory tests is not unusual because of biomass accumulation within the column.

Carbon adsorbents may catalyze biological processes, thereby increasing biological buildup. Therefore, the removal of organics in full-scale columns frequently is greater than that suggested by laboratory evaluation. After the predicted adsorptive capacity of the carbon is exhausted, biological activity often continues.

If the supply of dissolved oxygen in the wastewater is insufficient, anaerobic bacterial activity may occur. In the presence of oxygenated compounds (nitrates, sulfates, carbohydrates) and readily decomposable organic compounds, anaerobic bacteria cause the oxygen in these compounds to react with organics, producing gases such as nitrogen, hydrogen sulfide, and methane.

Carbon Transport

Spent carbon must be removed from carbon absorbers. In downflow contactors, removal provisions are straightforward because all of the carbon is removed at the same time and the column is refilled with fresh or regenerated carbon. Care should be used to prevent entry of the gravel or stone supporting media used in downflow contactors to the carbon transport system.

In upflow pulsed beds, only 5 to 25% of the total carbon column charge is removed at any given time to achieve maximum loading of the carbon with organic contaminants. Spent carbon is removed from the bottom of the vessel and a fresh carbon charge, equal in volume to the spent carbon removed, is added to the top of the vessel. This procedure ensures that fresh carbon is positioned to "polish" the effluent before it leaves the adsorber and that the carbon containing the most adsorbate of the most spent carbon is removed for regeneration. It is important to obtain a uniform withdrawal of carbon over the entire horizontal surface area of the carbon bed.

Carbon Regeneration

The carbon dosage or use rate for regeneration equipment sizing depends on the wastewater characteristics and the required effluent quality. Typical carbon dosages for municipal wastewater are shown in Table 10.1.

Table 10.1 Typical carbon dosages for various wastewater influents.

Prior treatment	Typical carbon dosage required/mil. gal column throughput,[a] lb/mil. gal[b]
Coagulated, settled, and filtered activated-sludge effluent	200–400
Filtered secondary effluent	400–600
Coagulated, settled, and filtered raw wastewater (physical–chemical)	600–1800

[a]Loss of carbon during each regeneration cycle is typically 5 to 10%. Makeup carbon is based on carbon dosage and the quality of the regeneration carbon.
[b]lb/mil. gal × 0.1198 = g/m^3.

Dewatering of the spent carbon slurry before thermal regeneration is accomplished in spent carbon drain vessels. Screens in drainage vessels allow transport water to flow from the carbon. Gravity drainage vessels dewater the carbon to 40 to 55% moisture content. Two drain bins are provided to permit continuous carbon feed to the furnace.

Partially dewatered carbon may be fed to the regeneration furnace with a screw conveyor equipped with a variable-speed drive to precisely control the rate of carbon feed. Carbon feed rate is controlled because carbon moisture content and quantities of adsorbed organics vary.

The anticipated carbon dosage governs the theoretical furnace capacity. For multiple-hearth furnaces, approximately 0.005 m^2 of hearth area is required for each kg (dry weight) of carbon per day (0.025 ft^2/d/lb). Actual furnace capacity includes an allowance for furnace downtime of approximately 40% beyond theoretical capacity.

Based on operating experiences at two full-scale facilities, the furnace should have provisions to add approximately 1 kg of steam per kg of carbon regenerated (1 lb/lb). Fuel requirements for the furnace are 7000 kJ/kg (3000 Btu/lb) of carbon when regenerating spent carbon for tertiary and secondary effluent applications. To this value, energy requirements for steam and an afterburner should be added if they are required.

Chemical Treatment

Chemicals are used for a variety of municipal treatment applications including enhancement of flocculation–sedimentation, solids conditioning, odor control, algae control, nutrient addition, activated-sludge bulking control, acid–base neutralization, precipitation of phosphorus, and disinfection.

Chemicals can also be used for precipitation of heavy metals, but metals are better controlled at the source through an industrial pretreatment program. Because flocculation, settling, solids conditioning, odor control, and disinfection are discussed in other chapters, chemical treatment in this chapter is limited to phosphorus precipitation and neutralization.

Phosphorus Precipitation

Phosphorus enters a wastewater treatment plant in three forms: (1) organic phosphorus found in organic matter and cell protoplasm, (2) as complex inorganic phosphates (polyphosphates), such as those used in detergents, and (3) as soluble inorganic orthophosphate.

During the treatment process, most of the organic phosphate and complex phosphates are converted to inorganic orthophosphate. In wastewater that has been treated by biological processes, most of the phosphorus is in the soluble form, although a small amount of insoluble organic phosphorus also exists in the form of cell protoplasm. For example, the average amount of orthophosphate discharged in the effluent of municipal trickling filter plants without chemical precipitation is approximately 8 mg/L phosphorus. In effluent from plants with chemical precipitation for phosphorus removal most of the remaining phosphorus is in the insoluble form (calcium, aluminum, or iron phosphate). Insoluble phosphorus compounds do not release phosphorus in other units of the treatment plant or in the receiving water body.

Phosphorus is an essential element in biological treatment. For activated sludge, the required minimum phosphorus/BOD_5 ratio has been estimated as 1:100, with similar ratios expected for other biological processes. Municipal wastewater has a BOD_5 in the range of 175 to 250 mg/L and a phosphorus content of 8 to 12 mg/L, thus exceeding the nutrient requirements for aerobic biological treatment.

Phosphorus Removal Methods

Phosphorus removal may be part of the primary, secondary, or tertiary treatment process. Physical–chemical treatment techniques include chemical precipitation and flocculation of phosphorus followed by settling, flotation, or filtration. Incidental uptake by biological processes, in general, accounts for relatively little phosphorus removal unless these systems are designed for phosphorus removal. Chapter 8 discusses biological phosphorus removal methods.

Precipitants

Phosphorus precipitation requires addition of a coagulant aid (flocculant) and a coagulant. Coagulants used for phosphorus precipitation are lime, alum, sodium aluminate, ferric chloride, and ferrous sulfate. In addition to its reactions with carbonate species, lime reacts with orthophosphate to precipitate hydroxyapatite as

$$5Ca^{2+} + 4OH^- + 3HPO_4^= \rightarrow Ca_5OH(PO_4)_3 + 3H_2O \tag{10.2}$$

The theoretical molar calcium/phosphorus ratio is 5:3. The above equation is representative, but not exact, because the composition of the apatite precipitate varies. As a result, the calcium/phosphorus mole ratio may vary from 1.3 to 2.0. Hydroxyapatite solubility decreases rapidly with increasing pH and, in general, phosphate removal increases with increasing pH. Essentially all orthophosphate converts to the insoluble form if the pH is greater than 9.5.

The actual pH required to precipitate a given amount of phosphate and the amount of lime required to raise the pH to the desired level vary with wastewater composition. These parameters should be determined by laboratory jar tests.

The chief variable that affects phosphorus removal by lime is wastewater alkalinity. Unless a high pH is used, waters with low alkalinity (150 mg/L or less) form a poorly settleable floc because of the small fraction of dense calcium carbonate ($CaCO_3$) precipitate. Sufficient quantities of calcium carbonate act as a flocculant that enhances settling of the hydroxyapatite. For wastewater with high alkalinity, a pH of 9.5 to 10 can result in excellent phosphorus removal.

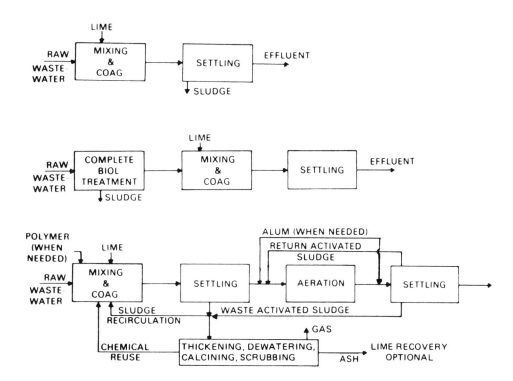

Figure 10.4 Various schemes of single-stage lime precipitation for phosphorus removal.

Magnesium hardness also affects the efficiency of phosphorus removal. At the higher pH range, magnesium hydroxide precipitates as

$$Mg^{2+} + Ca(OH)_2 \rightarrow Mg(OH)_2 + Ca^{2+} \quad (10.3)$$

This reaction begins at a pH of approximately 9.5 and is complete at a pH of 11. The magnesium hydroxide precipitate is gelatinous and removes fine suspended solids as it settles. However, these same gelatinous properties can impair solids dewatering.

Because chemical requirements impose the principal operational cost associated with phosphorus removal, proper chemical dosages are critical to process economy. The lime dosage required to reach a given effluent phosphorus level is, as a practical matter, independent of influent phosphorus concentration. The degree of phosphorus removal is a function of pH. Lime requirements to reach the required pH correlate more closely with wastewater alkalinity than with any other variable. For other phosphorus precipitants (alum or ferric salts), chemical requirements are proportional to influent phosphorus concentration. In contrast, phosphorus removal is decidedly nonstoichiometric with all chemicals.

Various lime precipitation schemes exist for phosphorus removal (Figure 10.4). Lime may be added before the primary settling tank in a biological treatment plant. Because the CO_2 produced in biological treatment can neutralize the carbonate and hydroxide alkalinity and lower the pH into the range for biological treatment, lime can be added so as to increase the pH to 10 to 11 at the primary clarifier. The maximum phosphorus insolubilization attained with lime addition is approximately 80%. At this limited pH, 2 to 3 mg/L of remaining soluble phosphorus is not unusual. Additional phosphorus may be removed, if necessary, by using aluminum or iron addition in biological reactors or the final sedimentation tank. The use of lime in primary treatment has the added advantage of increasing organics and suspended solids removal efficiencies in the primary sedimentation tank, thereby decreasing the load on the aeration system.

A second alternative consists of lime treatment following biological treatment. Phosphorus removal from the secondary effluent ensures enough phosphorus in the aeration stage to meet the nutrient demand of the biological floc. In addition, the biological system breaks down many of the complex phosphates to the more readily precipitated orthophosphate form. However, the high pH of returned solids may affect biological treatment.

Lime precipitation of phosphorus may require filtration to ensure continuous compliance with effluent requirements. Even with a process pH as high as 11.4, the effluent may contain some residual phosphorus. Although high pH values ensure that nearly all phosphorus is insolubilized, effluent total phosphorus depends on suspended solids removal efficiency. In cases where the precipitate floc is difficult to settle, granular-medium filtration can ensure an extremely high degree of suspended solids and phosphorus removal. A residual phosphorus concentration of 0.1 to 0.2 mg/L may be achieved with granular-medium filtration.

Phosphorus may be precipitated from wastewater by chemical reaction with most mineral salt coagulants, including various salts of iron and aluminum. The principal source of aluminum in phosphorus precipitation is aluminum sulfate, commonly known as alum. The reaction is

$$Al_2(SO_4)_3 \cdot (14H_2O) + 2H_2PO_4^- + 4HCO_3^- \rightarrow 2AlPO_4 + 4CO_2 + 3SO_4^{2-} + 18H_2O \tag{10.4}$$

The sulfate ion remains in solution and pH is depressed. From the above equation, the calculated weight ratio of alum to phosphorus is 9.6:1 (0.87 Al:1.0 P). More alum is actually required because of side reactions involving wastewater alkalinity and organic matter.

The solubility of aluminum phosphate is a function of pH. The most efficient chemical use occurs at a process pH near the range of minimum solubility, approximately 5.5 to 6.5. Because alum use results in a small pH depression and most existing treatment systems operate at near neutral wastewater pH, addition of alum almost automatically results in a pH in the range of minimum aluminum phosphate solubility. Sometimes excess alum is required to depress the pH sufficiently to reach the optimal pH range for phosphate removal. Excess alum then simply acts as an acid that might be replaced by another acid. Using alum following a lime treatment process provides an example of such alum applications.

Alum addition offers flexibility in the location of chemical application points. Alum can be added before the primary settling tank, in the biological reactor, or between aeration and final settling. Alum addition before the primary settling tank provides additional removal of suspended solids and organic matter in addition to phosphorus precipitation. These removals, however, result from reactions that compete with soluble phosphorus for available alum, thereby increasing the dosage required to obtain a given phosphorus residual level. In addition, some polyphosphates that are more difficult to treat may be present in raw wastewater.

Providing alum addition in the activated-sludge biological reactor allows use of the mixing already provided for that system. The best point of addition for alum in an activated-sludge plant is likely to be in the biological reactor effluent channel that carries mixed liquor to the final settling basin. Turbulence in this channel adequately mixes the chemical. The addition of alum after the biological system takes advantage of wastewater stabilization with the attendant hydrolysis of the complex phosphates to the more readily reacted orthophosphate form.

Most of the discussion concerning phosphorus precipitation with aluminum sulfate also applies to other mineral precipitants. Sodium aluminate can serve as a source of aluminum for the precipitation of phosphorus. Granular trihydrate is one commercial form of sodium aluminate. The reaction for phosphorus precipitation is

$$Na_2Al_2O_4 + 2H_2PO_4^- + CO_2 \rightarrow 2AlPO_4 + 2Na^+ + 4HCO_3^- \qquad (10.5)$$

Dissolved carbon dioxide or other acidity must be present to avoid a pH increase beyond the optimum zone. When this occurs, sodium aluminate alone is not satisfactory as a phosphate precipitant. In contrast to alum, which reduces pH, addition of sodium aluminate results in a slight rise in pH. Use of this chemical may be appropriate where wastewater pH is already low, and further depression must be avoided. From the above equation, the calculated sodium aluminate/phosphorus weight ratio is approximately 3.6:1. In practice, side reactions will require a higher dosage.

Both ferric and ferrous iron compounds may be used in the chemical precipitation of phosphorus. Although both types of compounds produce equivalent results, ferric chloride is used more often. The dominant reaction between ferric chloride and phosphorus is believed to be similar to that between phosphate and alum

$$FeCl_3 \cdot (6H_2O) + H_2PO_4^- + 2HCO_3^- \rightarrow FePO_4 + 3Cl^- + 2CO_2 + 8H_2O \qquad (10.6)$$

This equation indicates that an iron/phosphorus mole ratio of 1:1 is required. This corresponds to an iron/phosphorus weight ratio of 1.8:1. Like alum, greater amounts of ferric chloride are required to satisfy side reactions with alkalinity that produce $Fe(OH)_3$. Alkalinity reactions occur because the well-flocculating ferric hydroxide precipitate aids the settling of the colloidal ferric phosphate.

Experience has shown that efficient phosphorus removal requires the stoichiometric amount of iron to be supplemented by at least 10 mg/L of iron for hydroxide formation. Iron requirements for municipal wastewater are 15 to 30 mg/L as Fe (45 to 90 mg/L as $FeCl_3$) to reduce phosphorus by 85 to 90%. Dosages vary with the phosphorus concentration of the influent. The optimum pH range for iron precipitation of phosphorus is between 4.5 and 5.0. However, significant phosphorus removal occurs at pH values of 7 and somewhat above. As with the aluminum salts, iron salts may be added during primary, secondary, or tertiary treatment stages.

Both aluminum and iron salts, when used as phosphorus precipitants, increase dissolved solids in plant effluent. In plants with poor solids capture, ferric iron may impart a slight reddish color to the effluent.

Solids Considerations

Addition of mineral salts for phosphorus precipitation can significantly increase the quantity of solids generated because of production of metal-phosphate precipitates and metal hydroxides and improved suspended solids removal. Addition of metals upstream of the primary clarifier results in a primary sludge mass increase of 50 to 100%, almost equally attributable to improved suspended solids capture and additional chemical sludge generation. Overall plant solids mass increase is smaller because of reduced secondary sludge production from improved primary removals (for example, a 60 to 70% increase is typical across the entire plant).

For metal addition to secondary processes, waste activated sludge may increase by 35 to 45%, and the overall plant solids mass increase may be 5 to 25%. Metal addition to either primary or secondary treatment units not only increases solids mass but also solids volume because settled solids concentration in clarifiers may decrease by as much as 20%. Secondary clarifiers are especially sensitive to chemical phosphorous removal and must be oversized in order to take this solids load into account.

In the absence of definitive bench- or pilot-scale data, stoichiometric reactions of aluminum and ferric ions will provide a useful estimate of solids production. Overall reactions are shown below

$$Al^{3+} + 3H_2O \rightarrow Al(OH)_3 + 3H^+ \tag{10.7}$$

and

$$Fe^{3+} + 3H_2O \rightarrow Fe(OH)_3 + 3H^+ \tag{10.8}$$

Each mole of cation should react with three moles of water to produce one mole of metal hydroxide and three moles of hydrogen ions. Therefore, one milligram of alum, $Al_2(SO_4)_3 \cdot 14\ H_2O$, will react to produce 0.26 mg of insoluble aluminum hydroxide while consuming 0.5 mg/L of alkalinity as calcium carbonate. One milligram of ferric sulfate will produce approximately 0.5 mg of aluminum hydroxide and consume 0.75 mg of alkalinity. Alkalinity reductions are important design considerations for low alkalinity waters or nitrified effluent. During nitrification, significant alkalinity reductions occur and additional chemical treatment that further reduces alkalinity should be carefully evaluated. In denitrifying systems, however, almost half of the alkalinity used in nitrification is recovered or returned.

pH Adjustment

One of the most common types of chemical processes used in domestic wastewater treatment facilities is pH adjustment. Adjustment of pH, frequently used for coagulation and phosphorus precipitation, simply raises or lowers a pH to a more acceptable value. For example, wastewater that is excessively acidic or alkaline is objectionable in collection systems, treatment plants, and natural streams. Removal of excess acidity or alkalinity by chemical addition to provide a final pH of approximately 7 is called *neutralization*. Most effluents must be neutralized to a pH of 6 to 9 before discharge.

There are three critical components of any pH control system: mixing intensity or turnover time in the reactor, response time of the control system, and the ability of the chemical metering system to match process requirements. If any one of these components is not properly designed, significant problems in system performance can be anticipated.

Other factors that complicate the design of pH control systems include the amount of buffering capacity in wastewater, the change in mass flowrate of the hydrogen ion, variations in wastewater flowrate, and variations in wastewater temperature.

Before designing a pH control system, a thorough wastewater characterization study should be performed. If characteristics are not clearly defined, additional sampling and analysis should be considered. However, in almost all cases titration curves should be developed to provide definitive data with respect to chemical demands and consumption.

Methods used for pH adjustment are selected on the basis of overall cost because material costs vary widely as do equipment needs for different chemicals. The volume, kind, and quantity of acid or alkali to be neutralized or partially removed are also variables influencing the selection of a chemical agent.

Rapid Mixing

Associated with each type of chemical treatment process is a series of chemicals that can be used. In most cases, the success of the chemical process depends on mixing—the rapid dispersal of the chemical reagent throughout the waste stream.

Separate mix tanks are preferred for chemical mixing. If the process lacks a separate tank, the chemical must be added at a point that offers sufficient agitation and time for mixing. Pump suction and discharge lines have been used for this process.

Impeller Mixers

The most common type of mixing device for wastewater treatment is the rotating impeller mixer. Three groups of impeller mixers are used: paddles, turbines, or propellers. Of these, only turbine and propeller mixers are used for rapid mixing applications.

A *turbine impeller mixer* is submerged in a liquid and operates like a centrifugal pump without a casing. Most turbine mixers resemble multibladed paddle mixers with short blades turning at high speeds on a shaft located in the center of the mixing chamber. The diameter of the impeller is within a range from 30 to 50% of the diameter of the mixing vessel.

Propeller mixers have high-speed impellers and are used for thick solutions. Small propeller mixers revolve at full motor speed, 1750 rpm; larger mixers turn at 400 to 800 rpm. Propellers generate currents that are primarily axial and continue through the liquid in a given direction until deflected by the flow or wall of the mixing chamber. Propeller blades vigorously cut or shear the liquid. Propeller mixers are smaller in diameter than either paddle or turbine mixers, rarely exceeding 460 mm (18 in.) in diameter, regardless of the size of the mixing chamber. In a deep chamber, two or more propellers may be mounted on the same shaft; they direct the liquid in the same direction.

Other Mixing Devices

Mixing may be accomplished by several other devices, including baffled channels, hydraulic jump mixers, pneumatic mixing by the injection of compressed air, and in-line static mixing devices. Baffled channels and pneumatic mixing are better suited for flocculation operations than rapid mixing. In-line static mixers frequently are used for rapid mixing, but they have two disadvantages: head losses are up to 0.9 m (3 ft) and the mean temporal velocity gradient, G, cannot be changed to meet varying requirements but is a function of flowrate through the unit.

Fluid Regimes

The term "fluid regime" refers to the flow pattern and overall summation of the mass flow-shear relationships existing in a fluid in motion. The type of flow in a mixing chamber depends on impeller type; fluid characteristics; and tank, baffles, and mixer sizes and proportions. Fluid velocity at any point in the tank has three components; the overall flow pattern in the tank depends on the variations in these three velocity components from point to point.

Three orthogonal components—radial, longitudinal, and tangential—conveniently define the flow pattern. The radial component acts in a direction perpendicular to the shaft of the impeller; the longitudinal component acts in a direction parallel to the shaft; and the tangential or rotational component acts in a direction tangent to a circular path around the shaft.

The tangential component induced by the rotating impeller promotes rotational movement or vortexing around the impeller shaft. This vortexing impedes mixing by reducing the velocity of the impeller relative to the liquid. A preferable method to minimize this vortexing is to install baffles that impede rotational flow without interfering with radial or longitudinal flow. Simple and effective baffling may be achieved by installing vertical strips perpendicular to the wall of the tank. Four baffles suffice to prevent vortex formation, except in large mixing chambers. For turbine mixers, the width of the baffle need not exceed 8.3% of the tank diameter; for propeller mixers the baffle width should be 5.5% or less of the tank diameter. With side-entering, inclined, or off-center propellers, baffles are unnecessary.

If the combined effects of velocity components of the moving fluid produce laminar flow, no mixing occurs within the fluid except that caused by diffusion. If flow is turbulent, fluid particles move in all directions and mixing results primarily from connective displacement. The momentum transfer associated with such displacement generates strong shear stresses within the fluid. Almost all wastewater mixing regimes are in the turbulent flow range.

Design Considerations

The design of a functional and economically feasible rapid mixing system requires consideration of the following:

- Power requirements,
- Laboratory scale-up,
- Batch and continuous systems,
- Hydraulic retention time,
- Vessel geometry,
- High- and low-speed mixers,
- Propeller and turbine mixers,
- Mixer mounting,
- Top-entering turbines and side-entering propeller mixers, and
- Single-propeller and multi-propeller mixers.

Power requirements of an impeller mixer to maintain turbulent hydraulic conditions (Reynolds number greater than 10^5) can be estimated as (Rushton, 1952)

$$P = \rho K_T n^3 D^5 / g_c \tag{10.9}$$

where
    P = power requirement, N·m/s (ft-lb$_f$/s);
    ρ = mass density of the fluid, 1000 kg/m^3 (62.4 lb/ft^3) for water;

Table 10.2 Values of K_T used for determining impeller power requirements.

Impeller type	K_T
Propeller (square pitch, three blades)	0.32
Propeller (pitch of two, three blades)	1.00
Turbine (six flat blades)	6.30
Turbine (six curved blades)	4.80
Turbine (six arrowhead blades)	4.00
Fan turbine (six blades)	1.65
Flat paddle (two blades)	1.70
Shrouded turbine (six curved blades)	1.08
Shrouded turbine (with stator, no baffles)	1.12

K_T = constant;
n = impeller revolutions per second (s^{-1});
D = diameter of the impeller, m (ft); and
g_c = gravitational acceleration, $\dfrac{9.79 \text{ m·kg}}{\text{N·s}^2} \left(\dfrac{32.17 \text{ ft-lb}_m}{\text{lb}_f - \text{s}^2} \right)$.

K_T depends on the impeller shape and size, the number of baffles used to eliminate vortexing, and other variables not included in the power equation. Table 10.2 presents K_T values for mixing impellers rotating at the center line of cylindrical vessels with a flat bottom, four baffles at the vessel wall, baffle widths equaling 10% of the vessel diameter, liquid depth equal to one tank diameter, impeller diameter equal to 30% of the tank diameter, and the impeller positioned one diameter above the tank floor.

Several empirical parameters have been developed to describe performance characteristics of the rapid mixing unit operation. These include the power dissipation function, mean temporal velocity gradient, mixing opportunity parameter, and mixing loading parameter.

The power dissipation function, or power input per unit of mixer volume, provides a rough measure of the mixing effectiveness of the system because more power input creates greater turbulence and greater turbulence leads to better mixing. The mean temporal velocity gradient, G (s^{-1}), describes the degree of mixing of the system. As G increases, the degree of mixing increases. In domestic wastewater treatment, values of G range from 300 to 1500 s^{-1} for rapid mixing. The power dissipation function is expressed as

$$P/V = \mu G^2 \quad (10.10)$$

where
P = power, N·m/s (ft-lb$_f$/s);
V = mixing chamber volume, m^3 (ft^3);
μ = absolute viscosity of fluid, N·s/m^2 (lb-s/ft^2); and
G = velocity gradient (s^{-1}).

The *mixing opportunity parameter* is a ratio of power-induced rate of flow to hydraulic-induced rate of flow. In domestic wastewater treatment, this parameter ranges from 9000 to 180,000 (dimensionless) for rapid mixing. The mixing opportunity parameter is

$$G\bar{t} = (PV/\mu)^{0.5}/Q \qquad (10.11)$$

where $\bar{t}$ = hydraulic retention time of mixing basin (s) and Q = design flowrate (m³/s).

Batch-mixing systems are used in the makeup of chemical solutions. Selection of the proper chamber size depends, to a large extent, on the volume to be mixed and the time allowed for such mixing. For the makeup of specific chemical solutions requiring rapid mixing, the chemical manufacturer should be consulted to obtain the economical volume and suggested mixing times.

Continuous systems nearly always are required for mixing process waste streams. Where required in the treatment of domestic wastewater, multiple rapid-mixing chambers operated in parallel should be used. Hydraulic retention time in mixing reactors varies from 0.5 to 2 minutes. Undermixing results in inadequate dispersal of additives and uneven dosing. Overmixing may rupture wastewater solids already present in the waste stream or cause excessive dispersal of newly formed floc.

The shape and size of the tank often are dictated by process considerations. As a general rule, circular mixing tanks are more efficient for rapid mixing than square or rectangular tanks. For circular tanks, liquid depths equal to the tank diameter are a good practice. For tanks less than 4000 L (1000 gal) in capacity, portable or compact turbine mixers are most practical. For larger tanks, heavy-duty, top-entry turbine mixers are used. Squat tanks (i.e., tanks with top dimensions greater than liquid depth) are not recommended for mixing applications. For rectangular tanks, the mixers should be through one of the narrow walls of the chamber. For all tanks, one or more side-entry mixers are used in lieu of top-entry mixers with extremely long shafts.

Chemical Feed Systems

Feeding systems are necessary for the addition of reagents—solid, liquid, or gas—to the waste stream at a controlled rate. Design of a chemical feed system must consider the desired state of each chemical to be fed, the particular physical and chemical characteristics of the chemical, maximum and minimum wastestream flowrates, and the reliability of feeding devices.

Chemicals used in municipal plants may be received in either liquid or solid form. Coagulants in solid form are converted to solution or slurry before entering the waste stream. The dry feeder has numerous forms to handle wide ranges in chemical characteristics, feed rates, and degree of accuracy required. Solution feeding of coagulants depends primarily on liquid volume and viscosity.

Water-soluble coagulant aids, available as dry granular powders or concentrated liquids, are compatible with most dissolved inorganic salts at low concentrations found in tap water. Water for preparing flocculation solutions should have a low suspended solids content to avoid chemical sludge formation. Concentrated solutions of coagulants and flocculants should never be prepared consecutively in the same vessel unless residues are thoroughly removed between uses. Dry coagulant aids should be given enough time to solubilize completely because long, entwined molecules must hydrate fully before they will uncoil completely. Preparation time is decreased when dry particles initially are distributed evenly without large lumps. Then, only minimum agitation is required after dispersion to ensure a uniform solution. Flocculant solutions, unlike primary coagulants, are viscous and exhibit non-Newtonian flow. Ordinary regulating devices are inadequate and pumps and lines should not be sized on the basis of Newtonian flow properties.

The capacity of a chemical feed system is an important consideration for storage and feeding. Storage capacity designs must take into account the economy of quantity purchase versus the disadvantages of construction cost and chemical deterioration with time. Potential delivery delays and chemical use rates also merit careful consideration. Design of storage bins or tanks for chemicals must account for the angle of repose of the chemical and its necessary environmental requirements such as temperature and humidity. Size and slope of feeding lines and construction materials are other important considerations.

Chemical feeders must accommodate minimum and maximum feeding rates required. Manually controlled feeders have a common range of 20 to 1, but this range may be increased to approximately 100 to 1 with dual control systems. Chemical feeder control may be manual, automatically proportioned to flow, dependent on some form of process feedback, or a combination of any two of these. More sophisticated control systems are feasible if proper sensors are available. Standby units should be included for each type of feeder used. For proper operational flexibility, points of chemical addition and associated piping should be capable of handling all possible changes in dosing patterns.

A dry-feed installation consists essentially of a hopper, a feeder, and a dissolver tank. All three units are sized based on waste volume, treatment rate, and an optimum length of time for chemical feeding and dissolving. The best applications of dry-feed systems have high treatment rates, more stable chemicals, and more fluid materials. Less fluid materials can be handled, but feeder accessories are needed. Because a powdered or granular material may arch or bridge in a hopper, it needs vibration for continuous flow. To prevent flooding by some powders that are too free-flowing, a rotor below the hopper exit ensures flow control.

Dry feeders are either of the volumetric or gravimetric type. Most types of volumetric feeders are the positive-displacement type, incorporating some form of moving cavity of a specific or variable size. Accurate feeding depends on the chemical having a constant specific weight per unit volume. In operation, the chemical falls by gravity into the cavity and is then enclosed and separated from the hopper's feed. The size of the cavity and its rate of movement govern the material feed rate.

Solution Feed

Liquid-feed systems (Figure 10.5) are best applied for chemical treatment with lower treatment rates, less stable chemicals, chemicals that are more readily fed as a liquid to avoid handling of dusty or more dangerous chemicals, or materials available only as liquids.

Liquid-feed units include piston, positive-displacement diaphragm, and balanced diaphragm pumps and liquid gravity feeders (rotating dippers). The unit best suited for a particular application depends on feed pressure, chemical corrosiveness, treatment rate, accuracy desired, viscosity and specific gravity of the fluid, other liquid properties, and the type of control.

Piston-plunger and positive-displacement pumps are available with low to high capacities at pressures up to 41,000 kPa (6000 psig). Liquids with viscosities up to 10 N·s/m^2 (100 P) may be handled by the plunger pump, depending on the rate of feed.

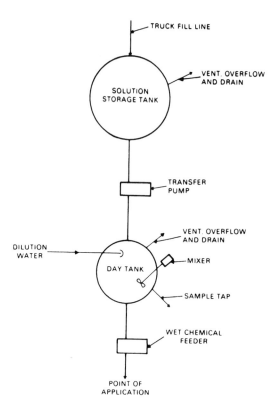

Figure 10.5 Typical feed system for wet chemicals.

Mechanically actuated diaphragm pumps are designed for low discharge pressures—less than 830 kPa (120 psig). Flow capacities of 0.076 L/s (72 gph) may be handled, although process fluids are limited to liquids with viscosities under 0.1 N·s/m^2 (1 P). Hydraulically actuated, balanced diaphragm pumps can operate against discharge pressures up to 21,000 kPa (3000 psig). They can handle capacities up to 0.4 L/s (400 gph), and liquids with viscosities of 1 N·s/m^2 (10 P) and higher can be handled at lower flowrates.

Membrane Processes

Process Description
Membrane processes can be pressure or vacuum driven or depend on electrical potential gradients, concentration gradients, or other driving forces. The focus of this section is on the predominant pressure-driven process, which include microfiltration (1 to 0.1 μm), ultrafiltration (1 to 0.01 μm), nanofiltration (0.5 to 0.05 μm), and hyperfiltration (<0.001 μm), commonly called reverse osmosis.

These four classes of membranes have been established primarily based on the size range of particles each type will remove from water or wastewater. This characteristic is described in terms of nominal pore size for micro- and ultrafiltration membranes and by molecular weight cutoff for nanofiltration and reverse osmosis membranes. Microfiltration and ultrafiltration membranes are used as filtration processes, whereas nanofiltration and reserve osmosis membranes can be classified as desalination processes.

Membrane processes were originally developed as desalination, or demineralization, processes. After the successful commercial application of membranes in reverse osmosis processes, membrane development moved into the areas of microfiltration and ultrafiltration. Microfiltration membranes have the largest pore size and are designed to remove large suspended particles such as colloids, bacteria, and cysts. Ultrafiltration membranes have a smaller pore diameter than microfiltration membranes and can remove viruses in addition to constituents removed by microfiltration membranes. In addition to constituents removed by microfiltration and ultrafiltration membranes, nanofiltration membranes can also remove large organic molecules. Reverse osmosis membranes can remove all of these contaminants in addition to ions.

Process Objectives
The objectives of membrane processes will vary considerably with the facility, location, and water quality goals desired. Membranes can be manufactured to reject many different species, and it is important that treatment objectives be determined early in the project development stage.

Pretreatment
Pretreatment requirements for various membrane processes vary tremendously and are membrane and application dependent. Because pretreatment for these processes is the most demanding, most of the following discussion is directed at reverse osmosis processes. Other processes are less demanding, but the pretreatment needs for each application should be carefully reviewed with anticipated membrane manufacturers.

To increase the efficiency and life of a membrane process, effective pretreatment of feedwater is required. Selection of proper pretreatment will maximize efficiency and membrane life by minimizing fouling, scaling, and membrane degradation. The net result of the above will be the optimization of product flow, salt rejection, product recovery, and operating costs.

Membrane Systems
Although natural semipermeable membranes displaying osmotic characteristics are common in living organisms, it was not until 1960 that a synthetic membrane with commercial possibilities became available.

Two properties are of paramount importance for a membrane. The *water permeability* determines the production rate of desalted water per unit membrane area, called *flux*. *Salt rejection* describes the capability of a membrane to retain dissolved solids. It is indicated by the *brine minimum product/brine* ratio, expressed as percent.

One of the primary difficulties facing the original design of a reverse osmosis module was the confinement of large hydraulic pressures associated with the process. Over the years, several novel systems have been developed to cope with this problem.

The first concept, know as *tubular configuration,* incorporated a porous wall tube with the membrane inserted or cast on the inner wall. The brine flow is axial within the tube, while the product water flow is radial through the membrane and the porous structure. Tubes are arranged in bundles, with each unit bundle being referred to as a module. All tubes of a module are connected in series so that the module has only one brine entrance and exit. The product water is collected in a trough or shell. Tubes are approximately 13 mm (0.5 in.) in diameter and are perforated or made from epoxy-bound fiberglass, woven nylon or polyester, or other porous materials of sufficient strength to withstand hydraulic pressure.

The second concept is the *spiral-wound* or *spiral-wrap configuration*. The process can best be understood if one visualizes a sealed envelope made from membrane with a porous backing material inside the envelope. As feed water pressure is applied to the outside of the envelope, product water traverses the membrane and collects in the porous material within the envelope. The product water is tapped off the envelope along one open edge that is glued to a plastic collection tube. Holes in the side of the tube provide entrance for product water. One or more envelopes are connected to the same tube. After the brine side separator screens are placed on each side of the envelope, the whole assembly is wrapped around the collection tube. The rolled-up unit (a cylindrical module) is then placed in a pressure vessel. The pressure vessel is a plastic-lined pipe with feed water entrance and brine exit at the ends. A product water exit is provided at the center of one end. Several modules are placed end to end in the pressure vessel with O-ring connectors used to fit product collection tubes together. A seal is provided between the outer wrap of each module and the inside of the pipe. Feed water entering the pipe is directed to the brine side separator screens in the module and the outer wrap seal prevents short-circuit brine flow around the module. Brine flows axially through the module by way of separator screens. The product collection tube is well-sealed so that contamination by saltwater cross leakage is not possible.

The third design is the *hollow fine fiber configuration*. As the name implies, a hollow fiber only 25 to 250 μm in diameter (approximately the size of a human hair) is manufactured. The fiber has a wall thickness of only 5 to 50 μm and is composed of unsupported membrane material. The use of unsupported membrane is possible because of the small diameter of the fibers (which experience relatively low stress, even under high pressure). In the original design, the brine flows externally to the fibers and the product flows through the fibers. Millions of fibers are looped into a U-shape and the ends are potted into a special epoxy resin, which serves as a tube sheet. The whole mass of fibers is then inserted to a containment vessel. Feed water, under pressure, enters one end of the vessel and exits at the other end. The flow of water through the shell side of the containment vessel is countercurrent to the flow of product water inside the hollow fibers. There are several reasons why brine pressure external to the fibers is preferable to internal pressure. First, the fiber wall will withstand greater pressure under compression than under tension. Second, the probability of the fibers clogging is less and the configuration is easier to clean. In addition, this design functions in a fail-safe mode. If a fiber fails, it is closed by external pressure, thereby preventing product water contamination.

One of the outstanding features of the hollow fine fiber concept is the extremely high membrane area that can be installed in a relatively small space. Packing densities as high as 40,000 m^2/m^3 (10,000 ft^2/ft^3) are possible compared with 300 to 1000 m^2/m^3 (100 to 300 ft^2/ft^3) for the spiral-wound configuration and 130 to 300 m^2/m^3 (40 to 100 ft^2/ft^3) for the tubular design. The hollow fine fibers are manufactured from polymid, polypropylene, and other organic materials. Their permeability is approximately two orders of magnitude less than typical cellulose acetate membranes. Thus, 0.03 m^3 (1 ft^3) of hollow fine fibers would have approximately the same capacity as would 0.03 m^3 of the cellulose acetate spiral-wrap configuration.

The tubular concept has one distinct advantage over other configurations. It has large, well-defined flow passageways and is, therefore, less affected by feedwater containing a high degree of particulate matter. In short, there is less tendency for clogging and channeling.

In addition, slimes and scales are relatively easy to remove by chemical cleaning. The disadvantage of the tubular design is the large number of tubes and fittings required per unit surface area. The tubular design also has a relatively high initial cost.

The spiral wound and hollow fine fiber concepts have the advantage of high membrane area packing density and a record of leakproof operation. It is essential when using these designs for reverse osmosis applications to provide for a high degree of prefiltration. Close attention must be paid to pretreatment in general, and slimes and other organic and inorganic formations must not be allowed to accumulate. The spiral wound concept permits easy field replacement of membranes. Similarly, the hollow fine fibers design lends itself to field replacement of fiber elements.

During operation, the water production rate per unit membrane area (flux) of a reverse osmosis plant tends to decline. It is important to keep this in mind and provide for additional capacity where constant water production rates are required. Flux decline has been traced to two principal causes: (a) membrane compaction and compression related to the sustained high-pressure fluid crushing porous substructures of cellulose acetate-type membranes and (b) membrane fouling by scale and contaminants. Salt rejection decline rates are not serious, especially if cross leakage from brine to product can be controlled. Systems now on the market indicate that mechanical and nonmechanical seals appear to be adequate.

Membrane Module Configuration

Because there are several types of membrane systems and many types of membrane materials, the designer has several different ways in which to configure membrane modules to accomplish desired treatment objectives.

Membrane Filtration

In a typical membrane filtration process for use in pretreating secondary effluent before demineralization, secondary effluent is pumped through a strainer before entering membrane modules. The membranes produce two streams: the filtrate, which would be pumped to the reverse osmosis process, and the backwash stream, which is returned to the headworks. The membranes in this example use compressed air for the backwash operation and the water removed from membrane modules is contained in the backwash water surge tank before return to the headworks.

Reverse Osmosis

In this system, the feedwater is filtered after pretreatment and boosted to the required system pressure by the high-pressure pump. The modules produce two process streams: the permeate, which is the product water from the process, and the concentrate (or reject), which is a waste stream. The ratio of permeate to reject is controlled by throttling with the concentrate control valve.

Reject and Brine Disposal

The disposal of wastewater brine presents significant engineering and economic problems. The old approach of freely discharging brine waste to streams, rivers, oceans, ponds, or underground reservoirs is carefully regulated in today's environment. Membrane process waste brines may vary from 5000 to 20,000 mg/L dissolved solids. Depending on the process, the effluent may also be heated, contain corrosion–erosion products from the plant, have a pH ranging from 4 to 9, be low in oxygen, and have varying degrees of turbidity. Membrane filtration processes may have reject streams with suspended solids concentrations varying from 50 to 1000 mg/L.

For high dissolved solids, demineralization-process waste streams, four possible methods of disposal are possible: direct discharge to surface water, deep-well injection, evaporation pond disposal, and evaporation to dryness and crystallization by a separate process.

Disposal to Surface Water

In general, direct discharge (without treatment) to a stream, lake, or other watercourse cannot be made without degrading water quality. Permitting of the discharge would require sufficient investigation and analysis of the effect of the discharge on water quality. Density differences between the discharge and the receiving water may dictate the use of three-dimensional water quality models.

Deep-Well Injection

Injection to subsurface strata is a promising method of disposal for waste brine. The oil and gas industry has used the method extensively. Such disposal is feasible only at locations where suitable underground formation for receiving the brine exist, and each potential site must be evaluated individually for establishing suitability. Drilling pond disposal wells is regulated by state agencies, and a properly designed system should be based on sound engineering and geologic principles and should place the waste where fresh water and natural resources will not be adversely affected.

Evaporation Ponds

Evaporation from surface ponds is another method of disposal of waste brine and this practice is subject to regulation to prevent pollution of underground and surface environments. Brine disposal ponds can be costly to permit and construct because of the liner systems required in many locations and the land costs associated with solar evaporation ponds. The use of brine disposal ponds will find application primarily in relatively warm, dry climates with high evaporation rates, level terrain, and low land costs.

Evaporation to Dryness and Crystallization

Evaporation to dryness and crystallization of the waste brine to disposable solid salts is probably the most expensive approach to the waste brine disposal problem. It would be used only if legal or site restrictions eliminated other disposal techniques or if a valuable byproduct could be recovered. This method is in use in process industries on products of greater value than are recoverable from saline water. The solid salts produced would either be sold or, more likely, would require transportation to a disposal area.

Conclusions

After the Civil War, the city of New Orleans found itself in an advantageous position regarding trade and commerce, and quickly grew into a bustling community. But its growth was hampered by the lack of clean drinking water. Several entrepreneurs decided that this was silly. The huge Mississippi River flowed right past the city. Why not tap this source for drinking water? So they set out to design a water treatment system that consisted mostly of granular sand filters for removing the mud from the river water. The filters were made of iron and were back-washable, looking much like the modern swimming pool filters. But what the engineers who designed these filters did not count on was the incredibly heavy silt load in the river water. The filters operated well for only a short time before they had to be backwashed, and there was not enough backwash water available to get all the mud out of the filters. The plant was a dismal failure and the investors all lost their money.

The reason the plant failed is that the engineers did not understand the proper function of a sand filter. It is not a roughing tool, but a polishing instrument. It is not a sieve, meant to take out the mud, but rather a filter that removes minute quantities of solids and produces a crystal clear effluent. The reason sand filters work well in both drinking water treatment plants as well as wastewater treatment plants is that the rest of the plant does the heavy lifting, producing an already highly treated water that is then further clarified by the filter.

References

Mattson, J. S.; Kennedy, F. W. (1971) Evaluation Criteria for Granular Activated Carbon. *J. Water Pollut. Control Fed.,* **43**, 2210.

Rushton, J. H. (1952) Mixing of Liquids in Chemical Processing. *Ind. Eng. Chem.,* **44**, 2931.

U.S. Environmental Protection Agency (1973) *Process Design Manual for Carbon Adsorption*; EPA-625/1-71-002a; Technology Transfer: Washington, D.C.

Symbols Used in this Chapter

c = unadsorbed concentration (kg/L)
D = impeller diameter (m)
G = velocity gradient (s^{-1})
g_c = gravitational constant
k = constant
K_T = constant
m = unit mass of adsorbing material (kg)
n = constant
n = impeller revolutions (s^{-1})
P = power requirement (n·m/s)
Q = flowrate (m^3/s)
$\bar{t}$ = hydraulic retention time (s)
V = volume (m^3)
x = impurity adsorbed (kg)
μ = viscosity (N·s/m^2)
ρ = density (kg/m^3)

11

Disinfection, Reoxygenation, and Odor Control

Disinfection

To reduce the risk of infection from pathogens in wastewater, disinfection is commonly practiced in wastewater treatment plants, and since 1972, is federally mandated. The effectiveness of disinfection is typically defined by a drop in fecal coliform concentration. Although coliforms are hardy and have other attributes that make them excellent indicators, some pathogens such as *Giardia* and *Cryptosporidium* can survive even when coliform organisms are killed and hence the coliform indicator is not foolproof.

In modern treatment plants, two methods of disinfection are commonly employed, chlorination and UV irradiation. Chlorine has historically been the most popular method, but its use has disadvantages. For example, chlorine may react with aquatic dissolved organic matter to form potentially carcinogenic halogenated byproducts. Chlorine can also kill aquatic organisms in the receiving waters—organisms that are important to the health of the aquatic ecosystem. Chlorine is also toxic and has to be carefully handled and controlled. Although there are no data to back this up, it is likely that during the past few decades, many more people have been killed while handling chlorine in wastewater treatment than have been fatally infected by pathogens in wastewater.

Discharge or receiving water requirements by the National Pollutant Discharge Elimination System permit system set limits such as 200 most probable number (MPN) of fecal coliform per 100 mL, 240 MPN total coliform per 100 mL, or 2.2 MPN total coliform per 100 mL. These limits vary depending on different state regulations and on characteristics of the particular receiving water body (for example, designated uses or dilution capacity).

Pathogens either die naturally or are destroyed in significant numbers in the course of wastewater treatment. Such reduction should be distinguished from purposeful disinfection, whereby physical and chemical facilities are installed at wastewater treatment plants for inactivating residual pathogens.

Sunlight is a natural disinfectant, principally acting as a desiccant. Irradiation by UV light intensifies disinfection and makes it a manageable undertaking. The most common source of UV light is a mercury vapor lamp constructed of quartz or a similar material that is transparent to intense and destructive invisible light of 253.7 nanometers (nm). The light is emitted by a mercury vapor arc. To ensure disinfection, the water must be free from suspended material and light-absorbing substances. Other forms of radiant or sonic energy are destructive to microorganisms, but they have yet to find engineering application in water disinfection.

Oxidizing chemicals—including halogens (chlorine, bromine, and iodine), ozone, potassium permanganate, hydrogen peroxide, hypochlorites of sodium and calcium, chlorine dioxide, and bromine chloride—can facilitate disinfection if organisms in water or wastewater are exposed to the proper dosage for the appropriate contact time. Pathogenic bacteria cannot survive long in highly alkaline or highly acidic waters (that is, at high [>11] or low [<3] pH values). Destruction of bacteria during lime softening is an example of pH effects.

Disinfection Kinetics

The two most common means of achieving wastewater disinfection are by chlorination and UV radiation. Disinfection, irrespective of the technique used, is a time-dependent process. The outcome of bacterial and viral destruction is the result of a series of physical, chemical, and biochemical actions that can be approximated by simple kinetic expressions. Although the kinetic descriptions are simple, applications are not universal. Site-specific conditions may create problems with precision and accuracy in the use of an empirical relationship that is effective at another site.

In addition to time (often referred to as *contact time*), disinfection also depends on the intensity or amount of the physical (as in UV light) or chemical (as in the case of the oxidants) entity applied to water and wastewater for the inactivation of microorgan-

Note on Terminology

Sterilization should not be confused with disinfection. *Sterilization* implies the destruction of all living things in a medium. The production of sterile water is confined to research, medical practices, and manufacture of pharmaceuticals. *Disinfection* is an operation by which living, potentially infectious organisms are killed or are rendered incapable of either reproduction or causing infection in humans. Disinfection can also be achieved through the application of heat, light, oxidizing chemicals, acids and alkalies, metal ions, and surface active chemicals.

isms. This amount is known as the *dose*. In the case of chemical disinfection, because the chemicals used are often oxidants, a portion of the dosed chemicals is consumed in reactions with reducing agents present in wastewater. Such reactions are fast. Hence, the full amount of the applied dose is not available for disinfection. The part of the chemical dose that is consumed in the extraneous chemical reactions is known as the *demand*. The part of the dosed chemical left after satisfaction of the demand and available for disinfection is known as the *residual*. Thus, kinetic expressions for microbial inactivation should be based on the residual and not the applied dose. In UV disinfection, the concepts of residual and demand do not apply.

In chlorine disinfection of wastewater, the concepts of dose, residual, and demand are complicated by the fact that chlorine reacts with ammonia, resulting in other disinfectants called *combined chlorine* (e.g., monochloramine). Such reactions are known to depend on the ratio of the chlorine dose to ammonia. The dose–residual relationship is known to be nonlinear and is described by the *breakpoint curve*. The kinetics of chlorine disinfection are also complicated by the fact that various species of chlorine resulting from the reactions with ammonia have varying inactivation potentials for microorganisms. Ammonia–nitrite–nitrate conditions in the effluent can also affect chlorine demand.

The information needed for designing a disinfection system includes knowledge of the rate of inactivation of the target organisms by the disinfectant. In particular, the effect of disinfectant concentration on the disinfection rate will determine the most efficient combination of contact time and disinfectant dose to use. The primary precepts of disinfection kinetics were first enunciated by Chick (1908), who recognized the similarity between microbial inactivation by chemical disinfectants and chemical reactions. He postulated that the rate of disinfection could be described as

$$-\frac{dN}{dt} = kN \tag{11.1}$$

where
- $-dN/dt$ = rate of decrease in organism population,
- k = organism die-off rate constant, and
- N = number of surviving organisms at any given time t.

The solution of this equation follows first-order kinetics and is often referred to as *Chick's law*. Divergences from exponential decay in disinfection are commonly observed, and it is recognized that many factors can cause these deviations (for example, changes in disinfectant concentration with time, differences in resistance between individual organisms of various ages in the same culture, existence of clumps of organisms, or occlusion of organisms by suspended solids).

Watson (1908) analyzed data with varying concentrations of disinfectant and demonstrated a definite logarithmic relationship between the concentration of disinfectant and the mean reaction velocity. He proposed the following equation to relate the organic die-off constant to the disinfectant concentration:

$$k = k'C^n \tag{11.2}$$

where
- k' = corrected die-off rate constant, presumably independent of C and N, the number of organisms,

C = disinfectant concentration, and
n = coefficient of dilution.

Combining these two equations

$$-\frac{dN}{dt} = k'NC^n \qquad (11.3)$$

The process of disinfection is influenced by temperature, and the Arrhenius equation can be used to predict temperature effects when direct heat-kill is not a significant factor

$$k'_T = k'_{20}\beta^{(T-20)} \qquad (11.4)$$

where
k'_T = rate constant at temperature T (°C),
k'_{20} = rate constant at 20 °C, and
β = empirical constant related to activation energy and universal gas constant.

Unfortunately, inactivation of organisms in batch experiments, even when the disinfectant concentration is kept constant, does not follow the predicted exponential decay pattern. Various attempts have been made to refine Chick's law or Chick–Watson models. Hom (1972) developed a flexible but highly empirical kinetic formulation based on the modification to the following form:

$$-\frac{dN}{dt} = k'Nt^mC^n \qquad (11.5)$$

where m = an empirical constant.

For changing concentrations of disinfectant, the observed disinfection efficiency is approximated by (Fair et al., 1968)

$$C^n t_p = \text{constant} \qquad (11.6)$$

where n = coefficient of dilution or a measure of the order of the reaction and t_p = time required to produce a constant percentage of die-off.

This observation has evolved into a Ct concept (concentration × time) that is currently being used in regulations for potable water treatment to ensure a certain percentage die-off of *Giardia*, viruses, *Cryptosporidium*, and other organisms. The percentage kills are being expressed in terms of log removals. In these and all other models of disinfection, the constants are functions of the nature of wastewater; chemical species; and, in the case of chlorine, the chlorine dose/ammonia ratio, all of which limit the usefulness of the models.

The kinetics of disinfection are also complicated by the fact that disinfectant-injured organisms, through repair mechanisms, can reactivate to a limited extent both in light and dark. This reactivation, sometimes referred to as *regrowth*, is mostly reported in the case of UV disinfection. As reactivation is never complete and only a fraction of affected organisms recover, the problem of reactivation may not be severe.

Protozoan parasites *Cryptosporidium* and *Giardia* are found outside their animal hosts as oocysts and cysts, respectively. Once inside the appropriate host, they infect the gastrointestinal tract and reproduce. Currently, it is not clear how enumeration of oocysts and cysts relates to their viability or potential infectivity. However, determination of cyst viability is difficult, requiring in vitro excystation.

Chlorine Disinfection

Chemistry of Chlorine Disinfection

Chlorine is the most widely used chemical for the disinfection of wastewater. It can be applied either as gaseous chlorine, sometimes referred to as elemental chlorine (Cl_2), or as a hypochlorite compound. Sulfur dioxide (SO_2) and other sulfur compounds, which react with excess chlorine, are used in the dechlorination of wastewater.

Elemental Chlorine

Chlorine is an element that occurs naturally only in a combined state. At standard conditions, chlorine, a member of the halogen family, is a greenish-yellow gas. When cooled and compressed to -34.5 °C (-30.1 °F) and 100 kPa (1 atm), respectively, it condenses to a clear, amber-colored liquid. Commercial chlorine is classified as a nonflammable, toxic, compressed gas. It is shipped in steel containers that are designed, constructed, and handled in accordance with strict government regulations.

In the gaseous state, chlorine is 2.5 times as heavy as air. In liquid form, chlorine is approximately 1.5 times as heavy as water. Liquid chlorine vaporizes rapidly. One volume of liquid yields approximately 450 volumes of gas. Thus 1.0 kg (2.2 lb) of liquid vaporizes to approximately 0.31 m^3 (11 ft^3) of gas. Chlorine is only slightly soluble in water with a maximum solubility at 100 kPa (1 atm) of approximately 10,000 mg/L (1%) at 9.6 °C (49.3 °F). The solubility of chlorine, like all gases, decreases with rising temperature. Practical solubility is approximately 50% of theoretical.

Chlorine is highly reactive and, under specific conditions, chlorine reacts with many compounds and elements, sometimes rapidly. Because of its affinity for hydrogen, chlorine removes hydrogen from some compounds, as in the reaction with hydrogen sulfide to form elemental sulfur (S) or the sulfate ion (SO_4^{-2}), depending on the chlorine/sulfur ratio and reaction conditions.

Chlorine gas is primarily a respiratory irritant and is highly toxic. A concentration greater than approximately 1.0 ppm (by volume) in air can be detected by most people because of its characteristic odor. Chlorine causes varying degrees of irritation of the skin, mucous membranes, and respiratory system. Liquid chlorine will cause skin and eye burns on contact.

Chlorine gas produces no known cumulative effects and complete recovery can occur following mild, short-term exposures. The current Occupational Safety and Health Act permissible exposure level is 0.5 ppm, the short-term exposure level is 1.0 ppm, and the immediately dangerous to life or health level is 30 ppm. Higher concentrations can be fatal.

Hypochlorites

Hypochlorites are salts of hypochlorous acid. Sodium hypochlorite (NaOCl) is the only liquid hypochlorite form in current use. There are several grades available. Calcium hypochlorite [$Ca(OCl)_2$] is the predominant dry form. Sodium hypochlorite, often referred to as liquid bleach, is commercially available only in liquid form, in concentrations between 5 and 15% available chlorine.

Sulfur Dioxide

Sulfur dioxide is commonly used for dechlorination. Sulfur dioxide is classified as a nonflammable, corrosive, and liquefied gas. In the gaseous state, sulfur dioxide is colorless with a suffocating, pungent odor and is approximately 2.25 times as heavy as air. Liquid sulfur dioxide is approximately 1.5 times as heavy as water. Commercially, sulfur dioxide is supplied as a pressurized, colorless, liquefied gas. The gas solubility in water is approximately 20 times greater than that of chlorine. In solution, sulfur dioxide hydrolyzes to form a weak solution of sulfurous acid (H_2SO_3). Sulfurous acid dissociates as

$$H_2SO_3 \rightleftharpoons H^+ + HSO_3^- \quad pK_a = 1.76$$
$$HSO_3^- \rightleftharpoons H^+ + SO_3^{-2} \quad pK_a = 7.19 \quad (11.7)$$

where the dissociation constants (pK_a) are at 25 °C (77 °F).

Chemistry of Chlorine in Water

When chlorine is added to water or wastewater, hydrolysis occurs, and a mixture of hypochlorous acid and hydrochloric acid is formed. The reaction is pH and temperature dependent, completed within milliseconds, and reversible. The hypochlorous acid formed is a weak acid and dissociates or ionizes to form an equilibrium solution of hypochlorous acid and hypochlorite ion (OCl^-). The equilibrium approaches 100% dissociation ($H^+ + OCl^-$) when the pH exceeds 8.5 and approaches 100% hypochlorous acid when the pH is less than 6.0.

$$Cl_2\,(g) + H_2O \rightleftharpoons HOCl + H^+ + Cl^- \quad pK_a = 7.54 \text{ at } 25\,°C\,(77\,°F)$$
$$HOCl \rightleftharpoons H^+ + OCl^- \quad pK_a = 7.54 \text{ at } 25\,°C\,(77\,°F) \quad (11.8)$$

Chlorine exists predominately as hypochlorous acid at pH levels between 2 and 6. At a pH of less than 2, hypochlorous acid reverts to chlorine; at a pH of greater than 7.8, hypochlorite ions are the predominate form.

Hypochlorite solutions of sodium hypochlorite and calcium hypochlorite also dissociate in water to form hypochlorite ions. The formation of both hydrogen and hypochlorite ions is pH dependent. Just as exhibited in the hydrolysis of chlorine gas, an equilibrium is established. The hypochlorite solutions formed are at an elevated pH. Chlorine tends to decrease the pH, while hypochlorites tend to raise the pH. Chlorine also reduces alkalinity by as much as 2.8 parts of chlorine per part of calcium carbonate ($CaCO_3$). An alkali is added to calcium and sodium hypochlorite solutions to enhance stability and slow decomposition. Thus, hypochlorite solutions can have pHs greater than 10. The treated wastewater is not affected by the pH of the chlorine solution added regardless of the source of the chlorine (chlorine gas or hypochlorites).

$$NaOCl \rightleftharpoons Na^+ + OCl^-$$

$$Ca(OCl)_2 \rightleftharpoons Ca^{+2} + 2OCl^- \quad (11.9)$$

The chlorine present as both hypochlorous acid and hypochlorite ions is defined as *free available chlorine*. Aqueous chlorine may also be referred to as free available chlorine, but its usefulness is not practical because of the low pH conditions involved.

Chloramines are products of the reactions between chlorine and ammonia-nitrogen found in wastewater. Ammonia-nitrogen may be present in appreciable amounts, 10 to 40 mg/L, in wastewater as either dissolved ammonia gas (NH_3), the ammonium ion (NH_4^+), or organic nitrogen compounds. The ammonia-ammonium equilibrium is pH dependent

$$NH_3\,(g) + H_2O \rightleftharpoons NH_4^+ + OH^- \quad pK_a = 9.24 \text{ at } 25\,°C\,(77\,°F) \qquad (11.10)$$

Three chloramines are formed—monochloramine (NH_2Cl), dichloramine ($NHCl_2$), and trichloramine or nitrogen trichloride (NCl_3)—in a stepwise process beginning with the reaction

$$NH_3 + HOCl \rightleftharpoons NH_2Cl + H_2O \qquad (11.11)$$

The hypochlorous acid then continues to react forming dichloramine

$$NH_2Cl + HOCl \rightleftharpoons NHCl_2 + H_2O \qquad (11.12)$$

and trichloromine

$$NHCl_2 + HOCl \rightleftharpoons NCl_3 + H_2O \qquad (11.13)$$

Chlorine Toxicity and Effects on Higher Organisms

Chlorine, when added to wastewater as a disinfectant, has varied effects both on downstream users of the receiving water body and on the living organisms that may be present in the waste effluent or the biota of the receiving water. In the waste effluent, cysts of *Entamoeba histolytica* and *Giardia lamblia* and eggs of parasitic worms are resistant to chemical disinfectants. Because of their high resistance to disinfection, cysts and ova can be removed from wastewater more effectively by methods other than chlorine disinfection.

A coagulation and settling process followed by filtration is the primary method of removing these organisms from potable water. Chlorination has limitations, and wastewater treatment operators should not try to accomplish by disinfection what can be accomplished more effectively and economically by other methods.

Aftergrowth

In the receiving stream, aftergrowths of some organisms occur after discharge of chlorinated effluents. The extent of these effects is primarily governed by the amount of biologically oxidizable material in the effluent. Aftergrowths observed in waters receiving chlorinated effluent are presumed to be a result of the destruction of large numbers of protozoa by chlorination. This permits subsequent multiplication of surviving bacteria unhampered by predatory protozoa such as ciliates and flagellates. Aftergrowths develop rapidly after chlorination, presumably because of the abundant food supply available for the small number of surviving organisms combined with the absence of predatory protozoa. When effluent is chlorinated, the greater the initial reduction in bacterial population, the longer the lag time will be before the multiplication of surviving organisms becomes apparent.

Dechlorination

Toxicity is related to the individual sensitivities of target organisms. A residual's toxicity is also related to the degree of chemical reactivity of the compound. Free chlorine residual, hypochlorous acid, and hypochlorite ion are more reactive compounds than are the combined residuals, mono- and dichloramine. Free chlorine residuals, hypochlorous acid and hypochlorite ion, exist only momentarily when added to wastewater containing ammonia-nitrogen. Because ammonia is a significant constituent of most effluents and chlorine is able to form chloramines, it follows that combined residuals, which are predominately monochloramine, will be responsible for most of the germicidal activity of chlorine present in chlorinated effluent.

Dechlorination, the removal of remaining chlorine, is required in most states. Stream standards that limit chlorine as total residual have been established in most parts of the country. Free and combined chlorine residuals can be effectively reduced by sulfur dioxide and sulfite salts. The sulfite ion reacts rapidly with free and combined chlorine. The sulfite ion is the active agent when sulfur dioxide or sulfite salts are dissolved in water. Their dechlorination reactions are identical. Sulfite reacts instantaneously with free and combined chlorine

$$SO_3^- + HOCl \rightarrow SO_4^{-2} + Cl^- + H^+$$

$$SO_3^{-2} + NH_2Cl + H_2O \rightarrow SO_4^{-2} + Cl^- + NH_4^+ \tag{11.14}$$

Reactions yield small amounts of acidity, which is neutralized by the alkalinity of the wastewater (2.8 mg of alkalinity as calcium carbonate is consumed per milligram of chlorine reduced). From the above equations, the amount of sulfur dioxide required per part of chlorine is 0.9, but the actual practice calls for the use of a 1:1 ratio.

Safety and Health

Chlorine gas has a detectable odor at low levels of concentration and has a greenish-yellow color at higher levels of concentration. At volumes of less than 0.1 ppm in air, chlorine gas is detectable only by instruments. The maximum contaminant level established for chlorine gas by the Occupational Safety and Health Administration (OSHA) is a 1-ppm, time-weighted average over 8 hours. Harmful effects of chlorine gas exposure begin to manifest at approximately 5 ppm and higher. Between 5 and 10 ppm, however, these effects (choking, coughing, watery eyes, mild skin irritation, and lung irritation) are acute and temporary. At higher concentrations, the effects become more long lasting and can result in death.

Shipment and Handling

Calcium hypochlorite is classified as a corrosive and rapid oxidant. Sodium hypochlorite is a corrosive agent. Chlorine and sulfur dioxide are nonflammable, corrosive, toxic, liquefied gases under pressure. Various sulfite solutions are classified as corrosive.

Facility Design

All handling facilities must be designed with adequate space for loading and unloading cylinders or 900-kg containers. Vehicles used to transport cylinders must be equipped with an upright cylinder rack, chain restraints for the cylinders, and a lift gate. If no lift gate is available on the vehicle, the facility should be constructed with a raised loading dock and interior ramp to allow the cylinders to be transported without lifting by hand.

A properly designed handling facility will have panic or escape hardware on all doors. All entrances from storage or use areas should be from outside the building, and all doors should be designed to open outward. Every room in a chlorine or sulfur dioxide facility in which a chlorine or sulfur dioxide gas container is stored or used must be negatively ventilated with one air change per minute and have a gas detector. Because these gases are heavier than air, an exhaust blower and screened floor vents are required. The gas detector should be interlocked with the fan and audible or visible alarms. Proper design would also permit observation of the gas detector from outside the storage room. Currently available gas detectors use remote sensors that permit the indicator and transmitter to be installed outside the storage area and as far as 300 m (approximately 1000 ft) away. In larger facilities the leak alarms are transmitted to the control building or to emergency response personnel.

If not equipped with a gas scrubber, the doors of the facility should have an electrical interlock that automatically turns on the lights and exhaust fan in the room before entry and when the doors are opened. A manual switch to operate the lights and fan should also be located near the doors on the outside wall. Some states require the ventilation fan to be interlocked with the leak detector to lock out operation of the fan to reduce dispersion of chlorine to the atmosphere. A gastight window should be installed in at least one of the doors of the facility to permit observation of the interior before entry. The ready availability of an exterior emergency shower and eyewash stations should be mandatory.

Many jurisdictions are requiring the installation of gas-scrubbing equipment for chlorine and sulfur dioxide storage rooms. Regulations are changing rapidly and are subject to wide variations in local interpretation. Planning and design of such facilities should be carefully coordinated with local authorities.

Appropriate OSHA-approved warning signs should be posted at the entrance and any other exposed side. The appropriate container repair kit and self-contained breathing apparatus should be located at a convenient external location.

Gas storage and use areas should be dedicated rooms. In these facilities, nothing should be stored and no work performed that is not related directly to handling chlorine or sulfur dioxide. Typical designs call for chlorine-feed equipment to be located in an additional dedicated room separate from the gas-storage room. Where evaporators are used, they are placed in the gas-storage room so that all the gas under pressure is located in a single room. Chlorine-storage and chlorine-feeder rooms must be well lighted, with adequate maneuvering areas. There should be no common drain, ventilation system, or doors between them. There should also be no doors between gas-storage or gas-handling spaces and other inhabited spaces (offices, shops, or other work spaces).

In any facility, chlorine or sulfur dioxide equipment must be located in isolated rooms separate from cylinder storage. Remote vacuum regulators should be used to convert gas to a vacuum, preferably at the source. This will maintain any gas piping in the equipment room under vacuum, which improves safety, and will permit the use of plastic piping such as schedule 80 polyvinyl chloride from the regulator to the feeding equipment and ejector. To facilitate operation, a gastight observation window should be provided in the common wall between the storage and vacuum regulator rooms.

Design and Selection of Equipment

Chlorinators and Sulfonators

Chlorine gas feeders are referred to as *chlorinators* to differentiate them from hypochlorinators, which feed a hypochlorite solution. Sulfur dioxide gas feeders are referred to as *sulfonators*.

A chlorinator has the following basic components: a vacuum regulator with an outside vent, feed-rate control, a Venturi-operated ejector with check valve, and an indicating flow meter. All chlorinator designs in current use consist of these components. Occasional use is made of direct-pressure feeders, but they are rare, particularly in wastewater disinfection. Direct-pressure feeders are used primarily for wastewater disinfection in remote areas or where there is an absence of electrical power preventing the development of a vacuum. The vacuum regulator may be found in many different configurations, depending on the manufacturer and the feed-rate capacity of the system. In wastewater disinfection systems, the vacuum regulator is most often located in the chlorine storage room, so that gas leaves the storage room under vacuum. This practice is true regardless of the capacity of the chlorinator and enhances the safety of the installation. The vacuum regulator is a diaphragm-operated device with one side of the diaphragm open to the atmosphere to permit venting should gas pressure suddenly develop. The other side of the diaphragm is connected to the vacuum source and linked to permit gas flow only under vacuum. This arrangement minimizes chances for chlorine leaks because a leak on the vacuum side causes chlorine flow to cease.

The feed-rate control device may be a simple, manually operated valve or an automatically controlled motor-driven valve. The gas-flow meter is calibrated to indicate an instantaneous feed rate in either metric units (grams or kilograms per hour) or English units (pounds per day).

The injector (or ejector) is a vacuum-producing device that consists of a nozzle and throat assembly and a check valve. The device produces a vacuum when a designed waterflow passes through a Venturi or orifice. When waterflow ceases, the check valve closes to prevent water from entering the chlorinator. To increase safety in handling chlorine and improve the speed of response to flow or residual changes, a longer gas vacuum line is preferred to a long solution line. Therefore, positioning the injector as close as possible to the point of addition should be strongly considered in each installation.

The injector is a critical component of the chlorinator for two reasons: first, its hydraulic component creates the vacuum under which the system operates; second, it mixes the gas with the makeup water to produce the solution that is injected to the wastewater. If the hydraulic conditions under which the injector must operate are not properly specified, the entire system may operate erratically or not at all. The design of the installation must consider the pressure at the point of injection and the supply pressure needed to the injector. A booster pump designed to provide correct waterflow and pressure must be chosen. The manufacturer of the injector can provide that information.

Chemical-Feed Pumps

Chemical-feed pumps used for the feeding of sodium or calcium hypochlorite are referred to as *hypochlorinators*. Like chlorinators, hypochlorinators come in many configurations, but the basic system components are similar in all cases. Sulfite solutions are also fed with chemical-feed pumps. Construction materials are changed to meet chemical-handling needs.

Sulfur Dioxide Feeders
Sulfur dioxide gas feeders (sulfonators) are similar in design and construction to chlorinators but sometimes are composed of different materials.

Feed Control Strategies
There are several ways to control the feed rate of chlorine gas or hypochlorite solutions: manual control, automatic flow proportioning or open-loop control, automatic residual or closed-loop control, or automatic compound-loop control, which combines flow and residual signals to vary the gas-feed rate. Flow proportioning is sometimes referred to as *feed-forward* control, and residual control is sometimes referred to as *feed-back* control.

In feed-forward control, the chlorine feed rate is in proportion to flow. Figure 11.1 illustrates the feedback loop for a flow-proportional chlorination. Flow-proportional chlorination in wastewater, however, is not always the best approach because the chlorine demand can vary not only with the flow but also with the nature of the wastewater.

Residual control, the feed-back approach, involves varying the chlorine feed rate based on deviation from a set-point on a controller. For systems in which the flowrate is nearly constant on a daily basis or is strictly seasonal, this type of control system works well. For systems in which the flowrate varies often and demand is variable as well, residual control may not be as effective as flow pacing because residual control systems do not react well to large variations in flowrate over short periods of time.

Compound-loop control provides the ability to take a flow and residual input to control gas feed. Flow is the primary drive, while residual is used to trim the gas feed. When the set-point in a compound-loop control system is further controlled automatically, the configuration is referred to as *cascade control*. Cascade control requires the use of another analyzer downstream of the compound-loop control analyzer. Output from the cascade control analyzer is used to regulate the compound-loop controller.

The choice of control strategy for a particular installation is based on regulatory requirements, existing facilities, wastewater treatment system design, economics, cost-effectiveness, and required system maintenance. The more complicated the selected control system, the more likely it is that service requirements will be more exacting and, therefore, will require more training and an increase in the skill level of operating personnel.

Reactor Design
Chlorine contact chambers are designed to be plug-flow reactors with a constant hydraulic residence time, typically 30 minutes, at average daily flow. In some instances, such as water destined for reclamation in California, the holding time is a minimum of two hours. Figure 11.2 shows a typical chlorine contact chamber. Contact chambers should have smooth corners to minimize dead space and short circuiting. A smooth surface will also guard against the potential of bacterial growth within protected grooves. The ability to remove accumulated suspended solids should be incorporated to the design. Most importantly, many evaluations of various designs have reaffirmed that long narrow channels with longitudinal baffles make the best contact chambers. This chamber attribute most likely influences the plug-flow nature of the reactor and the need for low velocities.

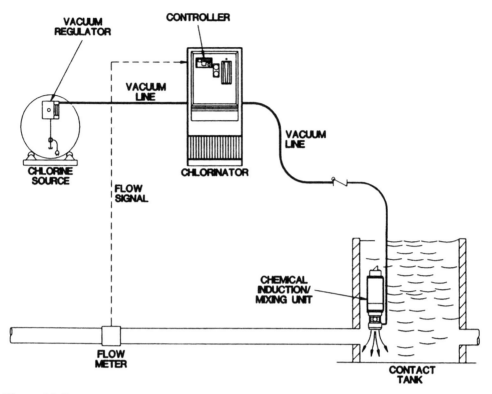

Figure 11.1 Flow-proportional chlorination.

The importance of initial mixing in chlorination is also important. This may be accomplished at the inlet zone using both static and mechanical mixers. In open-channel reactors, the designer must consider wind effects that may cause surface currents, which consequently may short-circuit and disrupt an acceptable plug-flow condition. When monochloramine is used as the disinfection agent, turbulence within the chamber should be minimized because it may cause back-mixing and reduce the concentration of volatile monochloramine.

Ultraviolet Disinfection

Chlorination has been the de facto choice for most wastewater disinfection operations since the early 1900s. Although chlorination is still used in the majority of disinfection applications, alternative processes are increasingly being selected. Ultraviolet irradiation has become the most common alternative to chlorination for wastewater disinfection.

The emergence of UV irradiation as an important wastewater disinfection alternative may be attributed to the drawbacks of conventional chlorination, improvements in UV technology, and advances in our understanding of the UV process. Primary problems associated with chlorination are effluent toxicity and safety. Free and combined chlorine elicit a toxic response in fish and daphnids at extremely low concentrations. Residual chlorine may be effectively eliminated by dechlorination (as is required in most new discharge permits), but effluent toxicity will remain in some cases. Factors that may

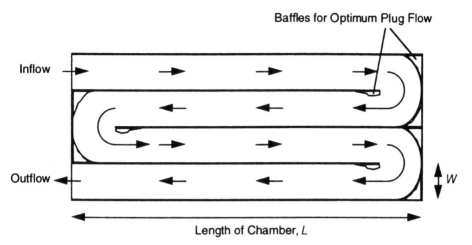

Figure 11.2 A typical chlorine contact chamber (from White, G. C. *The Handbook of Chlorination and Alternative Disinfectants,* 3rd ed. Copyright © 1992, Van Nostrand Reinhold: New York. This material is used by permission of John Wiley & Sons, Inc.).

contribute to toxicity following chlorination–dechlorination include chlorinated disinfection byproducts, active (+1 valent) chlorine not removed by dechlorination, and unoxidized ammonia (dechlorination restores chloramines to ammoniacal nitrogen).

Dechlorination and containment facility requirements have increased the cost of chlorine-based disinfection. At the same time, the development and application of open-channel, modular systems have reduced the cost of UV. Consequently, the costs of the two processes are comparable for new facilities.

The majority of UV disinfection systems today use an open-channel, modular design. Two principal lamp geometries have been adopted: (a) horizontal, uniform arrays with flow directed parallel to lamp axes and (b) vertical, staggered arrays with flow directed perpendicular to lamp axes (Figure 11.3). The horizontal lamp orientation has been adopted in the majority of applications.

Ultraviolet disinfection initially was used only on water of high quality (transmittance, dissolved solids, and so on) or in nonstringent, low-risk regulatory situations. With improvements in the technology and available equipment, higher degrees of treatment on poor-quality effluent is now possible. Ultraviolet disinfection is now routinely used for reclaimed water, direct-contact recreation receiving water, and similar sensitive applications.

Mechanism of Ultraviolet Disinfection

Ultraviolet irradiation is a physical disinfection process and, as such, has several fundamental characteristics that distinguish it from chemical disinfection processes (such as chlorination). Ultraviolet irradiation achieves disinfection by inducing photobiochemical changes within microorganisms. At a minimum, two conditions must be met for a photochemical reaction to take place

(1) Radiation of sufficient energy to alter chemical bonds must be available and
(2) Such radiation must be absorbed by the target molecule (organism).

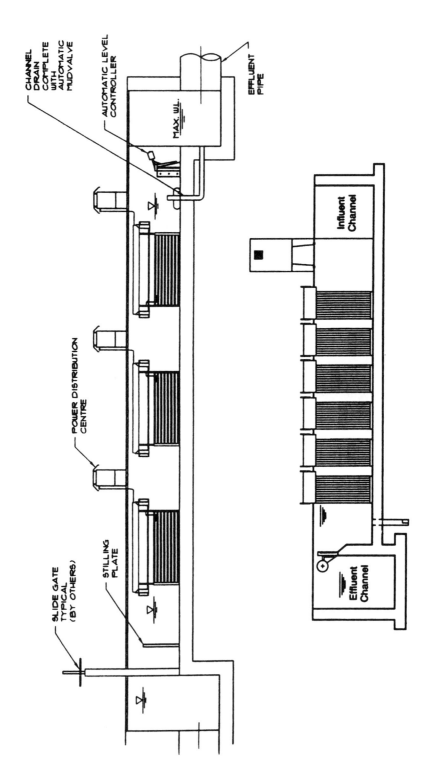

Figure 11.3 Open-channel UV disinfection systems: horizontal lamp configurations (top) and vertical lamp configuration (bottom).

In the majority of UV disinfection applications, low-pressure mercury arc lamps have been chosen as the source of UV radiation. Alternative sources of UV radiation are also being investigated for disinfection processes. In particular, medium-pressure mercury arc lamps have been used for disinfection in some applications. The output spectrum of these lamps is substantially different from the spectrum of conventional low-pressure lamps. Radiation is emitted from these lamps over a large fraction of the UV spectrum. As a result, responses of microorganisms to radiation from these lamps may be more complex than responses elicited by exposure to radiation from low-pressure lamps. Furthermore, a theoretical analysis of photobiochemical change induced by medium-pressure lamps is more complex because of the polychromatic nature of the radiant energy source. However, the fundamental operation of disinfection processes that use these lamps is conceptually similar to operation observed for conventional low-pressure mercury arc lamps.

Ultraviolet Inactivation Kinetics

In evaluating the rate of photobiochemical change induced by UV irradiation, it may be useful to view a UV photon as a "reactant". As in the case of strict chemical reactions, the rate of a photochemical reaction is governed by the availability of reactants. In a chemical reaction, reactant availability is quantified using concentration or activity. Intensity is the measure of radiation "availability" in a photochemical reaction. Therefore, knowledge of radiation intensity is required for estimating photochemical kinetics.

The experimental procedure used to evaluate inactivation dose-response behavior (kinetics) involves exposure of a microbial population to a measurable source of radiation for a known period of time followed by quantification of microbial viability. In most cases, the source of radiation is a collimated beam.

Because radiation energy may be absorbed by the medium in which microorganisms are suspended, the actual intensity within the medium will decrease from its top surface downward. Given this non-uniform intensity field, it may be necessary to calculate depth-averaged intensity within the entire irradiated volume. The variation in intensity with depth may be calculated by applying Beer's law

$$I_x = I_0\, e^{-\alpha x} \tag{11.15}$$

where

I_x = radiation intensity at depth x ($x > 0$, mW/cm^2);
x = depth (cm);
I_0 = incident radiation intensity (mW/cm^2); and
α = absorbance coefficient (cm^{-1}).

This equation may be integrated over the entire fluid depth ($0 < \alpha < H$) to yield a depth-averaged intensity for the entire reactor

$$I_{avg} = \frac{I_0}{H\alpha}(1 - e^{-H\alpha}) \tag{11.16}$$

where H = liquid depth in cm.

If the liquid within the reactor is kept shallow ($H < 1$ cm) and well-mixed, I_{avg} can be used to characterize the entire medium with little error. As a first approximation, inactivation kinetics in a well-mixed, shallow petri dish can be modeled as a first-order photobiochemical reaction

$$\frac{dN}{dt} = kI_{avg}N \tag{11.17}$$

where
- N = concentration of viable organisms (number of organisms/L);
- t = time (s);
- k = first-order inactivation constant (cm^2/mW·s); and
- I_{avg} = average radiation intensity (mW/cm^2).

The results of collimated beam experiments have been reported for many microorganisms. Generally speaking, viruses and bacteria are inactivated effectively by UV irradiation, whereas protozoan cysts and spore-forming bacteria are relatively resistant to UV inactivation. The UV dose required to meet coliform limitations will achieve better virus inactivation than the comparable chlorine dose.

Ultraviolet Disinfection System Design

At present, UV system design relies on a combination of past experience, pilot testing, and numerical modeling. Each factor is related, and the degree to which each is used often depends on the size of the system being considered, the budget, and the schedule. Use of a low-pressure lamp in open-channel configurations is conventional practice today and has received significant experience among many operating wastewater treatment plants and primary UV suppliers. In smaller systems, it is probably sufficient to base the design on expected wastewater characteristics and conventional practices, unless obvious differences from typical municipal wastewater exist. Acceptable numerical models can be used with appropriate default values for design parameters and coefficients. Redundancy should be incorporated to the design, and the sizing should be relatively conservative.

For the design of medium to large facilities, capital and operating costs can be substantial; in such cases, it is important to base the design sizing on relevant and site-specific wastewater characteristics. Pilot testing is recommended, particularly if advanced, nonconventional UV systems are being considered.

Wastewater Characteristics

Data should be collected to obtain a thorough characterization of effluent quantity and quality. For existing facilities, direct sampling and testing should be conducted and should address seasonal and diurnal variations. If the facility is new, an effort should be made to develop design-effluent characteristics from similar wastewater treatment plants and collection systems.

Critical data to be evaluated in design include flow, UV transmittance, suspended solids, and viable indicator organism (for example, coliform bacteria) concentrations. The UV transmittance of secondary effluent will be greater than 60% on a filtered basis, although lower values (on the order of 50%) have been observed. When collecting UV transmittance data, it is important to measure the parameter on both filtered and unfiltered samples. The unfiltered measurement presents a more representative estimate of the transmissibility through effluent and is critical to sizing of the system. The lower the transmittance, the greater will be the size requirements. In some cases, particularly at low transmittance levels, it may be necessary to reduce spacing of the lamps or consider using advanced higher intensity systems to overcome the lower transmissibility of the water. This is the case at transmittance levels of less than 50%.

Suspended solids will affect the transmittance of UV, occlude bacteria, and generally interfere with the UV disinfection process to a greater degree than encountered with chemical disinfection systems. This, in effect, establishes a limit of disinfection efficiency that can be accomplished by UV; this limit is a function of the particulate matter in the effluent. Certainly, one can expect this effect to vary and depend on the type of particulate matter and size distribution of particles. In waste streams containing a high degree of inorganic matter or naturally occurring soil solids (such as in combined sewer overflow and stormwater), concentrations of viable organisms associated with particulate matter may be relatively small. In cases of typical municipal wastewater and biologically treated wastewater, however, these concentrations can be significant and account for essentially all residual coliforms in the final effluent after clarification. For this reason, a high degree of filtration is required, including (in some cases) chemical coagulation of colloidal solids, to achieve high disinfection efficiencies.

Ultraviolet Disinfection Equipment

Original systems offered by vendors in the early 1980s consisted of enclosed chambers using either a submerged-lamp system or a noncontact lamp system. The technology evolved to a modular, submerged-lamp system installed in an open channel, which significantly improved system maintenance and afforded better hydraulics. The modular, open-channel UV system using a conventional low-pressure, mercury arc vapor lamp is currently the industry standard. An estimated 80% of all UV systems in operation today are open-channel, low-pressure lamp systems. A recent significant improvement to these systems was the development of the electronic ballast. Current emphasis in the market is research and development of alternate high-intensity UV sources, which fall into two basic categories: high-intensity, low-pressure lamp systems and medium-pressure lamp systems. Changes in lamp physics allow each of the new systems to provide similar germicidal performance, with substantial reductions in the number of lamps used (one-eighth to one-twentieth the number of lamps) compared with conventional low-pressure lamp systems.

Effluent Reoxygenation

Reoxygenation of treated wastewater effluent is often necessary if permit limits require high dissolved oxygen concentrations. Two principal categories of reoxygenation systems are cascade reoxygenation and mechanical or diffused air reoxygenation.

Cascade Reoxygenation

These systems rely on air and water interfaces at weir overflows, flumes, spillways, and similar hydraulic structures. Precise measurements of oxygen supplied (mass) or changes in dissolved oxygen are difficult to obtain and the process is difficult to control. These systems are not energy efficient, but they work well and cost little where hydraulic head is sufficient and the design cannot be modified to conserve the head or use it for other more efficient purposes.

In weir systems, reaeration occurs at the weir's crest during water surface formation and is enhanced by bubble entrainment and splashing in the lower pool. Studies by Nakasone (1987) indicate the reaeration over a single weir can be estimated by the *deficit ratio*, defined as

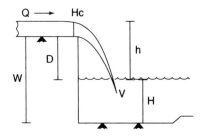

Figure 11.4 Graphical presentation of the terms in cascade reoxygenation equations (from Nakasone, H. [1987] Study of Aeration at Weirs and Cascades. *J. Environ. Eng.*, **113**, 64, with permission from the American Society of Civil Engineers).

$$r = \frac{C_s - C_o}{C_s - C} \quad (11.18)$$

where
 C_s = dissolved oxygen saturation concentration of the wastewater,
 C_o = dissolved oxygen saturation concentration after reoxygenation, and
 C = required dissolved oxygen concentration after reoxygenation.

$$\ln r_{20} = 0.0785(D + 1.5\,H_c)^{1.31}\, q^{0.428}\, H^{0.310} \quad (11.19)$$

for $(D + 1.5\,H_c) > 1.2$ m and $q\text{ ″ } 235$ m³/m·h
$$\ln r_{20} = 0.0861(D + H_c)^{0.861}\, q^{0.428}\, H^{0.310} \quad (11.20)$$

for $(D + 1.5\,H_c)\text{ ″ } 1.2$ m and $q > 235$ m³/m·h
$$\ln r_{20} = 5.39(D + 1.5\,H_c)^{1.31}\, q^{-0.363}\, H^{0.310} \quad (11.21)$$

for $(D + 1.5\,H_c) > 1.2$ m and $q > 235$ m³/m·h
$$\ln r_{20} = 5.92(D + H_c)^{0.816}\, q^{-0.363}\, H^{0.310} \quad (11.22)$$

where
 r_{20} = deficit ratio at 20 °C;
 D = drop height (m);
 H_c = critical water depth on the weir (m);
 q = discharge per width of weir (m³/m·h); and
 H = tailwater depth for downstream channels having horizontal beds (m).

The variables are defined in Figure 11.4.

Deficit ratios at other water temperatures (T) can be calculated as follows (Gameson et al., 1958):

$$\ln r_T = \ln r_{20}\,(1 + 0.0168\,[T - 20]) \quad (11.23)$$

where T = temperature (°C).

Another method used for determining required cascade height is based on the following equations developed by Barrett et al. (1960):

In customary U.S. units

$$h = r - \left[\frac{1}{0.11ab(1 + 0.045T)}\right] \quad (11.24)$$

In SI units

$$h = r - \left[\frac{1}{0.11ab(1 + 0.046T)}\right] \quad (11.25)$$

where
- h = height through which water falls, m (ft);
- r = deficit ratio at temperature T;
- C_s = dissolved oxygen saturation concentration of wastewater at temperature T (mg/L);
- C_0 = dissolved oxygen concentration of postaeration (mg/L);
- C = required final dissolved oxygen level after postaeration influent (mg/L);
- a = water quality parameter equal to 0.8 for a wastewater treatment plant effluent;
- b = weir geometry parameter (for a weir, b = 1.0; for steps, b = 1.1; for step weir, b = 1.3); and
- T = water temperature (°C).

A key element in the use of this method is the proper selection of the critical wastewater temperature that affects the dissolved oxygen saturation concentration, C_s.

Mechanical and Diffused Air Reoxygenation

These techniques encompass a wide range of mechanically assisted reoxygenation systems, including mechanical surface aeration (fixed, floating), jet diffusion, pump aerators, agitator-sparged systems, U-tube aerators, and diffused air (coarse, fine bubble, and so on) systems. These systems are amenable to engineering analysis and can be controlled to conserve energy, prevent overaeration, and reduce costs.

Relationship of Reoxygenation to Other Unit Processes

Reoxygenation should be the last step in the treatment process. Reoxygenation upstream of effluent filters, in the absence of compelling considerations to the contrary, is inadvisable because some filter designs allow negative heads to occur in filter media. High dissolved gases in the unfiltered effluent may escape from solution in the filter bed, producing bubbles and air binding. For nonnitrified effluents, nitrification by attached growths in the filter media can consume dissolved oxygen in the reoxygenated effluent, offsetting, at least partially, the reoxygenation.

Odor Control

Wastewater treatment facilities, regardless of how well designed they are, at one time or another can generate odors as byproducts of the wastewater treatment process. As public awareness increases, odor complaints plague municipal officials. As urban development spreads, wastewater treatment facilities that were once isolated often become surrounded by residential developments. As a result, a treatment facility may quickly become an unwanted neighbor even though it may have been around long before the residential development. To address increasing public concerns, local governmental agencies are promulgating odor-control ordinances with increasing frequency. Regulations that target the reduction of odors considered to be a nuisance are often vague and other regulations result in strict discharge limits imposed on point-source discharges from odor control facilities. Therefore, early in the design process, a design engineer should consult local and state rules and regulations to ensure that odor control is properly addressed to meet local conditions.

Sources of Odor

Odors prevalent at wastewater facilities generally result from inorganic gases and vapors. The most common inorganic vapors are hydrogen sulfide and ammonia. Odorous compounds, such as mercaptans, organic sulfides, and amines, are common byproducts of the decomposition of organic matter. Industrial wastes discharged to the municipal wastewater system can also generate significant odorous vapors, which, if emitted at high enough concentrations, may pose serious health and safety risks to plant personnel.

Common locations in wastewater treatment facilities where odors originate include collection systems, in-system pumping stations, preliminary and primary treatment systems, and solids-processing facilities. Of these locations, attention to odor control has primarily been focused at pumping stations, headworks facilities (wet wells, screening facilities, and grit chambers), and, more recently, solids-storage, thickening, stabilization, and dewatering facilities.

During the early stages of design of a wastewater treatment plant, a design engineer should identify the potential sources of odor and define local regulatory requirements. Once this is accomplished, methods of odor abatement or control can be identified, evaluated, and selected for inclusion in the design of the support system.

Table 11.1 lists some of the odorous compounds found in wastewater and their odor detection and recognition thresholds. Most of the odorous substances are gaseous under normal atmospheric conditions or at least have a significant volatility. The volatility is shown in the table as parts per million and is equal to the vapor pressure. The molecular weights of these substances typically range from 30 to 150. The lower the molecular weight of a compound, the higher its vapor pressure and potential for emission to the atmosphere. Substances of high molecular weight are less volatile and, thus, have a smaller likelihood for causing odor complaints.

These odorous compounds tend to follow Henry's law ($Hx = p$), which states that the partial pressure of the gas (p) above the liquid surface is directly proportional to the molecular concentration of the gas (x) dissolved in a liquid. Because Henry's constant (H) is known for many compounds dissolved in water, this is a useful relationship for estimating the concentration of these compounds in wastewater emissions.

Also, it is interesting to note that the reduced sulfur compounds, such as the mercaptans and organic sulfides, tend to be the most odorous because of their relatively low odor threshold concentrations. This also applies to the nitrogen-bearing amines, but to a lesser extent. Although hydrogen sulfide is considered to be the most prevalent odorous compound present in wastewater, it should not be presumed in every case that an odor problem is caused exclusively by hydrogen sulfide. Other odorous compounds, such as mercaptans, organic sulfides, and amines, may also be significant contributors.

Table 11.1 Odorous compounds in wastewater treatment (AIHA, 1989; Amoore and Hautala, 1983).

Compound name	Formula	Molecular weight	Volatility at 25°C, ppm (v/v)	Detection threshold, ppm (v/v)	Recognition threshold, ppm (v/v)	Odor description
Acetaldehyde	CH_3CHO	44	Gas	0.067	0.21	Pungent, fruity
Allyl mercaptan	$CH_2{:}CHCH_2SH$	74		0.0001	0.0015	Disagreeable, garlic
Ammonia	NH_3	17	Gas	17	37	Pungent, irritating
Amyl mercaptan	$CH_3(CH_2)_4SH$	104		0.0003	—	Unpleasant, putrid
Benzyl mercaptan	$C_6H_5CH_2SH$	124		0.0002	0.0026	Unpleasant, strong
n-Butyl amine	$CH_3(CH_2)NH_2$	73	93,000	0.080	1.8	Sour, ammonia
Chlorine	Cl_2	71	Gas	0.080	0.31	Pungent, suffocating
Dibutyl amine	$(C_4H_9)_2NH$	129	8000	0.016	—	Fishy
Diisopropyl amine	$(C_3H_7)_2NH$	101		0.13	0.38	Fishy
Dimethyl amine	$(CH_3)_2NH$	45	Gas	0.34	—	Putrid, fishy
Dimethyl sulfide	$(CH_3)_2S$	62	830,000	0.001	0.001	Decayed cabbage
Diphenyl sulfide	$(C_6H_5)_2S$	186	100	0.0001	0.0021	Unpleasant
Ethyl amine	$C_2H_5NH_2$	45	Gas	0.27	1.7	Ammonialike
Ethyl mercaptan	C_2H_5SH	62	710,000	0.0003	0.001	Decayed cabbage
Hydrogen sulfide	H_2S	34	Gas	0.0005	0.0047	Rotten eggs
Indole	$C_6H_4(CH_2)_2NH$	117	360	0.0001	—	Fecal, nauseating
Methyl amine	CH_3NH_2	31	Gas	4.7	—	Putrid, fishy
Methyl mercaptan	CH_3SH	48	Gas	0.0005	0.010	Rotten cabbage
Ozone	O_3	48	Gas	0.5	—	Pungent, irritating
Phenyl mercaptan	C_6H_5SH	110	2000	0.0003	0.0015	Putrid, garlic
Propyl mercaptan	C_3H_7SH	76	220,000	0.0005	0.020	Unpleasant
Pyridine	C_5H_5N	79	27,000	0.66	0.74	Pungent, irritating
Skatole	C_9H_9N	131	200	0.001	0.050	Fecal, nauseating
Sulfur dioxide	SO_2	64	Gas	2.7	4.4	Pungent, irritating
Thiocresol	$CH_3C_6H_4SH$	124		0.0001	—	Skunky, irritating
Trimethyl amine	$(CH_3)_3N$	59	Gas	0.0004	—	Pungent, fishy

Odor Control

Approaches to control odor emissions from wastewater facilities vary and include methods to inhibit the development of odorous vapor and the treatment of foul air streams.

Upstream Controls

One approach to abating odors at a wastewater treatment facility or in-system pumping station is to control the introduction of an odorous vapor or its formation in the collection system. Controlling the introduction of wastes that may present special odor problems requires that sewer use and pretreatment ordinances be established and, more importantly, be enforced. In addition, the collection system should be designed to minimize the potential for the wastewater to become anaerobic. To accomplish this, travel time should be minimized in the sewer while, at the same time, maintaining a velocity sufficient to scour deposits that may become a source of odor generation.

Because most collection systems are currently in place, it may not be possible to maintain "fresh" wastewater in the collection system. In cases where odors are generated within collection systems, various attempts, including oxygen addition and chemical addition, have been made to inhibit or control the formation of the odorous vapors. Some chemicals used are chlorine, hydrogen peroxide, and metal salts.

Chemical Additions

Chemical addition has been used at wastewater treatment facilities to control odors. This approach, similar to that used for upstream odor control, generally has been more successful because it is conducted under more controlled conditions. Chemicals historically used to control common odors (for example, hydrogen sulfide) are strong oxidants such as chlorine and hydrogen peroxide. The chemical addition is performed at headworks facilities where hydrogen sulfide is a prevalent problem. Prechlorination has been the most common chemical addition approach used. Ozonation has also been attempted, but with less success.

Adsorption Systems

One approach to the treatment of odorous air is to pass the foul air through an adsorptive medium to which odorous compounds adhere. Adsorptive media include granular activated carbon (either virgin carbon or chemically impregnated), activated alumina, activated aluminum with potassium permanganate, and silica gel.

Granular activated carbon is the most widely used adsorption system today. These systems include fans or blowers, adsorptive media beds (one or more stages), corrosion-resistant duct work, and necessary controls. Adsorptive media are regenerated by ambient temperature solubilization, in-place chemical regeneration, or using steam at high temperatures in a reducing atmosphere.

Biological Systems

Treatment of odorous air using biological systems has been tried for many years. The basic treatment process involves biological oxidation of odorous compounds. Biological approaches attempted include using bulk media filters (such as a soil filter), using odorous air as feed air to an activated-sludge treatment system, passing odorous air through a wastewater trickling filter, or using separate trickling filters dedicated to treat foul air. Many biological odorous air treatment systems are in use today.

Combustion Systems

Combustion, or high-temperature oxidation of odorous air, has been an effective way to destroy odorous compounds. Odorous air may be used to meet a portion of the combustion air of a residuals incinerator. Another approach is to use fume burners to destroy odorous compounds. In most cases, separate combustion systems for foul-air treatment have been too costly. In addition, if foul air is used as combustion air to an incinerator, some type of backup system is required for periods when the incinerator is not in use.

Ozonation

Because ozone is a powerful oxidant, ozone treatment systems have been used to treat odorous air. Controlling the dosage of ozone to ensure that overdosing does not occur is critical and expensive. Overdosing results in an odorous ozone offgas if an ozone destruction system is not used. Associated facilities include ozone diffusion and ozone off-gas destruction systems. Ozonation systems that have been installed have experienced serious mechanical problems. As a result, most installed systems are no longer in use.

Wet Scrubbers

In a treatment system that uses wet scrubber absorption, odorous air is brought into contact with a scrubbing liquid chemical. Odorous compounds in the foul air are absorbed in the scrubbing chemical and are thereby removed from the air. Strong oxidizing agents such as sodium hypochlorite and potassium permanganate are diluted and used as the scrubbing liquid chemical.

Two wet scrubber systems commonly used today are the wet packed-bed scrubber system and the mist scrubber system. The packed-bed scrubber consists of a contact chamber with inert packing material (to increase the exposed surface), scrubbing liquid, a recirculating system, an air blower, duct work, and controls. Odorous gases enter the scrubber and pass through the packed bed, where the odorous gases come in contact with the liquid-scrubbing chemical. The scrubbing liquid collects at the bottom of the contact chamber and is recirculated through the unit as required along with a makeup chemical.

Conclusions

Every culture has its stories and fables, often used for generations as educational tools. The morals from the stories help teach the new generation the wisdom of the elders. One such story in the Estonian culture is the story of a man and his bear. They were traveling along a hot, dusty road and the man decided to take a rest. They sat under a tree and the man went to sleep, telling the bear to keep watch. As the man was sleeping, the bear saw that a fly was buzzing around the man's head, and it finally landed on top of his head. The bear was worried that the fly would wake up the man and so, with all good intentions, he gave a mighty swat with his paw right on top of the man's head. He got the fly, but he also killed the man.

The moral of the story here is that sometimes the best of intentions can cause a great deal of harm. The same can be said about the disinfection of wastewater treatment plant effluents. Given the harmful nature of chlorine and the mischievous compounds it is associated with, does it make sense to continue using it to kill pathogens that do not seem to be causing much harm to anyone? Is the chlorination of wastewater treatment plant effluents like the bear's swat? Yes, we get the pathogens, but we cause a great deal more harm in the process.

References

American Industrial Hygiene Association (1989) *Odor Thresholds for Chemicals with Established Occupational Health Standards;* Akron, Ohio.

Amoore, J. E.; Hautala, E. (1983) Odor as an Aid to Chemical Safety: Odor Thresholds Compared with Threshold Limit Values and Volatiles for 214 Industrial Chemicals in Air and Water Dilution. *J. Appl. Toxicol.,* **3** (6), 272.

Barrett, M. J.; Gameson, A. L. H.; Ogden, C. G. (1960) Aeration Studies of Four Weir Systems. *Water Water Eng.,* **64** (9), 407.

Chick, H. (1908) An Investigation of the Laws of Disinfection. *J. Hyg. (G.B.),* **8**, 1908.

Fair, G. M.; Geyer, J. Ch.; Okun, D. A. (1968) Water Purification and Wastewater Treatment and Disposal. In *Water and Wastewater Engineering*; Wiley & Sons: New York; Vol. 2.

Gameson, A. L. H.; Vandyke, K. G.; Ogden, C. G. (1958) The Effect of Temperature on Aeration. *Water Water Eng.,* 489.

Hom, L. W. (1972) Kinetics of Chlorine Disinfection in an Eco-System. *J. Sanit. Eng. Div., Proc. Am. Soc. Civ. Eng.,* **98**, 183.

Nakasone, H. (1987) Study of Aeration at Weirs and Cascades. *J. Environ. Eng.,* **113**, 64.

Watson, H. E. (1908) A Note on the Variation of the Rate of Disinfection with Change in the Concentration of the Disinfectant. *J. Hyg. (G.B.),* **8**, 536.

White, G. C. (1992) *The Handbook of Chlorination and Alternative Disinfectants,* 3rd ed.; Van Nostrand Reinhold: New York.

Symbols Used in this Chapter

a = water quality parameter equal to 0.8 for a wastewater treatment plant effluent
b = weir geometry parameter (for a weir, b = 1.0; for steps, b = 1.1; for step weir, b = 1.3)
C_s = dissolved oxygen saturation concentration of wastewater at temperature T (mg/L)
C_0 = dissolved oxygen concentration of postaeration (mg/L)
C = required final dissolved oxygen level after postaeration influent (mg/L)
C = disinfectant concentration = drop height (m)
h = height through which water falls, m (ft)
H = liquid depth (cm)
H = tailwater depth for downstream channels having horizontal beds (m)
H_c = critical water depth on the weir (m)
k' = die-off constant
k'_T = rate constant at temperature T
I = radiation intensity (mW/cm^2)
I_{ave} = average radiation intensity (mW/cm^2)
I_x = radiation intensity at depth x (mW/cm^2)
m = constant
N = concentration of viable organisms (organisms/L)

q = discharge per width of weir (m³/m·h)
r_{20} = deficit ratio at 20 °C
r = deficit ratio = $C_s C_0 / C_s C$
T = water temperature (°C)
t_p = time required to produce constant die-off (h)
x = depth, m (ft)
α = absorbance coefficient (cm⁻¹)
β = constant

Production and Transport of Wastewater Sludge

Introduction

It is not realistically possible to completely separate liquid-handling processes from sludge- or residuals-handling processes. There are interrelationships between liquid and solids handling that must be considered in wastewater treatment plant design. The quantity, and characteristics, of solids generated is affected by how the liquid streams is processed. Biological nutrient removal processes produce secondary solids that are hard to dewater, for example. Similarly, the recycle streams from solids processing can affect the design of the liquid-handling processes. For example, centrifuge dewatering of anaerobically digested sludge produces a centrate with a high ammonia concentration that will increase the ammonia loading to a biological nitrogen removal process.

Sludge, and its characteristics, treatment, and final disposal, is covered in the next few chapters. Design engineers must understand that the quantities and characteristics of the sludges produced vary widely and in wastewater treatment are directly related to the characteristics of the plant influent and the type of treatment used to meet effluent guidelines.

> **Note on Terminology**
>
> The residuals from wastewater treatment have, since the evolution of the English language, been known as *sludge*. The name is translatable to many other languages, such as *Schlamm* in German, *osady* in Polish, and *solk* in Estonian. The unifying linguistic element seems to be the use of the "s" sound in combination with *l* or *g*, which appears to be particularly offensive to the human ear. All one has to do is say "sludge" in order for a person who knows nothing whatsoever of where this stuff comes from to perhaps be repulsed by it.
>
> Operators and public agencies, who were trying to find beneficial uses for wastewater sludges and often encountered severe public resistance to the disposal of sludge on land, worked with the Water Environment Federation (WEF) to seek a term to describe the "organic product of municipal wastewater treatment that can be beneficially reused" to distinguish it from that which cannot be land applied, sold as soil amendment, etc. After soliciting input from WEF members and a public relations firm, WEF came up with the word *biosolids*.
>
> One could argue that "a rose by any other name . . .", but the point is well taken. Public utilities do indeed have difficulty with the public in promoting beneficial use of wastewater sludge, and if changing names assists them in doing their work, then it is a useful change. The word itself, however, may not be a wise selection, because for one thing, there is no comparable word in other languages, since it is an artificially constructed word. So people who speak German or Polish or Estonian will not know what to make of the term biosolids. More than likely, they will continue to use their translation for sludge.
>
> In some texts, the term "solids" is used in preference to sludge. This is an inappropriate term because solids are contained in the sludge itself. That is, sludge is composed of liquids and solids. Other texts use the term "residuals", which is an umbrella term and could easily be substituted for sludge. Accordingly, in this text, sludge is a suspension of solids and liquids that is extracted from the liquid stream and that often requires further treatment prior to its disposal.
>
> In this book, the word "biosolids" is used when the treated solids meet beneficial use criteria as determined by compliance with the U.S. Environmental Protection Agency's (U.S. EPA's) 40 CFR Part 503 regulations.

Sludge Quantities

The quantity of sludge generated during the treatment of wastewater is one of the most important design parameters in wastewater treatment because it affects not only the sizing of processes for the treatment of the sludge, but also the liquid wastewater treatment processes. For example, the sizing of a secondary treatment process to maintain a desired solids retention time is governed by the quantity of sludge produced. The quantity of sludge generated also determines the size of all the processing equipment used for thickening, dewatering, stabilization, and incineration.

Estimating Sludge Quantities

While the importance of sludge production in wastewater treatment plant design has been generally recognized, a lack of understanding of the plant-to-plant variability and the difficulty in accurately estimating sludge quantities exists. It should be emphasized that the best source of information for estimating sludge reduction is plant-specific data that reflect the nature of the wastewater being treated and the type of wastewater treatment processes being used. In the absence of plant-specific data, general rules or more sophisticated models can be used; however, estimates obtained using such approaches may differ significantly from what may actually result. Consequently, conservative safety factors should be applied to an estimate developed from typical value models.

As an approximation, for typical wastewater of domestic origin, the total quantity of sludge produced is 0.25 kg/m^3 (1 dry ton/mil. gal), ranging from 0.2 to 0.3 kg/m^3 (0.8 to 1.3 dry ton/mil. gal) (Olstein et al., 1996). Sludge production is smaller for plants using processes such as digestion or heat treatment that destroy solids, and higher for plants with chemical addition. However, 0.25 kg/m^3 is a convenient benchmark for cursory comparisons.

A good approach to estimating sludge production is to develop a mass balance for the entire treatment plant that relates sludge production to design parameters for the wastewater treatment processes. Figure 12.1 illustrates a typical solids balance. The mass balance should show the key constituents, such as flow, suspended solids, and biochemical oxygen demand (BOD), and the process assumptions used in the mass-balance calculations. In the case of nutrient removal facilities, nitrogen and phosphorus should be added to the mass balance.

It is important to include recycle streams in the solids balance. This can be done in one of two ways. In the first approach, it is assumed that a fixed percentage of the solids or BOD is recycled from the downstream unit processes to the head of the plant. The solids balance can then be iterated until the recycled quantities assumed at the head of the plant become equal to the sum of the recycled quantities computed for each unit process.

An alternative approach is to estimate the net solids production from the plant based on historical data, the anticipated influent strength, or experience from similar facilities. The net solids production is then used to determine the amount of solids leaving the treatment plant. This is usually applied to the dry solids production following dewatering. The solids loading to a specific unit process can then be back-calculated from the mass balance.

For either of the aforementioned approaches, it is usually necessary to estimate separately the quantities of primary, secondary, and chemical solids. It is also important to take into account expected fluctuations in wastewater characteristics that can occur because of changes in industrial contribution, stormwater flows, seasonal variations in weather conditions, and expansion of the collection area. An understanding of expected peak solids production and diurnal variation is essential to the proper sizing of solids-processing systems.

Primary Solids Production

Most wastewater treatment plants use primary settling tanks to remove settleable solids from wastewater. Primary settling provides a relatively efficient method for reducing BOD loading to secondary treatment processes. The quantity of solids removed during primary settling is generally related to either the surface overflow rate or hydraulic

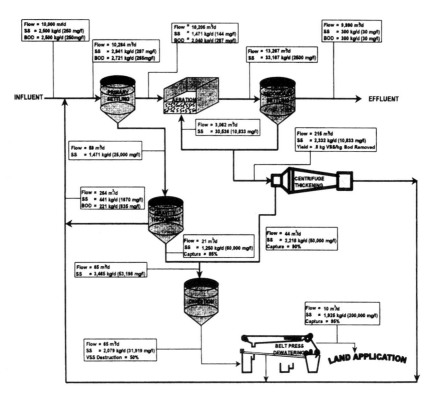

Figure 12.1 Typical solids balance in a wastewater treatment plant (BOD = biochemical oxygen demand, SS = suspended solids, VSS = volatile suspended solids).

retention time. Primary solids production can be related to hydraulic retention time as (Koch et al., 1990)

$$P = Q \times SS \times R \times C \qquad (12.1)$$

where
- P = primary sludge production, kg/d (lb/d);
- Q = plant influent flowrate, m³/d (mgd);
- SS = influent suspended solids (mg/L);
- C = conversion factor, 10^{-3} (8.34 if flow is in mgd and production is lb/d); and
- $R = \bar{t}\,(a + b\,\bar{t})$
  - $\bar{t}$ = hydraulic retention time (min),
  - a = constant (0.406 min), and
  - b = constant (0.0152).

This expression has been developed by fitting a curve to data, but it is important to realize that there is a high degree of variation about the predicted value, as shown in Figure 12.2. In fact, a rather poor correlation coefficient is often obtained if the removal efficiency is plotted versus the retention time.

The variability of sludge production is further illustrated by the results of a study of primary wastewater sludge production in Ontario, Canada. Figure 12.3 shows both

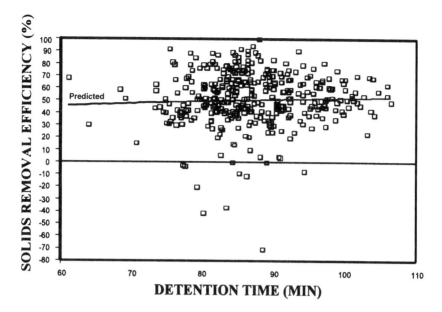

Figure 12.2 Typical primary settling tank performance for a single settling tank over many days of operation.

volume and mass of sludge produced in primary plants. While the mean sludge production is approximately 2 m³/1000 m³ wastewater treated, the production can be as high as 3 m³/1000 m³.

Secondary Solids Production

Secondary solids are biological solids that result from the conversion of soluble and colloidal wastes or substrates (measured as BOD or chemical oxygen demand [COD]) to microorganisms or biomass. Secondary solids also include some of the particulate matter that is not removed by primary sedimentation but becomes incorporated to the biomass. Secondary solids are produced by treatment processes such as activated sludge, biological nutrient removal, trickling filters, rotating biological contactors (RBCs), and other attached-growth systems. The quantity of secondary solids produced is a function of many factors, such as the efficiency of the primary treatment process, the ratio of suspended solids to 5-day BOD (BOD_5), the amount of soluble wastes (soluble BOD or COD) present in the wastewater, and the design parameters of the secondary treatment process. For the activated-sludge process, the amount of time that secondary solids are retained in the process (that is, the solids retention time) will have a significant effect on the amount of secondary solids produced because the longer the solids are retained, the more endogenous decay or self destruction of the biomass that occurs.

Temperature also has an effect on secondary solids production. At higher temperatures, the solids production should be decreased by a higher growth rate and more endogenous respiration.

The final determinant is the solids retention time. A low solids retention time (sludge age) will result in reduced production in the summer, while a shorter solids retention time will increase solids in the summer.

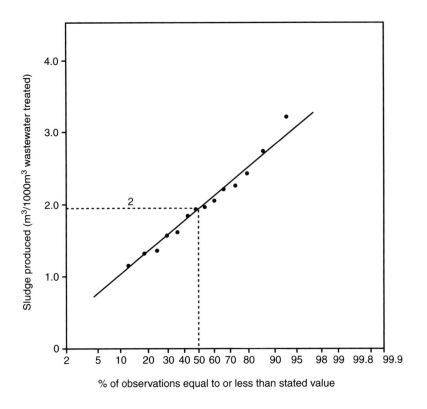

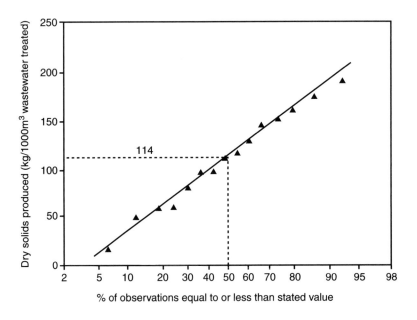

Figure 12.3 Variability in the production of raw primary sludge in Ontario wastewater treatment plants (Antonic et al., 1974).

> **Note on Terminology**
> The phrases *total yield* (including effluent solids) and *net yield* (excluding effluent solids) have been used by some designers to distinguish between these two yield expressions. This is the preferred definition of the phrases net yield and gross yield, although in related literature these phrases have been used interchangeably (along with the phrases *observed yield* and *apparent yield*), adding to the confusion when trying to compare plants.

The kinetic relationship between secondary solids production and solids retention time can be theoretically expressed using an equation of the following:

$$Y_{obs} = \frac{Y_t}{1 + k_d(\theta_c)} \qquad (12.2)$$

where
- Y_{obs} = observed yield (kg biomass produced/kg substrate used);
- Y_t = true yield (g biomass/g substrate);
- k_d = endogenous decay rate (d^{-1}); and
- θ_c = solids retention time (d).

The substrate concentration is represented by either BOD$_5$ or COD. The substrate concentration is expressed as grams of BOD$_5$ or COD consumed by the process, although some designers prefer to express the substrate as the concentration applied rather than removed. The biomass concentration is expressed either in terms of suspended solids or volatile suspended solids (volatile solids is approximately 70 to 80% of suspended solids). The yield can also include effluent biomass as well as the biomass that is wasted from the secondary treatment process.

The ranges listed in Table 12.1 have been reported for the true yield and endogenous decay coefficients and are expressed as net production (that is, kilograms of suspended solids versus kilograms of substrate removed). In all cases, the solids lost in the effluent are not counted in the yield calculations, although they should be for completeness.

A curve in Figure 12.4, labeled "Monod" after the scientist who pioneered application of Michaelis–Menten enzyme kinetics to microbial growth (Monod, 1949), represents fitting eq 12.2 to the data using linear regression techniques. The data in Figure 12.5 are presented to emphasize the variation typical of operating plants. In many plants, it is difficult to see a clear relationship between solids retention time and solids production. In fact, some references have reported that solids production does not appear to be affected by solids retention time (Wilson et al., 1984; Zabinski et al., 1984). However, when multiple plants are plotted, a relationship between yield and solids retention time usually becomes apparent.

Table 12.1 Typical range of values for sludge yield coefficients.

Coefficient	Basis	Range	Typical
Y_t	g VSS/g BOD$_5$	0.4–0.8	0.6
Y_t	g VSS/g COD	0.25–0.4	0.4
k_d	d^{-1}	0.04–0.15	0.10

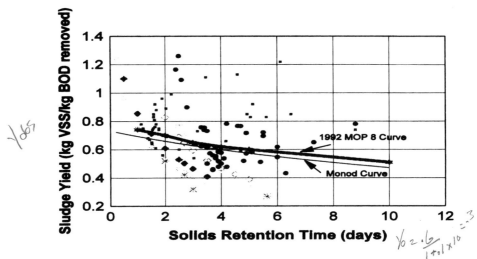

Figure 12.4 Secondary sludge yield for an activated-sludge system in a plant with primary settling.

The values of the coefficients should also vary with temperature. For a given solids retention time, the growth rate and amount of endogenous respiration should increase with temperature, resulting in lower solids production. In practice, although the temperature of the wastewater varies over the year, the effect on solids production may be masked because many plants adjust solids retention time on a seasonal basis and see little difference in solids production throughout the year.

Another approach sometimes used to estimate secondary solids production is to separate the production into three terms as shown in eq 12.3 (Koch et al., 1990)

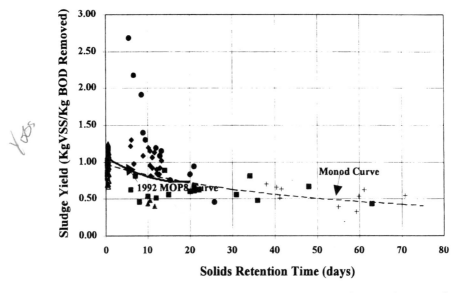

Figure 12.5 Secondary sludge yield in an activated-sludge plant without primary settling.

12-8 | Wastewater Treatment Plant Design

$$\text{Net production (kg/d [lb/d])} = I + a(\text{VSS}) + b(\text{SBOD}_5) \quad (12.3)$$

where
- I = inerts removed, kg/d (lb/d);
- a = volatile solids coefficient;
- b = soluble BOD coefficient;
- VSS = volatile suspended solids removed, kg/d (lb/d); and
- SBOD_5 = soluble BOD_5 removed, kg/d (lb/d).

The volatile suspended solids coefficient varies between 0.6 and 0.8, and the soluble BOD coefficient varies between 0.3 and 0.5. Both coefficients decrease with increasing solids retention time and are dimensionless.

All of the approaches discussed above can be used to estimate secondary solids production from suspended-growth activated-sludge systems. A similar approach can be applied to attached-growth secondary treatment systems such as rotating biological contactors and trickling filters. In general, solids production for attached-growth systems can be expressed by the following equation:

$$\text{Net production (kg/d [lb/d])} = Y[(\Delta \text{BOD}_5) + k_d(\text{AB})] \quad (12.4)$$

where
- Y = observed yield (kg biomass/kg substrate);
- k_d = endogenous decay rate (kg biomass/kg biomass·d);
- ΔBOD_5 = BOD_5 removed, kg/d (lb/d); and
- AB = attached biomass, kg (lb).

The yield coefficient has units similar to the coefficient for BOD removed for activated-sludge systems (that is, grams of VSS per grams of BOD_5 removed). The attached-biomass coefficient has the same units as the biomass coefficient for activated-sludge systems. The amount of attached biomass is usually directly related to the amount of surface area available to support the attached growth; in addition, plots of solids production per unit of surface area versus BOD_5 removal per unit of surface area can be used to derive plant-specific kinetic constants from plant data. The values of the yield coefficient for attached-growth systems are similar to those for suspended-growth systems; however, the values of the attached-biomass endogenous decay rate coefficients tend to be higher to reflect the longer effective solids retention time with attached systems. Typical values of the attached-biomass coefficient for the attached-growth system range from 0.03 to 0.40.

Combined Solids Production

In plants that do not use primary treatment, a combined secondary and primary sludge is generated. The quantity of combined sludge can be estimated by adjusting the yield coefficients used to estimate secondary solids production to account for the additional solids. The amount of secondary solids production is significantly increased for those plants that do not use primary treatment. In such plants, the primary solids become an inseparable portion of the biomass or secondary solids product. Figure 12.5 is a plot of secondary solids production for plants that do not use primary treatment (Koch et al., 1997; Schultz et al., 1982). The values of the yield and endogenous decay rate coeffi-

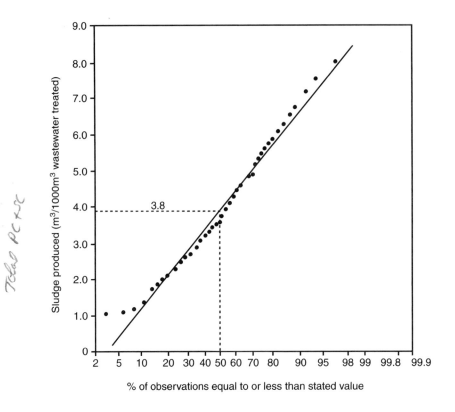

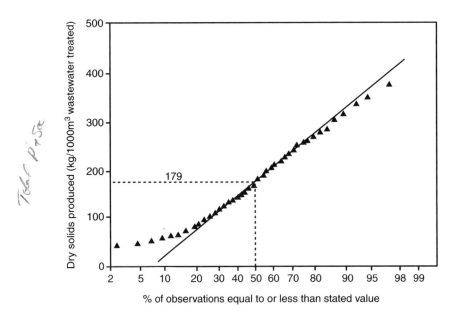

Figure 12.6 Variability in total sludge production in secondary wastewater treatment plants in Ontario (Antonic et al., 1974).

cients for the Monod curve shown in Figure 12.5 are 0.975 kg VSS/kg BOD_5 removed and 0.0177 d^{-1}, respectively. The high solids retention times represent extended aeration plants and oxidation ditch plants. Although a trend in reduced solids production versus solids retention times can be seen, it is important to note that there are several plants that experience solids production levels that are significantly above the curve.

A survey of a number of plants in Ontario, Canada, found high variations of sludge production in secondary treatment plants. The total sludge production numbers in these plants (primary sludge plus waste activated sludge) are shown in Figure 12.6. As with primary sludge production, the sludge volume and mass produced in conventional activated-sludge plants varies significantly, ranging from 1.0 m^3/1000 m^3 wastewater treated, to over 8 m^3/1000 m^3. This is an impressive range that emphasizes the need for careful consideration in the design of sludge handling facilities.

Chemical Solids Production

The quantity of chemical solids produced can be estimated from the anticipated chemical reactions. The addition of chemicals can increase solids production in direct proportion to the amount of chemical added. Competing reactions, however, must be considered. For example, the addition of ferric chloride will generate more solids than what would be estimated by recognizing the formation of ferric chloride to form ferric hydroxide. The ferric hydroxide preferentially reacts with phosphate, yielding higher levels of solid precipitates than the hydroxide reaction. Generally, a yield of 1 g of additional solids per 1 g of ferric chloride added is a good approximation. Similarly, lime addition can significantly increase solids production.

Sludge Characteristics

An engineer must know the characteristics and quantities of solids to be handled when designing conveyance, conditioning, and thickening or dewatering facilities. Numerous references can assist the designer in obtaining detailed information about sludge sources, characteristics, and quantities. Because this text focuses on the design of facilities, however, only those topics that significantly affect the design process are discussed. Characteristics of solids depend on the percentage of industrial wastes, ground garbage, sidestreams in the wastewater, the use of chemical precipitants and coagulants, process control, peak loads and weather conditions, and the unit processes being used at a wastewater treatment plant. Types of solids include primary, secondary, mixed primary, chemical sludge, and biosolids.

Primary Sludge

Primary sludge consists of organic solids, grit, and inorganic fines. Primary sludge is pumped to a downstream process for additional thickening, stabilizing, and dewatering; disposal or reuse then follows. The composition of primary sludge varies widely from day to day and even hour to hour within a given plant and from plant to plant. Table 12.2 presents a summary of the typical composition of raw primary sludge.

The total solids concentration varies depending on how the sludge collection and removal system is operated. If sludge is removed from the primary settling tank at a high rate, a lower solids concentration can be expected. Some plants remove the solids at a low rate and essentially use the primary settling tanks as gravity thickeners.

Plants that receive combined sanitary sewage and stormwater or have a high contribu-

Table 12.2 Primary sludge characteristics.

Parameter	Concentration (dry-weight basis)
Total solids	2.0–8.0
Total volatile solids, % of TS	60–80
Grease, % of TS	5.0–8.0
Phosphorus, % of TS	0.8–2.8
Protein, % of TS	20–30
Cellulose, % of TS	8–15
Nitrogen, % of TS	1.5–4.0
pH	pH 5.0–8.0

tion of infiltration and inflow can have a great variation of volatile solids concentration and quantity of primary sludge. In addition, plants with inadequate grit removal can also have increased primary sludge production. In these cases, VSS fractions can decrease to approximately 60%. The lower VSS concentration reflects a higher proportion of inorganics and grit in the primary solids. The heavy metal content of primary sludge varies depending on the types of industries contributing flow to the plant and is higher for plants that add inorganic chemicals, such as ferric chloride or lime, in primary tanks.

Secondary Sludge

The phrase *secondary sludge* refers to biological sludges that are generated from the conversion of soluble wastes in primary effluent to biomass and particles escaping primary treatment. Secondary sludge is produced by treatment processes such as activated sludge, trickling filters, and rotating biological contactors. Although numerous variations are possible, all secondary sludges result from aerobic biological treatment of wastewater. Generally, biological sludges are more difficult to thicken or dewater than primary sludges and most chemical sludges. Table 12.3 shows a typical composition of secondary sludge characteristics.

High-rate (high F/M) plants tend to have higher nitrogen levels in the sludge than conventional activated-sludge plants because of the lower level of endogenous respiration. For plants that use biological processes for phosphorus removal, higher phosphorus content in sludges can be expected. Heavy metal concentrations are determined by the industrial contribution to the plant and the result of chemical phosphorus precipitation if this is used.

Table 12.3 Secondary sludge characteristics.

Parameter	Concentration (dry-weight basis)
Total solids, %	0.4–1.2
Total volatile solids, % of TS	60–85
Grease, % of TS	5–12
Phosphorus, % of TS	1.5–3.0
Protein, % of TS	32–41
Nitrogen, % of TS	2.4–7.0
pH	pH 6.5–8.0

Combined Sludge

Generally, sludges produced from combinations of primary and secondary solids have properties that are closer to secondary sludge, although they are affected by the respective composition of each type. For purposes of thickening and dewatering, it is axiomatic that when two sludges are combined, the result will exhibit the dewatering characteristics of the poorer sludge. Accordingly, design engineers usually keep sludges apart and do not try to save money by combining them, which typically results in greater operating expenses.

The two chemical characteristics of concern in sludges produced in wastewater treatment plants (primary plus biological sludges) are heavy metals and nutrients. The metals limit the disposal of the sludge on farmland and the nutrients affect their market value. These characteristics are further discussed in Chapter 16.

Chemical Sludge

Chemical sludges result from the addition of aluminum or iron salts and lime to improve suspended solids removal or to precipitate phosphorus. Wastewater chemistry, pH, mixing, reaction time, and opportunities for flocculation all affect chemical sludge characteristics. Generally, the addition of lime improves thickening and dewatering performance of chemical sludges. The heavy metal content of chemical solids is typically higher because of the additional precipitation of heavy metals with the iron and aluminum and the addition of heavy metals contained in the coagulant.

Liquid Sludge Storage

Sludge storage is an important, integral component of every wastewater solids treatment and disposal system. Sludge storage provides the following benefits: equalizes flow to downstream processes such as thickening or dewatering; allows liquid sludge accumulation during scheduled or unscheduled times when solids-processing facilities are not operating; allows more uniform feed rate to enhance thickening, conditioning, and dewatering operations; and permits scheduling flexibility and process optimization for thickening and dewatering operations.

Sludge can be stored within wastewater treatment process tankage, solids treatment process systems, or in separate specially designed tanks. Storage can be short term (for example, in wastewater clarification basins or in solids thickening tanks). Such storage, however, is limited and is used mostly by small treatment plants where storage time may vary from a few hours to 24 hours. The amount of storage necessary depends on the redundancy of the sludge-handling equipment. If thickeners or dewatering equipment is down, and there is not redundancy, the sludge must be stored somewhere. Large treatment plants often use aerobic digesters, anaerobic digesters, facultative lagoons, and other solids treatment processes with long detention times to provide short-term storage. In cold climates, storage of up to six months may be necessary because of the odiferous nature of liquid sludge, such storage facilities can cause severe public relations problems if not properly designed with covers and means for controlling odors.

Liquid Sludge Transport

Sludge pumping systems are an important consideration in wastewater treatment plants, particularly those with average flows greater than about 4000 m^3/d (approximately

1 mgd). Typical needs of pumping systems include transporting solids from the following: primary and secondary clarifiers to thickening, conditioning, or digestion facilities; thickening and digestion facilities to dewatering operations; biological processes for recycle or further treatment; and degritting facilities to temporary storage. Even in smaller plants, sludge transport may require careful consideration to ensure adequate velocities without undersizing the piping and increasing the risk of line blockage.

Although specifying a single type of pump to satisfy the handling of all solids within a plant might seem advantageous, the wide range of conditions imposed on such service exceeds the capabilities of a single type of pump. Fortunately, many pumps are available to the design engineer.

Flow and Head Loss Characteristics

Flow characteristics (rheology) of wastewater sludge cannot be simply defined. Rheological properties vary widely from process to process and from plant to plant. Because rheological properties directly influence pipeline friction losses, head loss characteristics also vary extensively.

Design Approach

Minimizing pumping distance and applying a conservative multiplier to head losses calculated for equivalent flows of water is the traditional approach to designing sludge pumping and piping systems. However, this approach can be inaccurate. For short pumping distances, the inaccuracies often lack practical consequence. Two types of sludges falling under the classification are dilute sludges and thickened sludges because these act as liquids.

Dilute Sludges

Settled sludge from clarifiers is typically relatively dilute. For activated sludge, concentrations of 1.2 to 1.5% are typical. At these concentrations, data indicate that the head loss essentially is equal to that of water for velocities greater than 0.3 to 0.6 m/s (1 to 2 ft/s) (Mulbarger, 1997). At these velocities, the slurry is in the turbulent flow regime and essentially behaves the same as water. At lower velocities, the flow becomes laminar, and head losses will increase sharply. Therefore, whenever possible, pumping systems for dilute sludges should be designed to maintain a minimum velocity of 0.6 to 0.75 m/s (2 to 2.5 ft/s), thus ensuring turbulent flow.

Thickened Sludges

The concentration at which sludge can be defined as "thickened" varies with the type of sludge and the preceding treatment processes. Thickened sludges range from a readily pourable to viscous material, but will not have any free water apparent. At low Reynolds numbers (less than 2000), the sludge will be flowing in the laminar zone, and experimental evidence suggests that Figures 12.7 and 12.8 are useful for estimating the headlosses for sludges of various concentrations in 15- to-20 cm (6- to 8-in.) diameter pipes. At higher velocities (higher Reynolds numbers), the sludge will be in turbulent flow. At that point, head losses can be calculated using the Hazen–Williams formula. The value of C in the Hazen–Williams equation should be assumed to be 140 for normal conditions and 100 for worst-case design conditions.

If the curves in Figures 12.7 and 12.8 are used, pumps and motors should be designed and selected to operate satisfactorily over the entire head loss range from

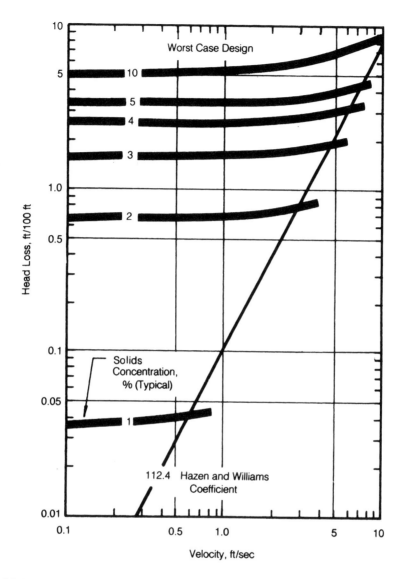

Figure 12.7 Frictional head losses for a 6-inch diameter sludge force main (in. × 25.4 = mm; ft × 0.3048 = m) (Mulbarger et al., 1981).

"water" to "worst-case" design. These changes in head affect centrifugal pumps much more than positive-displacement-type pumps. If centrifugal pumps (for example, recessed-impeller centrifugal pumps) are used, motors should also be checked for overloading if the operating head is significantly less than the design head. When the pump operates beyond the right terminus of its characteristic curve, it is in a "runaway" condition. Under this condition, motors may be overloaded. A designer should also recognize that a routine sludge can occasionally exceed the worst-case head loss curve, so oversizing of motors may be advisable in some instances.

In addition to careful consideration of head loss, the nature of the process receiving the pumped sludge must be considered. Many positive-displacement pumps that are suitable

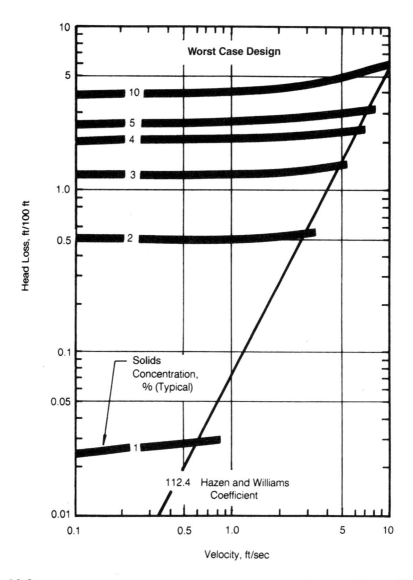

Figure 12.8 Frictional head losses for an 8-inch diameter sludge force main (in. × 25.4 = mm; ft × 0.3048 = m) (Mulbarger et al., 1981).

for pumping thickened residuals produce a pulsating flow. This may not be acceptable for downstream processes that have a chemical addition paced proportional to the flow or that depend on a steady flow to maintain proper operation, such as centrifuge.

Kinetic Pumps

Kinetic (or dynamic) pumps are pumps in which energy is continuously added to the fluid to increase the velocity within the pump to a value greater than the velocity at the discharge. The velocity reduction that occurs at the discharge results in an increase in pressure. These pumps should be used only for dilute sludges. Several types of these pumps and their common applications are described in the following sections.

Solids-Handling Centrifugal Pumps

Centrifugal pumps encompass a wide range of designs. With the exception of special designs, such as the recessed impeller, use of centrifugal pumps is restricted in most common cases to relatively dilute (less than 1% solids), trash-free sludges. Conventional centrifugal pumps are commonly applied for transport of return activated sludge because of the pump's high volume and excellent efficiency.

Centrifugal pumps are not recommended for pumping primary sludge, primary scum, or thickened sludge applications. There are two factors that contribute to the failure of standard solids-handling centrifugal pumps in this application. First, there is no method to ensure that the thickened solids will be positively drawn to the pump by the pump suction. Secondly, the system head curve varies significantly depending on the solids concentration, causing variations in the liquid flowrate and in the amount of power required by the centrifugal pump.

Recessed-Impeller Pumps

The recessed-impeller pump is also known as a *torque flow pump*, *vortex pump*, or *shear-lift pump*. This pump has a standard concentric casing with an axial suction opening and a tangential discharge opening. The impeller, recessed into the pump casing, can be the open or semiopen type, with either straight radial blades or blades tapered to the shaft.

For sludge pumping applications, these pumps have fully recessed, open impellers. The rotating impeller creates a spiraling vortex field in the sludge within the casing; this vortex moves sludge from the centrally located suction to the tangentially located discharge. Most of the sludge does not pass through the impeller vanes. Thus, abrasive contact between sludge solids and the impeller is minimized, and large solids are easily passed.

Recessed-impeller pumps are well suited to raw sludges with solids concentrations of approximately 2.5% or less. When digested solids (biosolids) are pumped, these units will work well at solids concentrations up to approximately 4%. Although thicker sludges can be pumped, the varying friction losses cause erratic flowrates and heavy radial thrusts on the pump shaft. The various positive-displacement pumps perform better than recessed-impeller pumps for most thickened sludge applications. If recessed-impeller pumps are used, flow meters and variable-speed drives should be provided to maintain a relatively constant flow. The heaviest possible shafts and bearings should also be specified. In addition, the pumps should be horizontally mounted to simplify maintenance, with adequate cleanouts and flushing connections provided.

Screw–Combination Centrifugal Pumps

Screw–combination centrifugal pumps are a combination of screw feeder and a normal centrifugal impeller. These pumps have a high efficiency with low net positive suction head requirements. In addition, the corkscrew action of screw impellers may improve the handling of thicker sludges by providing a more positive feed to the suction.

Grinder Pumps

Special combination centrifugal pump grinders are also available. These pumps combine a hardened steel cutting bar with a typical centrifugal vortex-type pump. These pumps can be used as digester recirculation pumps and are effective in preventing rag balls. However, operating experience indicates that such pumps require as much maintenance as grinders.

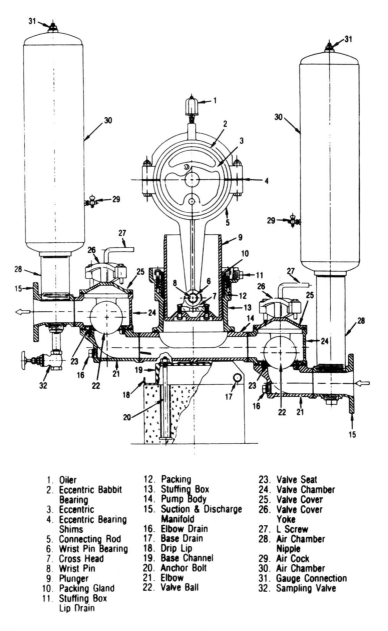

Figure 12.9 Details of a positive-displacement plunger pump.

1. Oiler
2. Eccentric Babbit Bearing
3. Eccentric
4. Eccentric Bearing Shims
5. Connecting Rod
6. Wrist Pin Bearing
7. Cross Head
8. Wrist Pin
9. Plunger
10. Packing Gland
11. Stuffing Box Lip Drain
12. Packing
13. Stuffing Box
14. Pump Body
15. Suction & Discharge Manifold
16. Elbow Drain
17. Base Drain
18. Drip Lip
19. Base Channel
20. Anchor Bolt
21. Elbow
22. Valve Ball
23. Valve Seat
24. Valve Chamber
25. Valve Cover
26. Valve Cover Yoke
27. L Screw
28. Air Chamber Nipple
29. Air Cock
30. Air Chamber
31. Gauge Connection
32. Sampling Valve

Positive-Displacement Pumps

Several types of positive-displacement pumps and their common applications are described in the following sections.

Plunger Pumps

Plunger pumps are positive-displacement units with pistons driven by either an exposed eccentric crank shaft or a walking beam. They are available in simplex, duplex, triplex,

Figure 12.10 Progressing cavity pump.

and quadplex configurations. Plunger pumps have an output of 150 to 225 L/min (40 to 60 gpm) per plunger and can develop up to 70 m (230 ft) of discharge head. These pumps are designed for an efficiency of 40 to 50%. This allows a power reserve to overcome changes in pumping head. Advantages of plunger pumps include the ability to handle heavy solids concentrations (up to 15%); wide choice of capacities; low pumping rates can be used with large port openings; positive delivery unless some object prevents the ball check valves from seating; constant but adjustable capacity, regardless of large variations in pumping head; and they can operate for a short period of time under no-flow conditions, such as those that may occur with a plugged suction line, without damage. Adjusting the length of the stroke allows the output of the pump to be changed, but the pumps operate more satisfactorily at or near full stroke: a variable-pitch V-belt drive or some type of variable-speed drive commonly is provided for pumping capacity control. Plunger pump construction is illustrated in Figure 12.9.

Plunger pumps can operate with up to 3 m (10 ft) of suction lift, although suction lifts can reduce the solids concentration that can be pumped. Using a pump with suction pressure higher than the discharge is impractical because flow would be forced past the check valves. The use of special intake and discharge air chambers reduces noise and vibration and dampens pulsations of intermittent flow.

Progressing Cavity Pumps

When used in applications similar to those of plunger pumps, the progressing cavity pump operates more cleanly and discharges a smoother flow. In addition, the pumps can provide a consistent flow despite changes in discharge head. However, improper selection and design of these pumps can lead to excessive maintenance problems and costs. A concern that should be guarded against is operation of these pumps in no-flow conditions. If a no-flow condition occurs, damage to the stator can quickly occur. Good design includes the use of pressure-release valves and a feed back loop to protect the pump.

The progressing cavity pump (see Figure 12.10) uses a worm-shaped metal rotor that turns in an eccentric motion inside a pliable elastomeric stator. The stator has an axial pitch approximately 50% that of the rotor. The different axial pitches provide sealing lines that move down the pump as the rotor turns. Between the sealing lines, cavities progress axially and move the sludge from the suction end toward the discharge end.

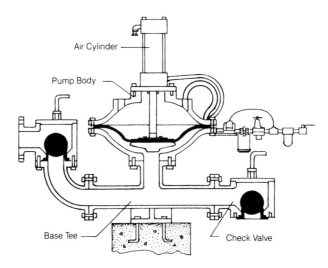

Figure 12.11 Air-operated diaphragm pump.

Some slippage flow at the sealing lines between the stator and rotor occurs as the stator wears; the additional slippage causes further wear. Slippage can be effectively minimized by using sufficient cavities (multistage construction) to limit the pressure difference across the sealing lines.

Because the elastomeric stator is relatively soft and subject to wear from excessive grit, progressing cavity pumps should preferably be used in facilities with good grit removal facilities. They should not be used to transport grit. It is also important to minimize the rotational speed of the pump rotor. In some applications (particularly with variable-speed drives), it may be prudent to select a pump larger than required for the design flow to ensure that the pump can meet the design flow requirements after some degree of stator wear has occurred.

An advantage of the progressing cavity pump is that the stator tends to act as a check valve, preventing backflow through the pump under most conditions. Therefore, in many cases, a check valve is not required. If the static backpressure on the pump is high (greater than 50 m), either a check valve or an anti-reverse ratchet should be provided. If stator wear is expected, as in applications with a significant grit concentration, a check valve is a more appropriate device. Although the check valve may be eliminated in most applications, it is important to include isolation valves on both the suction and discharge side of the pump. These will allow the pump to be removed from service for routine maintenance.

Air-Operated Diaphragm Pumps

The most common application of air-operated diaphragm pumps is pumping sludge from primary settling tanks and gravity thickeners. These pumps provide a relatively simple means of pumping thickened residuals, allowing them to handle grit with minimum wear. However, depending on downstream processes, pulsation in flow may not be acceptable.

In air-operated diaphragm pumps (see Figure 12.11) compressed air flexes a membrane that is pushed or pulled to contract or enlarge an enclosed cavity. In most operations, this pump has a single-chambered, spring-return diaphragm pump, an air-pressure regulator, a solenoid valve, a gauge, a muffler, and a timer. Unless compressed air is already planned for the facility, providing this service can significantly increase the cost of the pump as well as sufficient space for disassembly. The air exhaust from these valves is also noisy, which may create problems depending on the plant location.

Rotary Lobe Pumps
A rotary lobe pump is a positive-displacement pump that uses multilobed, intermeshed rotating impellers. Like the progressing cavity pump, these pumps have the benefit of offering a relatively smooth flow of liquid. They also do not require check valves in many applications with low-to-moderate discharge static heads. However, as with the progressing cavity pump, it is important to provide both suction and discharge isolation valves. In addition, because pumping efficiency is dependent on the maintenance of close tolerances between the rotating lobes, it is important to remove all large or abrasive material upstream of the pumps. Therefore, rotary lobe pumps are more suitable in applications with efficient grit removal and should not be used for the transport of grit. Their use has been most successful with waste activated sludge.

Pneumatic Ejectors
Pneumatic ejectors lack rotating elements and electric motors. They have a receiving container, inlet and outlet check valve, air supply, and liquid level detector. When the sludge reaches a preset level, air is forced into the container and the sludge is ejected. After the discharge cycle, the air supply cuts off and sludge flows through the inlet to the receiver. This results in a sudden blast of flow and can harm downstream processes.

Peristaltic Hose Pumps
Although more widely applied in the industrial sector, peristaltic hose pumps have also been occasionally used for wastewater sludge pumping. The pumps, available in capacities of 36 to 1250 L/min (1 to 330 gpm) and heads up to approximately 152 m (approximately 500 ft), have self-priming capabilities. As a positive-displacement pump, the output of the pump is directly proportional to speed at either high or low discharge pressures; thus, a metering capability is provided. These pumps are relatively simple devices that require only common tools and basic mechanical skills for assembly, servicing, and repair. The hoses, however, are prone to breakage and must often be replaced.

Reciprocating Piston Pumps
Reciprocating piston pumps are useful and cost effective in special applications where dewatered cake must be transported to cake-storage or -loading facilities. These piston pumps are not used ahead of the dewatering process. However, because these units can achieve discharge pressures up to 1.5×10^4 kPa (2200 psi), they are the primary choice for long-distance pumping of thickened sludge. With such high potential discharge pressures, it is important to properly design downstream piping systems.

Other Pumps

Air-Lift Pumps
Air-lift pumps are frequently used for return activated sludge and similar applications in smaller treatment plants where a precisely controlled flowrate is not required. An air-lift

Figure 12.12 Archimedes screw pumps used for return activated sludge.

pump has an open riser pipe, the lower end of which is submerged in the liquid to be pumped. When an air supply tube introduces compressed air at the bottom of the pipe, the resultant mixture of air bubbles and liquid rises; the liquid is pushed up and out through the riser pipe.

The pumping capacity of an air-lift pump can be varied by increasing the air supply rate to an optimum amount of air. Because the pump discharges a liquid–air mixture, an increase in air supply beyond the optimum value causes a decrease in the volume of liquid that is discharged. Although these pumps are capable of pumping against high heads, their application in wastewater treatment is high volume and low head. Lifts are typically less than 1.5 m (5 ft) because of the need for at least 70% air-release submergence. The primary advantages of air-lift pumps are the absence of moving parts and the simplicity of construction and usage. The air supply arrangement governs the capability of handling solids.

Archimedes Screw Pumps

The Archimedes screw pump (see Figure 12.12) is occasionally used for return activated sludge pumping applications. This pump has either an open design, for lifts up to 9 m (30 ft), or an enclosed design, for lifts up to 12 m (40 ft) or more. The Archimedes screw pump, a positive-displacement-type pump, automatically adjusts its rates of discharge in proportion to the depth of liquid in the inlet chamber. This characteristic gives the Archimedes screw pump an inherent variable flow capacity without the motor speed controllers needed for centrifugal or recessed-impeller pumps.

The principal disadvantage of the Archimedes screw pump is its space requirements. The screw pumps can warp from thermal expansion caused by exposure to the sun if left unoperated for extended periods or if not equipped with cooling water sprays and covers. At the opposite extreme, offline units may experience freezing during cold weather periods. Archimedes screws also require continuous lubrication of the bearings. Another disadvantage is the aeration of the return activated sludge that often occurs with these systems. In some instances, the use of Archimedean screw pumps for return activated sludge was discontinued because the high dissolved oxygen content of the return activated sludge was interfering with the biological nutrient removal process. Their use in biological nutrient removal plants for lifting return activated sludge is therefore questionable.

Long-Distance Pipelines

Pumping of sludge for several thousand meters has been successfully accomplished at many locations throughout the United States. However, special design criteria must be developed to minimize the potential for operating problems. Sludge characteristics, such as viscosity, solids percentage, and type of sludge, should be carefully determined. The effects of flow velocity on sludge viscosity and pipe friction losses should also be studied.

Common Design Deficiencies in Pumps and Piping

Several design errors in pumping and piping systems are particularly noteworthy.

- Incorrect calculation of friction head and inadequate allowance for variations that occur during operation. Some procedures are simple and easy to use but yield incorrect results.
- Poor suction conditions for thickened sludge. Because thixotropy and plasticity can affect friction, a good design includes a straight, short suction pipe to a pump in a deep room to allow substantial positive suction pressure.
- High maintenance costs are particularly common for progressing cavity pumps that should be operated at limited speeds and pressures per stage. An incorrectly applied pump or a pump of inadequate mechanical quality will have a high maintenance cost.
- Buried or encased elbows—grit slurries and sometimes sludge that has supposedly been degritted can wear out elbows.
- Inadequate allowance for flushing and cleaning lines—many sludges form grease deposits or scale in pipes.
- Sludge withdrawal from two or more tanks at the same time with one pump. Valves should be provided so that a pump can draw from tanks sequentially, or, if possible, a pump should be provided for each tank with interconnections to be used when pumps are out of service.
- Pipeline route with high spots that trap air or gas. Air-relief valves in sludge service are too troublesome to be a suitable alternative to avoiding high spots.
- Improper selection of valves. Valves should preferably be ball or V-port valves with 100% clear waterway area, or at the very least plug valves with a minimum 80% clear waterway area.
- Lack of rupture disks or other pressure-relief devices between isolation valves.
- Lack of sampling ports.

Standby Capacity

Because primary and secondary sludge pumping are critical functions, either dual units or units performing dual duty typically provide standby capacity. In addition, primary sludge pumps usually provide standby service scum pumping. Standby capacity for activated-sludge recirculation is mandatory because an interruption can quickly impair effluent quality.

Dewatered Sludge Storage

Dewatered cake storage provides similar benefits for downstream disposal alternatives (for example, trucking to land application sites, composting, and incineration) to sludge storage provided for thickening and dewatering operations. Storage evens out fluctuations, enhances operational flexibility and reliability, and allows accumulation while downstream operations are out of service.

Bins or hoppers are commonly used to store dewatered cake; they can be made any size. Existing facilities have individual hoppers ranging from a few cubic meters to 380-m^3 (500-yd^3) capacity. Multiple hoppers, which are typical at most dewatered cake-storage facilities, provide firm capacity during hopper maintenance or inspection. Hoppers are typically covered and equipped with duct work, blowers, and scrubbers to control odors. They are also equipped with specially designed screw augers at the bottom of the bin to control discharge to pumps, conveyors, or trucks for further transport or handling.

Dewatered cake can also be stored in open or enclosed concrete pads. Access for rubber tire bucket loading equipment can be provided, along with methods for dealing with stormwater (if the pad is open) and odors.

Dewatered Sludge Transport

Modern dewatering operations can produce cake with 15 to 40% solids content or greater, depending on the conditioning chemicals and type of dewatering equipment used. These cakes range in consistency from that of pudding to damp cardboard; moreover, they will not flow by gravity in a pipe or channel from the dewatering equipment. Instead, they must be transported from the dewatering equipment by mechanical conveyors such as flat, troughed or corrugated belts; screw augers; gravity drops in which dewatering equipment positioned directly above a storage hopper allows the cake to drop directly into it; and pumping using positive-displacement pumps. In many plants, inadequate shielding has resulted in horrendous housekeeping problems when the sludge splashes and spatters over floors, walls, equipment, and even operators.

Pumps

Progressing cavity pumps and hydraulically driven reciprocating piston pumps have been used to pump dewatered cake in lieu of transporting cake by belt or screw conveyors. Compared with conveyors, pumps offer the advantages of odor control (because the dewatered cake travels in an enclosed pipe), elimination of spills, and reduced maintenance. In addition, dewatered cake pumps require significantly less space than any of the other conveyance options and are therefore suitable for retrofit of conveyor systems in existing buildings with space constraints. Noise levels can also be reduced in some cases. Depending on the application, pumps may need to be designed for high pressure in the discharge piping. Pumps often require more energy expenditure per volume of cake transported than conveyors.

The hydraulic characteristics of dewatered cake (that is, over 15% solids) have not been extensively studied or widely reported. Dewatered cake exhibits both plastic and thixotropic behavior. For a Bingham plastic, there is a minimum shearing stress that is required to initiate flow. For thixotropic materials, the shear rate decreases with time at a constant rate of shear. These two behaviors complicate hydraulic design. Nonetheless, dewatered cake can be pumped, although head losses for most dewatered cake applications are high, often in the range of 1380 to 6900 kPa (200 to 1000 psig). Constrictions, such as smaller-than-line-size valves and short radius bends, should be avoided in discharge piping. The piping should be of sufficient size so that theoretical cake flow velocities never exceed 0.15 m/s (0.5 ft/s), with maximum velocities of 0.08 m/s (0.25 ft/s) preferred, especially in cases where dewatered cake solids concentrations exceed 30%. The pipe should be designed to allow flushing and pigging of the line.

Progressing Cavity Pumps

Progressing cavity pumps have been used with limited success for pumping dewatered cake. Two problems that seem to plague this application is bridging, where the pump begins to run dry and burns out, and clogging in the throat section. These pumps should be limited to wet cake with solids concentrations of approximately 20% or less, limited to pump rotational speeds to 100 rpm or less, and limited to the pressure per stage to 52 kPa (75 psi).

Hydraulically Driven Reciprocating Piston Pumps

Hydraulically driven reciprocating piston pumps use a twin screw auger feeder. The principal advantage of these pumps is that they can move dewatered cake with up to 20% solids. Drier cakes are moved more economically with conveyors. Hydraulically driven reciprocating piston pumps have a higher purchase cost than other pumps, but they can convey dewatered cake at solids concentrations that other pumps (i.e., progressing cavity) cannot.

Hydraulically driven reciprocating piston pumps are available from manufacturers in a range of capacities and discharge pressure capabilities, up to 1500 L/min (400 gpm) and 13,800 kPa (2000 psi), respectively. They also have a wide range of suction feed hopper-to-pump configurations. Hydraulic power units are also available in broad output ranges, depending on the required discharge line pressure.

The phrase *volumetric efficiency* can be described as the ratio of the volume of solids pumped per piston stroke to the total volume displaced per piston stroke. If a hydraulically driven reciprocating piston pump is pumping water or dilute sludges (that is, 1 to 4% solids), this ratio would essentially be 100% because the wastewater is nearly incompressible. This behavior is typical of a true Newtonian fluid. Dewatered cake, however, neither physically resembles a true Newtonian fluids nor behaves like one. It contains air, other entrained or dissolved gases, and concentrated organic material, which are all compressible. The cake, therefore, tends to compress as the piston begins applying pressure. Until the cake is squeezed against the downstream resistance, the cake does not move forward relative to the pumping cylinder. Once the cake does move forward, the piston has already displaced a certain volume of the cylinder. This displaced volume is lost and accounts for a small part of the volumetric efficiency. The simple inability to completely fill the cylinder with each return stroke of the piston, however, accounts for most of the loss in volumetric efficiency. Even with the slight pressure of 34 to 103 kPa (5 to 15 psi) provided by a twin screw auger or conical plow feeder plus a partial vacuum in the cylinder, the dewatered cake resists moving into the cylinder bore. This resistance increases as cake dryness increases, with a corresponding drop in volumetric efficiency. Regardless of the actual reasons for loss of volumetric efficiency, this loss must be accounted for when sizing these pumps. No theoretical model exists for predicting volumetric efficiency, although it typically ranges from 60 to 90%. Once the characteristics of the cake are known or can be estimated, manufacturers should be consulted for recommendations regarding a typical volumetric efficiency to apply. As a rough estimate, a conservative value of 70% can be used for pumping dewatered cake in the range of 20 to 30% solids.

Conveyors

Conveyors are used to move wet or dry solids materials that are not easily pumped, such as primary grit, screenings, dewatered cake, and other treatment plant solids. Generally, conveyors for municipal plants are either belt conveyors or screw conveyors.

Belt Conveyors

Belt conveyors move material that is loaded on top of a moving flexible belt. This type of belt is supported on rollers spaced 0.9 to 1.5 m (3 to 5 ft) apart on the carrying side and approximately 3 m (10 ft) apart on the noncarrying, or return, side of the conveyor. The rollers on the carrying side are called *load-side rollers*; on the noncarrying, or return, side the rollers are called *idlers*. To increase solids-carrying capacity, load-side rollers may be troughed or angled to form a concave carrying surface. One or more drive drums or pulleys connected to a motor through a belt or chain drive provide the belt-driving force. For simple conveyor systems, the drive pulley, which is usually single, is located at the discharge end or head of the belt. The pulley at the solids-loading end is referred to as the *tail pulley*. The belt must have a device for maintaining a minimum tension to reduce sag between carrying idlers, provide contact force, and prevent slippage at the drive pulley. Such devices can include a weighted pulley (gravity take-up), a spring-loaded pulley, or a screw adjustment for pulley position. The screw take-up on the tail pulley, the least costly device, is typically used for conveyors less than 90 m (300 ft) long.

The following design guidelines can assist in designing the conveying of solids:

- Belt transfer points should have minimum drop heights and be equipped with skirtboards with wipers to minimize splashing and spillage.
- Belt cleaning is potentially troublesome. Counterweighted rubber scrapers below the head pulley have been ineffective and have required intensive maintenance. Multiple-finger-type scrapers with adjustable tensions are suggested. An alternative often not mentioned is a water spray followed by a rubber scraper if the water can be easily collected and disposed.
- Accessories, such as snubber or counterweight pulleys that contact the dirty side of the belt, should be avoided. Snubbers are pulleys positioned to increase the angle of contact between the belt and the drive pulley. This increases the area for friction and reduces drive slippage. Instead of gravity take-ups, manual screw take-ups and, if necessary, multiple shorter belts should be used to avoid snubbers and gravity take-ups.

- Housekeeping facilities such as adequate hose stations, oversized floor or paving drains with exaggerated grades below the conveyor, and nonskid tread plates instead of grating should be provided.

Screw Conveyors

Screw conveyors move material by the pushing action of a helical blade, or "flight". Screw conveyors can either have their flights attached to a center shaft or consist of a shaftless design. The screw is mounted in a U-shaped trough or enclosed in a tubular housing. Flights are manufactured in a wide variety of designs that can be tailored to an application. The shaft can be tapered and the flights can be tapered in radius; they can have full or partial cross sections, or both. The horizontal distance between flight blades, called the *pitch*, can vary along the shaft length. The material can be mixed or folded by two flights or blades that are cut, folded, or otherwise shaped. In shafted screw conveyors, a drive mechanism turns the center shaft that is supported by end bearings and intermediate hanger bearings as necessary to reduce shaft deflection.

The design approach for screw conveyors resembles that of belt conveyors. The material properties, volume, and variability must first be defined. Capacity of the conveyor is a direct function of the screw speed, the flight size or diameter (assuming shaft size remains fixed), and the amount of trough loading. Conveyor flight pitch and any folding, cutting, or other special flight designs also affect capacity. Some screw conveyor manufacturers provide tables and charts in their catalogs to determine preliminary sizes of their conveyors for a range of these variables.

Screw conveyors should be avoided if the sludges contain sticks, large objects, or rope-type materials. Enclosed-tube screw conveyors can reduce or eliminate spillage and housekeeping problems but are slightly more susceptible to jamming and are more difficult to access for maintenance. For sticky solids such as dewatered cake, intermediate support or "hanger" bearings should be avoided because they could cause plugging as material packs against them. Larger shafts, heavier shaft-wall thicknesses, or both allow greater screw lengths between support bearings; generally, enlarging the shaft is more effective than increasing the shaft-wall thickness. If intermediate bearings are unavoidable, the flight design near the hanger can be modified to minimize the packing problem.

Dried Sludge Transport

Belt Conveyors

Belt conveyors are one of the most widely used and efficient means of transporting bulk materials. To transport material, material is loaded onto the top of the belt at one end through a loading chute and is discharged to a chute at the opposite end. Belt conveyors range in width from 35 to 150 cm (14 to 60 in.). The smallest belt conveyors accommodate capacities up to 76 kg/s (300 ton/hr), traveling at speeds up to 100 m/min (325 ft/min). The largest belt conveyors operate at speeds of up to 200 m/min (650 ft/min), handling capacities up to 1135 kg/s (4500 ton/hr). Belt conveyors can travel long distances at high speeds and are also appropriate for short distances and small capacities. Other conveying devices are preferred for distances less than 6 m (20 ft).

Because they provide a gentle, efficient, and durable means of conveying, belt conveyors are appropriate for transporting dried solids. In addition, they are one of the best types of conveyors to prevent material degradation. When installing a belt conveyor, the belt should be angled to form a trough. The trough presents a carrying surface that will not be amenable to material rolling off the belt.

Conveyors used for transporting dried solids are typically no larger than 100 cm (40 in.) wide. While belt conveyors are preferred for horizontal transport applications involving dried solids, other conveying devices should be considered when large elevation gains are required. Typically, belt conveyor inclination angles exceeding 20° are not recommended for dried solids, particularly dried wastewater sludge pellets, which tend to be free flowing. If belt conveyor inclination angles exceeding 20° are required to transport dried materials, sidewalls, cleats, or cover belts are recommended.

There are several important material characteristics and design criteria that should be considered in belt conveyor design. The Conveyor Equipment Manufacturers Association's *Belt Conveyors for Bulk Materials* (1979) includes procedures for establishing design and other criteria and belt sizing.

Screw Conveyors

Screw conveyors are particularly well suited to transporting dried materials and are available in shafted and shaftless or centerless varieties. These conveyors use a spiral helix mounted on a shaft or a self-supporting helix to move material within a covered trough. Screw conveyors are among the most economical types of conveyors for moving dried biosolids. Inlet and discharge openings may be located where needed. A complete screw conveyor unit is supported at the ends, loading points, and by either feet or saddles at intermediate points.

Standard screw conveyor sizes range from 150 mm (6 in.) to 600 mm (24 in.) in diameter. Screw conveyors are installed horizontally or at inclines of less than 45°. New designs of shaftless screw conveyors can be used to convey dried solids vertically. Standard lengths (3 to 4 m) of shafted screw sections are coupled together using bolts; in addition, internal intermediate hangers provide support, maintain alignment, and serve as bearing surfaces. Shaftless screw spirals are furnished in continuous lengths exceeding 6 m and rely on polyethylene liners or steel wear bars for intermediate support of the spiral. Torsional capacity of the center shaft and couplings (shafted screw conveyors) or spiral (shaftless screw conveyors) are limiting factors of a conveyor's overall length. The length of a screw conveyor is typically limited to 14 m, unless exceptional design considerations are addressed.

Drag Conveyors

Drag, or *en masse*, conveyors are also acceptable equipment for transporting dried solids. Drag conveyors have a wide range of uses in numerous industries. Drag conveyors have a long history of conveying materials such as rock salt, coal, wood chips, sawdust, sludge, grit, and logs. These conveyors are highly adaptable and can be customized to convey most, if not all, wastewater treatment plant solids.

Drag conveyors are recognized by their slow moving chain-and-flight assembly that typically pushes material along a steel pan or trough. The trough cross section is rectangular or U-shaped. The flights used to drag the solids are bolted directly to the drag conveyor's chain. Most drag conveyors are installed in horizontal conveying configurations. Special designs, however, permit their installation in vertical and "Z"-lift applications. A Z-lift is an installation in which the drag conveyor starts out horizontal, bends vertically, and then bends back horizontally.

Drag conveyors are used in many processes such as primary clarification, grit removal, dewatered cake transport, composting, and other solids-processing operations. Drag conveyors also may be used as live bottom feeders in the bottom of hoppers.

Drag conveyors are exceptionally strong and robust conveyors. The lengths of drag conveyors are generally only limited by the strength of the conveyor chain and the weight of the material being conveyed. Practical length limits of drag conveyors are approximately 40 m on horizontal and 10 to 15 m on vertical.

Bucket Elevators

Bucket elevators are simple and dependable units for vertically conveying dry materials. A bucket elevator consists of a series of buckets mounted on a belt or chain within a housing. The buckets are filled with material at the base of the unit and discharge material at the top. They are available in a wide range of capacities (10 to 350 kg/s [4 to 140 ton/hr]) and are completely enclosed to prevent the escape of dust and odors. There are three basic types of bucket elevators: centrifugal discharge, positive discharge, and continuous.

Centrifugal discharge elevators are the most common type of bucket elevator and are best suited for handling fine, free-flowing materials that can be dug from the elevator boot at the base of the unit. Loading is simple and fewer buckets are required for this type of elevator. The buckets, mounted either on a chain or belt, travel at speeds sufficient to discharge the material by centrifugal force as they pass around the head pulley or sprocket (up to 90 m/min [300 ft/min]).

Pneumatic Conveyors

Pneumatic conveying involves moving material in a pipeline using air. There are two categories of pneumatic conveying: dilute phase and dense phase. *Dilute phase* refers to systems that use conveying air at low pressure (less than 100 kPa [14.5 psig]) and have low material/air ratios. *Dense phase* refers to systems that use air at pressures higher than 300 kPa (45 psi) and have high material/air ratios.

In dilute-phase pneumatic conveying systems, the conveying air velocity is high enough to suspend the material in the air stream, in the range of 20 to 40 m/s (65 to 130 ft/s). These systems use either a positive or negative pressure to push or pull the material through the conveying line. Dilute-phase conveying systems use rotary airlocks to feed material to the conveying line, and positive-displacement blowers or fans to supply the conveying air. The system incorporates a large volume of air to move a small amount of material, with material/air ratios of less than 5:1 (5 kg material/kg air).

With dense-phase pneumatic conveying systems, the air velocity is too low to suspend the material in the air stream. Material settles in the conveying line, and as the pressure increases, the material forms a plug that is pushed by the conveying air. Average air velocities for dense-phase systems are less than 2.5 m/s (8 ft/s), and material/air ratios are quite high (up to 100 kg material/kg air). Dense-phase pneumatic conveying systems require a pressure tank and a high-pressure air compressor that can provide 350 to 700 kPa (50 to 100 psig) of air for conveying. The material flows into the pressure tank or transporter by gravity until a high level is reached. The transporter inlet valve closes, the vessel is pressurized using the compressed air, and the material flows out of the vessel to the conveying line.

Dilute-phase systems are ideal for nonabrasive, powdered materials and short conveying distances. Although capital costs for dilute-phase conveying equipment are low, energy costs can be quite high because of the large conveying air requirements. Although dilute-phase systems are successfully used for conveying a wide variety of bulk solids, they may not be suitable for dried solids used for beneficial reuse that can be abrasive and degrade with exposure to high-velocity air. Dilute-phase conveying may be used for lime, sawdust, and other chemicals.

The design of a pneumatic conveying system depends on parameters such as material bulk density, capacity and flowrate, and equivalent length of conveying line. Pressure systems are used when high flowrates are needed (greater than 9000 kg/h [2.0×10^4 lb/h]). Vacuum systems are used when flowrates do not exceed 7000 kg/h and the equivalent length is less than 300 m.

Conclusions

It takes a certain amount of courage to spend one's professional career moving sludge from one place to another. "So, little girl, what does your father do?"

But moving sludge is one of the greatest and most daunting of technical challenges. Not only is sludge thixotropic and non-Newtonian, it also is dynamic in its properties. Give it a minute, and it will change right in front of your eyes. So designing a pumping system that responds to the ever-changing nature of the material you are pumping, even if you knew how to design for a constant and perceptible feed, is a challenging task. The engineer who designs such systems no doubt spends a lot of sleepless nights wondering if it will work. And when it does, the joy has to be tempered by professional pride. After all, you cannot show the client that you did not really know if it was going to work! It is no wonder that engineers engaged in sludge work are a tight and mutually supportive fraternity.

References

Antonic, M.; Hamoda, M. F.; Cohen, D. B.; Schmidtke, N. W. (1974) *A Survey of Ontario Sludge Disposal Practices*; Project No. 74-3-19; COA Research Report; Toronto.

Conveyor Equipment Manufacturers Association (1979) *Belt Conveyors for Bulk Materials*, 2nd ed.; CBI Publishing Co.: Boston, Massachusetts.

Koch, C.; et al. (1990) Spreadsheets for Estimating Sludge Production. *Water Environ. Technol.*, **2**.

Koch, C.; Lee, J. S.; Bratby, J. R.; Barber, D. B. (1997) A Critical Evaluation of Procedures for Estimating Biosolids Production. Presented at the Water Environment Federation/American Water Works Association Water Residuals and Biosolids Management: Approaching the Year 2000 Specialty Conference, Philadelphia, Pennsylvania.

Monod, J. (1949) The Growth of Bacterial Cultures. In *Annu. Rev. Microbiol.*, **3**, 371.

Mulbarger, M. C.; Copas, S. R.; Kordic, J. R.; Cash, F. M. (1981) Pipeline Friction Losses for Wastewater Sludges. *J. Water Pollut. Control Fed.*, **53**, 1303.

Mulbarger, M. C. (1997) Selected Notions about Sludges in Motion, and Movers. Presented at the Central States Water Environment Association Education Seminar, Madison, Wisconsin.

Olstein, M.; et al. (1996) *Benchmarking Wastewater Treatment Plant Operations;* Interim Report; Water Environment Research Foundation: Alexandria, Virginia.

Schultz, J., Hegg, B. A.; Rakness, K. L. (1982) Realistic Sludge Production for Activated Sludge Plants without Primary Clarifiers. *J. Water Pollut. Control Fed.*, **54**, 1355.

Wilson, T.; et al. (1984) Operating Experiences at Low Solids Retention Time. *Water Sci. Technol.* (G.B.), **16**, 661.

Zabinski, A.; Srinivasaraghavan, R. (1984) Low SRT—An Operator's Tool for Better Operation and Cost Savings. Presented at the 57th Annual Conference of the Water Pollution Control Federation, New Orleans, Louisiana.

Symbols Used in this Chapter

a	= constant (0.406 min)
AB	= attached biomass
b	= constant (0.0152)
C	= conversion factor, 10^{-3} (8.34 if flow is in mgd and production is lb/d)
k_d	= endogenous decay rate (d^{-1})
k_d	= endogenous decay rate (g biomass/g biomass·d)
P	= primary sludge production, kg/d (lb/d)
Q	= plant influent flowrate, m³/d (mgd)
R	= $\bar{t}(a + b)$
SS	= influent suspended solids (mg/L)
$\bar{t}$	= hydraulic retention time (min)
Y	= observed yield (kg biomass/kg substrate)
Y_{obs}	= observed yield (kg biomass produced/kg substrate used)
Y_t	= true yield (kg biomass/kg substrate)
ΔBOD_5	= BOD removed
θ_c	= solids retention time (d)

13
Sludge Conditioning

Introduction

Conditioning refers to chemical or thermal treatment used to improve the efficiency of a thickening or dewatering processes. Chemical conditioning involves use of inorganic chemicals, organic polyelectrolytes, or both. Thermal conditioning can be accomplished by heating the sludge, or by allowing it to freeze and then thaw. Thickening and dewatering of wastewater sludges—particularly those containing solids from biological treatment systems—is often not practical without conditioning.

Historically, the first chemical conditioners were ferric chloride and lime. The use of *organic polyelectrolytes*, or *polymers*, in municipal wastewater treatment plants was introduced during the 1960s and these were rapidly adopted for both thickening and dewatering processes. The primary advantage of polymers compared with inorganic chemicals is that polymers do not significantly increase the amount of solids production.

The conditioning method must be compatible with the proposed methods of dewatering and ultimate disposal. Some dewatering methods such as centrifugation use pressure to compact the solids, whereas filter presses work by allowing water to pass through the void spaces. A single type of conditioning agent cannot be expected to be useful for all applications.

Production of a cake suitable for incineration increases constraints on the conditioning system. For incineration to be economical, cake solids should be maximized, but not at the expense of the volatile fractions. It is easy to achieve cake solids of 30 to 40% by adding large amounts (50 to 100%) of substances such as fly ash prior to dewatering; however, this only represents a real improvement if the mixture has changed the sludge to allow more water to drain from it. If the increase in cake solids is only from adding massive amounts of dry solids that are not combustible, then the heating value of the cake will actually decrease. High levels of inorganics in the cake also cause production of excessive ash, which ultimately must be disposed of. Bulking materials such as ash are not conditioners but may assist in reducing the overall water content of the mixture. Such additions may be necessary if the sludge is to be disposed of in a landfill.

Where landfill or land application is the ultimate disposal option, the cake may have to meet certain levels of biological and physical stability. The addition of a sufficiently high lime dosage may accomplish biological stability while fly ash or other bulking materials may provide the physical stability. Mechanical strength can be measured by a slump test similar to that used for concrete. It is becoming increasingly common for local or state authorities to require solids concentrations of 35 to 40% before allowing sludge to be co-disposed of with municipal solid waste in landfills.

Chemical Conditioning

Inorganic Chemicals

Conditioning with inorganic chemicals is principally associated with recessed chamber filter presses and vacuum filters, although they have also been used for belt filter presses. Inorganic chemicals used in conditioning are ferric chloride and lime. The quantities of inorganic chemicals required for conditioning are greater than that of organic polymers and this becomes important in ultimate disposal. The use of ferric chloride and lime, for example, can add as much as 20 to 40% more mass (and associated volume) to the amount of cake requiring disposal. This additional quantity can reduce the ultimate life of a landfill or land-application site and increase transportation costs.

A second consideration is the effect of the conditioning agents on the heat value of the sludge when incineration or some form of heat recovery is used. Ferric chloride and lime add precipitates to the total solids that do not burn. Nevertheless, when evaluating alternative conditioners, it is important to note that it is the ratio of water to dry volatile solids in the dewatered cake—not the presence or absence of chemical precipitates—that determines cake combustibility. In addition, ultimate disposal by composting could affect the type of conditioning chemical used.

Lime

Lime is used in some wastewater treatment plants for pH control. It is commercially available in two dry forms: pebble quicklime and powdered hydrated lime. In either form, lime is caustic, tends to dust when powdered, and is prone to precipitate out when slurried, forming a calcium carbonate scale on conveying vessels.

As a conditioner, lime is used to raise the depressed pH caused by ferric chloride addition. Besides pH control, lime provides a certain degree of odor reduction because sulfides are converted in solution from hydrogen sulfide to the sulfide and bisulfide ion, which are nonvolatile at alkaline pHs. At higher doses, lime can affect stabilization (for example, at disinfection levels greater than pH 12). An additional benefit is the formation of calcium carbonate and calcium hydroxide precipitates. These precipitates improve dewatering ability by acting much like a bulking agent, increasing porosity while resisting compression. Some dissolved calcium hydroxide also is available at high pH levels.

Ferric Chloride

Ferric chloride reacts with the bicarbonate alkalinity in sludge to form ferric hydroxide complexes that act as aggregates. In general, ferric chloride dosage ranges from 2 to 10% of solids and lime ranges from 5 to 40%, both of which are based on dry solids. Activated sludges alone require higher ferric chloride dosages, anaerobically digested sludge requires midrange dosage, and fresh raw primary sludges require lower dosages.

Ferric chloride is commonly used to flocculate wastewater solids. Ferric chloride is sold commercially in liquid form as a 30 to 35% ferric chloride solution. In colder climates, shipping strength is reduced to avoid crystalline hydrate formation on cold rail cars. As a liquid, ferric chloride is corrosive; consequently, care is essential for its proper handling and storage.

Ferric chloride coagulation is sensitive to pH and works best above pH 6. Below pH 6, floc formation is weak and dewaterability is sometimes poor. For this reason, lime is used to adjust the pH for effective ferric chloride use and optimum dewatering. Commercial ferric chloride liquid is an orange-brown color, with average concentrations between 30 and 35% by weight in water. At 30 °C (86 °F) and a specific gravity of 1.39, a 30% ferric chloride solution contains 1.46 kg (3.24 lb) of ferric chloride.

The majority of wastewater sludges cannot be adequately conditioned with ferric chloride without lime (Christensen and Stulc, 1979; Christensen et al., 1976; Stulc and Christensen, 1976; Webb, 1974). Instead of coagulation by adsorption of positively charged complexes onto negatively charged colloids, the coagulation seems to result from a double layer compression by a high Ca^{2+} concentration of negatively charged solids particles and $Fe(OH)_2(s)$ colloids. The key requirements are a pH of 11 to 12, a high Ca^{2+} concentration (10^{-2} M), and the presence of solid ferric species. Ferric should be added before lime and the points of addition should be separate. In ferric–lime conditioning, lime doses as quicklime typically run 2 to 4 times the ferric chloride doses to reach pH 11 to 12.

Aluminum Salts

The primary differences between aluminum and ferric chemistry are the relative solubilities of aluminum species above pH 7 and the relative insolubility of ferric iron above pH 7. The practical significance is that ferric hydroxide is insoluble at the highest pH values used in conditioning with ferric and lime (that is, pH 12 to 12.5), whereas aluminum hydroxide is quite soluble above pH 10. Aluminum salts, therefore, are unlikely to be effective with the same lime doses often used with iron salts and are seldom used for sludge conditioning.

Organic Polymers

Organic chemicals used for conditioning are primarily long-chain, water-soluble, synthetic organic polymers. Polyacrylamide, the most widely used polymer, is formed by the polymerization of backbone monomer acrylamide. Polyacrylamide is nonionic. To carry a negative or positive electrical charge in aqueous solution, polyacrylamide must be combined with anionic or cationic monomers. Because most solids carry a negative charge, the cationic polymers are the most commonly used polymers for conditioning. Polymers are further categorized by molecular weight (from 0.5 to 18 million), charge density (from 10 to 100%), and active solids levels (from 2 to 100%). Polymers are also categorized by the form when received (i.e., dry, liquid or solution, emulsions, or gel).

High-molecular-weight and long-chain polymers are highly viscous in liquid form, extremely fragile, and difficult to mix into aqueous solution. Unmixed polymers appear as "fish eyes" in a diluted solution. As the molecular weight increases, so does the difficulty in mixing and diluting the polymer.

Since the introduction of secondary treatment, polymers for conditioning have become attractive because they do not appreciably add to the volume to be disposed of, do not lower the fuel value to be incinerated, are safer and easier to handle than inorganic chemicals, and result in easier maintenance than inorganic chemicals that require frequent cleaning of equipment, typically with acid baths. However, polymers, are not completely stable and can plasticize at high temperatures, and polymers are slippery when spilled on floors, creating severe hazards to plant operators.

Polymer Charge

Chemical reactions involving polymers are similar to those for inorganic chemicals (that is, they neutralize surface charges and bridge the particles). Neutralization of the negative electrical charges of the particle by the positive charge of the polymer leads to a reduction of the electrostatic repulsion between particles and thus encourages aggregation. In polymer bridging, the long-polymer molecule attaches itself by adsorption to two or more particles at the same time. Flocs formed by particle bridging tend to have a higher resistance to shear than flocs formed by charge neutralization.

Charge is developed by ionizable organic constituents distributed throughout different sites on the polymer molecule. Measuring the charge of a specific polymer under field conditions is nearly impossible. Consequently, the polymer's relative charge (sometimes called the *application charge*) can be used to measure charge capability of a particular polymer.

For anionic and nonionic polymers, charge in application does not significantly change because usual levels of dissolved materials present do not overcome the ionic equilibrium among anionic-charged particles in the sludge. For cationic products, charge neutralization brought about by the influence of water alkalinity counter-ions depletes the cationic-charged species of the polymer. This effect usually causes deteriorating charge levels with the passage of time. Some polymers seem to be more charge-stable than others; however, polymer-charge stability usually is one of the most competitive trade secrets of manufacturers.

Most polymer manufacturers use the phrase *relative charge* to describe the measured titratable charge level of their products under specified test conditions. Consequently, comparative charge levels among different manufacturers may be practically meaningless. Thus, the user should be wary of claims relating to charge in applications without on-site testing under controlled conditions. In many cases, charge is not the sole governing criterion that determines a polymer's effectiveness in a given application.

Polymer Molecular Weight

Molecular weight of a polyelectrolyte is a rough indication of the length of the polymer chain. Molecular weight also affects other product attributes such as solubility, viscosity, and charge density in aqueous solution. Although there are exceptions, in general, lower molecular weight products tend to be more soluble, less viscous, and have higher charge density in water. Products encountered in conditioning practice can be categorized as low, intermediate, and high molecular weight.

Low-molecular-weight polyelectrolytes often are described as *primary coagulants*, a term usually reserved for products ranging from 2.0×10^4 to 1.0×10^5 mol wt. These products are water soluble and are typically marketed in concentrations of 30 to 50%. They have low viscosities (that is, close to the viscosity of water) and can be easily diluted and mixed with water at the application point.

These low-molecular-weight coagulants are useful for clarification applications where there are many small dispersed particles to be destabilized and settled. They are common in oily waste and biological waste treatment applications where low concentrations of solids are being treated. They also are sometimes used as the first part of a two-polymer program in which high-charge density is required to break the suspension.

Intermediate-molecular-weight cationic, nonionic, and anionic products are available as solutions and in dry and liquid-latex emulsion forms. Most are modified or substituted polyacrylamide-based polymers with molecular weights of 3.0×10^5 to several million. Although it is difficult to generalize about the entire class of intermediate-molecular-weight products, most products in this category require wetting (mixing activation to disperse the polymer) and aging to develop full-product activity in application.

Solutions of intermediate-molecular-weight products are more viscous than lower-molecular-weight products. In fact, product handling of feeding characteristics limits commercial solutions of these products to 1% (dry solids basis) or less. Consequently, supplemental dilution water is usually needed to improve polymer disbursement in the sludge being conditioned.

Intermediate-molecular-weight products are common in thickening and dewatering applications, especially those with high concentrations of secondary (biological) sludges. Many of the most popular cationic flocculants, along with some nonionics used in conditioning, fall into this category. Nearly all charge variations are available in the intermediate-molecular-weight range.

High-molecular-weight polymers can be cationic, anionic, and nonionic. These products are available as liquid viscous solutions, latex emulsions, or dry powder. Moreover, their molecular weights vary from 2×10^6 to more than 12×10^6. Solubility and viscosity considerations typically dictate the solution concentrations available in these high-molecular-weight products. Product solutions are made up at 0.5 to 1.0% solids concentration and allowed to age for several hours before additional dilution at the application point.

Polymer Forms

Polymers are available in two different physical forms: dry and liquid. Dry polymers can be delivered in a microbead, gel powder, or gel log. Liquid polymers can be delivered as a solution, emulsion, or mannich form. All dry and liquid polymers can be prepared with three charge types—cationic, anionic, and nonionic—and can be purchased with a wide array of molecular weights, charge densities, and active solids levels. The form, charge, and activity level of the polymer can greatly affect how they react with sludge.

The *activity* of a polymer relates to the percent of the molecular weight that is available to react with and flocculate the solids particles. The activity level can greatly vary with the form of the polymer. Polymer dosing criteria are stated in grams of active polymer per kilograms of dry solids, allowing polymers having different activity levels to be compared on an equivalent basis. For example, a polymer having an activity of 9% will require 10 times more grams of bulk polymer than a polymer that has an activity of 90%.

Dry polymers can have an activity level as high as 94 to 100%. The shelf life of dry polymers is typically 2 years, but wet and humid conditions should be avoided in the storage area because polymers will tend to cake and deteriorate.

Most dry polymers are difficult to dissolve. To make up a working solution, an *eductor* is used as a prewetting device to disperse the polymers in the water. The solution is slowly mixed in a mixing tank until the dry polymer particles are dissolved, and then aged in accordance with manufacturer recommendations. This aging time (30 minutes to 2 hours) allows the polymer particles to unfold into long chains. Once the dry polymer is diluted, the solution is only stable for approximately 24 hours. The quality of water used to dissolve the dry polymer particles is important. Water that is hard or high in free chlorine can cause the solution to deteriorate within a few hours. Water with a free-chlorine level greater than 0.5 mg/L will begin to have an adverse effect.

Gel polymers are high-molecular-weight monomers that are produced with gamma radiation instead of a chemical catalyst, as with other forms of polymers. Gels are typically shipped in log form with a diameter of 230 mm (9 in.) and a length of 510 mm (20 in.), and are 30 to 33% active solids. Gels require special makeup equipment.

Emulsions are dispersions of polymer particles in a hydrocarbon oil or light mineral oil. Surface active agents are typically applied to prevent separation of the polymer–oil phase from the water phase. Emulsion polymers must be mixed to prevent the separation of oil and water. With emulsions, it is possible to produce high molecular weight and maintain an activity level of 30 to 50% without producing high viscosity. The apparent viscosity of emulsion polymer in its as-delivered state ranges from 300 to 5000 cps. The shelf life of emulsion polymers is typically 6 months to 1 year. The initial breaking of the emulsion and aging, critical for optimum emulsion polymer performance, can be accomplished with a static mixer, high-speed mixer, or wet dispersal unit. Emulsion polymers can have higher molecular weights and higher charges than dry polymers but do not have the operational problems of dry polymers. Disadvantages of emulsion polymers are the potential for oil–water separation and the higher cost per volume of active material.

A *mannich polymer* contains 3 to 8% active polymer. It is produced by using a formaldehyde catalyst to promote the chemical reaction to create the organic compound. Because vapors from formaldehyde pose a safety hazard and can be carcinogenic, mannich polymers should be carefully stored and only used in well-ventilated areas. Characteristics of mannich polymers include their high viscosity range (from 5.0×10^4 cps to more than 1.5×10^5 cps) and extreme difficulty in pumping. Mannich polymers also have a relatively short shelf life. This type of polymer, however, can be quite effective and economical for large treatment plants, with economy directly related to the shipment cost of the polymer.

Dosage Optimization

Nearly any thickening or dewatering performance can be enhanced through the use of one type of polymer or another. Under normal design conditions, gravity thickeners and dissolved air flotation thickeners do not require polymer addition. They have polymer

Table 13.1 Typical dosages of polymer for thickening municipal wastewater sludges (U.S. EPA, 1979; WPCF, 1987, 1988).

Application	Sludge type	Polymer dosage, g/kg
Gravity thickening	Raw primary	2–4
	Raw (P^a + WAS^b + TF^c)	0.8
	Raw WAS	4.3–5.6
Dissolved air flotation	WAS (oxygen)	5.4
	WAS	0–14
	P + TF	0–3
	P + WAS	0–14
Solid-bowl centrifuge	Raw WAS	0–3.6
	Anaerobically digested WAS	2–7.2
Rotary drum	WAS	6.8
Gravity belt	Digested secondary	5

[a]Primary.
[b]Waste activated sludge.
[c]Trickling filter.

feed points for troubleshooting or upset conditions. In upgrades of existing gravity thickeners and dissolved air flotation thickeners, polymers are used to increase the solid and hydraulic-loading rates to the units and improve solids capture. Table 13.1 summarizes polymer dosage for various thickening applications and Table 13.2 reviews the experience of polymer conditioning for dewatering.

Table 13.2 Typical dosages of polymer for dewatering municipal wastewater sludges (U.S. EPA, 1979; WPCF, 1983, 1988).

Application	Sludge type	Polymer dosage, g/kg
Belt filter press	Raw (P^a + WAS^b)	2–5
	WAS	4–9
	Anaerobically digested (P + WAS)	6–10
	Anaerobically digested primary	4–7
	Raw primary	2–3
	Raw (P + TF^c)	3–6
Solid-bowl centrifuge	Raw primary	0.5–2.3
	Anaerobically digested primary	2.7–5
	Raw WAS	5–10
	Anaerobically digested WAS	1.4–2.7
	Raw (P + WAS)	2–7
	Anaerobically digested (P + WAS + TF)	5.4–6.8
Vacuum filter	Raw primary	1–5
	Raw WAS	6.8–14
	Raw (P + WAS)	5–8.6
	Anaerobically digested primary	6–13
	Anaerobically digested (P + WAS)	1.4–7.7
Recessed chamber filter press	Raw (P + WAS)	2–2.7

[a]Primary.
[b]Waste activated sludge.
[c]Trickling filter.

Selecting the right dosage of a chemical conditioner is critical to optimum performance. Dosage affects not only the dryness of the cake, but also the solids capture rate and solids disposal costs. Dosage is determined from pilot-plant tests, bench tests, and online tests. Because sludge characteristics periodically change, the dosage should periodically be reevaluated.

Numerous laboratory tests are available to determine conditioning effectiveness and dewaterability. These test objectives encompass the following: an evaluation of various conditioning and dewatering chemicals to provide the best dewaterability, development of design criteria for pilot- or full-scale dewatering unit processes, comparison and evaluation of different conditioning techniques, and the use of different conditioning techniques in controlling the operation of the dewatering unit processes.

Jar Test

The jar test is a visual observation of the size of flocs produced when various quantities of the conditioner are mixed with samples of the sludge. The test can be used to screen the different types of conditioner and to determine the effects of different dosages of a specific conditioner. Jar testing, the simplest type of testing, is often used for preliminary estimates of conditioner amounts and costs.

The jar test is probably the easiest method used to evaluate chemical conditioning because chemical conditioners can be compared and the best dose can be determined by treating 1-L samples at different conditioner concentrations. Following a short period of rapid mix, the sludge is flocculated for a few minutes and then allowed to settle. Liquid clarity and floc structure appearance indicate the ability of the chemical to condition. There is no standard jar test.

Capillary Suction Time Test

The capillary suction time (CST) test (Baskerville et al., 1978; Vesilind, 1988) indicates the time (in seconds) required for a small volume of filtrate to be withdrawn from conditioned sludge when subjected to the capillary suction pressure of dry filter paper.

The CST test is a rapid and simple method of screening dewatering aids. It relies on the capillary suction of a piece of thick filter paper to draw out the water from a small sample of conditioned sludge. The sample is placed in a cylindrical cell on top of chromatography grade filter paper. The time it takes for the water in a sludge to travel 10 mm in the paper between two fixed points is recorded electronically as CST. Capillary suction time is measured after sludge is mixed with varying conditioner dosages.

A typical CST for an unconditioned sludge is approximately 200 seconds or more. Sludges that hold water more tenaciously may exhibit CST values in thousands of seconds. A conditioned product that will readily dewater should yield a CST value of 10 seconds or less to produce a good cake either from filter presses or centrifuges.

The disadvantage of the CST (especially in comparison with the specific resistance to filtration) is that the test is specific only to the sludge being tested. While it gives comparative data, it is not a fundamental measure of sludge dewaterability (Vesilind, 1988). This problem can be solved by defining another term, *filterability*,

$$\chi = \left[\frac{1}{(D_2^2 - D_1^2)}\left(\frac{PA}{\pi d}\right)\right]\left[\frac{t}{\mu C}\right] \qquad (13.1)$$

where
- χ = filterability, m⁵/N·kg (ft⁴-s²/lb²);
- D_1, D_2 = diameters of sensor locations 1 and 2, m (ft);
- d = filter paper depth, m (ft);
- P = capillary suction of paper, m (ft);
- A = area of the bottom of the collar, m² (ft²);
- μ = viscosity of filtrate, N·s/m² (lb/s/ft);
- C = solid concentration of sludge, kg/m³ (lb/ft³); and
- t = CST as measured with any comparable CST apparatus (s).

The filterability, χ, is independent of the physical properties of the measuring device. Filterability is plotted versus polymer dose and the dose at which filterability is at the minimum is defined as the optimum polymer dose.

Standard Shear Test
The standard shear test (IAWPRC, 1981) assesses the physical strength of the conditioned flocs. The sludge is subjected to high-speed stirring for various periods. The change in CST—brought on by the different degrees of imposed shear—is then measured. A strong floc shows relatively little change in CST after stirring, whereas a weak floc shows a significant increase in CST after even a short period of stirring.

Buchner Funnel Test
The Buchner funnel test (Vesilind, 1979) is used to determine the specific resistance of a sludge to filtration or dewatering. The specific resistance is calculated by comparing the relationship between time and filtrate volume. By plotting specific resistance as a function of conditioner dose, a tester can determine the optimum dose of conditioner.

The Buchner funnel test provides useful data for all dewatering devices. The test is based on using the volume of filtrate obtained in a uniform filtration period as the measure of drainability or filterability. The laboratory apparatus required to perform the Buchner funnel test consists of a 9-cm Buchner funnel, a 250-mL lipless graduated cylinder filtrate receiver, a timer, and a vacuum pump.

Specific resistance of a mixture represents the relative resistance offered to drainage of its liquid component. Typical specific resistance values for municipal sludges are from 3 to 40 × 10¹¹ m/kg for conditioned digested wastewater solids and 1.5 to 5 × 10¹⁴ m/kg for primary sludge. Specific resistance can be calculated as

$$R = \frac{2bA^2P}{\mu C} \qquad (13.2)$$

where
- R = specific resistance (m/kg);
- b = slope of t/V versus V (s/m⁶);
- V = volume of filtrate collected (m³);
- t = time from start (s);
- A = area of filter (m²);
- P = test pressure (N/m²);
- μ = viscosity (N·s/m²); and
- C = mass of dry solids per unit volume of filtrate (kg/m³).

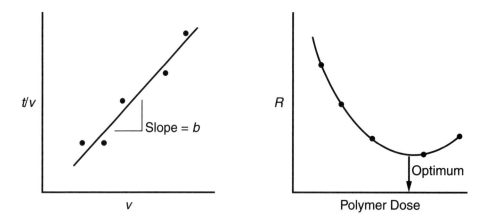

Figure 13.1 Typical results from a specific resistance to filtration test; establishing the slope *b* and calculating the specific resistance to filtration *R* using eq 13.2.

If a series of conditioners must be tested for their efficacy in dewatering a particular sludge, the best dosage (i.e., the dosage at which the specific resistance has its lowest value as shown in Figure 13.1) for each of the conditioners may be obtained and compared. Based on the cost of the conditioning chemicals tested and their optimum doses, the cost for conditioning a given sludge quantity with each of the conditioners tested can be evaluated and compared.

Sludge Compactability

This test, which uses a laboratory centrifuge, is useful in determining the effects of centrifugal acceleration, detention time, and conditioning chemicals and dosages applied on the cake solids concentration. The test is also useful in screening various conditioning chemicals for their performance. The test involves the spinning of a standard volume of conditioned solids with a given chemical at a specific dose for various times at different centrifugal forces. At the end of each spinning period, the centrate is decanted and the solids concentration is determined in both the cake and centrate. The compactability is defined as the increase in cake solids concentration with increasing centrifugal force (Vesilind, 1979).

Sludge Consistency

Although the bench-scale centrifuge test can screen conditioners for their performance based on the percentage of solids recovered and cake solids concentration, it does not give an indication of the consistency or firmness of the cake. These are important operating attributes because they are related to the ease with which sludge moves through the centrifuge (by way of scrolls and efficient discharge from the centrifuge. The *penetrometer test* can be used to measure this sludge property. The sludge is first centrifuged in a test tube and then a small rod is dropped into it. The extent of the penetration of this rod into the compacted sludge is a measure of its consistence. A fluffy biological sludge will have poor consistency (the rod will drop all the way through the sludge) while a raw primary sludge will frequently have excellent consistency (supporting the rod without penetration) (Vesilind, 1979).

Thermal Conditioning

Thermal conditioning is a phrase that refers to the simultaneous application of heat and pressure to sludge to enhance its dewaterability without the addition of conditioning chemicals. During thermal conditioning, the heat lyses the cell walls of microorganisms contained in biological sludges, releasing bound water from the particles. This process further hydrolyzes and solubilizes hydrated particles in biological sludges and, to a limited degree, organic compounds in primary sludges. Conventional mechanical dewatering devices can then readily separate the released water from the solid particles as long as the sludge has sufficient fibrous solids for cake structure. The two basic modifications of thermal conditioning employed in wastewater treatment are heat treatment and low-pressure oxidation. Research has shown little difference between the cake characteristics produced by the two thermal conditioning methods (Marshall and Gillespie, 1974).

A schematic diagram of the low-pressure oxidation system is shown in Figure 13.2. This system pumps macerated sludge at approximately 2760 kPa (400 psi) through a heat exchanger that raises the ambient temperature of the sludge to between 150 and 160 °C (300 and 320 °F). The preheated sludge and air mixture enters the bottom of the reactor where auxiliary heat from injected steam raises the sludge's temperature to between 165 and 176 °C (330 and 350 °F). After "cooking" for 15 to 30 minutes, the conditioned sludge travels back through the heat exchanger, where it cools to between 45 and 55 °C (110 and 130 °F) (U.S. EPA, 1985; WPCF, 1988). The conditioned sludge then discharges to a decant tank for gravity thickening and further processing. In the decant tank, vapors generated in the process are separated from the mixture and processed to remove odorous compounds. Suspended solids are concentrated in the underflow. Supernatant flows from the decant tank travel to either the main wastewater plant or to separate sidestream treatment facilities. The chemical oxygen demand of the supernatant is high and can increase the biochemical oxygen demand loading on the plant by as much as 50%.

Conventional mechanical dewatering devices with thermal conditioning may produce a cake of 30 to 40% solids. Higher concentrations of 40 to 55% solids may be obtained with filter presses. These concentrations exceed the cake solids obtainable from conventionally dewatered chemically conditioned cake using belt presses or centrifuges.

Odors associated with thermal conditioning are some of the most problematic and persistent of those encountered in wastewater treatment. The principal sources of odors include vents, decant tanks, and mechanical dewatering areas of the thermal conditioning process. All off-gases from the process require collection and treatment. Failure to do so will result in obnoxious odors and associated neighbor complaints. Such complaints have caused several installations to shut down. The most reliable method of odor control is off-gas combustion at 816 °C (1500 °F). Combustion is most effective in cases where a furnace exists on-site.

Freeze–Thaw Conditioning

For many years, wastewater treatment plant operators in northern climates observed that sludge left out over winter drained quite readily after the spring thaw. When sludge freezes the water molecules will form crystals which, during their growth, exclude the solids, pressing them together into granules. Upon thaw these granules dewater with little resistance.

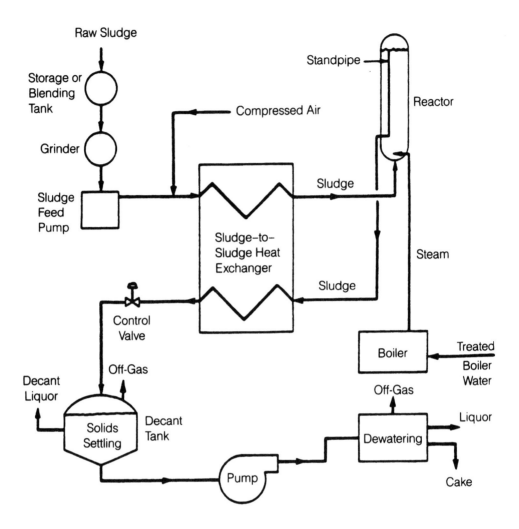

Figure 13.2 Flow diagram of a low-pressure oxidation process.

Freeze–thaw conditioning has been effectively applied to the conditioning of water treatment plant sludges such as waste alum sludge. The most effective means of using freeze–thaw is to construct a *freezing bed*, developed by Martel (1989). The freezing bed is similar in many ways to the sand drying bed, except that the sludge is placed on the bed in layers of less than 20 cm (approximately 6 in.) and allowed to freeze. When the weather warms, the bed thaws and drains naturally into perforated pipes under the bed.

Freeze–thaw conditioning is not quite as effective with biological sludges as it is with inorganic sludges because the particles do not stick together strongly and will break up under shear such as pumping. Nevertheless, the sludge-freezing bed conditioning and dewatering has been effectively applied in numerous locations in cold regions.

Conclusions

Reviews of designs by design teams are effective means of catching problems before the facility is even constructed. Will the operator be able to safely climb in (and out of) the tank? Is the valve within reach or is it in such a location that nobody can get to it without the assistance of Superman? Are any of the pipes taking up the same space? Experienced engineers learn to ask the right questions and often have a checklist they go through to study a plan. Often the advice of an experienced operator on the review team can be invaluable.

One such question is mass balances. In any given process the mass balance is absolutely necessary to understand what is going on. One example of where running mass balances has been invaluable is in the evaluation of "sludge additives". For the past 100 years, operators and engineers have thought that adding a bulky material to sludge will assist in its dewatering, and they have tried almost anything possible, from coal to sand to bark to shredded newspapers. In every case they have found that the dewatered sludge cake was always significantly drier with the additive and recommended that the plant operator begin to use the material to mix with the sludge prior to dewatering.

The problem was, of course, that the bulk materials were invariably drier than the sludge, and the product was therefore drier. A mass balance would show that the effect of the bulking agent was negligible. In fact, the same dryness would have been achieved in the dewatered sludge cake if the bulking agent were added *after* the sludge was dewatered. All the bulking agent was doing was adding dry mass without adding water, and producing a larger amount of sludge requiring disposal.

References

Baskerville, R. C.; et al. (1978) Laboratory Techniques for Predicting and Evaluating the Performance of a Filterbelt Press. *Filtr. Sep.*, 445.

Christensen, G. L.; Stulc, D. A. (1979) Chemical Reactions Affecting Filterability in Iron-Lime Sludge Conditioning. *J. Water Pollut. Control Fed.*, **51**, 2499.

Christensen, G. L.; Elliott, W. R.; Johnson, W. K. (1976) Interactions Between Sludge Conditioning, Vacuum Filtration, and Incineration at the Metropolitan Wastewater Treatment Plant. *J. Water Pollut. Control Fed.*, **48**, 1955.

International Association for Water Pollution Research and Control (1981) *Sewage Sludge II: Conditioning, Dewatering and Thermal Drying*. Manual of British Practice in Water Pollution Control, Maidstone, Kent, G.B.

Marshall, D. W.; Gillespie, W. J. (1974) Comparative Study of Thermal Techniques for Secondary Sludge Conditioning. *Proceedings of the 29th Industrial Waste Conference, Purdue University*, West Lafayette, Indiana.

Martel, C. J. (1989) Development and Design of Sludge Freezing Beds. *J. of Environ. Eng., ASCE*, **115**, 788.

Stulc, D. A.; Christensen, G. L. (1976) *Chemical Reactions Affecting Filtration in Iron-Lime Sludge Conditioning*; Research Report of the Metropolitan Waste Control Committee; University of Minnesota: St. Paul.

U.S. Environmental Protection Agency (1979a) *Chemical Aids Manual for Wastewater Treatment Facilities*; EPA-430/9-79-018; Washington, D.C.

U.S. Environmental Protection Agency (1979b) *Chemical Primary Sludge Thickening and Dewatering*; EPA-600/20-79-055; Cincinnati, Ohio.

U.S. Environmental Protection Agency (1979c) *Evaluation of Dewatering Devices for Producing High-Solids Sludge Cake*; EPA-600/2-79-123; Cincinnati, Ohio.

U.S. Environmental Protection Agency (1979d) *Process Design Manual for Sludge Treatment and Disposal;* EPA-625/1-79-011; Municipal Environmental Research Laboratory; Office of Research and Development; Cincinnati, Ohio.

U.S. Environmental Protection Agency (1979e) *Review of Techniques for Treatment and Disposal of Phosphorus-Laden Chemical Sludges;* EPA-600/2-79-083; Cincinnati, Ohio.

U.S. Environmental Protection Agency (1985) *Heat Treatment/Low Pressure Oxidation: Design and Operational Considerations;* EPA-430/9-85-001; Washington, D.C.

Vesilind, P. A. (1979) *Treatment and Disposal of Wastewater Sludges;* Ann Arbor Science Publishers, Inc.: Ann Arbor, Michigan.

Vesilind, P. A. (1988) The Capillary Suction Time as a Fundamental Measure of Sludge Dewatering. *J. Water Pollut. Control Fed.*, **60**, 2.

Water Pollution Control Federation (1983) *Sludge Dewatering;* Manual of Practice No. 20; Washington, D.C.

Water Pollution Control Federation (1988) *Sludge Conditioning;* Manual of Practice No. FD-14; Alexandria, Virginia.

Webb, L. J. (1974) A Study of Conditioning Sewage Sludges with Lime. *Water Pollut. Control*, **73**, 192.

Symbols Used in this Chapter

A = area of the bottom of the collar, m^2 (ft^2)
A = area of filter (m^2)
b = slope of t/V versus V (s/m^6)
C = mass of dry solids per unit volume of filtrate (kg/m^3)
C = solid concentration of sludge, kg/m^3 (lb/ft^3)
D_1, D_2 = diameters of sensor locations 1 and 2, m (ft)
d = filter paper depth, m (ft)
P = capillary suction of paper, m (ft)
P = test pressure (N/m^2)
R = specific resistance (m/kg)
t = CST as measured with any comparable CST apparatus (s)
t = time from start (s)
V = volume of filtrate collected (m^3)
μ = viscosity, $N \cdot s/m^2$ (lb/s/ft)
χ = filterability, $m^5/N \cdot kg$ ($ft^4 \cdot s^2/lb^2$)

14

Sludge Thickening, Dewatering, and Drying

Introduction

Wastewater treatment plants commonly use thickening unit processes for the concentration of either combined or separate solids streams. Thickening reduces the volumetric loading to, and increases the efficiency of, subsequent solids processing steps. When more water is to be removed from sludge than is possible by thickening methods, dewatering is necessary. Even more water is removed in drying, which uses thermal energy.

The ease with which different sludges thicken or dewater varies widely. A waste activated sludge, for example, is difficult to dewater, whereas a well-digested primary sludge can be concentrated quite readily. Much of this variability results from the kinds of solids and the way in which the water is bonded to the solids. In waste activated sludge, much of the water is difficult to remove because it is attached to the bacterial cells or tied up chemically in the cell structures.

Not only does ease of solids concentration vary, but the solids differ, often with the same sludge, depending on how the material has been processed and managed before the thickening or dewatering. The degree of solids concentration achieved also depends on the temperature of the sludge, its pH, and the chemical and physical makeup of the solids. In short, the variables affecting the ease of water removal are many and to date it has not been possible to relate the physical, chemical, and biological properties of a sludge to its ease of yielding the entrapped water.

> **Note on Terminology**
>
> Sludge, the residues of wastewater treatment, is made up of *solids* and *liquid*. The solids are suspended in the liquid at various concentrations, from a low of less than 1% solids in a waste activated sludge, to perhaps as high as 5% solids from the primary settling tank. Because the *suspended solids* concentration is so much higher than the *dissolved solids* concentration, the solids in sludge management are assumed to be all suspended solids.
>
> Solids concentration is expressed either as mg/L or percent solids. The conversion from mg/L to percent is tricky. The first term, mg/L, is on a mass per volume basis, while the second term, percent, is on a mass per mass basis. To make the conversion, it is necessary to assume that the density of the solids is equal to that of water. This is a reasonable assumption for biological sludges, but not so good for heavy (especially some industrial) sludges. Nevertheless, the conversion from one to the other is
>
> $$1\% \text{ solids} \approx 10{,}000 \text{ mg/L}$$
>
> The percent designation is reasonable for sludge management work because the test is invariably run by first weighing the wet sludge sample, drying it, and weighing the remaining dry solids. The solids concentration is then expressed as
>
> $$\text{Percent solids} = \frac{\text{Dry solids}}{\text{Wet solids}} \times 100$$
>
> In most cases, the solids concentration of the sludge as extracted from the liquid treatment train is not sufficiently high and must be increased. That is, some of the liquid must be separated and removed. There are several terms in common usage that describe this separation. *Thickening* is the process of removing some of the water from sludge, with the thickened sludge retaining its liquid character. That is, the sludge still acts like a liquid. Methods of thickening, such as gravity thickening, will not produce solids concentrations much greater than 5% solids. *Dewatering* removes more water, and a dewatered sludge behaves like a solid. These sludges are difficult to pump but can be managed with solids conveying equipment such as conveyor belts. Most sludges attain characteristics of solids at approximately 20% solids. Some dewatering devices such as modern centrifuges can attain solid concentrations as high as 40%. Drying is the process of removing water from sludge by evaporation using thermal means. This can be natural drying, as in sand beds, or artificial drying such as drum dryers, which can attain solids concentrations approaching 100%.

Gravity Thickening

A gravity thickener operates in a similar manner to a settling tank. Solids settle to the bottom under the force of gravity, and the separated water flows up and over weirs at the top. In gravity thickening, solids are concentrated by the gravity-induced settling and compaction of solids. A typical gravity thickener is shown in Figure 14.1.

Gravity thickeners are most successfully used for thickening primary and lime sludges, although they have also been successfully used for trickling filter sludges, activated sludges, and anaerobically digested sludges. Primary and lime sludges settle quickly and achieve a high underflow concentration without chemical conditioning. The

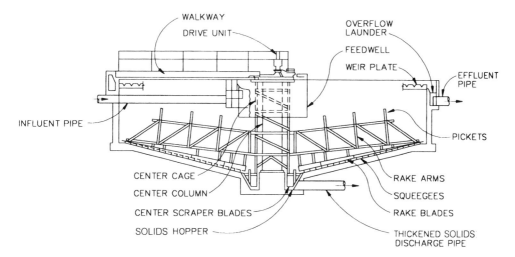

Figure 14.1 Typical gravity thickener.

presence of biological solids, particularly waste activated sludge, results in lower capture rates and lower underflow solid concentrations. In addition to concentration, gravity thickening also provides some equalization and storage of solids that may be of benefit to downstream operations.

The most common design for a gravity thickener is a circular tank with a side water depth of 3 to 4 m (10 to 14 ft). Gravity thickener tanks range up to 21 to 24 m (70 to 80 ft) in diameter. Tanks larger in diameter that are used for thickening organic sludges can cause gasification and flotation problems because of anoxic and anaerobic activity caused by increased solids detention time. The floor slope of gravity thickeners, typically 1:6 to 1:4, is steeper than that of primary settling tanks. This increased floor slope deepens the solids blanket in the center area, permits minimum solids detention while maximizing the depth of the solids over the withdrawal pipe, and reduces raking transport problems.

Combination clarifier–gravity thickener units are circular settling tanks with a deepened center section that functions as a gravity thickener. Rectangular tanks are seldom used as combined units because of difficulties associated with solids removal.

Theory of Gravity Thickening

The gravity thickening tank is divided into two areas, with the top area being governed by the settling of individual particles. There is then a sharp and sudden increase in solids concentration, known as a *sludge blanket*. The sludge blanket defines the level at which the particles are in solid-solid contact. Settling in this *zone settling* area is governed not by the settling velocity of the particles, but rather the rate at which the solids can compress to expel the entrapped water.

The theory of continuous thickening assuming that the rate of particle movement in the zone settling area is governed only by the solids concentration and assumes that zone settling prevails throughout the entire concentration range. Under zone settling, the settling rate of an ideal suspension depends on the solids concentration only, a concept originally developed by Coe and Clevenger (1916).

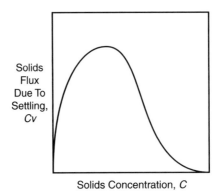

Figure 14.2 Solids flux curve resulting from a settling test in a cylinder.

When solids are settling in a liquid, the *solids flux* (mass of dry solids passing a given point per time per area of the cylinder) due to settling can be expressed as Cv, where C = solids concentration and v = velocity of the solids. Note that C is mass/unit volume and v is length/time, so that Cv = mass/time/unit area (e.g., lb solids/min·ft^2). The solids flux due to settling can be estimated using batch tests in which a given solids concentration is placed in a cylinder and the height of the interface measured with time. This *interface settling velocity* is assumed to be a characteristic of that solids concentration only. Multiplying the settling velocity (v) times the initial concentration (C) of the sludge in the cylinder yields solids flux due to settling (Figure 14.2).

In a continuous thickener, solids are transported downward not only by settling, but also by the withdrawal of solids from the bottom of the tank. The total solids flux for any solids concentration in the thickener is therefore

$$G = Cv + Cu \qquad (14.1)$$

where
- G = total solids flux at some solids concentration C (kg/m^2·d);
- C = solids concentration (kg/m^3);
- v = settling velocity at solids concentration C (m/d, as measured in a batch thickening test); and
- u = downward velocity caused by solids withdrawal (m/d).

Any solids concentration, C, can occur at any point in the continuously operating thickener. The downward movement of the solids (solids flux) is due to both the solids settling ($C \times v$) and the withdrawal of sludge as underflow ($C \times u$). Both of these fluxes, the one due to settling and the one due to the extraction of solids from the bottom of the thickener, can be plotted as shown in Figure 14.3. Within the thickener, as the solids concentration changes from the feed solids to the underflow solids, the greatest flux the thickener is able to maintain is the limiting flux, G_L, which occurs at concentration C_L in Figure 14.3.

Using the principle of materials balance, the flux within any part of the thickener operating at steady state must be the same at all times. Thus, the total flux times the

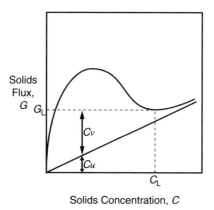

Figure 14.3 Total solids flux curve, with flux due to settling (Cv) and flux due to withdrawal from the bottom of the thickener (Cu).

thickener area must be the same as the underflow concentration times the underflow flowrate, or

$$GA = Q_u C_u \tag{14.2}$$

where
- G = solids flux (kg/m²·s);
- A = surface area of the thickener (m²);
- Q_u = underflow flowrate (m³/s); and
- C_u = underflow solids concentration (kg/m³).

Note that the units are (kg/m²·d) × (m²) = (m³/s) × (kg/m³), or kg/d = kg/d.

The downward velocity, u, created by the underflow is

$$u = \frac{Q_u}{A} = \frac{GA}{C_u A} = \frac{G}{C_u} \tag{14.3}$$

Substituting into eq 14.1, the general solid flux equation, and rearranging gives

$$G = \frac{v}{\dfrac{1}{C} - \dfrac{1}{C_u}} \tag{14.4}$$

If C in the above equation is the concentration of the limiting flux G_L, then the velocity is the settling velocity at the limiting concentration, or v_L yielding.

$$G_L = \frac{v_L}{\dfrac{1}{C_L} - \dfrac{1}{C_u}} \tag{14.5}$$

where G_L = limiting flux (kg/m²·d) and v_L = settling velocity at limiting concentration C_L (m/d).

Table 14.1 Typical gravity thickening data for various types of sludges (U.S. EPA, 1979).

Type of sludge	Feed solids concentration, solids, %	Expected underflow concentration, solids, %	Unit solids loading, kg/m³·d
Separate sludges			
Primary (PRI)	2–7	5–10	100–150
Trickling filter (TF)	1–4	3–6	40–50
Rotating biological contractor (RBC)	1–3.5	2–5	35–50
Waste activated sludge (WAS)			
WAS—air	0.5–1.5	2–3	20–40
WAS—oxygen	0.5–1.5	2–3	20–40
WAS—(extended aeration)	0.2–1.0	2–3	25–40
Anaerobically digested solids from primary digester	8	12	120
Thermally conditioned sludge			
PRI only	3–6	12–15	200–250
PRI + WAS	3–6	8–15	150–200
WAS only	0.5–1.5	6–10	100–150
Tertiary sludge			
High lime	3–4.5	12–15	120–300
Low lime	3–4.5	10–12	50–150
Iron	0.5–1.5	3–4	10–50
Other sludges			
PRI + WAS	0.5–1.5	4–6	25–70
	2.5–4.0	4–7	40–80
PRI + TF	2–6	5–9	60–100
PRI + RBC	2–6	5–8	50–90
PRI + iron	2	4	30
PRI + low lime	5	7	100
PRI + high lime	7.5	12	120
PRI + (WAS + iron)	1.5	3	30
PRI + (WAS + alum)	0.2–0.4	4.5–6.5	60–80
(PRI + iron) + TF	0.4–0.6	6.5–8.5	70–100
(PRI + iron) + WAS	1.8	3.6	30
WAS + TF	0.5–2.5	2–4	20–40
Anaerobically digested PRI + WAS	4	8	70
Anaerobically digested PRI + (WAS + iron)	4	6	70

Once the limiting flux for a thickener is known, the required area to achieve the desired underflow concentration C_u can be calculated as

$$A_L = \frac{Q_o C_o}{G_L} \qquad (14.6)$$

where
 Q_o = influent flowrate (m³/min);
 C_o = influent solids concentration (mg/L); and
 A_L = limiting area, or the minimum area necessary for a thickener to achieve an underflow concentration of C_u (m²).

The limiting flux, G_L, is also called *unit solids loading* in the thickener design. Note again that for a thickener operating at steady state, $Q_o C_o = Q_u C_u = Q_i C_i$.

Design of Gravity Thickeners

Experience has shown that wastewater solids thickening characteristics vary considerably not only among different types of solids, but among the same types of solids from different locations. These variations can be caused by a wide range of factors such as variations in the physical properties of solids particles, the type and volume of industrial wastes treated, the type and operating conditions of wastewater treatment processes, or solids handling practices before thickening. Consequently, it is desirable to design a thickening facility based on the criteria developed from some test program. Such a design approach may not be practicable in many cases, however, because the solids in question may not be available for testing. In this case, thickener design can be accomplished using available performance data from similar thickening operations.

When designing a thickener, sufficient tank area and depth should be provided for both thickening and clarification functions. The most important aspect of thickener design, however, is the establishment of the area required to achieve a desired degree of thickening. Under normal operating conditions, the clarification requirements seldom become a governing factor in thickener design, but must still be checked as part of the thickener design.

Area Determination for Solids Thickening

The key design variables in gravity thickening is the surface area. Two approaches to design are used in practice—one based on prior experience, and one on laboratory testing of the sludge to be thickened.

The first approach is to use prior experience. Table 14.1 provides typical data for various types of municipal wastewater sludge. The unit solids loading, defined as the limiting solid flux above, is the quantity of solids allowable per unit area of thickener per unit time (kilograms per square meter per day) to achieve the indicated underflow solids concentration. The required area is determined by dividing the expected daily solids loading (with dimensions such as kilograms per day) by a reasonable value of unit solids loading for the particular type of solids in question.

The second approach to thickener design is the use of the solids flux theory presented in the preceding section. This method requires that the relationship between the settling flux and solids concentration be determined. Such a relationship can be developed from the results of batch settling tests.

The batch settling test (zone settling rate test in *Standard Methods*) is commonly conducted in a transparent cylinder equipped with a stirring mechanism. The recommended size of the cylinder is at least 1 m high and 100 mm in diameter, with a peripheral stirrer speed of 10 mm/s or less (APHA et al., 1998). The stirrer, which consists of one or more thin rods positioned near the internal wall of the settling column, should be extended over the entire depth of solids. The importance of using a large settling column and the stirring mechanism is well documented (Dick, 1972; Vesilind, 1974b).

The graphical method of Yoshioka et al. (1957) may be used to determine the area required to accomplish a desired degree of thickening. As shown in Figure 14.4, a line is drawn as a tangent to the settling flux curve. The intercept on the abscissa of this line is the underflow solids concentration, and the intercept on the ordinate is the limiting

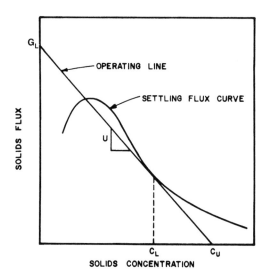

Figure 14.4 Graphical method for determining limiting solids flux and underflow concentration.

solids flux, or the maximum solids flux (or unit solids loading) that can be transported to the bottom of the thickener.

In this procedure, the operation of a thickener is assumed to be strictly one dimensional. One-dimensional operation requires that solids be uniformly distributed horizontally at the feed level and that removal of thickened underflow produce equal downward velocities throughout the cross section of the tank. These conditions generally cannot be met in full-scale thickeners because of the relatively small feed well and central withdrawal of thickened solids.

Other Design Considerations

Although the area requirements for solids loading usually far outweigh the area requirements for overflow rate, the area requirements for settling must be considered. Maximum overflow rates of 15.5 to 31.0 $m^3/m^2 \cdot d$ (380 to 760 gpd/ft^2) are used for primary sludges, whereas overflow rates in the 4 to 8 $m^3/m^2 \cdot d$ (100 to 200 gpd/ft^2) range are more applicable for secondary sludges.

The inlet to the thickener should be designed to minimize turbulence in the feed well. Most circular thickeners employed for the treatment of domestic wastes use bottom-feed inlets to a center thickener feed well. Feed to this system flows vertically and then laterally with a low level of turbulence. In most industrial applications and in some domestic wastewater treatment cases, other configurations such as overhead feed are used.

Gravity thickening mechanisms (see Figure 14.1) are sometimes provided with pickets to help release water from the solids. The design may vary, depending on the nature of the solids to be handled. Reports of the effects of pickets on the mechanism of solids consolidation have been varied. Many carefully carried out studies have produced contradictory results. If rotated at a high enough speed, picket rakes can actually hinder the thickening process (Vesilind, 1968; Jordan and Scherer, 1970). When solids are producing sufficient gas to prevent subsidence, pickets may provide a channel for the release of gas and, under this circumstance, may enhance thickening. Some thickener

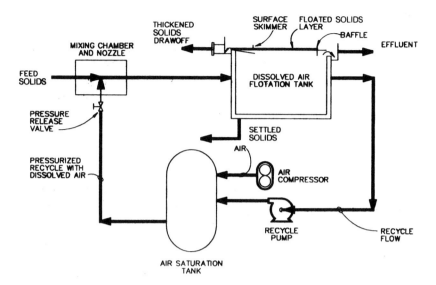

Figure 14.5 Schematic of dissolved air flotation thickener.

mechanisms can provide sufficient agitation without pickets, perhaps explaining why beneficial effects are usually not achieved.

Dissolved Air Flotation Thickening

Dissolved air flotation (DAF) thickening refers to the process of solids-liquid separation caused by the introduction of fine gas (typically air) bubbles to the liquid phase. The bubbles attach to the solids, and the resultant buoyancy of the combined solids-gas matrix causes the matrix to rise to the surface of the liquid where it is collected by a skimming mechanism. Dissolved air flotation thickening is commonly used for waste activated sludge, aerobically digested solids, and contact-stabilized, modified activated, or extended-aeration solids without primary settling. Dissolved air flotation is generally not used for primary or trickling filter sludges because gravity settling for these solids is more economical than DAF.

Figure 14.5 presents a schematic of a typical DAF thickener system. The primary components of a DAF thickener are the pressurization system and the DAF tank. The pressurization system has a recycle pressurization pump, an air compressor, an air saturation tank, and a pressure-release valve. Dissolved air flotation thickener tanks, either rectangular or circular, are equipped with both surface skimmers and bottom solids removal mechanisms. The surface skimmers remove floated solids from the tank surface to maintain a constant float blanket depth. The bottom mechanism removes the nonfloatable heavier solids that settle to the bottom of the tank.

Dissolved air flotation tanks are baffled and equipped with an overflow weir. Clarified effluent passes under an end baffle (rectangular units) or peripheral baffle (circular units) and over the weir to an effluent launder. The weir controls the liquid level within the tank with respect to the float collection box and helps regulate the capacity and performance of the unit.

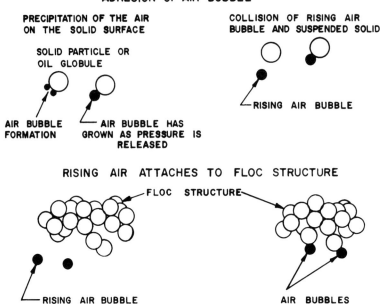

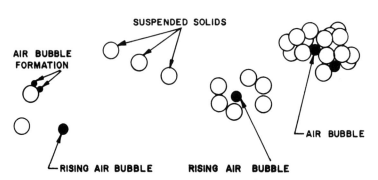

Figure 14.6 Solid-bubble contacting mechanisms in dissolved air flotation.

Dissolved air flotation thickening was introduced in the United States in the 1930s. Initially, it was primarily used in the paper industry for treating white water. Since then, DAF thickening has been used for specialty waste treatment applications.

Flotation can be employed in both liquid clarification and solids concentration applications. Flotator liquid effluent (known as *subnatant*) quality is the primary performance factor in clarification applications. These applications include flotation of refinery, meat-packing, meat-rendering, and other oily wastewaters. Float-solids concentrations are the primary performance criteria in solids concentration flotation applications. Concentration applications include the flotation of waste solids of biological, mining, and metallurgical processes.

Theory of Dissolved Air Flotation

Dissolved air flotation is used to float particles with specific gravities much greater than 1.0 because the bulk specific gravity of the aggregated air and solids is reduced to less than 1.0 by the adhesion of low-density gas (usually air) bubbles to the aggregate material. The gas-solids contacting sequence can occur by adhesion, trapping, adsorption, or other mechanisms. These mechanisms, in addition to chemical and electrostatic effects, can alter the collision efficiency of particles with gas bubbles (Flint and Howarth, 1971; Roberts et al., 1978).

Gas bubbles can adhere to a solid particle or oil globule by precipitation or collision mechanisms. Bubbles can be trapped in a floc structure as the bubbles rise through the liquid media. Alternatively, gas bubbles can be absorbed or adsorbed in a floc structure as the floc structure is formed (see Figure 14.6). The adhesion, trapping, and adsorption sequences provide a means of reducing the aggregate specific gravity of a solids-gas matrix to less than 1.0. Once the aggregate specific gravity is substantially less than 1.0, the material can be rapidly floated and removed from the liquid phase.

The size of the bubbles generated in a DAF system significantly affects flotation performance. The bubbles generated in a DAF system are microscopic in size, ranging from 10 to 100 μm. When gas bubbles are released from a pressurized liquid, as is the case with DAF systems, individual gas bubbles are not distinguishable to the naked eye and impart a milky white appearance to the depressurized liquid.

Design of Dissolved Air Flotation Thickeners

Ideally, the dissolved air flotation, or DAF, systems should be designed using pilot plants. These pilot units should operate with the same feed material as full-sized equipment. The gas bubbles created in a pilot-unit system should be the same size as those created in full-sized equipment; in addition, the pressurization system must operate at the same pressure in both systems. The loading rate and air/solids ratio chosen for use during pilot-unit tests must be applicable to full-sized equipment.

The *hydraulic loading rate* refers to the sum of the feed and recycle flowrates divided by the net available flotation area. Dissolved air flotation thickeners are designed for hydraulic loading rates of 30 to 120 $m^3/m^2 \cdot d$ (0.5 to 2 gpm/ft^2), with a suggested maximum daily hydraulic loading of 120 $m^3/m^2 \cdot d$, assuming no use of conditioning chemicals. The additional turbulence in flotators caused when the hourly hydraulic loading rate exceeds 5 $m^3/m^2 \cdot h$ may hinder the establishment of a stable float blanket and reduce the attainable float-solids concentration. Deterioration in solids capture can also result as increased turbulence forces the flow regime away from plug flow and more toward mixed flow. The addition of a polymer flotation aid typically is required to maintain satisfactory performance at hourly hydraulic loading rates greater than 5 $m^3/m^2 \cdot h$.

The solids loading rate of a DAF thickener is generally denoted in terms of weight of solids per hour per effective flotation area. Without any chemical conditioning, the range of loading rates for DAF thickening waste activated sludge is approximately 2 to 5 $kg/m^2 \cdot h$ (0.4 to 1 $lb/hr/ft^2$) to produce a thickened float layer of 3 to 5% total solids. With the addition of polymer, the solids loading rate to a DAF thickener typically can

Figure 14.7 Gravity belt thickener.

be increased 50 to 100%, with up to a 0.5 to 1% increase in the thickened-solids concentration.

The air/solids ratio is perhaps the single most important factor affecting DAF performance. It refers to the weight ratio of air available for flotation to the solids to be floated in the feed stream. Reported ratios range from 0.01:1 to 0.4:1; adequate flotation is achieved in most municipal wastewater thickening applications at ratios of 0.02:1 to 0.06:1. Pressurization system sizing depends on many variables, including design solids loading, pressurization system efficiency, system pressure, liquid temperature, and concentration of dissolved solids. Pressurization system efficiencies differ among manufacturers and system configurations and can range from as low as 50% to more than 90%.

Gravity Belt Thickening

Gravity belt thickening is a solids-liquid separation process that was made popular in the 1980s and relies on coagulation and flocculation of solids in a dilute slurry and drainage of free water from the slurry through a moving fabric-mesh belt. A gravity belt thickener is a mechanical filtration device that accomplishes this solids-liquid separation (see Figure 14.7).

The gravity belt thickener is a modification of the upper gravity drainage zone of the belt filter press. It is used to thicken solids of initial concentrations as low as 0.4% and up to 4 to 8% with a polymer addition of 1.5 to 5 g/kg (3 to 10 lb/ton) on a dry-weight basis. It also achieves greater than 95% solids capture. Applications include thickening of waste activated sludge, primary sludge, anaerobically or aerobically digested solids, and blended solids from wastewater treatment plants. Gravity belt thickeners are also used to thicken alum and lime sludges from water treatment plants. The process is polymer dependent to achieve any of the thickening criteria established for solids capture. One notable feature of the gravity belt thickener is its use of relatively open-mesh filter belts that require high polymer dosages. The open fabric is used to achieve higher thickening rates. However, without chemical addition, the solids generally cannot be concentrated and the open-mesh fabric permits excessive solids losses.

Theory of Gravity Belt Thickening

Biological solids are difficult to thicken because water is retained within the floc structure, adsorbed onto the particles, and contained within the individual biological cells. Particle size also affects the ability of the solids to be thickened: the smaller the particle size, the greater the particle surface area of a given volume. Effects of smaller particles include greater repulsive force between particles caused by larger negatively charged surface area, greater frictional resistance to water movement, and greater attraction of water to particle surface.

Primary sludge, which has larger particles, is easier to thicken than biological sludge because it has fine colloidal particles, their biological cells are mostly water, and the formation of floc traps the water. Digested solids are more difficult to thicken than raw primary sludge because fine particles are created in the breakdown of organic particles in the stabilization process. Difficult-to-thicken solids can result in a higher dose of polymer, reduced overall solids capture, and reduced thickened-solids concentration.

The gravity belt thickener is dependent on coagulant addition to concentrate solids. Without proper coagulant addition, the process will fail. Coagulant type and dosage are important operating variables. Cationic (positively charged) polymers are used to neutralize the negative charge of biological solids. The type and amount of polymer used depends on the type of solids to be thickened and the floc characteristics.

In most systems, polymer is added to the solids in the feed pipe using a full-circumference injection ring (see Figure 14.8). The feed pipe downstream of the polymer addition point, as well as the solids feed well, provide the detention time and low turbulence conditions necessary to facilitate the development of floc particles. Some manufacturers recommend that the polymer be added immediately upstream of the feed well. The solids flow from the inlet feed well down the gently sloped inlet ramp, with guide veins and baffles to uniformly disperse the solids across the width of the continuously moving fabric belt. Free water released from the solids drains through the belt while the solids remain on top of the fabric. Approximately 80% of the water removed drains through the belt during the first meter of travel after discharge from the feed chute. The solids form a mat on the belt, which can limit the release of additional free water. Most manufacturers use plows or vanes that ride on the surface of the belt to fold or turn over the solids and expose clean areas of the fabric belt and allows any remaining free water to drain (see Figure 14.9).

An adjustable pitch ramp or variable-height dam on the discharge end causes the solids to pool and roll backward, increasing the retention time on the drainage belt and gently imparting additional energy to the solids. The pooling and backward-rolling action release more free water. The solids then travel over the ramp into a hopper and is pumped to downstream stabilization, transport, or disposal processes.

Solids that cling to the belt as they move under the ramp are scraped from the belt by a spring-tensioned doctor blade. The belt is then washed with high-pressure spray water

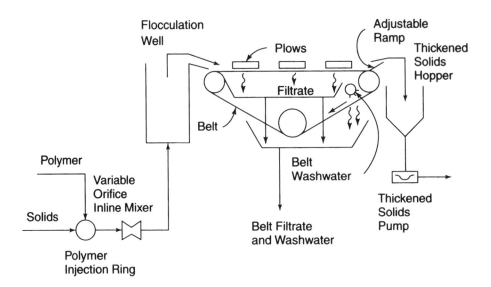

Figure 14.8 Gravity belt thickener schematic.

to remove excess polymer and solids from the pores of the fabric belt. The belt then travels to the influent end to receive a fresh charge of sludge. Free water draining through the belt and wash water from the belt cleaning process are conveyed to liquid processes for additional treatment. The solids concentration of the combined wash water and filtrate flow typically is less than 400 mg/L for waste activated sludge thickening.

Figure 14.9 Movement of solids on gravity belt thickener to allow water to drain through fabric belt.

Table 14.2 Typical gravity belt thickener performance (from Ashbrook-Simon-Hartley, Aquabelt Operations & Maintenance Manual. Copyright © 1992, Ashbrook Corporation, Houston, Texas, with permission).

Type of biosolids	Initial concentration, %	Solids loading, kg/m·h	Polymer dosage, g/kg	Final concentration, %
Primary (P)	2–5	900–1400	1.5–3	8–12
Secondary (S)	0.4–1.5	300–540	3–5	4–6
(50% P)/(50% S)	1–2.5	700–1100	2–4	6–8
Anaerobic (50% P)/(50% S)	2–5	600–790	3–5	5–7
Anaerobic (100% S)	1.5–3.5	500–700	4–6	5–7
Aerobic (100% S)	1–2.5	500–700	3–5	5–6

Design of Gravity Belt Thickeners

The gravity belt thickener has demonstrated good performance with many types of wastewater solids. Operating experience has shown that the process is less affected by plant operating problems than many other thickening processes. Difficult-to-thicken solids only require modification of the polymer dosage and hydraulic and solids loading rates. With these minor modifications, thickened solids concentration and solids capture remain high.

Pilot testing is now primarily done to evaluate various manufacturers' machines to determine compliance and cost-effectiveness in competitive bids. Because of the good performance and flexibility of the machine, pilot testing is not needed for most wastewater sludges. Pilot testing is needed in cases where a unique thickening application is being considered or where physical viewing of the machine and the results is needed to gain acceptance. Typical laboratory testing consists of sending a sample of the solids to the manufacturer for jar tests and free-drainage tests to determine polymer type and dosage. Such tests provide some preliminary information, but full-scale trials are always necessary to confirm the laboratory results.

As with other wastewater thickening processes, performance is specific to the particular solids being thickened. Expected performance and design criteria can, however, be approximated from similar full-scale installations. Table 14.2 presents performance and design criteria for various municipal wastewater solids.

Other Methods of Thickening

Rotary Drum Thickening

A *rotary drum thickener*, also called a *rotary screen thickener*, functions similarly to a gravity belt thickener. In both devices, free water drains through a moving porous media while flocculated solids are retained on the media. The rotary drum thickener consists of an internally fed rotary drum with an integral internal screw for transporting thickened solids or screenings out of the drum. The drum rotates and is driven by a variable- or constant-speed drive. Figure 14.10 shows a typical rotary drum thickener.

Rotary drum thickeners can be used as a prethickening step before belt filter press dewatering and are used in small- to medium-sized plants for waste activated sludge thickening. They are well suited to thickening high-fiber sludges, such as those in the pulp and paper industry, and either raw or digested municipal solids that contain a significant primary sludge fraction. They are also used for removing screenings from

Figure 14.10 Rotary drum thickener (side panel removed) (MacConnell et al., 1989).

raw wastewater or primary sludge. Their success with municipal waste activated sludge is variable and highly dependent on actual solids characteristics. Similar to gravity belt thickening, the addition of large amounts of polymer can be a concern in thickening because of floc sensitivity and shear potential in the rotating drum.

The rotary drum thickener is most applicable in small- to medium-sized wastewater treatment plants for thickening waste activated sludge because the largest unit has a capacity of approximately 1100 L/min (300 gpm). It is attractive for use because of its efficient space requirements, low power usage, moderate capital cost, and ease of enclosure, which improves housekeeping and odor control. Table 14.3 shows some operating performance data for a rotary drum thickener.

Centrifuges

The *conveyor* or *solid-bowl centrifuge* is a useful tool for thickening sludge. It is a versatile process and in many plants centrifuges are used for both thickening and

Table 14.3 Typical rotary drum thickener performance.

Type of solids	Feed, % TS[a]	Water removed, %	Thickened solids, %	Solids recovery, %
Primary	3.0–6.0	40–75	7–9	93–98
WAS[b]	0.5–1.0	70–90	4–9	93–99
Primary and WAS	2.0–4.0	50	5–9	93–98
Aerobically digested	0.8–2.0	70–80	4–6	90–98
Anaerobically digested	2.5–5.0	50	5–9	90–98
Paper fibers	4.0–8.0	50–60	9–15	87–99

[a]TS = total solids.
[b]WAS = waste activated sludge.

dewatering. The use of solid-bowl centrifuges is most often in dewatering sludge, and thus this device is discussed in detail in the next section of this chapter.

Centrifugal Dewatering

The separation of a liquid-solids slurry during centrifugal dewatering is analogous to the separation process in a gravity thickener. In a centrifuge, however, the applied force can be 500 to 3000 times the force of gravity. Separation results from the centrifugal force—driven migration of the suspended solids particles through the suspending liquid toward or away from the axis of rotation of the centrifuge, depending on the density difference between the liquid and solid phases. The increased settling velocity imparted by the centrifugal force and the short settling distance of the particles accounts for the comparatively high capacity of centrifugal equipment.

Centrifuges have been used to thicken waste solids since the early 1920s. The solid-bowl conveyor centrifuge units have found the widest application in dewatering. Variables affecting sedimentation in these units are grouped into the following three basic categories: performance, process, and design. Performance is measured by the concentration achieved in the underflow or thickened solids product and the solids recovery obtained while operating at that underflow concentration. The recovery is calculated as the ratio of the dewatered-cake dry solids to the feed dry solids, typically expressed as a percentage. Using the commonly measured solids concentrations, the recovery is calculated as

$$R = \frac{C_d(C_f - C_c)}{C_f(C_d - C_c)} \times 100 \tag{14.7}$$

where
- R = recovery percent;
- C_d = solids concentration of the dewatered sludge cake (mg/L or % by weight);
- C_f = solids concentration in the feed (mg/L or % by weight); and
- C_c = solids concentration in the centrate (mg/L or % by weight).

The second variable needed to assess centrifuge performance is the sludge cake solids concentration. A centrifuge can be operated so that it can produce a high level of solids recovery, but produce a wet cake. Or, it can be operated so it produces a dry sludge cake, but this would be at the sacrifice of solids recovery. The objective, developed through judicious manipulation of process variables and sludge conditioning, is to produce a high recovery while also producing a cake high in solids.

Process variables that affect sedimentation include the feed flowrate, rotational speed of the centrifuge, differential speed of the conveyor relative to the bowl, depth of the thickening zone, chemical use, and the physicochemical properties of the solids and suspending liquid (such as particle size and shape, particle density, temperature, and liquid viscosity). These variables are the tools that the wastewater treatment operator must use to make the most of centrifuge performance. The design, or fixed parameters, differ for each centrifuge, and include maximum operating speed, feed-inlet design, and centrifuge geometry.

The two basic configurations of solid-bowl conveyor centrifuges are countercurrent and cocurrent (see Figure 14.11). The primary differences between thickening and

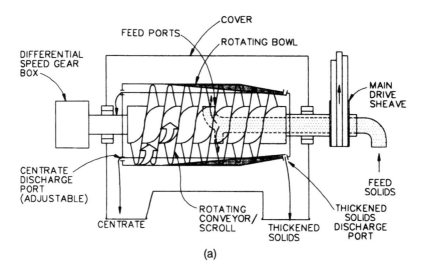

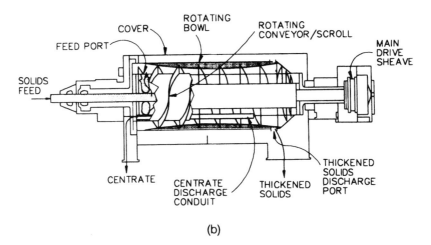

Figure 14.11 Schematic of two solid-bowl centrifuge configurations: (a) countercurrent flow and (b) cocurrent flow.

dewatering centrifuges are the configuration of the conveyor (scroll) and the location and configuration of the solids discharge ports.

Theory of Centrifugation

The *solid-bowl conveyor centrifuge*, often referred to as a *continuous decanter scroll* or *helical screw-conveyor centrifuge*, consists of a solid-walled imperforated bowl that generally employs a horizontal axis of rotation. The bowl is of a compound design in that it combines conical and cylindrical sections. Centrifugal force causes the liquid surface to essentially be parallel to, and equidistant from, the axis of rotation. The solids-discharge ports at one end of the bowl conventionally are at a smaller radius than the liquid-discharge ports at the other end, although the reverse can be true if internal disc barriers are used.

The solids-feed slurry is introduced through a concentric tube to an appropriate point within the bowl. The liquid depth in the centrifuge is determined by the discharge-weir elevation relative to the bowl wall. The weirs are adjustable. Because they are denser, the solids are settled against the bowl wall and are continuously transported to the opposite end of the bowl (conical end) by a helical screw conveyor that extends the full length of the bowl. The differential speed between the bowl and the conveyor is produced by mechanical, hydraulic, or electric means.

The two most important criteria for the successful operation of a centrifuge relate to solids separation (hydraulic or clarification capacity) and solids removal (cake-conveying capacity). The hydraulic capacity for the solid-bowl centrifuge is based on the "sigma concept", originally developed by Ambler (1952). The settling capacity of a machine can be described using the following equation:

$$\Sigma = 2\pi l \frac{\omega^2}{100g} (0.75\ r_2^2 + 0.25 r_1^2) \tag{14.8}$$

where
 Σ = theoretical hydraulic capacity (cm²);
 l = effective clarifying length of centrifuge bowl (cm);
 ω = angular velocity of centrifuge bowl (rad/s);
 g = acceleration from gravity (m/s²);
 r_1 = radius from centrifuge centerline to the liquid surface in the centrifuge bowl (cm); and
 r_2 = radius from centrifuge centerline to the inside wall of the centrifuge bowl (cm).

A simplified equation, applicable only to a solid-bowl centrifuge, is (Vesilind, 1974b)

$$\Sigma = \frac{V\omega^2}{g \ln\left(\frac{r_2}{r_1}\right)} \tag{14.9}$$

where V = centrifuge pool volume (cm³).

The hydraulic capacity of two different size, but geometrically similar, centrifuges to dewater a specific sludge can be estimated as

$$\frac{Q_1}{\Sigma_1} = \frac{Q_2}{\Sigma_2} \tag{14.10}$$

where
 Q_1 = flow to machine number 1 (L/min);
 Σ_1 = sigma for machine number 1 (cm²);
 Q_2 = flow to machine number 2 (L/min); and
 Σ_2 = sigma for machine number 2 (cm²).

If a pilot-scale machine is named "machine number 1", the sigma value Σ_1 for this machine can be calculated. If the machine is now run at a desired performance at some hydraulic feed rate Q_1, and the sigma value Σ_2 for the large-scale machine is calculated, then the flowrate that will produce the same performance in the larger machine (2) can be calculated as Q_2.

Table 14.4 Factors affecting centrifugal thickening and dewatering.

Basic machine design parameters	Adjustable machine and operational features	Solids characteristics

Flow geometry
  Countercurrent
  Cocurrent
  Internal baffling
Bowl/conveyor geometry
  Diameter
  Length
  Conical Angle
  Pitch and lead
  Maximum pool depth
Solids and flocculant feed points
Maximum operating speed
	Bowl speed
	Bowl and conveyor
	  differential speed
	Pool depth and volume
	Feed rate
	  Hydraulic loading
	  Solids loading
	Flocculant use
		Particle and floc size
		Particle density
		Consistency
		Viscosity
		Temperature
		Sludge volume index
		Volatile solids
		Solids retention time
		Septicity
		Floc deterioration

The second function the solid-bowl machine has to fulfill is to get the dewatered sludge out of the machine. The ability of a machine to do that can also be estimated using pilot-plant data and the "beta" concept, originally suggested by Vesilind (1974a). Scale-up of the solids-handling (cake-conveying) capacity is based on the assumption that when the ratio of solids discharge rate to the theoretical solids-handling capacity is the same for two geometrically similar—but different-sized—machines, then the performance of the two centrifuges will be similar for a given solids feed. The parameter that estimates this ability is called beta and is defined as

$$\beta = \Delta\omega \, (SN\pi Dz) \qquad (14.11)$$

where
- β = beta (m³/h);
- $\Delta\omega$ = differential speed (rad/s);
- S = pitch of the conveyor blades (distance between the blades [m]);
- N = number of leads on the conveyor (number of conveyor blades wrapped around the conveyor);
- D = bowl diameter (m); and
- z = pool depth (m).

All of the variables in the beta equation relate to the actual centrifuge hardware and can be measured, and thus a beta value can be calculated for any solid-bowl machine. The ability of any one machine (1) to process solids can be used to estimate the solids handling capability of another machine (2), by relating the solids loading using the beta values, as

$$\frac{U_1}{\beta_1} = \frac{U_2}{\beta_2} \qquad (14.12)$$

where
- U_1 = solids loading on machine number 1 (kg/h);
- β_1 = beta value for machine number 1 (m³/h);

U_2 = solids loading on machine number 2 (kg/h); and
β_2 = beta value for machine number 2 (m³/h).

Design of Centrifugal Dewatering

When a sample of the sludge to be centrifuged is not available, as would occur for new wastewater treatment plants, the design of the centrifuge is simply a purchase based on the best estimate of the vendor. This is not unreasonable because the operator has great leeway in the operation of the machine once installed. Table 14.4 lists the significant design and operating variables that influence the operation of a horizontal solid-bowl centrifuge. A desirable characteristic of the centrifuge is that its performance—as measured by cake solids and solids capture—can be adjusted to desired values by modifying such control variables as feed flowrate, bowl and conveyor differential speed, conditioning chemical (polymer) use, and pool depth. Process design criteria generally are defined by establishing certain solids characteristics and relating their significance to the application of the centrifugal thickener and the effects of these characteristics on centrifuge performance. Solid-bowl conveyor centrifuges are quite versatile in their application and can be used for thickening a variety of wastewater streams. The majority of applications have been for thickening air or oxygen waste activated sludge. The hydraulic loadings to the centrifuge control the liquid-phase residence time. Unfortunately, specific design criteria are not possible because of the many variations in solids characteristics and centrifuge design.

Solids concentration also is an important factor in determining the operating parameters of the centrifuge. It determines the specific solids input load applied to the centrifuge (kilograms per day of dry solids), as well as the dewatered (cake solids) output volume (cubic meters per day). These measurements assist a centrifuge designer in determining the proper machine design parameters; they also assist an operator in adjusting machine and process variables (for example, conveyor differential and feed rates) to balance load demands.

If a sample of the sludge to be centrifuged is available, either laboratory or pilot tests can be conducted to estimate the performance of the machine. Laboratory tests would include centrifugal dewatering in test tubes as discussed above. The most important variable would be the sludge consistency as measured by the penetration test. A sludge that has little body of consistency, where the penetrometer rod sinks right through the sludge, is a poor candidate for centrifugal dewatering and would probably yield poor performance.

If enough sludge is available for conducting pilot tests, then the sigma and beta relationships discussed above can be used. A smaller geometrically similar machine is run in order to obtain its best performance (including all the operating variables as well as the chemical conditioning of the sludge) and then this performance can be expected in the larger machine if the sigma and beta relationships are adhered to. That is, suppose the sigma and beta values are calculated for the model (machine 1) and for the prototype (machine 2), then the flowrate to the larger machine that would, on the basis of clarification, produce the same result is

$$Q_2 = Q_1 \frac{\Sigma_2}{\Sigma_1} \qquad (14.13)$$

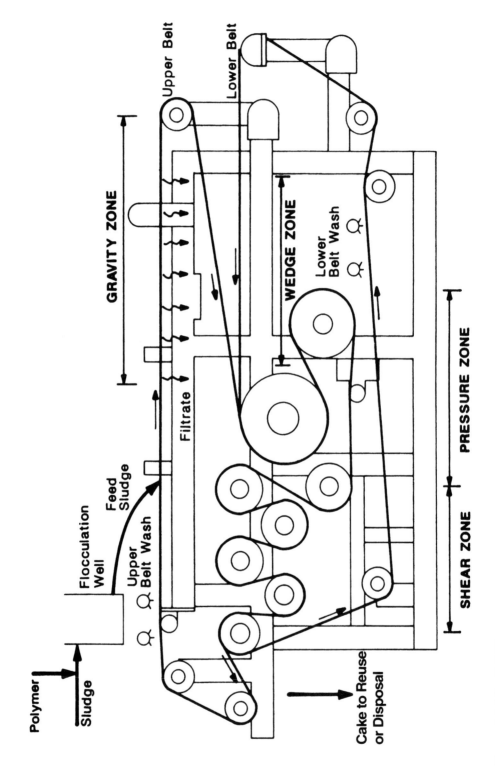

Figure 14.12 Belt filter press.

Similarly, the solids throughput or solids handling capacity can be estimated by calculating the beta terms for both the model (1) and the prototype (2) and the maximum solids the machine can be expected to handle is calculated as

$$U_2 = U_1 \frac{\beta_2}{\beta_1} \qquad (14.14)$$

One of these terms would be the limiting design parameter.

Belt Filter Dewatering

Belt filters have double moving belts to continuously dewater through a combination of gravity drainage and compression. Because of its low capital and operating cost, and its reliability in producing a well-dewatered sludge, the belt filter is now widely used all over the world.

Theory of Belt Filter Dewatering

Sludge is sequentially dewatered in the belt filter process through the following three operational stages: chemical conditioning, gravity drainage to a nonfluid consistency, and compaction in a pressure and shear zone. Figure 14.12 presents a simplified belt filter schematic that shows these three stages.

Chemical conditioning by organic polymer is the most common method of preparing the feed for gravity and pressure dewatering on the belt filter. Polymer can be added to a separate flocculation tank located upstream of the press or to the feed piping leading to the belt unit. The polymer must be intimately and thoroughly mixed with the feed to prepare for further dewatering. Proper mixing is an essential step in belt filter operation.

The dewatering operation of the belt filter begins when the polymer-flocculated sludge enters the gravity drainage section. This section has a continuous porous belt that provides a large surface area through which gravity drainage occurs. A distribution system evenly applies the mixture onto the gravity feed belt and the filtrate from the gravity zone is collected and piped to a drain system. The thickened sludge then moves into the compression stage where it is squeezed between the two porous belts. Increasing pressure and compression occur between the two belts because of compressive and shear forces as the sludge winds its way through the press. The pressure increase begins in the wedge zone where the two belts are brought back together following the gravity dewatering zone and continues to increase as the sludge enters the high-pressure or drum-pressure stage of the belt filter. Tension on the belts develops a squeezing action on the cake sandwich as the belts proceed around the several drums or rollers (of varying diameters) to maximize shearing action. As the sludge moves through the press, the decreasing diameters of the rollers progressively increase the pressure. The final dewatered cake is removed from the belts by doctor blades.

Design of Belt Filter Dewatering

Many variables affect the performance of the belt filters, including sludge characteristics; method and type of chemical conditioning; pressures developed; machine configuration, including gravity drainage; and belt speed. Although belt filter press performance data indicate significant variations in the dewatering capability of different types of sludges, the press typically is capable of producing a dewatered cake that is 18 to 25%

Table 14.5 Typical performance of a belt filter.

Type of sludge	Dry feed solids, %	Loading per meter belt width		Dry polymer,[a] g/kg dry solids	Cake solids, %	
		L/min	kg/h		Typical	Range
Raw primary (P)	3–7	110–190	360–550	1–4	28	26–32
Waste activated sludge (WAS)	1–4	40–150	45–180	3–10	15	12–20
P + WAS (50:50)[b]	3–6	80–190	180–320	2–8	23	20–28
P + WAS (40:60)[b]	3–6	80–190	180–320	2–10	20	18–25
P + Trickling filter (TF)	3–6	80–190	180–320	2–8	25	23–30
Anaerobically digested:						
P	3–7	80–190	360–550	2–5	28	24–30
WAS	3–4	40–150	45–135	4–10	15	12–20
P + WAS	3–6	80–190	180–320	3–8	22	20–25
Aerobically digested:						
P + WAS, unthickened	1–3	40–190	135–225	2–8	16	12–20
P + WAS (50:50), thickened	4–8	40–190	135–225	2–8	18	12–25
Oxygen activated WAS	1–3	40–150	90–180	4–10	18	15–23

[a]Polymer needs based on high molecular weight polymer (100% strength, dry basis).
[b]Ratio is based on dry solids for the primary and WAS.

solids for typical combined primary and secondary sludges. Solids recovery, including backwash-water solids, ranges from 80 to 95%. Table 14.5 lists typical performance data for the belt filter press dewatering of sludges from municipal wastewater treatment plants in the United States.

Although there is considerable experience and operational performance data for belt filters on a variety of sludge types, performance does vary. Depending on the specific application, reliance on historical performance data may be inadequate. For example, if a thermal process, such as drying, is to be used after dewatering, the percent solids achieved in the dewatering process can have a significant effect on the overall process costs. In these situations, performing a more detailed evaluation of belt filter dewatering is recommended.

The best method for evaluating belt filter performance on a specific material is to dewater the sludge using a pilot-scale test unit. Several belt filter manufacturers have mobile trailer-mounted pilot units that can be brought to a plant site for testing. Most of these units are small production machines and will provide performance comparable to larger models. Data to be collected as part of the pilot test include hydraulic and solids loading rates, polymer type and usage, percent solids, and solids recovery.

Whenever possible, polymer dosages and feed rates should be specifically optimized. Specific resistance or capillary suction time tests can be used to compare the filtration characteristics of different sludges and to determine the optimum coagulation requirements.

When evaluating the performance of the belt filter, like that of any other dewatering process, quantity and quality of the filtrate and backwash and their effects on the wastewater treatment system should be considered. Sidestream biochemical oxygen demand (BOD_5) varies from 150 to 300 mg/L, and suspended solids vary from 400 to 800 mg/L.

Handling or throughput capacity of a belt filter is the primary design criterion for sizing a dewatering facility. As with centrifuges, throughput capacity is either hydraulically limited or solids limited, depending on the feed solids concentration.

The flow-limiting characteristics of the belt filter typically are a result of withdrawal rates from settling tanks or thickeners and associated pump capacities. Nominal feed rates to belt presses per unit of belt width range from 3 to 4 L/m·s (14 to 19 gpm/ft), although belts have satisfactorily handled higher rates. Solids loading rates, which often limit performance, range between 0.10 and 0.20 kg/m·s (5 to 10 lb/ft-min).

Filter Press Dewatering

The primary advantage of a filter press system is that it often produces cakes that are drier than those produced by other dewatering alternatives. In cases where cake solids content must be greater than 35%, filter presses can be a cost-effective dewatering option. Filter presses also have adaptable operation to a wide range of solids characteristics, acceptable mechanical reliability, comparable energy requirements to vacuum-filter dewatering systems, and high filtrate quality that lowers recycle stream treatment requirements.

The significant disadvantages of filter presses are their high capital cost, the substantial quantities of conditioning chemicals or precoat materials that are required, periodic adherence of cake to the filter medium that requires manual removal, and relatively high operating and maintenance costs. The pressure filter press system remains more expensive than other dewatering alternatives; however, when disposal requirements dictate drier cakes, pressure filter presses have often been proven cost effective because of the lower disposal costs associated with drier cakes. Pressure filter presses have been shown to be cost effective when the solids cake must be incinerated. Often, the increased dry solids content of the filter press cake (which increases the proportion of volatile matter to water content) enables autogenous combustion in incinerators, thus reducing the need for other fossil fuels such as natural gas or fuel oil.

Theory of Filter Press Dewatering

A filter press consists of a number of plates that are rigidly held in a frame to ensure alignment and are pressed together either hydraulically or electromechanically between a fixed and moving end (see Figure 14.13). As shown in Figure 14.14, the plates have a drainage surface, drainage ports for the discharge of filtrate, and a large centralized port for solids feed. A filter cloth covers the drainage surface of each plate and provides a filter medium. A closing device presses and holds the plates close together while feed solids are pumped into the press through the inlet port at pressures of 700 to 2100 kPa (100 to 300 psig). The filter medium captures the solids and permits filtrate to drain through the plate drainage channels. A backing cloth or underdrainage, typically made of a rigid cloth or polyvinyl chloride, is sometimes used to keep the cloth separated from the drainage channels or pipes and the drainage ports during the high-pressure cycle. The backing cloth is the size of the recessed portion of the plate and is held in place by pins. Solids collect in the chambers until a practical low feed-rate limit is reached (typically 5 to 7% of the initial flowrate) and the filter cycle is terminated. At this time, the filter press feed pump is stopped and the individual plates are shifted and cakes discharged.

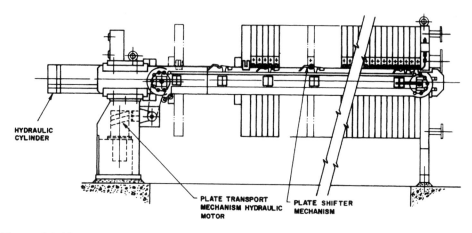

Figure 14.13 Filter press.

Solids conditioning, generally required to produce a low-moisture cake, involves adding lime and ferric chloride, polymer, or polymer combined with either inorganic compound to the solids before filtration. Using only polymer for solids conditioning decreases performance, but this is offset by savings of chemical costs, reduced ammonia odors, and reduced overall volume of cake produced. One of the problems with only using polymer is cake release from the cloth during the discharge cycle, and ferric chloride can be used to enhance cake release. A disadvantage of using ferric chloride with polymer is severe corrosion of piping and press. This does not occur with lime and

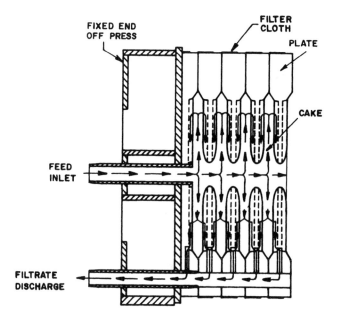

Figure 14.14 Typical filter press recessed plate.

ferric chloride conditioning because the lime neutralizes the corrosive action of the ferric chloride. Installations that intend to use polymer and ferric chloride should provide rubber lining on all metallic surfaces to protect against the corrosive nature of ferric chloride.

Design of Filter Press Dewatering

Evaluation of pressure filter presses is easy because most manufacturers have small skid-mounted units that are simple to test and operate. The pilot pressure filter press is much smaller than pilot-scale centrifuges and belt presses. Difficulties in pilot testing of pressure filter presses arise because of the variety of conditioning agents and filter cloths available.

Scaleup is a critical factor when testing pressure filter presses. With a pilot system, a thin cake coupled with a small recessed chamber (400 to 800 mm [16 to 30 in.] square) can lead to high expectations of dewatering capability. Pilot testing should use minimum 500-mm (20-in.) square plates and multiple chambers (typically four to eight) with a chamber thickness the same as that planned in the full-scale installation.

Drying Bed Dewatering

Sand Drying Beds

Sand drying beds have been used for more than 100 years. If well designed and properly operated, they are less sensitive to influent solids concentration and can produce a drier product than most mechanical devices. Particularly suited to small facilities, they can be successfully used in wastewater treatment of all sizes and in widely varying climates.

Compared with mechanical dewatering, sand drying is a more labor-intensive process and requires more land. The high capital and operating costs of mechanical systems have caused designers to take a second look at sand drying in cases where adequate land is available and environmental conditions are acceptable. This trend has been accompanied by growing concerns of groundwater contamination. The additional costs of bed lining and groundwater quality monitoring of bed systems may make mechanical dewatering more cost effective for all but small plants.

Sludge on sand beds is primarily dewatered by drainage and evaporation. Removing water from the sludge by drainage is a two-step process. Initially, the water is drained into the sand and removed through the underdrains. This step, a few days in duration, continues until the sand becomes clogged with fine particles or until all the free water has drained away. Once a sludge supernatant layer has formed, decanting removes surface water. This step is especially important for removing rainwater, which slows the drying process if it is allowed to accumulate on the surface. Decanting may also be necessary for removal of free water released by chemical treatment. Drying in beds can also be enhanced by use of auger mixing vehicles in paved beds. Water remaining after initial drainage and decanting is removed by evaporation.

Drying beds can be categorized as (1) conventional rectangular beds with side walls enclosing layers of sand and gravel, with underdrainage piping to carry away the liquid with or without provisions for mechanical removal, and with or without a roof or a greenhouse-type covering; (2) paved rectangular drying beds with a center sand-drainage strip with or without heating pipes buried in the paved section, and with or without

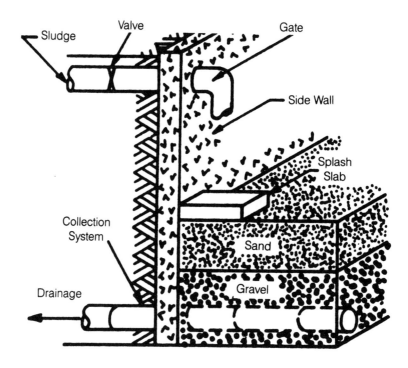

Figure 14.15 Schematic of a section of a typical sand drying bed.

covering to prevent incursion of rain; (3) wedge-wire drying beds with a wedge-wire septum, including provision for an initial flood with a thin layer of water followed by introduction of liquid sludge on top of the water layer, controlled formation of cake, and mechanical cleaning; and (4) rectangular vacuum-assisted beds with provisions for application of a vacuum as a motive force to assist gravity drainage.

Each drying bed is typically designed to hold, in one or more sections, the full volume of sludge to be removed from the digester or aerobic reactor at one drawing. Structural elements of the bed include the side walls, underdrains, gravel and sand layers, partitions, decanters, sludge distribution channel, runway and ramps, and possibly bed enclosures (see Figure 14.15).

Construction above the sand surface should include an embankment or vertical wall with above-sand freeboard of 0.5 to 0.9 m (20 to 36 in.). Underdrains, which typically are constructed of perforated plastic pipe or vitrified clay tile, are sloped toward a primary collecting pipe or outlet drain. The primary underdrain pipes should be no less than 100 mm (4 in.) in diameter and have a minimum slope of 1%. Spacing should range from 2.5 to 6 m (8 to 20 ft) and should take into account the type of sludge-removal vehicles to be used to avoid damage to the underdrain. Gravel layers are graded to an overall depth of 200 to 460 mm (8 to 18 in.), with the relatively coarser materials at the bottom. The gravel particles range from 3 to 25 mm (0.1 to 1.0 in.) in diameter. Sand depth varies from 200 to 460 mm (9 to 18 in.). However, a minimum depth of 300 mm (12 in.) is suggested to secure a good effluent and to reduce the frequency of sand replacement caused by losses from cleaning operations. A good-quality sand exhibits the following characteristics: particles that are clean, hard, durable, and free

Table 14.6 Design criteria for sand drying beds, using anaerobically digested sludge without chemical conditioning.

Initial sludge source	Uncovered beds Area, m²/cap	Uncovered beds Solids loading, kg/m²·a	Covered bed area,[a] m²/cap
Primary			
Imhoff and Fair (1940)	0.09	134	
Rolan (1980)	0.09–0.14		0.07–0.09
Walski (1976)			
N45°N latitude	0.12		0.09
Between 40 and 45°N	0.1		
S40°N latitude	0.07		0.05
Primary plus chemicals			
Imhoff and Fair (1940)	0.2	110	
Rolan (1980)	0.18–0.21		0.09–0.12
Walski (1976)			
N45°N latitude	0.23		0.173
Between 40 and 45°N	0.18		0.139
S40°N latitude	0.14		0.104
Primary plus low-rate trickling filter			
Quon and Johnson (1966)	0.15	110	
Imhoff and Fair (1940)	0.15	110	
Rolan (1980)	0.12–0.16		0.09–0.12
N45°N latitude	0.173		0.145
Between 40 and 45°N	0.139		0.116
S40°N latitude	0.104		0.086
Primary plus waste activated sludge			
Quon and Johnson (1966)	0.28	73	
Imhoff and Fair (1940)	0.28	73	
Rolan (1980)	0.16–0.23		0.12–0.14
Walski (1976)			
N45°N latitude	0.202		0.156
Between 40 and 45°N	0.162		0.125
S40°N latitude	0.122		0.094
Randall and Koch (1969)	0.32–0.51	35–59	

[a] Only area loading rates available for covered beds.

from clay, loam, dust, or other foreign matter; a uniformity coefficient that is not more than 4.0, but preferably under 3.5; and an effective size of sand grains that is between 0.3 and 0.75 mm (0.01 and 0.03 in.).

For manual removal of sludge in smaller plants, the drying bed typically is divided into sectional areas approximately 7.5-m (25-ft) wide; some mechanical removal methods have wider areas. The width should be designed to accommodate the removal method used (for example, multiples of loader bucket width and span of vacuum removal system).

Per-capita area criteria commonly used for sizing sand drying beds are shown in Table 14.6. These criteria are largely based on empirical studies of primary sludges conducted in the early 1900s by Imhoff and Fair (1940), who recommended a range from 0.1 to 0.3 m²/cap (1.0 to 3.0 ft²/cap), depending on the type and solids concentration of the sludge applied to the bed. Other important factors, such as the applied sludge depth and number of yearly applications, were also considered in this pioneering work. In attempting to uniformly apply these criteria, however, many design engineers fail to adequately consider the basic parameters of the original work. Changes in characteristics and quantity of sludge produced per person make current use of these criteria highly questionable.

Other Types of Drying Beds

Other types of drying beds include paved drying beds, wedge-wire drying beds, and vacuum-assisted drying beds. Because few plants use these methods, only limited information on their performance is available.

Paved Drying Beds

The primary advantages of paved drying beds are that front-end loaders can be more easily used for cake removal, auger mixing vehicles can speed up drying, and bed maintenance is reduced. Most beds are rectangular in shape and are 6 to 15 m (20 to 50 ft) wide by 20 to 45 m (70 to 150 ft) long, with vertical sidewalls. Often, paved drying beds are designed with sand-filled gutters in the middle to assist in drainage.

Wedge-Wire Drying Beds

Wedge-wire drying bed systems have been used since the early 1970s. In a wedge-wire drying bed, slurry is spread onto a horizontal, open drainage media in a way that yields a clean filtrate and provides a reasonable drainage rate. The bed has a shallow, rectangular, watertight tank fitted with a false floor of wedge-wire panels with slotted openings of 0.3 mm (0.01 in.). Caulking where the panels abut the walls will make the false floor watertight. An outlet valve to control the rate of drainage is located underneath the false floor.

Vacuum-Assisted Drying Beds

The principal components of vacuum-assisted drying beds are a bottom ground slab with a layer of stabilized aggregate several millimeters thick that supports the rigid multimedia filter top (this space is also the vacuum chamber and is connected to a vacuum pump) and a rigid multimedia filter top, which is placed on the aggregate support. Sludge spreads onto the filter surface by gravity flow at a rate of 57 L/min (150 gpm) and to a depth of 300 to 750 mm (12 to 30 in.). Polymer is injected to the sludge in the inlet line. Filtrate drains through the multimedia filter into the space containing the aggregate and then to a sump. From the sump, a level-actuated submersible pump sends the filtrate back to the plant. After the sludge is applied and allowed to gravity-drain for approximately 1 hour, the vacuum system is started and maintains a vacuum at 34 to 84 kPa (10 to 25 in. Hg) in the sump and under the media plates.

Other Dewatering Methods

Reed Beds

Reed beds combine the action of conventional drying beds with the effects of aquatic plants on water-bearing substrates. While conventional drying beds are used to drain more than 50% of the water content from the sludge, the resulting residue must be hauled away for further treatment or disposal at designated sites. When the drying beds are constructed in a specific manner and then planted with reeds of the genus *Phragmites communis*, further desiccation results from the demand for water by these plants. To satisfy this demand, the plants continually extend their root system into the sludge deposits. Additionally, this extended root system causes the establishment of a rich microflora that feeds on the organic content of the sludge. This microflora is also partly kept aerobic by the action of the plants. Degradation by the microflora is so effective that eventually up to 97% of the solids are converted into carbon dioxide and water with a corresponding volume reduction.

The reed-bed treatment system generally consists of a composition of parallel basins that are rectangular in shape with concrete sidewalls. The bottom of each bed is lined and provided with two underdrains. In addition, a 230-mm (9-in.) layer of 19-mm (0.75-in.) washed river gravel is topped with a 102-mm (4-in.) layer of filter sand. The reeds are planted in the gravel, with 11 plants per square meter (1 plant per square foot) of filter area. A freeboard of 1.0 to 1.5 m (3.5 to 5 ft) is often provided depending on storage design requirements. Basins are cyclically loaded. In each cycle, the first basin is loaded over a 24-hour period and then allowed to absorb the loading over a 1-week resting period before the cycle is repeated (Banks and Davis, 1983).

Hydraulic design loadings for sludge concentrations of 3 to 4% solids are 0.0042 $m^3/m^2 \cdot h$ (2.5 gal/d/ft^2) or 35 $m^3/m^2 \cdot y$ (86 gal/yr/ft^2). At this loading rate, approximately 1.0 m (3.5 ft) of product at a moisture content of 70% will accumulate over 10 years. When solids are removed for disposal, the top layer of sand will also be removed and must be replaced. Generally, the root system remaining in the gravel bed will allow for regeneration of the reed plants without the need for replanting. These systems are most applicable in climates where winter temperatures ensure at least one prolonged frost. The planted reeds are harvested in the fall once they have become dormant.

Lagoons

Lagoons, which are natural or artificial earth basins, can be used for both drying and storage. A few communities have used some form of lagoon system, reportedly with favorable results, although the use of lagoons typically has been started as a temporary expedient to handle sludge volumes in excess of the original plant design. Climatic conditions have a decided effect on the functioning of a lagoon, with warmer, arid climates producing the best results. However, cases have been reported where lagoons have successfully been used for wastewater dewatering in the far north.

Operation of lagoons generally involves the pumping of liquid during a period of several months or more into the lagoon. The pumped sludge is stabilized before application to minimize odor problems. The supernatant is continuously or intermittently decanted from the lagoon surface and returned to the wastewater treatment plant. The dewatered material is, after some time, removed with mechanical equipment.

The location of natural or artificial lagoons for dewatering must be considered before the use of lagoons is selected as a treatment method. The proposed site should be sufficiently removed from dwellings and other areas where odors would produce problems. Because they go through a series of wet and dry conditions, large lagoons can produce nuisance odors.

The use of deep lagoons for drying raw sludge has resulted in severe odor problems, but the use of shallow lagoons for drying generally has not produced odors more intense than those sometimes experienced with conventional sand drying beds. However, odors produced from lagoons used for storage of solids can be more of a problem because wet treated municipal solids retain a higher moisture content far longer than do solids treated on conventional sand drying beds. Odors can vary from a gas- and tar-like odor produced by a well-digested material to the putrid odors produced from the decomposition of raw sludge.

If the subsurface soil is permeable, the potential for groundwater contamination exists. Clay or membrane liners and underdrain systems can minimize this potential contamination.

Vacuum Filters

Vacuum filters were the first mechanical devices for sludge dewatering but have now been almost totally replaced by other processes such as belt filters. Some older plants still have vacuum filters, but they are seldom operated. The most common vacuum filter consists of a rotating cylindrical drum that is partially submerged in a vat of sludge. The cylindrical drum is mounted on a structural framework with a horizontal axis of rotation. The surface of the drum is covered with a porous medium, the selection of which is based on its dewatering characteristics. The lowest part of the drum is submerged in a vat of sludge that is to be dewatered.

The operating performance of vacuum filters is measured by filter yield, the efficiency of solids capture, and the filter-cake characteristics. Each of these criteria is important, but any one may be particularly significant at a given plant. Filter yield, measured in $kg/m^2 \cdot h$ ($lb/hr/ft^2$), is the most common measure of filter performance.

The efficiency of solids capture is measured as the percentage of feed solids that is recovered in the filter cake. Solids capture is influenced by filter media, vacuum pressure, and feed solids characteristics. Solids-capture efficiency, which ranges between 85 and 99%, can be an important consideration in the cost of treating recycle streams.

Characteristics of the filter cake, such as moisture content and heat value, constitute a third measure of performance. Cake solids content varies from 12 to 50% by weight, depending on the type of conditioning, filter cycle time, and submergence. Most cakes will fall in the 15 to 25% range. When large doses of inorganic conditioners are used, cake solids contents may appear high because of the added solids from the conditioners.

Thermal Drying

Thermal drying involves the application of heat to evaporate water. Thermal drying reduces the moisture content to a level below that achievable by conventional mechanical dewatering methods. Compared with these methods, the advantages of thermal drying include reduced transportation costs, further pathogen reduction, improved storage capability, and marketability (provided the metal concentrations are low). Thermally dried cake can be easily marketed as a fertilizer or soil conditioner. It is also acceptable for landfill disposal or incineration.

The drying rate depends on the internal mechanism of liquid flow and the external mechanism of evaporation. During the drying process, a temperature gradient develops from the heated surface inward, causing moisture to migrate from within the wet sludge to its surface by the mechanisms of diffusion, capillary flow, and internal pressures generated by shrinkage during drying. These internal drying mechanisms must be understood to predict the probable behavior of the sludge during the drying process. The transfer of heat from the heat transfer medium to the wet sludge raises the temperature of the sludge and evaporates water at the solid surface. External conditions affecting the transfer of heat or the drying process include temperature, humidity, rate and direction of gas flow, the exposed surface area, physical form of sludge, agitation, detention time, and the method of support employed during drying. Understanding these external conditions and their effects is necessary when investigating the drying characteristics of sludge, choosing the correct dryer, and determining the optimal operating conditions. Although the internal and external mechanisms occur simultaneously, either mechanism may limit the drying rate.

The three general stages of thermal drying include the warm-up stage, the constant-rate stage, and the falling-rate stage. During the warm-up stage, the sludge temperature and the drying rates increase to the steady-state conditions of the constant-rate stage. The warm-up stage typically is short and results in little drying. During the constant-rate stage, interior moisture replaces the external moisture as it evaporates from the saturated surface of the sludge. The transfer of heat to the evaporating surface controls the drying rate—similar to the evaporation of water from a pool of liquid. The constant-rate stage, which generally is the longest stage, results in most of the drying. The drying rate, essentially independent of the internal mechanism of liquid flow, depends on the following three external factors: heat- or mass-transfer coefficient, area exposed to the drying medium, and temperature and humidity differences between the drying medium and the wet surface of the sludge. Finally, during the falling-rate stage, external moisture evaporates faster than it can be replaced by internal moisture. As a result, the exposed surface is no longer saturated, latent heat is not transferred as rapidly as sensible heat is received from the heating medium, the temperature of the sludge increases, and the drying rate decreases. The transition point between the constant-rate and falling-rate stages is called the *critical moisture*.

Theory of Thermal Drying

The classification of dryers is based on the predominant method of transferring heat to the wet solids. These methods are convection, conduction, radiation, or a combination of these.

In *convection* (direct) *drying systems*, the wet sludge directly contacts the heat-transfer medium, typically hot gases. Convection heat transfer is expressed mathematically as

$$q_{conv} = h_c A (T_g - T_s) \qquad (14.15)$$

where

q_{conv} = convective heat transfer, kJ/h (Btu/hr);
h_c = convective heat-transfer coefficient, kJ/m²·h/°C (Btu/hr/ft²/°F);
A = area of wetted surface exposed to gas, m² (ft²);
T_g = gas temperature, °C (°F); and
T_s = temperature at sludge gas interface, °C (°F).

The convective heat-transfer coefficient can be obtained from the dryer manufacturer or from pilot studies.

In *conduction* (indirect) *drying systems*, a solid retaining wall separates the wet sludge from the heat-transfer medium, typically steam or another hot fluid. The mathematical expression for conduction heat transfer is

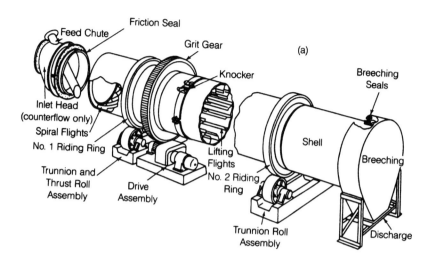

Figure 14.16 Rotary dryer.

$$q_{cond} = h_{cond}A(T_m - T_s) \qquad (14.16)$$

where

q_{cond} = conductive heat transfer, kJ/h (Btu/hr);
h_{cond} = conductive heat-transfer coefficient, kJ/m²·h/°C (Btu/hr/ft²/°F);
A = area of heat-transfer surface, m² (ft²);
T_m = temperature of heating medium, °C (°F); and
T_s = temperature of sludge at drying surface, °C (°F).

The conductive heat-transfer coefficient, a composite term, includes the effects of the heat-transfer surface films of the sludge and the medium. The coefficient may be obtained from the dryer manufacturer or from pilot studies.

In *radiation* (infrared) *drying systems*, infrared lamps, electric resistance elements, or gas-heated incandescent refractories supply radiant energy that transfers to the wet sludge and evaporates moisture. Radiation heat transfer is expressed as

$$q_{rad} = C_s As(T_r - T_s) \qquad (14.17)$$

where

q_{rad} = radiation heat transfer, kJ/h (Btu/hr);
C_s = emissivity of the drying surface (dimensionless);
A = sludge surface area exposed to radiant source, m² (ft²);
s = Stefan–Boltzman constant, 4.88×10^{-8} kcal/m²·h/K (1.73×10^{-9} Btu/hr/ft²/°R);
T_r = absolute temperature of radiant source, K (°R); and
T_s = absolute temperature of the sludge drying surface, K (°R).

The emissivity of the drying surface may be obtained from the dryer manufacturer or from pilot studies.

Design of Thermal Drying Systems

Thermal dryers are grouped into four categories: direct, indirect, combined direct–indirect, and infrared.

Direct Dryers

Direct (convection) dryers that have been successfully employed for drying municipal wastewater sludge include the flash dryer, rotary dryer, and fluidized-bed dryer.

The *flash dryer*, or pneumatic conveyor dryer, consists of a furnace, mixer, cage mill, cyclone separator, and vapor fan. The mixer blends wet sludge with dried cake to achieve a feed mixture with 40 to 50% moisture.

A *rotary dryer* consists of a cylindrical steel shell that is rotated on bearings and typically is mounted with its axis at a slight slope from the horizontal (see Figure 14.16). Rotary dryers include both single-pass and triple-pass systems. A mixer blends wet sludge with previously dried cake, thereby reducing the moisture content of the feed material to 30 to 50% and dispersing the cake. The blended product continuously enters the upper end of the rotary dryer along with hot furnace gases at a temperature ranging from 260 to 482 °C (500 to 900 °F). The mixture and hot gases are conveyed (usually cocurrently) to the discharge end of the dryer. During conveyance, axial flights along the slowly rotating interior wall of the dryer pick up and cascade the sludge through the dryer. This creates a thin sheet of falling particles, which directly contact the hot gases and rapidly dry. The exhaust gases exit the dryer at 66 to 105 °C (150 to 220 °F), and travel to air pollution control equipment for odor and particulate removal. Rotary dryers have successfully dried raw primary and secondary sludge mixtures, waste activated sludge, and digested primary sludge. The dried product is amenable to handling, storage, and marketing as a fertilizer or soil conditioner.

The *fluidized-bed dryer* consists of a stationary, vertical chamber with a perforated bottom through which hot gases (typically air) are forced or induced by a blower arrangement. Figure 14.17 shows a typical diagram of a fluidized-bed dryer system. A screw conveyor feeds the wet sludge through an air lock. In the dryer, the heated gas fluidizes or suspends the sludge and sand bed. The heated gases from the system furnace enter the plenum chamber and a perforated plate at the base of the dryer evenly distributes the gases across the body of the dryer. This produces a high level of intermixing and an intimate gas solids contact, resulting in high mass and heat transfer between the solid and gas phases. The dry cake exits the dryer by overflowing from the chamber through a pipe or adjustable weir with a rotary air lock. Vent gases exhaust to a cyclone separator or other air pollution control equipment.

Indirect Dryers

Indirect (conduction) dryers that have been employed for drying municipal sludge include the paddle dryer, hollow-flight dryer, disk dryer, and multiple-effect evaporation dryer.

Paddle, hollow-flight, and *disk dryers* consist of a stationary horizontal vessel, or trough, with a jacketed shell through which a heat-transfer medium (typically steam) circulates. The vessel, or trough, contains a rotating agitator assembly with a series of agitators (disks, flights, or paddles) mounted on a rotating shaft (rotor). The rotor and agitators (usually hollow) allow the heat-transfer medium to circulate through the hollow core. The heat-transfer medium (typically steam) flows to the disks (and shell) and is discharged as condensate after transferring its available energy to the product. Other

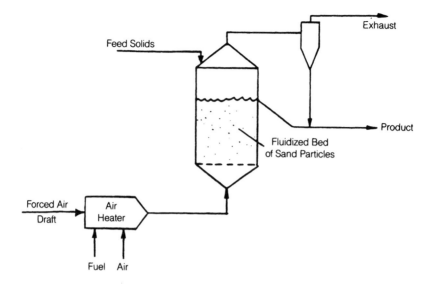

Figure 14.17 Flow diagram of a fluidized-bed sludge drying system.

thermal media, such as hot water or oil, may also be used. However, the dryer is internally constructed in a manner different from that required for steam. The agitators not only transport the sludge through the unit, but also provide an additional heat-transfer surface that contacts the sludge. Fully circular hollow disks provide a smooth heat-transfer surface with no leading edges. This construction technique minimizes fouling, product accumulation in dead spots, and wear. Disks are fabricated of carbon steel or stainless steel, and many dryers are supplied with a combination of both. The units can also be constructed with various special alloys if highly corrosive elements are present. A schematic diagram of a paddle or hollow-flight dryer is shown in Figure 14.18.

Direct–Indirect Dryers

The direct–indirect dryer, which is a jacketed trough or vessel, uses larger volumes of hot gases (typically air) as a heating medium or sweep gas than the indirect dryer. This mode of operation lowers the boiling point of the volatile fraction and causes a fluidizing effect, thus improving the heat-transfer coefficient between the sludge and heating surfaces. Another advantage of this mode of operation is that the direct and indirect heating can be varied to minimize energy consumption and maximize the drying rate.

Infrared Dryers

Municipal wastewater sludge drying applications of infrared (radiant) dryers usually involve combustion (that is, an infrared furnace or multiple-hearth furnace).

Conclusions

Engineering decision making has two aspects: technical and ethical. Technical questions are questions about how to achieve a goal by the most effective and efficient means, and

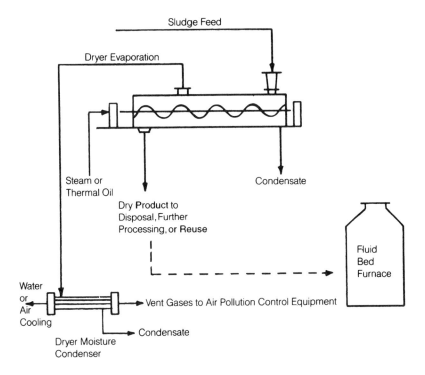

Figure 14.18 Flow diagram of a paddle or hollow-flight sludge drying system.

there is usually (though not always) a state-of-the-art answer on which the acknowledged experts in the field agree—a right answer. That is, one piece of dewatering equipment might be just the right application for a client. But this may not be true in all cases.

Let us use an example to illustrate a point. Suppose you own a software company and your group signed a contract to write a complex data-managing system coupled with geographic information system tracking. Your program will establish a person's location at any time if the person uses a cell phone. You do not know who your client is, but the money is good and you are well along in writing the software. You then discover that this program is to be used by a particularly cruel dictator who wants it to track dissidents and others he considers dangerous to his regime. You find out that this dictator has been routinely arresting and torturing these people, and many have disappeared without a trace. Your program will help the dictator be more efficient in fighting his opponents.

Does it matter to you whether you work on this project or not—because someone else surely will? What about your obligations to your employees, including the ones you hired specifically for the job? How important are your principles: are you prepared to run the risk that your company will get a reputation for unreliability, and lose business?

Sludge dewatering is not quite as dramatic as the above example, but the essence is often the same. Let us look at another case. Suppose you are working for a manufacturer of dewatering equipment and you go out on a client sales visit. Your boss makes a pitch to the potential customer, selling a piece of equipment that you know is not at all in the best interest of the customer (and the public that eventually will pay the price of an unusable piece of very expensive equipment). Does it matter whether or not you say anything (and risk losing both the sale and your job)? What are your obligations, as an engineer, to the public who is buying the equipment? How important are your principles?

Gunn (Vesilind and Gunn, 2002) suggests a test to determine if, after due consideration, you have done the right thing. He asks whether, after the decision is made, you can look yourself in the eye for a full minute in the evening and feel proud (or at least not ashamed) of the decision. Alternatively, suppose the essence of your decision appeared on the front page of the *New York Times* tomorrow morning. Would you feel proud of what you have done?

References

Ambler, C. M. (1952) Evaluation of Centrifuge Performance. *Chem. Eng. Prog.*, 48.

American Public Health Association; American Water Works Association; Water Environment Federation (1998) *Standard Methods for the Examination of Water and Wastewater*, 20th ed.; Washington, D.C.

Ashbrook-Simon-Hartley (1992) Aquabelt Operations & Maintenance Manual. Houston, Texas.

Banks, L.; Davis, S. (1983) Desiccation and Treatment of Sewage Sludge and Chemical Slimes with the Aid of Higher Plants. Presented at the 15th National Conference on Municipal and Industrial Sludge Utilization and Disposal.

Coe, H. S.; Clevenger, G. H. (1916) Methods for Determining the Capacities of Slime Settling Tanks. *Trans. Am. Inst. Min. Eng.*, 55, 356.

Dick, R. I. (1972) Gravity Thickening of Waste Sludges. *Proc. Filtr. Soc., Filtr. Sep.*, 9, 177.

Flint, L. R.; Howarth, W. J. (1971) The Collision Efficiency of Small Particles with Special Air Bubbles. *Chem. Eng. Sci. (G.B.)*, 26, 1155.

Imhoff, K.; Fair, G. M. (1940) *Sewage Treatment*; Wiley & Sons: New York.

Jordan, V. J., Jr.; Scherer, C. H. (1970) Gravity Thickening Techniques at a Water Reclamation Plant [Part I]. *J. Water Pollut. Control Fed.*, 42, 180.

MacConnell, G. S.; Harrison, D. S.; Mousavipour, F.; Kirby, K. W.; Lee, H. (1989) Full Scale Testing of Centrifuges in Comparison with DAF Units for WAS Thickening. Presented at 62nd Annual Conference of the Water Pollution Control Federation, San Francisco, California.

Quon, J. E.; Johnson, G. E. (1966) Drainage Characteristics of Digested Sludge. *J. Sanit. Eng. Div., Proc. Am. Soc. Civ. Eng.*, 92, 4762.

Randall, C. W.; Koch, C. T. (1969) Dewatering Characterisitics of Aerobically Digested Sludge. *J. Water Pollut. Control Fed.*, 41, 215.

Roberts, K. L.; et al. (1978) Dissolved Air Flotation Performance. *Proceedings of the 33rd Industrial Waste Conference, Purdue University*, West Lafayette, Indiana; p 194.

Rolan, A. T. (1980) Determination of Design Loading for Sand Drying Beds. *J. N.C. Sect., Am. AWWA, N.C. Water Pollut. Control Assoc.*, L5, 25.

U.S. Environmental Protection Agency (1979) *Process Design Manual for Sludge Treatment and Disposal*; EPA-625/6-85-010; Municipal Environmental Research Laboratory, Office of Research and Development: Cincinnati, Ohio.

Vesilind, P. A. (1968) The Influence of Stirring in the Thickening of Biological Sludge. Ph.D. Thesis, University of North Carolina, Chapel Hill.

Vesilind, P. A. (1974a) Scale-Up of Solid Bowl Centrifuge Performance. *J. Environ. Eng.*, 100, 479.

Vesilind, P. A. (1974b) *Treatment and Disposal of Wastewater Sludges*; Ann Arbor Science Publishers, Inc.: Ann Arbor, Michigan.

Vesilind, P. A.; Gunn, A. S. (2002) *Engineering, Ethics, and the Environment*; Cambridge University Press: New York.

Walski, T. M. (1976) Mathematical Model Simplifies Design of Sludge Drying Beds. *Water Sew. Works*, **123**, 64.

Yoshioka, N.; et al. (1957) Continuous Thickening of Homogeneous Flocculated Slurries. *Chem. Eng.* (Jpn.), **21**, 66.

Symbols Used in this Chapter

- A = area of thickener (m^2)
- A = area of wetted surface exposed to gas, m^2 (ft^2)
- A = sludge surface area exposed to radiant source, m^2 (ft^2)
- C_s = emissivity of the drying surface (dimensionless)
- C_d = solids concentration of the dewatered sludge cake (mg/L or % by weight)
- C_f = solids concentration in the feed (mg/L or % by weight)
- C_c = solids concentration in the centrate (mg/L or % by weight)
- C = solids concentration (kg/m^3)
- C_o = influent solids concentration (mg/L)
- C_L = limiting solids concentration (mg/L)
- D = bowl diameter (m)
- G = total solids flux at some solids concentration C ($kg/m^2 \cdot d$)
- g = acceleration from gravity (m/s^2)
- h_c = convective heat-transfer coefficient, $kJ/m^2 \cdot h/°C$ ($Btu/hr/ft^2/°F$)
- l = effective clarifying length of centrifuge bowl (cm)
- N = number of leads on the conveyor (number of conveyor blades wrapped around the conveyor)
- Q_o = influent flowrate (m^3/min)
- Q_1 = flow to machine number 1 (L/min)
- Q_2 = flow to machine number 2 (L/min)
- Q_u = underflow flowrate (m^3/h)
- Q_o = inflow rate (m^3/h)
- q_{conv} = convective heat transfer, kJ/h (Btu/hr)
- q_{rad} = radiation heat transfer, kJ/h (Btu/hr)
- R = percent recovery
- r_1 = radius from centrifuge centerline to the liquid surface in the centrifuge bowl (cm)
- r_2 = radius from centrifuge centerline to the inside wall of the centrifuge bowl (cm)
- S = pitch of the conveyor blades (distance between the blades [m])
- s = Stefan–Boltzman constant, 4.88×10^{-8} $kcal/m^2 \cdot h/K$ ($1.73 \, 10^{-9}$ $Btu/hr/ft^2/°R$)
- T_r = absolute temperature of radiant source, K (°R)
- T_s = absolute temperature of the sludge drying surface, K (°R)
- T_g = gas temperature, °C (°F)
- T_s = temperature at sludge gas interface, °C (°F)
- u = downward velocity caused by solids withdrawal (m/d)
- U_1 = solids loading on machine number 1 (kg/h)
- U_2 = solids loading on machine number 2 (kg/h)
- V = centrifuge pool volume (cm^3)

v = settling velocity at solids concentration C (m/d)
v_L = limiting settling velocity (m/h)
z = pool depth (m)
β_1 = beta value for machine number 1 (m³/h)
β_2 = beta value for machine number 2 (m³/h)
$\Delta\omega$ = differential speed (rad/s)
ω = angular velocity of centrifuge bowl (rad/s)
Σ = theoretical hydraulic capacity (cm²)
Σ_1 = sigma for machine number 1 (cm²)
Σ_2 = sigma for machine number 2 (cm²)

Sludge Stabilization

Introduction

Stabilization processes treat the solids generated in the primary treatment process, converting them to a stable (that is, not readily putrescent) product for use or disposal. Further, stabilization processes reduce odor and pathogens in the sludge, thus providing a more attractive and safer product for beneficial use. The four most common stabilization processes used today are anaerobic digestion, aerobic digestion, composting, alkaline stabilization, and combustion.

Anaerobic digestion is perhaps the most widely used of the processes in wastewater treatment plants. It produces relatively stable biosolids at a moderate cost and, as an added benefit, produces methane gas. This gas can be used to heat the digester, facility, or both, and, in larger treatment plants, may even be used for generation of electricity. Some of the disadvantages of this process include the following: high initial cost; a significant amount of mechanical equipment (particularly where digester gas is used); a strong sidestream (supernatant); the necessity for heat input to maintain the desired temperature; and the tendency toward process upset as a result of poor mixing, overloading, lack of temperature control, and heavy metals or other toxic agents in the feed.

Aerobic digestion is found in smaller wastewater treatment plants that treat less than approximately 20,000 m^3/d (5 mgd). It is a power-intensive process (because of power requirements for oxygen transfer) compared with anaerobic digestion, but is less expensive at the outset. Aerobic digestion is less complex operationally, and is sometimes not even a separate process. In many extended aeration facilities, such as oxidation ditches, the solids retention time (SRT) is maintained long enough to provide at least partial digestion by endogenous respiration in the aeration system.

Autothermal thermophilic aerobic digestion is a stabilization process that operates at thermophilic temperatures of 40 to 80 °C (104 to 176 °F) and additionally uses aerobic microorganisms to avoid methane gas production, in conjunction with facultative and anaerobic microorganisms for stabilizing organic wastes.

Composting is often employed to prepare sludge for use as a soil amendment and conditioner rather than as a fertilizer. Feed to the composting process can be either raw or digested solids. The final compost mixture should have a minimum solids content of 40%. A bulking agent is added to increase solids content, provide the desired structural properties of the mass, and to promote adequate air circulation. Composting is a labor-intensive process because of the addition of the bulking agent, turning of the windrows (for windrow composting), and recovery of the bulking agent. The potential for odors exists, especially at poorly designed and operated sites. Additionally, there is a possibility that pathogens will be spread through dust.

Lime stabilization of wastewater sludge is used to produce a soil amendment and conditioner. Some of the disadvantages are that the lime-treated sludge can become unstable if the pH drops after treatment and regrowth of biological organisms occurs, and the added lime increases the mass of solids to be used or disposed of.

Combustion of sludge is the ultimate in stabilization. Many of the concerns in sludge disposal, such as the prevalence of pathogenic organisms, are eliminated and the sludge becomes a solid product with no value that must be disposed of. The success or failure of sludge combustion is often dependent on the ability to dewater sludge to its autothermal combustion point, reducing or completely eliminating the need for auxiliary fuel.

Anaerobic Digestion

The purpose of anaerobic digestion is to reduce pathogens, reduce biomass quantity by partial destruction of volatile solids, and produce usable gas as a byproduct. Anaerobic digestion requires both proper design and careful operation, and continues to be one of the most widely used processes for the stabilization of sludge. The complexity of anaerobic digestion arises from process sensitivity and interactions of components that make up the complete system.

Anaerobic Digestion Theory

Microbiology and Biochemistry

Anaerobic digestion occurs as the result of a complicated set of chemical and biochemical reactions. Reactions occur within the context of a complex ecosystem involving many types of bacteria, with each type providing a unique and indispensable bio-transformation. Figure 15.1 shows a simplified representation of the anaerobic digestion process, which is divided into the following three stages: hydrolysis, formation of soluble organic compounds and of short-chained organic acids, and methane formation.

In the first stage (hydrolysis), the proteins, cellulose, lipids, and other complex organics are made soluble. In the second stage (acid formation), the products of the first

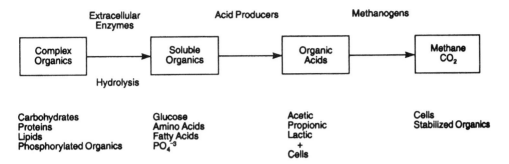

Figure 15.1 Microbiological pathway of anaerobic digestion.

stage are converted to complex soluble organic compounds including long-chained fatty acids; these soluble organic compounds are then broken down to short-chained organic acids (known as *acidification*). In the third stage (methane formation, or *methanogenesis*), the organic acids are converted to methane and carbon dioxide.

The overall extent of stabilization by anaerobic digestion is measured by the amount of volatile solids destruction that occurs through the digester. Because anaerobic digestion is biologically mediated and depends on the growth of microorganisms, complete volatile solids destruction does not occur, with 40 to 65% being typical. Lower percentage destruction occurs with biological solids or solids containing significant concentrations of materials that are difficult to degrade. High-percentage destruction is achieved when the digestion of primary sludge contains easily degraded materials such as simple carbohydrates, complex carbohydrates (cellulose), proteins, and lipids (grease). Studies by Eastman and Ferguson (1981) indicate that for complex particulate substrates, such as wastewater residuals, hydrolysis is the rate-limiting step for the acid-forming phase. After these complex organic substances are solubilized, they are immediately converted to volatile acids and do not accumulate.

In addition to the rate and extent of digestion, gas production and characteristics of the gas produced vary depending on the nature of the solids being digested. For example, destruction of 0.454 kg (1 lb) of chemical oxygen demand (COD) produces 0.350 standard m^3/kg (5.62 standard ft^3/lb) of methane (Parkin and Owen, 1986), measured at 101 kPa (1 atm) and 20 °C (68 °F). The COD content of biodegradable solids per unit of mass varies, with lipids and proteins having a higher COD content than simple and complex carbohydrates. (The chemical oxygen demand, COD, is used to express the substrate concentration, instead of the more familiar BOD, or biochemical oxygen demand, because the process is anaerobic, and BOD is an aerobic test. It makes more sense therefore to use the more precise COD parameter instead of the BOD.) As a result, complete digestion of lipids and proteins produces more methane than does complete digestion of simple and complex carbohydrates. The concentration of carbon dioxide in the digester gas also varies with the type of solids being digested.

Variables Affecting Anaerobic Digestion

Of the many environmental factors that affect the rates of the three anaerobic digestion reactions, the most important are solids retention time, effectiveness of mixing, hydraulic retention time, temperature, pH, and the presence of toxic materials.

Uniformity of environmental conditions (particularly the aforementioned factors) within the anaerobic digester is critical and largely determines the maximum possible rate of digestion and the potential for digester upsets. The degree of mixing determines the uniformity of conditions in the digester.

Solids and Hydraulic Retention Times

The sizing of anaerobic digesters is based on providing sufficient retention time in these well-mixed reactors to allow significant volatile solids destruction to occur. Sizing criteria, expressed either as solids retention time (days, calculated as the mass of solids in digester divided by the mass of solids removed per day), and the hydraulic retention time (days, the working volume divided by the volume sludge fed to the digester per day). If there is no change in solids concentration within the digester, the hydraulic retention time is the same as the solids retention time.

The solids and hydraulic retention times and the extent of each of the three reactions occurring during anaerobic digestion (hydrolysis, acid formation, and methane formation) are directly related: an increase in solids retention time increases the extent of each reaction; a decrease in solids retention time decreases the extent of reaction. There is a minimum critical retention time for each reaction, and if this is not provided, the bacteria cannot grow rapidly enough to remain in the digester, the reaction mediated by those bacteria will cease, and the digestion process will fail.

Temperature

Temperature is important in determining the rate of digestion, particularly rates of hydrolysis and methane formation. From a design standpoint, both the design operating temperature and the ability to maintain that temperature within close tolerances are important. The design operating temperature establishes the minimum solids retention time required to achieve a given amount of volatile solids destruction. Most anaerobic digestion systems are designed to operate in the mesophilic temperature range (approximately 35 °C [95 °F]). Some systems have been designed to operate in the thermophilic temperature range (approximately 55 °C [131 °F]). Advantages claimed for thermophilic digestion include improved dewaterability, increased pathogen destruction, and increased scum digestion.

While selection of the design operating temperature is important, maintaining a stable operating temperature in the digester is more important. This is because bacteria involved (particularly methane formers) are sensitive to temperature changes. Temperature changes greater than 1 °C/d can result in process failure. A good design avoids a change in digester temperature greater than 0.5 °C/d. Temperature changes must be held to less than 1 °C/d. This is a critical consideration when determining feed schedules.

pH

Anaerobic bacteria, particularly the methane formers, are also sensitive to pH. Optimum methane production occurs when the pH level is maintained between pH 6.8 and 7.2. Acid formation continuously occurs during the digestion process and tends to lower the digester pH. However, methane formation also produces alkalinity, primarily in the forms of carbon dioxide and ammonia. These materials buffer changes in pH by combining with hydrogen ions.

A reduction in digester pH (such as overloading the digester, which leads to an increase in acid formation) inhibits methane formation. As acid production continues, methane and alkalinity formation are further inhibited, possibly leading to process failure. Mixing, heating, and feed-system designs are important in minimizing the potential for this type of process upset. Provisions for the addition of chemicals, such as lime, sodium bicarbonate, or sodium carbonate, to neutralize excess acid in an upset digester should be included.

Toxic Materials

If concentrations of certain materials, such as ammonia, heavy metals, light metal cations, and sulfide, sufficiently increase in anaerobic digesters, they can create unstable conditions within the digester. Toxic conditions can occur as a result of a sudden change in digester operation, such as overfeeding or excessive addition of chemicals, or as a result of a shock loading of these materials in the plant influent. Table 15.1 presents toxic and inhibitory concentrations of selected inorganic materials. Table 15.2 presents toxic and inhibitory concentrations of selected organic materials in anaerobic digestion.

The most common effect of excess concentrations of these materials in the digester is inhibition of methane formation. This leads to volatile acid accumulation, pH depression, and digester upset, as previously discussed. Concentrations of dissolved forms of some of these materials can be controlled by chemical addition. The most common example of this is control of sulfide using iron salts, discussed later in this chapter.

Applicability

Anaerobic digestion may be considered beneficial for stabilization when the volatile solids concentration is 50% or higher and if no biologically inhibitory substances are present or expected. Digestion of primary solids results in better solids-liquid separation characteristics than activated sludge. Combining components will result in settling characteristics, which are better than activated sludge but not as good as primary sludge alone. Chemical residuals containing lime, alum, iron, and other substances can be

Table 15.1 Toxic and inhibitory concentrations of selected inorganic materials in anaerobic digestion (from Parkin, G. F.; Owen, W. F. [1986] Fundamentals of Anaerobic Digestion of Wastewater Sludge, *J. Environ. Eng.*, 112, 5, with permission from the American Society of Civil Engineers).

Substance	Moderately inhibitory concentration, mg/L	Strongly inhibitory concentration, mg/L
Na^+	3500–5500	8000
K^+	2500–4500	12,000
Ca^{++}	2500–4500	8000
Mg^{++}	1000–1500	3000
Ammonia-nitrogen (pH dependent)	1500–3000	3000
Sulfide (un-ionized gas)	200	200
Copper (Cu)	—	0.5 (soluble) 50–70 (total)
Chromium VI (Cr)	—	3.0 (soluble) 200–250 (total)
Chromium III	—	180–420 (total)
Nickel (Ni)	—	2.0 (soluble) 30.0 (total)
Zinc (Zn)	—	1.0 (soluble)

Table 15.2 Toxic and inhibitory concentrations of selected organic materials in anaerobic systems.

Compound	Concentration resulting in 50% activity, mM
I-Chloropropene	0.1
Nitrobenzene	0.1
Acrolein	0.2
I-Chloropropane	1.9
Formaldehyde	2.4
Lauric acid	2.6
Ethyl benzene	3.2
Acrylonitrile	4
3-Chlorol-1,2-propandiol	6
Crotonaldehyde	6.5
2-Chloropropionic acid	8
Vinyl acetate	8
Acetaldehyde	10
Ethyl acetate	11
Acrylic acid	12
Catechol	24
Phenol	26
Aniline	26
Resorcinol	29
Propanol	90

successfully digested if the volatile solids content remains high enough to support the biochemical reactions and no toxic compounds are present. If an examination of past sludge characteristics indicates wide variations in quality, anaerobic digestion may not be feasible because of its inherent sensitivity to changing substrate quality.

One of the advantages of anaerobic digestion is that energy (more than that required by the process) is produced. Methane can be used to heat and mix the reactor and excess methane can be used to heat space or produce electricity, or be used as engine fuel to power blowers. However, gas cleaning systems are often required to render the digester gas useful for these purposes.

The disadvantages associated with anaerobic digestion are that the digester is easily upset by unusual conditions and erratic or high loadings, and is slow to recover. The process requires close operational control. Large reactors are required because of the slow growth of methanogens and required solids retention times increasing capital costs.

During digestion, heavy metals are concentrated in the sludge, possibly causing a failure to meet regulations, and therefore anaerobic digestion can restrict disposal alternatives. The resultant supernatant sidestream adds to the loading of the wastewater treatment plant. It is high in biochemical oxygen demand (BOD), COD, suspended solids, phosphorus, and ammonia-nitrogen. Cleaning operations are difficult and dangerous because of the closed vessel. Internal heating and mixing equipment can have significant problems as a result of corrosion and wear in harsh, inaccessible environments. External heating systems similarly can clog and need constant maintenance. The sludge dewatering characteristics are not often enhanced by anaerobic digestion, and its heating value is reduced for incineration purposes.

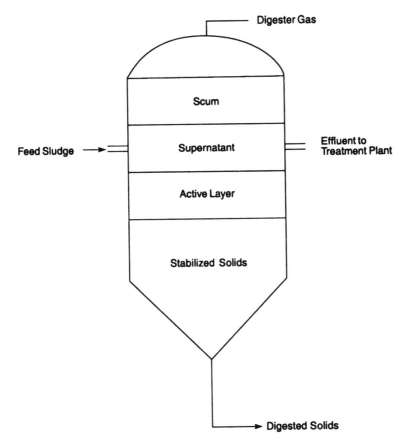

Figure 15.2 Low-rate anaerobic digestion (U.S. EPA, 1979).

For the operator, the possibility of explosion exists as a result of inadequate operation and maintenance, leaks, or carelessness. Gasline condensation or clogging can cause maintenance problems and there are comparatively high maintenance requirements because of potential scaling, scum buildup, and grit accumulation.

Process Variations

Three process configurations for anaerobic digestion in common use are low-rate, high-rate, and two-stage digestion. In addition, anaerobic digestion can also be operated in two temperature regimes—mesophilic (30 to 38 °C [86 to 100 °F]) and thermophilic (50 to 60 °C [122 to 140 °F]).

Low-Rate Digestion

Low-rate digesters are the oldest anaerobic stabilization systems and are also called *standard-rate digesters* or *conventional anaerobic digesters*. Figure 15.2 illustrates the low-rate digester. The digester consists of a cylindrically shaped tank with a sloping bottom and a flat or domed roof. No mixing is provided in this system.

Because no mixing occurs, stabilization in low-rate systems results in a stratified condition within the digester. Methane gas accumulates in the headspace of the tank and is drawn off for storage or use. Scum accumulates on the liquid or supernatant

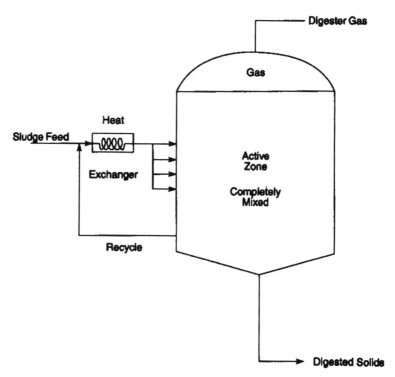

Figure 15.3 High-rate anaerobic digestion.

surface. The supernatant is drawn off and recycled either to the primary clarifier or to the secondary treatment process. The supernatant contains high ammonia and phosphorus concentrations. The stabilized solids settle to the tank bottom for removal and further processing.

Low-rate digestion is characterized by intermittent feeding, low organic loading rates, no mixing other than that caused by rising gas bubbles, large tank size because of the small effective volume, and detention times of 30 to 60 days. Grit and scum layers will accumulate on the bottom and top of the tank, respectively, thereby decreasing the effective volume. An external heat source may or may not be present to increase the digestion rate. Conditions are not maintained for optimum digestion. This type of digestion has traditionally been considered only for small plants, less than approximately 4000 m^3/d [1.0 mgd]). Low-rate digesters are seldom built today.

High-Rate Digestion

High-rate digesters (Figure 15.3) are characterized by supplemental heating and mixing, uniform feeding rates, and sludge thickening before digestion. In properly designed and operated digesters, these factors result in uniform conditions throughout most of the digester. As a result, the tank volume required for adequate digestion is reduced and the stability of the process is improved.

Heating the sludge digestion increases the microorganism growth rate, the digestion rate, and gas production. High-rate anaerobic digesters are operated at mesophilic and thermophilic temperature ranges. Thermophilic digestion may offer several advantages over mesophilic digestion, including increased reaction rates that can result in smaller digester volumes, increased destruction of pathogens, and better dewatering characteristics. Limitations of the process include extreme sensitivities of the organisms to the defined temperature range, a higher net energy input (and, therefore, higher operation costs) compared with the mesophilic process, and the production of digested sludge with a more offensive odor.

Several heating methods have been used for anaerobic digesters, including steam injection, internal heat exchangers, and external heat exchangers. External heat exchangers are the most popular because of their flexibility and ease of maintaining the heating surfaces. Internal coils can foul because of caking and, as a result, will have to be removed, or the digester will have to be emptied to clean them. Hot water supply temperatures to heat exchanger surfaces are maintained between approximately 50 and 62 °C (120 and 150 °F). At temperatures higher than 62 °C (150 °F), caking on the heat exchanger surfaces becomes more likely. The use of steam injection results in dilution. A considerable amount of water is necessary for steam production. Costs for raw water and softening add to the cost of steam injection.

Auxiliary mixing of the digester contents reduces thermal stratification, dispersing the substrate for better contact with the active biomass, and reducing scum buildup. Mixing also dilutes any inhibitory substances or adverse pH and temperature feed characteristics, thereby increasing the effective volume of the reactor.

Continuous or regular intermittent feeding is beneficial to digester operation because it helps maintain steady-state conditions within the digester. The methanogens are sensitive to changes in substrate levels. Uniform feeding and multiple feed-point locations in the tank can alleviate or reduce shock loading to those microorganisms. Excessive hydraulic loading should be avoided because it decreases detention time, dilutes the alkalinity necessary for buffering capacity, and requires additional heat energy.

Two-Stage Anaerobic Digestion

Two-stage digestion is an expansion of the high-rate digestion technology that divides the functions of fermentation and solids-liquid separation into two separate tanks in series (Figure 15.4). The first tank is a high-rate stabilization system, while the second is for solids-liquid separation. The second reactor does not have mixing or heating facilities unless it is also used to provide standby digester capacity. It may serve several other functions such as providing storage capacity and insurance against short circuiting of the process.

Anaerobically digested solids may not settle well, resulting in a supernatant containing a high concentration of suspended solids that can be detrimental to the liquid wastewater treatment system when recirculated. Several reasons for poor settling characteristics include incomplete digestion in the primary digester (which generates gases in the secondary digester and causes floating solids) and fine-sized solids that have poor settling characteristics. This latter case is associated with secondary or tertiary solids, including chemical solids.

Two-stage digesters using both thermophilic digestion (see below) followed by mesophilic digestion are in developmental stage and may prove to have superior operational characteristics.

Thermophilic Anaerobic Digestion

Digesters can be operated in the thermophilic temperature range (that is, 55 to 65 °C [130 to 150 °F]). Design criteria and system performance for thermophilic digestion are

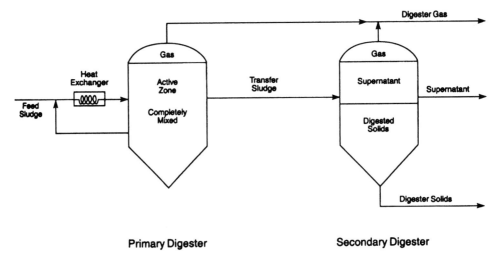

Figure 15.4 Two-stage, high-rate anaerobic digestion (U.S. EPA, 1979).

somewhat different than those for mesophilic digestion. Thermophilic digestion is thought to achieve higher volatile solids reduction per unit volume of digester than mesophilic digestion. Because the temperature is higher, however, the heating costs will be higher for thermophilic digestion.

Design Criteria

Volatile Solids

The anaerobic digestion process reduces the quantity of volatile solids. The degree of destruction is one indicator for the operator to determine if the digester is functioning properly. A typical high-rate digester will reduce the volatile solids by 40 to 60%. Each digester will display its own maximum level of destruction depending on feed characteristics. Between 60 to 80% of the volatile solids fed to the digester are biodegradable, unless industrial wastes or other abnormal characteristics are present.

The amount of volatile solids destroyed is a function of both temperature and solids retention time. The overall volatile solids destruction potential can vary as a function of the liquid treatment process that generates it. For example, when comparing extended aeration and high-rate processes, the biodegradable fraction of the volatile solids of waste activated sludge generated would be lower for the extended aeration process, given the effect of endogenous respiration. Therefore, under identical conditions, the digested solids from the high-rate system would show a greater volatile solids destruction (hence higher gas production).

Pathogen Reduction

Anaerobic stabilization also accomplishes substantial reductions in pathogen concentration (Farrell et al., 1986; Garber, 1982; Lee et al., 1989; Storey, 1987; Stukenberg et al., 1992).

Table 15.3 Typical gas production rates (Buswell and Neave, 1939).

Material	Specific gas production per unit mass destroyed	
	m³/kg	Methane content, %
Fats	1.2–1.6	62–72
Scum	0.9–1.0	70–75
Grease	1.1	68
Crude fibers	0.8	45–50
Protein	0.7	73

Gas Quality and Quantity

The quality and quantity of digester gas produced can also be used to evaluate digester performance. Gas production is directly related biochemically to the amount of volatile solids destroyed and is expressed as volume of gas per unit mass of volatile solids destroyed. This specific gas production rate is different for each organic substance in the digester. Table 15.3 gives the specific gas production for several organic substances. The range of gas production varies from approximately 1.2 to 1.5 m³/kg (20 to 25 ft³/lb) of volatile solids destroyed for fats to 0.7 m³/kg (12 ft³/lb) of volatile solids destroyed for proteins and carbohydrates. A typical municipal anaerobic digester handling primary and waste activated sludge should produce approximately 0.8 to 1 m³ of gas/kg volatile solids destroyed (13 to 18 ft³/lb). The amount of gas produced is a function of temperature, solids retention time, and the volatile solids loading.

Typical digester gas exhibits a heat content between 20 and 25 MJ/m³ (500 and 700 Btu/ft³). An average value of 25 MJ/m³ (640 Btu/ft³) has been used for design. This represents a heating value that is approximately two-thirds that of pure methane gas (39 MJ/m³ [960 Btu/ft³]).

Anaerobic Digester Design

The first important design consideration is to determine the correct reactor volume to ensure sufficient stabilization of the feed and methane gas production. Other decisions, such as mixing method, heating requirements, and energy recovery, are discussed later. This discussion is oriented toward two-stage mesophilic digestion. Additional design considerations are required for thermophilic or two-phase digestion, discussed later in this chapter.

Design Data

The data necessary for designing an anaerobic digester include the quality and quantity of feed solids to be digested, including all of the solids (mass per day) produced by primary, secondary, and tertiary (where applicable) treatment processes. Additional information is necessary on the percentage of solids, the percentage of volatile solids, and the ratio of primary to secondary solids to be fed to the digester. The total solids produced can be theoretically calculated based on the solids balance or extrapolated from existing operating data.

Table 15.4 Typical design parameters for low-rate and high-rate digesters (Burd, 1968).

Parameter	Low rate	High rate
Solids retention time, days	30–60	15–20
Volatile suspended solids loading, kg/m^3·d (lb/ ft^3/d)	0.64–1.6 (0.04–0.1)	1.6–3.2 (0.10–0.20)
Volume criteria, m^3/cap ft^3/cap		
Primary sludge	0.06–0.08 (2–3)	0.03–0.06 (1.3–2)
Primary sludge + trickling filter sludge	0.11–0.14 (4–5)	0.07–0.09 (2.6–3.3)
Primary sludge + waste activated sludge	0.11–0.17 (4–6)	0.07–0.11 (2.6–4)
Combined primary + waste biological sludge feed concentration, % solids-dry basis	2–4	4–6
Anticipated digester underflow concentration, % solids-dry basis	4–6	4–6

Design Parameters

Design of anaerobic digesters has been based on solids retention time, organic loading rate (volatile suspended solids per volume), and volume per capita. Design parameters for low- and high-rate digesters are detailed in Table 15.4. For estimating feed volumes at domestic plants in the absence of operating data, the volume figures per capita can be used. Low-rate digesters are organically loaded at rates equal to approximately 0.5 to 1.5 kg/m^3 volatile suspended solids (VSS)·d (0.04 to 0.1 lb VSS/d/ft^3). High-rate digesters with mixing and heating are characterized by organic loading rates of 2 to 3 kg/m^3 VSS·d (0.1 to 0.2 lb/d/ft^3).

Solids retention times are approximately 30 to 60 days for standard or low-rate digestion and 15 to 20 days for high-rate digestion at mesophilic temperatures. For two-stage digestion, the solids retention time is calculated only for the first reactor in the system because the second tank is used mostly for solids settling and storage, as well as gas storage. A minimum solids retention time is essential to the anaerobic digestion process to ensure that the necessary microorganisms are being produced at the same rate as they are wasted daily.

A rational approach to the selection of a design solids retention time is based on growth kinetics (Parkin and Owen, 1986). The minimum solids retention time, or that solids retention time on the edge of process failure, can be estimated from

$$\theta_{min} = \frac{YkS_e}{K_s + S_e} - b \qquad (15.1)$$

where

  θ_{min} = limiting solids retention time (SRT) for required digester performance (d);
  Y = yield of anaerobic organisms resulting from growth (g VSS/g COD destroyed);
  k = maximum specific substrate use rate (g COD/g VSS·d);
  S_e = concentration of biodegradable substrate in digested sludge (therefore, in digester) [COD/L]);
    = $S_O (1 - E)$, where S_O = concentration of biodegradable substrate in the feed sludge, g COD/L, and E = digestion efficiency, fraction of substrate destroyed;
  K_s = half saturation concentration of biodegradable substrate in feed sludge (g COD/L); and
  b = endogenous decay coefficient (d^{-1}).

Table 15.5 Estimated volatile solids destruction in anaerobic digesters.

	Digestion time, days	Volatile solids destruction, %
High rate (mesophilic range)	30	65.5
	20	60.0
	15	56.0
Low rate	40	50.0
	30	45.0
	20	40.0

Values for constants in the above equation have been proposed for typical municipal primary sludge within a temperature range of 25 to 35 °C (77 to 95 °F). The following proposed values (Parkin and Owen, 1986) are based on laboratory experiments:

$k = 6.67$ g COD/g VSS·d (1.035^{T-35});
$K_s = 1.8$ g COD/L (1.112^{T-35});
$b = 0.03$ d^{-1} (1.035^{T-35});
$Y = 0.04$ g VSS/g COD removed; and

where T = temperature (°C).

Volatile Solids Loading and Destruction

The phrase *volatile solids loading* refers to the mass of volatile solids (VS) added to the digester each day divided by the working volume of that digester (kg VS/m³·d [lb VS/d/ft³]). Loading criteria typically are based on sustained loading conditions (peak month or peak week solids production), with provisions for avoiding excessive loading during shorter time periods. A typical design sustained peak volatile solids loading rate is 1.9 to 2.5 kg VS/m³·d (0.12 to 0.16 lb VS/d/ft³). The upper limit on the volatile solids loading rate is determined by the rate of accumulation of toxic materials, particularly ammonia, or washout of methane formers. A limiting value of 3.2 kg VS/m³·d (0.20 lb VS/d/ft³) is often used. Excessively low volatile solids loading rates can result in designs that are expensive for both construction and operation. Construction is expensive because of the large tankage volume provided. Operation can be expensive because gas production rates may not be sufficient to provide the energy required to maintain the desired operating temperature in the digester. A low volatile solids loading rate (less than approximately 1.3 kg VS/m³·d [0.08 lb VS/d/ft³]) while maintaining the design solids retention time (or the hydraulic retention time) indicates that thickening before digestion may be cost effective. Table 15.5 shows the effect of digestion time on volatile solids destruction in both high-rate and low-rate digesters.

Gas Production

The specific gas production at municipal plants can be estimated by using the relationship of approximately 0.8 to 1.1 m³/kg (13 to 18 ft³/lb) of volatile solids destroyed. The greater the percentage of fats and grease, the higher the expected specific gas production as long as adequate solids retention time and mixing are provided because these materials are the slowest to metabolize. The total gas volume produced is as follows:

$$G_v = (G_{sgp})V_s \qquad (15.2)$$

where
- G_v = volume of total gas produced, m³ (ft³);
- V_s = volatile solids destroyed, kg (lb); and
- G_{sgp} = specific gas production, taken as 0.8 to 1.1 m³/kg (13 to 18 ft³/lb) of volatile solids destroyed.

The total amount of methane produced can be estimated from the amount of organic material removed each day by the following relationship:

$$G_m = M_{sgp}(\Delta OR - 1.42 [\Delta X]) \qquad (15.3)$$

or

$$G_m = (\Delta COD_R)(13.1 \text{ MJ/kg COD}_R)/38 \text{ MJ/m}^3 \text{ CH}_4 \qquad (15.4)$$

where
- G_m = volume of methane produced, m³/d (ft³/d);
- M_{sgp} = specific methane production per mass of organic material (COD) removed, m³/kg (ft³/lb);
- ΔOR = organics (COD) removed daily, kg/d (lb/d);
- ΔX = biomass produced kg/d (lb/d); and
- ΔCOD_R = COD removed across digester kg/d (lb/d).

Because digester gas is approximately two-thirds methane, the total digester gas produced is equal to 1.5 (G_m). Expected methane concentrations for different digesters can range from 45 to 75% in volume. Carbon dioxide concentrations range from 25 to 55%, by volume.

Gas Collection and Storage

Digester gas produced from the anaerobic processing of sludge is collected either for use or burning to avoid odor. As it rises through the sludge, gas is collected above the digestor's liquid surface and is released. The digester gas is then piped to heating or power equipment for immediate use, a gas holder for later use, or a waste-gas burner.

The digester gas collection and distribution system must be maintained under positive pressure to avoid the possibility of an explosion if the gas should be inadvertently mixed with ambient air. Mixtures of air and digester gas containing methane concentrations ranging as low as 5% are potentially explosive. Gas storage, piping, and valve arrangements should be designed such that when the digester sludge volume changes, the gas—not air—will be drawn into the digester and will not be lost by displacement.

The anaerobic digestion process produces gas at a fluctuating rate. Therefore, when digester gas is to be used as fuel for plant operations, some method of gas storage is necessary as a balance between supply and demand. This requires a storage device with an adjustable capacity that will provide a uniform pressure for gas use equipment.

Two types of gas storage tanks in use are the gravity-type tank and the pressure-type tank. Low-pressure, floating gas-holder digester covers are gravity-type tanks that are designed to float entirely on the gas produced. These are variable-volume, constant-pressure tanks. Roller guides and positive stops are integral to the cover to provide minimum resistance to motion and an upward travel limit.

Pressure-type tanks typically are spheres that hold the gas at pressures from 140 to 700 kN/m² (20 to 100 psi), with the average ranging between 140 and 350 kN/m² (20 and 50 psi). Gas may be pumped to the pressure storage tank by a digester-gas-operated compressor.

Mixing

Most manufacturers of digester mixing equipment will suggest the appropriate type, size, and power level of mixing equipment depending on the digester volume and geometry. These suggestions are based on in-house studies and successful experiences of other similar installations. Anaerobic digesters can be mixed with gas, mechanical, or pumped mixing systems (various mixing systems have different advantages and disadvantages). Selection of a mixing system is based on costs, maintenance requirements, process configuration, and the screenings, grit, and scum content of the feed. Suggested parameters for sizing digester mixing systems include unit power, velocity gradient, unit gas flow, and digester volume turnover time. These four parameters are related and can be used to equate manufacturers' recommendations.

The *unit power* is defined as delivered motor watts per cubic meter (horsepower per 1000 ft³) of digester volume. Actual energy applied, viscosity, and digester configuration are not accounted for. Several values have been suggested for unit power selection. These values range from 5.2 to 40 W/m³ (0.2 to 1.5 hp/1000 ft³) of reactor volume.

The velocity gradient parameter as a measure of mixing intensity was suggested by Camp and Stein (1943). It is expressed as the following equation:

$$G = (W/\mu)^{1/2} \tag{15.5}$$

where

G = root-mean-square velocity gradient (s⁻¹);
W = power dissipated per unit volume, Pa·s (lb-s/ft²); and
μ = absolute viscosity, Pa·s (l-s/ft²) (for water, 720 Pa·s at 35 °C [1.5×10^{-5} lb-s/ft² at 95 °F]).

and

$$W = Z/V \tag{15.6}$$

where Z = power and V = tank volume, m³ (ft³).

Note that pressure Pa (pascals) = N/m² = Joules/m³, and Pa·s = (Joules·s)/m³.

The power can be determined from

$$Z = 2.40\ P_1\ (Q)\ \ln P_2/P_1) \tag{15.7}$$

where

Q = gas flow, m³/s (ft³/min);
P_1 = absolute pressure at surface of liquid, Pa (psi); and
P_2 = absolute pressure at depth of gas injection, Pa (psi).

These equations can be used as an approximation of the necessary power and gas flow of compressors and motors for a gas-injection system. Viscosity is a function of temperature, solids concentration, and volatile solids concentration. As the temperature increases, viscosity decreases; as the solids concentration increases, viscosity increases. In addition, as the volatile solids increase to more than 3.0%, viscosity increases. Appropriate values of the root-mean-square velocity gradient are 50 to 80 s^{-1}. The lower values can be used for a system using a single gas port, or where grease, oil, and scum are suspected as potential problems.

By rearranging the preceding equations, the unit gas flow relationship to the root-mean-square velocity gradient can be solved by

$$Q/V = G^2\mu/(P_1) \ln P_2/P_1 \qquad (15.8)$$

Gas-injection systems using draft tubes are a common type of digester mixing system in use today. They can provide enough mixing to ensure a completely mixed process. A draft-tube gas-recirculation system consists of a series of large-diameter tubes into which digester gas is released, causing the biomass to rise and mix as it approaches the liquid surface of the tank. The number of draft tubes in a digester depends on the digester dimensions. More than one draft tube is provided for digesters with diameters greater than 18 m (60 ft). Compressed gas is released inside the draft tube through top-entering lances or laterally through the draft-tube wall near the bottom of the unit. The draft tube can be installed on the roof or mounted on the digester floor using supports. The system is supported by the required compressors and controls. The three types of compressors used are the rotary lobe, rotary vane, and liquid ring. The draft tubes are constructed of steel plate; typical diameters range from 0.5 and 1.0 m (20 and 40 in.). A heating jacket can be installed around the draft tube to provide mixing and heating.

Mechanical stirring systems use rotating impellers to mix the digester contents. The mixers can be either low-speed turbines or high-speed propeller mixers installed in draft tubes. The draft tubes can be either internally or externally mounted on the digester. The flow pattern for mechanically stirred and pumped mixing systems is from top to bottom. This is in contrast to gas systems, which mix from bottom to top. Disadvantages of mechanical stirring systems include sensitivity to liquid level and clogging of impellers with rags.

In *pumped mixing systems*, externally mounted mixing pumps withdraw biomass from the top center of the tank and re-inject it through nozzles tangentially mounted at the bottom of the tank. Scum breaker nozzles are also provided at the liquid surface for intermittent use in breaking up scum accumulations. A high-flow, low-head solids-handling pump, such as an axial-flow, mixed-flow, or screw centrifugal pump, is used. Pumps often are belt-driven to permit speed changes if digester solids contents vary.

Heating
Maintaining a constant temperature in the digester improves the performance of the process. Rapid changes in temperature can lead to process upset. Controlling the temperature near an optimum value maximizes the rate of digestion, thereby minimizing the required digester volume. To maintain digester temperature at a constant value and near optimum, heat is added to raise the temperature of incoming sludge and offset the heat loss from the digester to the surrounding ground and air. The heat addition required to raise the temperature of the incoming sludge can be estimated as

$$Q_1 = W_f C_p (T_2 - T_1) \qquad (15.9)$$

where
- Q_1 = heat required, kJ/d (Btu/d);
- W_f = feed weight, kg/d (lb/d);
- C_p = specific heat of water, 4.2 kJ/kg·°C (1.0 Btu/lb/°F);
- T_1 = temperature of sludge entering digester, °C (°F); and
- T_2 = temperature of product leaving digester, °C (°F).

The heat addition required to make up for heat losses from the digester to its surroundings can be estimated by the following equation:

$$Q_2 = UA(T_2 - T_1) \qquad (15.10)$$

where
- Q_2 = rate of heat addition required to compensate for losses from the digester (kg cal/h);
- U = heat-transfer coefficient (kg cal/m²·h in °C);
- A = surface area of digester over which losses occur (m²);
- T_2 = temperature of sludge in digester (°C); and
- T_1 = temperature of surroundings (°C).

Losses across various surfaces of the digester (floor, walls exposed to earth, walls exposed to air, and cover) are generally calculated separately, then added to estimate the total rate of heat loss from the digester. In making these calculations, the temperature in the digester and the temperature of the surroundings (air, wet earth, frozen earth, and dry earth) must either be known or estimated.

Values in Tables 15.6 and 15.7 can be used to estimate the components of heat loss from the digester. Table 15.6 presents heat-transfer coefficients for various construction materials; Table 15.7 presents heat-transfer coefficients for various tank components.

Where the walls or covers are constructed of two or more materials, an effective heat-transfer coefficient can be calculated as

$$1/U_e = 1/U_1 + 1/U_2 + \ldots \qquad (15.11)$$

where U_e = effective heat-transfer coefficient and $U_{1,2,\ldots}$ = effective heat-transfer coefficients of individual materials. In calculating heat loss, it is generally assumed that all of the digester contents (gas and sludge) are at the same temperature. The temperature of the surroundings, T_2, is the temperature of either the air or the earth to which the surface is exposed.

When calculating heating requirements, a range of possible operating conditions should be considered. The capacity of the heat-transfer system should be based on the maximum probable sludge feed rate during sustained minimum temperature conditions. Calculations are based on maximum weekly sludge production during minimum weekly temperature conditions. A heating system with adequate turndown facilitates operation during average and minimum heating requirement conditions. Heat-transfer requirements also include the efficiency of the heat exchanger, which may range from 60 to 90%.

Heat-transfer coefficients for external heat exchangers range from 0.9 to 1.6 kJ/m²·°C.

Table 15.6 Heat-transfer coefficients for various construction materials (Avallone and Baumeister, 1978; *Handbook,* 1974; U.S. EPA, 1979).

Material	(kg·cal·m/m^2·h·°C)[a]	(Btu-in./hr/ft^2/°F)[b]
Concrete, uninsulated	0.25–0.35	2.0–3.0
Steel, uninsulated	0.65–0.75	5.2–6.0
Mineral wool insulation	0.032–0.036	0.26–0.29
Brick	0.35–0.75	3.0–6.0
Air space (between materials)	0.02	0.17
Dry earth	1.2	10
Wet earth	3.7	30

[a]Divide by thickness (m) to obtain kg·cal/m^2·h·°C.
[b]Divide by thickness (in.) to obtain Btu/hr/ft^2/°F.

Transfer coefficients for internal coils used for heating range from 85 to 450 kJ/m^2·°C (15 to 80 Btu/hr/ft^2/°F) depending on the solids content of the biomass.

Chemical Requirements

Chemical feed systems sometimes become necessary because of the changing quality and quantity of the feed. Changes in alkalinity, pH, sulfides, or heavy metal concentrations may necessitate incorporating chemical addition into the total process. The ability to feed certain chemicals, such as sodium bicarbonate, ferrous chloride, ferrous sulfate, lime, and alum, should be considered during the early design phase. Standby chemical metering pumps may be sufficient to satisfy operator needs. Several points of application and the associated piping necessary to accomplish feeding should be evaluated. Although pumps and other chemical feed equipment need not be installed initially, provisions to allow their future incorporation (such as pipe taps and blank flanges) should be considered during initial design. Sufficient areas should also be provided for the storage of bulk chemicals.

Table 15.7 Heat-transfer coefficients for various tank components (Avallone and Baumeister, 1978; *Handbook,* 1974; U.S. EPA, 1979).

Tank component	Typical heat-transfer coefficient[a]	
	kg-cal/m^2·h·°C	Btu/hr/ft^2/°F
Fixed steel cover, 6 mm (0.25 in.)	100–120	20–25
Fixed concrete cover, 280 mm (9 in.)	1.0–1.5	0.20–0.30
Concrete wall, 370 mm (12 in.)		
Exposed to air	0.7–1.2	0.15–0.25
Plus 25 mm (1 in.) air space and 100 mm (4 in.) brick	0.3–0.5	0.07–0.10
Concrete floor, 370 mm (12 in.)		
Exposed to dry earth, 3 m (10 ft)	0.3	0.06
Exposed to wet earth, 3 m (10 ft)	0.5	0.11

[a]Smaller values indicate greater insulating capacity.

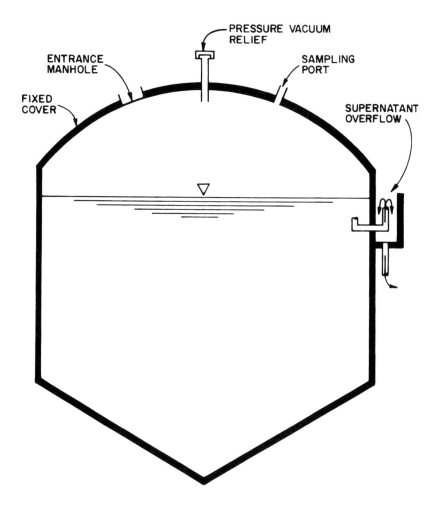

Figure 15.5 Fixed digester cover.

Digester Covers

Digester designs use covered tanks to collect gas, reduce odors, stabilize internal digester temperatures, and maintain anaerobic conditions. In addition, the cover can support mixing equipment and provide access to the tank interior. Two types of covers are available: fixed and floating.

A fixed digester cover and its appurtenances are illustrated in Figure 15.5. These are either flat or dome-shaped and are fabricated from reinforced concrete or steel. Reinforced concrete roofs are sometimes lined with polyvinyl chloride liners or steel plated to contain gas. Digester gas accumulates in the dome and is withdrawn for use or storage. Problems associated with a fixed cover can occur, especially with the introduction of air resulting in a potentially explosive gas mixture, or with the production of positive or negative pressures in the tank. This situation arises when feed and discharge rates are dissimilar, the supernatant lines clog, or pressure- or vacuum-relief valves fail.

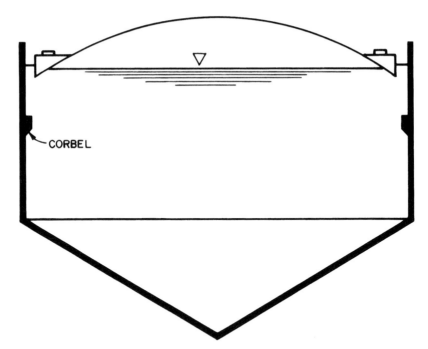

Figure 15.6 Floating digester cover.

Floating-type covers are divided into the following two classifications: those that rest on the liquid surface and those that rest on side skirts and float on the gas (gas holders). The Wiggins-type floating cover (see Figure 15.6) may have a large area above the liquid surface available for gas collection. To accomplish this, a buoyant force is developed on the outside rim of the cover that acts as a pontoon. The Downes-type floating cover (see Figure 15.7) is constructed to limit the space above the liquid surface by increasing the area of cover contact with the surface. Weights are primarily used to increase the flotation weight of the cover for increased gas pressure or to balance offset loads on the cover created by cover-mounted equipment. A common application for floating-type covers is on primary digesters to facilitate process control by separating filling and drawing operations and to provide for better scum control through scum submergence. A disadvantage of floating covers is their potential to tilt in cases of severe foaming.

Gas-holder covers are designed to provide additional gas storage space. Gas-storage space allows for variations in gas production and use rates. Gas-holder covers are modified floating covers designed to float on gas rather than on liquid (see Figure 15.8). Modifications include an extended rim skirt to contain the gas and a special guidance system to provide stability as the cover floats on gas. This type of cover must be designed to operate under lateral wind loads and resultant overturning forces.

Tank Configuration and Construction

Anaerobic digesters have been built in rectangular, square, cylindrical, and egg-shaped configurations. Rectangular tanks are used at locations where site availability is a problem. They are the least expensive to construct but are difficult to operate because of the tendency for dead zones to develop from poor mixing characteristics.

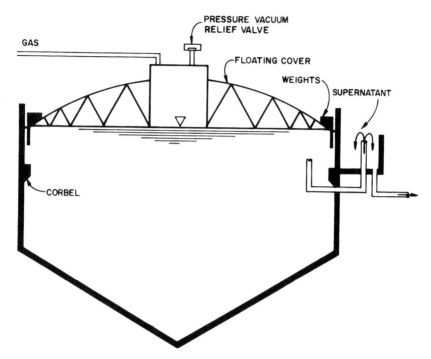

Figure 15.7 Cylindrical tank with Downes-type floating cover.

A common configuration is a low vertical cylinder with a conical floor. These circular tanks are constructed of reinforced concrete, with vertical sidewall depths ranging from 6 to 14 m (20 to 45 ft) and diameters ranging from 8 to 40 m (25 to 125 ft). Conical-shaped bottoms, with slopes varying between 1:3 and 1:6, are preferred for cleaning purposes. Slopes greater than 1:3, although desirable for grit removal, are difficult to construct and difficult to work on during cleaning. However, milder slopes are too flat to provide any significant improvement over flat-bottom designs. The floors can either contain one central withdrawal pipe or be divided into pie-like sections, each equipped with a separate withdrawal pipe (waffle-bottom digester). The latter are more costly to construct than traditional conical designs but may reduce cleaning costs and frequency. Where necessary, cylindrical digesters have been insulated using brick veneer and an air space, earth fill, polystyrene plastic, fiber glass, or insulation board.

An improved tank configuration is the egg-shaped digester (see Figure 15.9). The steeply sloped top and bottom cones help eliminate scum and grit buildup problems, in turn eliminating or reducing the necessity for digester cleaning. Mixing requirements of an egg-shaped digester are considerably lower than for a conventional shallow cylindrical vessel. In the latter, most of the mixing energy applied goes to maintaining grit in suspension and to control scum buildup.

There are three basic types of mixing systems used with egg-shaped digesters: unconfined gas mixing, mechanical mixing using an impeller and a draft tube arrangement, and external pumped recirculation (see Figure 15.10). Most egg-shaped digesters are also fitted with a gas lance or hydraulic jet near the bottom of the lower cone section to facilitate occasional stirring of grit that may have accumulated. Any or all of the mixing

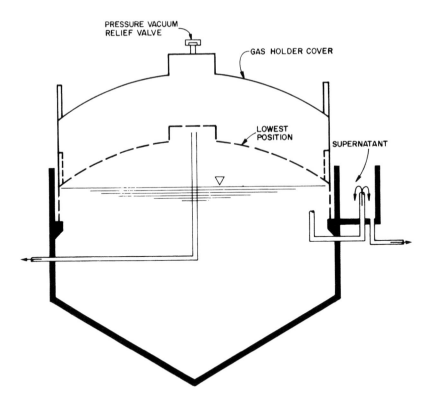

Figure 15.8 Floating gas-holder cover.

systems may be provided in any one digester, and all can be operated during any one day, although gas mixing and mechanical draft-tube mixing are seldom practiced at the same time. Egg-shaped digesters can be constructed of steel or concrete. The outside surfaces are insulated and covered with aluminum cladding for protection and aesthetics (Figure 15.11).

Effect of Digestion on Dewatering and Dewatering Recycle Streams
Anaerobic digestion decreases the mass of sludge requiring dewatering. However, the product that is produced is ordinarily more difficult to dewater than undigested sludge, primarily because digestion reduces flocculent characteristics and increases the concentration of dispersed, nonflocculent particles. Dewatering of anaerobically digested sludge produces a high-strength liquid recycle stream. Of the recycle stream, the most important constituent requiring subsequent treatment is total Kjeldahl nitrogen (primarily in the form of ammonia) and, to a lesser degree, BOD and suspended solids. Hydrogen sulfide can also be important.

Anaerobic digestion converts 50 to 60% of the particulate Kjeldahl nitrogen contained in the sludge into ammonia. Most of this ammonia is contained in the liquid recycle stream from the dewatering process. This can significantly affect liquid process design and operation, particularly if the plant has ammonia or nitrogen removal requirements or otherwise must be operated to nitrify.

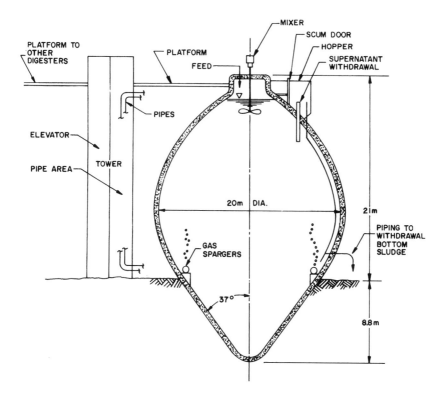

Figure 15.9 Egg-shaped anaerobic digester.

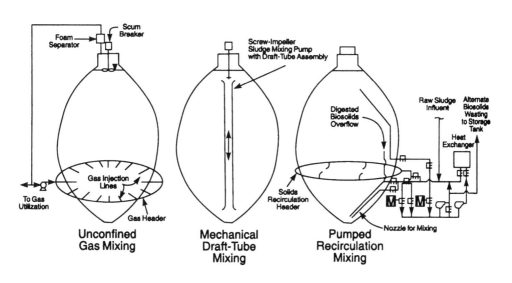

Figure 15.10 Mixing systems for egg-shaped anaerobic digesters.

Chapter 15 Sludge Stabilization

Figure 15.11 Egg-shaped digester installation.

Hydrogen sulfide in the recycle stream can also cause operating problems in downstream biological treatment units. Excessive growth of sulfur-oxidizing bacteria has been observed, particularly in fixed-film systems in plants where large recycle flows from anaerobic digestion occur.

Aerobic Digestion

Stabilization during the aerobic digestion process occurs from the destruction of degradable organic components and the reduction of pathogen organisms by aerobic, biological mechanisms. The aerobic digestion process is a suspended-growth biological treatment process and is based on biological theories similar to those of the extended aeration modification of the activated-sludge process.

The objectives of the aerobic digestion process, which can be compared with those of the anaerobic digestion process, include production of a stable product by oxidizing organisms and other biodegradable organics, reduction of mass and volume, reduction of pathogen organisms, and conditioning for further processing. Advantages of the aerobic process compared with anaerobic digestion are

- Production of an inoffensive, biologically stable product;
- Lower capital costs;

- More simple operational control with reductions in volatile solids concentrations slightly less than those achieved in the anaerobic digestion process;
- Safer operation with no potential for gas explosion and less potential for odor problems; and
- Discharge of a supernatant with a 5-day BOD (BOD_5) concentration less than that found in the anaerobic process. In addition, it is less prone to upsets and less susceptible to toxicity.

The primary disadvantage attributed to the aerobic digestion process is the higher power cost associated with oxygen transfer. Recent developments in the aerobic digestion process, such as highly efficient oxygen-transfer equipment and research into operation at elevated temperatures, may reduce this concern. Other disadvantages cited include the reduced efficiency of the process during cold weather, the inability to produce a useful byproduct, such as methane gas, from the anaerobic process, and the mixed results achieved during mechanical dewatering of aerobically digested solids.

Aerobic Digestion Theory

Aerobic digestion is based on the biological principle of endogenous respiration. Endogenous respiration occurs when the supply of available substrate (food) is depleted and microorganisms begin to consume their own protoplasm to obtain energy for cell maintenance reactions.

During the digestion process, cell tissue is oxidized aerobically to carbon dioxide, water, and ammonia or nitrates. Because the aerobic oxidation process is exothermic, a net release of heat occurs during the process. Although the digestion process should theoretically go to completion, in actuality only 75 to 80% of the cell tissue is oxidized. The remaining 20 to 25% is composed of inert components and organic compounds that are not biodegradable. The material that remains after the full completion of the digestion process exists at such a low energy state that it is essentially biologically stable. Consequently, it is suitable for a variety of disposal options.

The aerobic digestion process actually involves two steps: direct oxidation of biodegradable matter and subsequent oxidation of microbial cellular material by organisms. These processes are described by

$$\text{Organic matter} + NH_4^+ + O_2 \xrightarrow{\text{bacteria}} \text{Cellular material} + CO_2 + H_2O \quad (15.12)$$

$$\text{Cellular material} + O_2 \xrightarrow{\text{bacteria}} \text{Digested sludge} + CO_2 + H_2O + NO_3^- \quad (15.13)$$

These equations describe first the oxidation of organic matter to cellular material and then how this material is subsequently oxidized to digested solids. The process represented by the second equation is typical of the endogenous respiration process and is the predominant reaction in aerobic digestion systems.

Because of the need to maintain the process in the endogenous respiration phase, aerobic digestion is used to stabilize waste activated sludge. The inclusion of primary solids in the process can shift the overall reaction because they contain little cellular material. Most of the organic and particulate material in primary sludge constitutes an external food source for the active biomass contained in the biological sludge. Therefore, longer retention times are required to accommodate the metabolism and cellular growth that must occur before endogenous respiration conditions are achieved.

Using the formula $C_5H_7NO_2$ as representative of the cell mass of a microorganism, the stoichiometry of the aerobic digestion process can be represented as

$$C_5H_7NO_2 + 5O_2 \rightarrow 5CO_2 + 2H_2O + NH_3 + \text{Energy} \quad (15.14)$$

$$C_5H_7NO_2 + 7O_2 \rightarrow 5CO_2 + 3H_2O + NO_3^- + H^+ + \text{Energy} \quad (15.15)$$

The first equation represents a system designed to inhibit nitrification; nitrogen appears in the form of ammonia. The stoichiometry of a system in which nitrification occurs is represented by the second equation, where nitrogen appears in the form of nitrates.

The destruction of organic matter containing nitrogen results in the formation of one mole of alkalinity per mole of nitrogen released. This is because the nitrogen is released as ammonia (NH_3), which combines with carbon dioxide to form ammonium bicarbonate. Thus, if one begins with organic nitrogen it is possible to nitrify and denitrify with no net destruction of alkalinity (Daigger and Bailey, 2000).

Practically, approximately 50% of the alkalinity consumed by nitrification can be recovered by denitrification. If excessive pH depression is a problem (as a result of alkalinity consumption by nitrification), it may be possible to control this problem by periodic denitrification or by the addition of lime or sodium bicarbonate. Denitrification can be accomplished by periodically turning off the aerators while continuing to mix the digester if the facility is designed with a draft-tube aerator containing an air sparger.

Nitrification during the aerobic digestion process increases the concentration of hydrogen ions and subsequently decreases pH if there is insufficient buffering capacity in the sludge. As in the activated-sludge process, approximately 7 kg of alkalinity is destroyed per kilogram of ammonia oxidized (7 lb/lb). The pH may drop as low as pH 5.5 during long aeration times; however, the aerobic digestion process does not seem to be adversely affected.

Approximately 1.5 kg of oxygen is required per kilogram of active cell mass (1.5 lb O_2/lb) in the nonnitrifying system, whereas 2 kg of oxygen per kilogram of active cell mass (2 lb O_2/lb) is required when nitrification occurs. Actual oxygen requirements for the aerobic digestion process depend on factors such as operating temperature, inclusion of primary solids, and the solids residence time in the activated-sludge system.

Aerobic Digestion Design

Numerous variables govern the design of conventional aerobic digestion systems—those systems operating at temperatures between 20 and 30 °C that use air as the oxygen source for biological activity. The variables include desired reduction in volatile solids, quantities and characteristics of the influent, process operating temperature, oxygen-transfer and mixing requirements, tank volume and detention time, and method of system operation.

Reduction in Volatile Solids

The primary purpose of the aerobic digestion process is to produce sludge amenable to various disposal options. Reductions ranging from 35 to 50% in the levels of volatile solids are obtainable through the aerobic digestion process. Recall from the discussion on stabilization, however, that the reduction in volatile solids during the aerobic digestion process may not be a valid indication of stabilization.

Feed Quantities and Characteristics

Because the mechanism of the aerobic digestion process is similar to the activated-sludge process, the same concerns regarding variations in influent characteristics and levels of biologically toxic materials apply, although a dampening effect will occur as a result of upstream treatment processes. The accumulation of heavy metals in activated sludge by precipitation and adsorption that can occur at a pH level greater than pH 7 can result in toxicity in the aerobic digester from resolubilization of the heavy metals under low-pH conditions present in the digestion process.

Operating Temperature

The operating temperature of the aerobic digestion system is a critical parameter in the process. A frequently cited disadvantage of the aerobic process is the variation in process efficiency that results from changes in operating temperature. Changes in operating temperatures are closely related to ambient temperatures because most aerobic digester systems use open tanks.

In colder climates, diffused aeration is used for aerobic digestion because this reduces surface turbulence and hence heat loss. Aerobic digestion systems are operated in the temperature ranges from 10 to 40 °C. The effects of temperature can be estimated by

$$(K_d)_T = (K_d)_{20} \beta^{T-20} \tag{15.17}$$

where
K_d = reaction rate constant (time);
β = temperature coefficient; and
T = temperature (°C).

The reaction rate constant indicates the destruction rate of volatile solids during the digestion process. An increase in the reaction rate constant occurs with an increase in the temperature of the system and implies an increase in the digestion rate. Temperature coefficients ranging from 1.02 to 1.10 have been reported, with a temperature coefficient of 1.05 as the average.

Oxygen-Transfer and Mixing Requirements

The biological reaction that occurs during the aerobic digestion process requires oxygen for the respiration of cellular material in activated sludge and, in the case of mixtures with primary solids, the oxygen needed to convert organic matter to cellular material. In addition, proper operation of the system requires adequate mixing of the contents to ensure proper contact of oxygen, cellular material, and organic matter (food source). Because the introduction of oxygen to maintain the biological process provides a mixing action, these parameters are interrelated.

In aerobic digestion systems that strictly treat biological solids, the need for adequate mixing will govern the capacity of the oxygenation equipment. Systems treating primary and biological mixtures require additional oxygen for the biological oxidation process, and in many cases, this requirement will govern mixing equipment size.

Tank Volume and Retention Time Requirements

The required volume of an aerobic digestion system is governed by the retention time necessary to achieve a desired reduction in volatile solids. Retention time required to achieve a 40 to 45% reduction in volatile solids ranges from 10 to 12 days at an operating temperature of approximately 20 °C. The total required aeration time is significantly dependent on operating temperature and biodegradability of the sludge. Significant deterioration occurs in dewaterability of aerobically digested sludge in systems with unusually long detention times. However, many state agencies adopted the U.S. Environmental Protection Agency's (U.S. EPA's) standards for aerobic digestion that requires a retention time of 60 days at 15 °C and 40 days at 20 °C and focuses on pathogen reduction rather than on volatile solids reduction.

One equation for estimating the required volume of a continuously operating aerobic digester is

$$V = Q_i(X_i + YS_i)/X(K_d P_v + 1/\theta_c) \tag{15.18}$$

where
- V = volume of the aerobic digester (L);
- Q_i = influent average flowrate (L/d);
- X_i = influent suspended solids (mg/L);
- Y = portion of the influent BOD consisting of raw primary solids (%);
- S_i = influent digester BOD_5 (mg/L);
- X = digester suspended solids (mg/L);
- K_d = reaction rate constant (d^{-1});
- P_v = fraction of volatile suspended solids in the digester (% of suspended solids); and
- θ_c = solids retention time (d).

The term YS_i can be disregarded if no primary solids are included in the load to the aerobic digester. This equation should not be used to compute digester volumes in systems where significant nitrification will occur.

Benefield and Randall (1980) note that a portion of the volatile solids are non-biodegradable and a portion of the nonvolatile solids are solubilized. They suggest that the required digester hydraulic retention time can be calculated as

$$\bar{t} = (X_i X_e) + (Y)(S_a)/(K_d)(D)(X_{oad})(X_i) \tag{15.19}$$

where
- $\bar{t}$ = required digester hydraulic retention time (d);
- X_i = suspended solids concentration in the influent (mg/L);
- X_e = suspended solids concentration in the effluent (mg/L);
- Y = yield coefficient for organic content of the primary solids;
- S_a = ultimate BOD of the primary solids in the digester;
- K_d = reaction rate constant for the biodegradable portion of the active biomass (d^{-1});
- D = biodegradable active biomass in the influent that appears in the effluent (%); and
- X_{oad} = percentage of active biomass that is biodegradable in the influent.

If the aerobic digestion system is treating only biological solids, the yield coefficient for organic content of the primary solids and the ultimate BOD of the primary solids in the digester are equal to zero, and the other parameters are representative of the secondary sludge. Treatment of a mixture of primary and secondary sludge requires the inclusion of the yield coefficient for organic content of the primary solids and the ultimate BOD of the primary solids in the digester to reflect the conversion of organic matter to cellular material. The inclusion of additional factors for the primary component is based on the premise that the presence of these solids slows the destruction rate of cellular material, while the organic matter is converted to cellular material. The other terms in the equation should reflect the characteristics of the mixture.

Process Variations

Several variations to the standard mesophilic air oxygenation aerobic digestion system have been investigated in recent years. The more notable variations are high-purity oxygen aeration, autothermal thermophilic digestion, and low-temperature digestion.

High-Purity Oxygen Aeration

This modification of the aerobic digestion process substitutes high-purity oxygen for air. High-purity-oxygen systems are relatively insensitive to changes in ambient air temperatures because of the increased rate of biological activity and the exothermic nature of the process.

While one variation of this modification uses open tanks, aerobic digestion using high-purity oxygen is done in closed tanks similar to those used in the high-purity-oxygen activated-sludge process. The use of the high-purity-oxygen aerobic digestion system in enclosed tanks will result in higher operating temperatures because of the exothermic nature of the digestion process.

The primary disadvantage of this modification is the increased cost associated with oxygen generation. As a result, high-purity-oxygen aerobic digestion is cost effective only when used in conjunction with the comparable activated-sludge process. The use of high-purity oxygen also decreases the amount of carbon dioxide that would be added in the standard air-oxygenated system. Consequently, neutralization may be required to offset the reduced buffering capacity of the system.

Low-Temperature Aerobic Digestion

The operation of aerobic digestion systems at lower temperature ranges (less than 20 °C) has been studied to provide better operational control for small package-type wastewater treatment plants in northern climates. Investigations have shown that the solids retention time must be increased as operating temperatures decrease to maintain an acceptable level of suspended solids reduction (Koers and Mavinic, 1977; Mavinic and Koers, 1979).

Autothermal Thermophilic Aerobic Digestion

Jewell and Kabrick (1978) were the first to use the term *autothermal thermophilic aerobic digestion*, or *ATAD*. The process often achieves cost and space savings over other conventional process such as anaerobic or conventional aerobic digestion. With the U.S. EPA regulations on sludge disposal it became competitive with other processes as a result of its ability to produce a pathogen-free sludge. Such a product has potentially unrestricted use and wider marketing opportunities than unpasteurized sludge.

Theory of Autothermal Thermophilic Aerobic Digestion

The bulk of the heat in this process comes, as the name implies, from the energy released from oxidation of the volatile components of the sludge. Fats provide the highest heat release at 4.3×10^4 kJ/kg oxidized volatile solids followed by protein at 2.4×10^4 kJ/kg, carbohydrates at 1.9×10^4 kJ/kg and biomass at 1.7×10^4 kJ/kg. The mix of these organics in the sludge feed will affect the efficiency of the process to generate heat, the digestion time needed, the supplemental mixing energy needed and the odor compounds generated. The process therefore is not completely autothermal as the name implies.

It the process is a batch operation, temperatures can exceed 50 °C for a short period. If the system is semicontinuous batch operation or a series of reactors, the thermophilic temperatures will be restricted to the last or middle reactors. All of the reactors and reaction times are not thermophilic.

The process is not completely aerobic and includes fermentation and anaerobic as well as aerobic respiration. The addition of the air provides the oxygen or electron acceptor for the primary heat production and the formation of carbon dioxide and water. Fermentation and anaerobic oxidation are necessary prerequisite biochemical reactions that provide the low molecular weight substrate for the aerobic process.

Microorganisms in the feed sludge are killed by the high temperature before being reassimilated and undergoing further endogenous oxidation. The digestion does not have to go to completion if only 38% destruction of volatile solids is required, as specified by U.S. EPA regulations. Undigested organic biomass will therefore remain in the final product. The product is characterized primarily as cellulose and contains incomplete oxidized products, but in contrast to other mesophilic digestion processes, the product is a complete transformation of the original sludge.

Design of Autothermal Thermophilic Aerobic Digestion

The autothermal thermophilic aerobic digestion process must be integrated to the complete wastewater treatment process that includes complementary influent screening, primary sludge collection, biological sludge collection, thickening, sludge storage, dewatering, and final reuse programs as well as the plant odor control strategy.

The feed to the digesters should be above a minimum organic strength to ensure heat production. This is typically measured in chemical oxygen demand and total solids. A prethickener is needed to achieve the minimum concentrations necessary. Total volatile solids concentration may be used as a surrogate measure for chemical oxygen demand, and total solids is useful as a measure for mixing energy needs. A minimum total volatile solids of 3 to 4% (COD concentrations of 40 to 60 g/L) and a total solids of less than 7% are typical low and high boundaries for operation. Higher or lower concentrations of either require increasingly higher mixing energy.

Because of the higher temperatures, substrate diffusion, solubility of solids, and rates of absorption are increased, and viscosity and gas solubility are decreased, which lessens the reaction time to achieve stabilization. The reactors can be one-sixth to one-half the volume of conventional aerobic or anaerobic digesters.

Selection of the reactor number depends on the seasonal fluctuation in hydraulic loading and the selected operation of the digestion process. Unless a prefeed storage tank is available to equalize sludge flow, peak week or peak day loading should be used to size the reactors and select the required number. Where large fluctuations occur, such as at resort areas, three or more reactors are desirable. A normal design for peak conditions

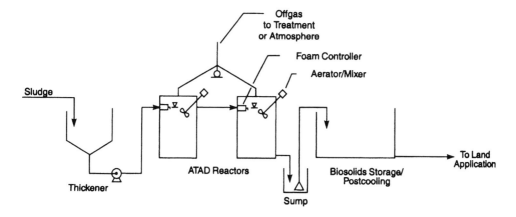

Figure 15.12 Typical autothermal thermophilic aerobic digestion system.

is to ensure the hydraulic residence time is at least above six days. Longer periods may be needed depending on the waste character or desired final product. A hydraulic residence time of 10 days is a good design value for average conditions. A heat exchanger is needed for cold climates to ensure the feed sludge is above 20 °C. Caution is needed in designing for cold-thickened waste biological sludge because viscosity is several orders of magnitude above that of water.

Following digestion, a final storage tank may be needed. Size and number depend on the disposal or use plans and the dewatering plant operation. The final reactor may function as a heat source for preheating the feed sludge in a multireactor configuration. Where the sludge is to be dewatered the solids should be cooled to 30 to 40 °C to reduce vapor loss and odor, and to potentially improve dewatering efficiency.

The autothermal digestion system has four styles of operation. Operation is simplest for a continuous or semicontinuous lineup of reactors and becomes more complex for a full or semibatch lineup. A continuous system allows displacement of solids from one reactor to another, as shown in Figure 15.12. This system will not meet U.S. EPA pathogen reduction regulations because a possibility exists that a pathogen could short-circuit the process. A semicontinuous operation requires storage tanks upstream and downstream to store sludge before daily feed periods and after transfers. A full batch process requires 6 to 10 days of storage upstream of a single reactor facility. Storage may be reduced by one day per reactor added. By contrast, the semibatch process incorporates the storage to the lineup and allows continuous flow to the first reactor and batch operation to the downstream reactors. The latter process allows variable liquid depth. Otherwise, semicontinuous or batch reactors are required to be taken on- or off-line to meet variations in flow.

Equipment includes aspirator-mixers, pump-aerator combinations and diffused air systems. To avoid early failure, all operating systems and equipment requiring servicing should not interfere with or be installed in the reactor vessels. All equipment and piping must be erosion–corrosion resistant. Foam control systems are necessary and should be backed up. Tanks may be protected steel or concrete and should be insulated throughout. Insulation efficiency will depend on the climate. Automation can control operation within 1 °C of set point in the pasteurizing vessel (Kelly and Wong, 1997). Aeration and mixing energy will depend on loading and hydraulic residence time. Air must be introduced as a fine-bubble mixture to achieve the high oxygen transfer needed. In a quiescent tank, fine-bubble airflow of less than 0.1 m³/h air/m³ of reactor volume in a liquid stream only is needed to float a 4% sludge feed. In the reactors under high mixing energy, floatation has been observed at over 2 m³/h air/m³ of reactor volume. Airflow requirements will depend on COD oxidation needed but is 0.5 to 1.5 m³/h air/m³ of reactor volume. The first reactor has the highest air requirements and should be sized to achieve a 30% COD reduction.

The reaction rate for a continuously stirred reactor is modeled by

$$S_{in}/S_{out} = [1 + K_T V_e/Q]^{-N} \qquad (15.20)$$

where
- S_{in} = COD or VSS into the first reactor (mg/L);
- S_{out} = COD or VSS out of the reactors (mg/L);
- $K_T = K_{20}\, \beta^{T-20}$;
- β = temperature coefficient (1.03);
- T_{20} = 0.025 d⁻¹;
- V_e = effective volume of each reactor (m³);
- Q = daily sludge feed rate (m³/d); and
- N = number of reactors in series, assuming reactors are of equal volumes.

Where volumes are different, extend $1 + K_T V_e/Q$ to add other reactors as multipliers and drop the exponent N.

Air requirements will depend on the efficiency of the fine-bubble aeration equipment but may be assumed at 50 to 70%. For an 8-day hydraulic retention time and a 43% COD reduction and assuming a COD/VSS ratio of 1.86:1, the volumetric airflow rate required is calculated at approximately 0.6 to 1.0 v/v-h. A higher airflow rate can be used but increases heat loss, especially where the reactor operates at thermophilic temperatures because much of the heat will be lost through latent heat of evaporation. Coarse-bubble aeration should not be used for this reason. Oxygen is assumed to be 23.2% of air by weight and air is 1.3 kg/m³.

Mixing energy is needed to provide a maximum interaction between the thermophiles and the substrate. It will also provide supplemental heat. A minimum shear gradient G of 450 s⁻¹ for complete mix requires a power density of at least 100 W/m³ of reactor volume at 50 to 60 °C.

Operation of Autothermal Digestion Plants

Operating temperatures in the reactors will vary depending on the operating strategy to be achieved. For pasteurization, the central or final reactor of a multistage facility must be held above 55 °C. The initial reactor may vary upwards from 45 °C.

When proteins and fats in a concentrated waste stream are fermented and oxidized, the products are essentially carbon dioxide and water as well as incompletely oxidized odorous and non-odorous compounds. The odorous compounds are protein building blocks and lower-molecular-weight fatty acids. The protein building blocks are known to be sulfur-based compounds including hydrogen sulfide, mercaptans and methlylated sulfur compounds, and nitrogen-based compounds including amines and ammonia. The lower-molecular-weight fatty acids include acetate, propionate, butyrate, and valerate. Other odorous compounds include alcohols, double alcohols, benzenols, aldehydes, ketones, esters, and other incompletely oxidized products. These are naturally occurring substances but are found in high concentrations. Because under normal reactor conditions the odor precursors are continuously being formed, the digestion would need to essentially be taken to completion to eliminate odor. This is rarely practical and therefore a containment and treatment system is essential.

Nitrifying tower oxidation, ozone, UV, two- and three-stage wet chemical scrubbers, activated carbon dry scrubbers, and soil biofilters have all been used to treat the off gas. They have all proven successful but like all systems are limited to operational capacities and system reliability. The advantage of this system is that the reactors are entirely contained and odorous off gases can be removed and treated. Odors can be significant, however, and the treatment requires appropriate consideration.

After digestion, cooling is sometimes necessary for effective thickening and enhanced supernatant quality. Twenty days of cooling is recommended. The feed must contain a minimum volatile solids content of 25 g/L (2.5%) and 40 g/L COD. Volatile solids contents less than this amount will have difficulty achieving autothermal conditions. Feed from treatment plants without primary clarifiers operating with food/microorganism ratios (F/M) as low as 0.1:0.15 is still suitable for the autothermal digestion process. However, retention times of 15 days or less should be maintained so volatile solids reduction via endogenous respiration in the biological reactors is kept to a minimum. The process can be used on low F/M feed solids but may require higher feed concentration or, if the heat content of the feed is low, an external preheat source via heat exchangers within the reactors.

Either the raw wastewater or the solids should undergo fine screening (6- to 12-mm spacing [0.25- to 0.5-in. spacing]) to eliminate inerts, plastics, and rags from the reactor. Macerating has been used, although it is not recommended as a replacement for screening. In cases where heat exchangers are used, maceration may be considered as a supplement to a good prescreening facility. Good grit removal from the wastewater is also necessary to reduce wear on aeration devices and grit buildup in the reactors. Retention times of 4 to 30 days have been reported (Kelly, 1991).

Composting

Composting is a biochemical method of stabilization that prepares wastewater residuals for beneficial use as a soil conditioner. It is a self-heating process that destroys pathogens and produces a material similar to soil humus. Well-stabilized compost can be stored indefinitely and has minimal odor even if rewetted. Compost is suitable for a variety of end uses, such as landscaping, topsoil blending, potting, and growth media, and can be distributed to the general public for gardening. It can also be used in agriculture for soil erosion control, improvement of soil physical properties, and in revegetation of disturbed lands. Local markets may be developed in urban and nonagricultural areas as well as in agriculture and mine revegetation.

During composting, a medium is used that is dry enough to provide pore spaces with free air but wet enough to sustain biological activity—35 to 50% solids. Porosity is provided by mixing dewatered cake with a bulking agent or amendment, such as wood chips. Because of the bulking agent, the volume of compost product is equal to or greater than the volume of dewatered cake. For a given amount of dry solids, the volume of material to be composted increases with the decreasing percentage of solids because of the greater amount of moisture.

Raw, digested, or chemically stabilized solids can be composted. The process has also been proven effective for organic residuals from the paper, pharmaceutical, and food-processing industries. The bulking agent or amendment can be a wide variety of materials including other wastes such as wood wastes, yard waste, paper, agricultural residue, wood ash, and animal bedding.

Composting is a simple process that can be performed outdoors in most climates. Because of a desire to more efficiently operate the process, control odors, and reduce operating costs, many facilities are constructed under structures, in fully enclosed buildings, or in entirely mechanized facilities.

Theory of Composting

Microbiology

Three primary categories of microorganisms involved in the composting process are bacteria, actinomycetes, and fungi. The bacteria are responsible for a significant portion of organic matter decomposition. Initially, at mesophilic temperatures (lower than 40 °C [104 °F]), bacteria metabolize carbohydrates, sugars, and proteins. At thermophilic temperatures (higher than 40 °C), they decompose protein, lipids, and the hemicellulose fractions. In addition, they are responsible for much of the heat energy produced.

Fungi are also present at both the mesophilic and thermophilic temperature ranges. Chang (1967) indicated that mesophilic fungi metabolize cellulose and other complex carbon sources. Their activity is similar to the actinomycetes. Both the fungi and actinomycetes are found in the exterior portions of compost piles. Golueke (1977) suggests that this phenomenon is related to the aerobic nature of the organisms because most of the fungi and actinomycetes are obligate aerobes.

Energy Balance

Heat is generated by the conversion of organic carbon to carbon dioxide and water vapor. The fuel is provided by the rapidly degraded volatile fraction of the solids supplemented by the amendment. Heat is primarily removed by evaporative cooling through aeration and agitation. To a lesser extent, heat is lost at the pile surface. The process temperature will not rise if the rate of heat loss exceeds the rate of heat generation. Haug (1980) concludes that if W, defined as

$$W = \frac{\text{Water evaporated}}{\text{Volatile solids destroyed}} \quad (15.21)$$

is below 8 to 10, sufficient energy should be available for heating and evaporation. If W exceeds 10, the mix will remain cool and wet. This generalization is based on heat of vaporization and does not consider the effect of ambient conditions on evaporation and surface cooling.

Carbon/Nitrogen Ratio

Microorganisms use carbon and nitrogen in proportions fixed by the composition of the microbial biomass. The ideal ratio of available carbon to nitrogen is in the range of 25:1 to 35:1. If the carbon/nitrogen ratio is less than 25:1, excess nitrogen will be released as ammonia, resulting in a loss of nutrient value and the emission of ammonia odor. If the carbon/nitrogen ratio exceeds 35:1, organic material will break down more slowly and remain active well into the curing stage. The carbon/nitrogen ratio of wastewater residuals is in the range of 5:20. The amendment and bulking agent provide supplemental carbon, improving both the energy balance and the carbon/nitrogen ratio of the mixture.

The calculation of the carbon/nitrogen ratio is complicated by the fact that some of the carbon becomes available more slowly than the nitrogen. If wood chips are used as a bulking agent, only a thin surface layer of the wood provides available carbon. Carbon in sawdust is more rapidly available.

Process Objectives

The primary objective of composting is to produce a fertilizerlike product that can be beneficially used. The compost must meet regulatory and public health requirements and be attractive for some end use. This primary objective is met through the following process objectives: pathogen reduction, maturation, and drying.

Pathogen Reduction

Pathogenic organisms found in wastewater residuals fall into five primary groups: bacteria, viruses, protozoa cysts, helminthic (parasitic worm) ova, and fungi. The first four groups are often termed *primary pathogens* because they can invade normally healthy persons and cause diseases. The last group, the fungi, are referred to as *secondary pathogens* because they only infect persons with weakened respiratory or immune systems.

Elevated temperature is one of the most effective methods of destroying pathogens. The temperature within a composting pile or vessel may not be uniform because of variations in heat loss, mix characteristics, and airflow. Composting in cases where temperatures reach the thermophilic range should eliminate practically all viral, bacterial, and parasitic pathogens. However, some fungi, such as *Aspergillus fumigatus*, are thermotolerant and therefore survive the composting process.

Maturation

The term *maturation* refers to the conversion of the rapidly biodegradable components in the organic material and amendment to substances similar to soil humus that slowly decompose. Compost that is insufficiently mature will reheat and generate odors in storage and upon rewetting. It may also inhibit seed germination by generating organic acids and inhibit plant growth by removing nitrogen as it decomposes in the soil. The term *stability* in composting refers to the reduction in the rate of microbial degradation of the biodegradable components in the mixture as the compost matures.

There are a number of testing methods and standards for measuring compost stability or maturity. At the present time, no single test is universally accepted. The standards associated with each test are still tentative, and much work needs to be done to correlate test results with odor generation and plant growth. A complete assessment of maturity may require more than one type of test.

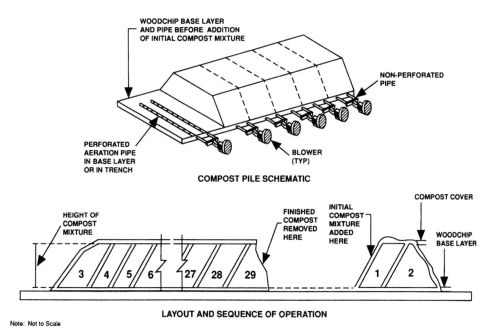

Figure 15.13 Aerated static pile composting system.

Drying
Water vapor is removed during composting, increasing the solids content of the mixture from approximately 40 to 55%. Drying is critical in processes that include screening because screens do not perform well on compost with less than 50 to 55% solids. Drying is achieved by providing sufficient aeration or agitation to make the most of the removal of water vapor.

Design of Composting Systems

Process Alternatives
Two basic options exist for composting of sludge: windrow (turned or aerated) and mechanical composting. Windrow composting without aeration requires periodic turning in order for the piles to remain aerobic. The aerated pile composting system (Figure 15.13) is a fundamental improvement over the turned windrow. The air blown into the static pile and the amount of air controls the temperature within the windrow. The disadvantages of this system are the amount of land area required and the problem with odor production. A typical mechanical composting system is shown in Figure 15.14. There are many types of mechanical composting systems available, but the cost is generally prohibitive and the ultimate use of the compost has to be firmly established before an investment in such a unit can be justified.

Bulking Agents and Amendments
The aerated static-pile process and some in-vessel processes require mixes with sufficiently high porosity to be aerated with low-pressure blowers. For these processes, bulking agents are required. Examples include wood chips, chipped brush, chipped waste lumber, and shredded tires. Leaves and paper have been used but are not effective.

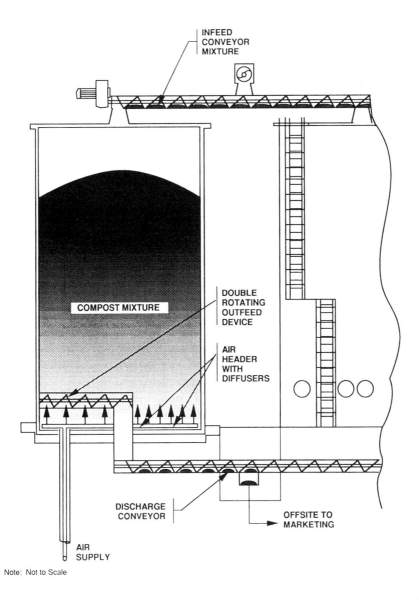

Figure 15.14 Vertical plug-flow composting reactor.

Bulking agents adjust the solids content of the mixture, provide supplemental carbon to adjust the carbon/nitrogen ratio and energy balance, and provide structural integrity to maintain porosity as the mixture is stacked.

A number of mixing technologies are used in composting operations. Front-end loaders have been used but result in poor mixing. Batch mixers are mobile or stationary, self-unloading, and are well suited for small- and medium-sized facilities. For each batch, the correct proportions of dewatered cake and bulking agents are loaded in and mixed by internal flights. In *continuous mixers*, the dewatered cake and bulking agents are continuously fed to the mixer from live-bottom bins and automatically propor-

Chapter 15 Sludge Stabilization

tioned. Pug mills and plough mixers are examples of commonly used continuous mixers. This system is more complex than the batch mixer but well suited to medium- and large-sized facilities. Windrow turners, specially built for composting, are most suitable for windrow composting operations.

Temperature Control and Aeration

Aeration removes heat and water vapor and supplies oxygen to the microorganisms. As the rate of airflow is increased in a forced aeration system, the pile temperature decreases and the rate of water vapor removal increases. Agitation quickly releases heat and water vapor and also enhances aeration by improving porosity. Without sufficient aeration, the pile temperature may exceed 70 °C (158 °F), which is detrimental to microbial activity. The optimum temperature range for volatile solids destruction is 40 to 50 °C (approximately 105 to 120 °F). In addition, 40 to 50 °C is optimal for the removal of water vapor because high rates of airflow are required to maintain low temperatures in a highly active process. To ensure pathogen destruction, the temperature should be more than 55 °C for some specified time (often about two weeks), depending on the type of composting process used.

Retention Time

The time period required to stabilize the organic material is divided between an active composting stage and a curing stage. Aerated composting requires 20 days of aeration followed by 30 days of unaerated curing. Most horizontal agitated-bed systems are designed for 21 days of aerated composting followed by curing. However, other in-vessel systems use shorter active composting times (often 14 days).

Alkaline Stabilization

In recent years, a number of advanced alkaline stabilization technologies have emerged that either use chemical additives in addition to, or instead of, lime, or involve special equipment or processing steps. These processes all claim advantages over traditional lime stabilization, including enhanced pathogen control and a more publicly acceptable product. The product of advanced alkaline stabilization technologies is sometimes referred to as *artificial soil* because it has been successfully used as a soil substitute. This section discusses traditional and advanced alkaline stabilization technologies.

Lime is the most widely used and one of the lowest cost alkaline materials available in the wastewater industry. Lime has been used for reducing odors in privies, increasing pH in stressed digesters, removing phosphorus in advanced wastewater treatment, treating septage, and conditioning before and after mechanical dewatering.

The purposes of alkaline stabilization may include the following: to substantially reduce the number and prevent the regrowth of pathogenic and odor-producing organisms, thereby preventing any health hazard associated with the biosolids; to create a stable product that can be stored; and to reduce the short-term leaching of metals from biosolids not incorporated with natural soil.

The alkaline stabilization process is a simple one. An alkaline chemical is added to raise the pH of the feed, and adequate contact time is provided. At pH 12 or higher, with sufficient contact time and homogeneous lime and feed mixing, pathogens and microorganisms are either inactivated or destroyed. Chemical and physical characteristics of the solids produced are also altered by the reactions with the alkaline material. The chemistry of the process is not well understood, although it is believed that some complex molecules are split by reactions such as hydrolysis and saponification (Christensen, 1982).

Both small and large treatment plants have used lime stabilization as the primary stabilization process. However, lime stabilization is more common for small facilities. It is more cost effective than other chemical stabilization options. Some large plants have used lime stabilization as an interim process when the primary stabilization process (such as anaerobic or aerobic digestion) has been temporarily out of service. Lime stabilization has also been used to supplement the primary stabilization process during peak sludge production periods.

Theory of Alkaline Stabilization

Lime stabilization depends on maintaining the pH at a high enough level for a sufficient period of time to inactivate the microorganism population of the sludge. This halts or substantially retards the microbial reactions that can otherwise lead to odor production and vector attraction. The process can also inactivate viruses, bacteria, and other microorganisms that are present.

The lime stabilization process involves a variety of chemical reactions that alter the chemical composition of the sludge. The following equations, simplified for illustrative purposes, show the types of reactions that occur:

Reactions with inorganic constituents include

Calcium: $$Ca^{2+} + 2HCO_3^- + CaO \rightarrow 2CaCO_3 + H_2O \quad (15.22)$$

Phosphorus: $$2PO_4^{-3} + 6H^+ + 3CaO \rightarrow Ca_3(PO_4)_2 + 3H_2O \quad (15.23)$$

Carbon dioxide: $$CO_2 + CaO \rightarrow CaCO_3 \quad (15.24)$$

Reactions with organic constituents include

Acids: $$RCOOH + CaO \rightarrow RCOOCaOH \quad (15.25)$$

Fats: $$Fat + CaO \rightarrow Fatty\ acids \quad (15.26)$$

Initially, lime addition raises the pH of the sludge. Then, reactions occur such as those in the aforementioned equations. If insufficient lime is added, the pH decreases as these reactions take place. Therefore, excess lime is required.

Biological activity produces compounds such as carbon dioxide and organic acids that react with lime. If biological activity in the sludge being stabilized is not sufficiently inhibited, these compounds will be produced, reducing the pH and resulting in inadequate stabilization.

Given sufficient information about a particular sludge, it theoretically would be possible to calculate the lime requirement to raise the pH to a given value.

If quicklime (or any compound high in quicklime) is added to sludge, it initially reacts with water to form hydrated lime. This reaction is exothermic and releases approximately 15,300 cal/g·mol (2.75×10^4 Btu/lb/mol). The reaction between quicklime and carbon dioxide is also exothermic, releasing approximately 43,300 cal/g·mol (7.8×10^4 Btu/lb/mol).

These reactions can result in substantial temperature rise, particularly in sludge cake with a low moisture content. For example, adding 45 g (0.1 lb) of quicklime per gram of solids to a cake containing 15% solids concentration can result in a temperature increase of more than 30 °C (50 °F)

$$(0.1 \text{ lb CaO}) (1 \text{ lb mol}/56 \text{ lb}) (27{,}500 \text{ Btu/lb mol}) = 49 \text{ Btu} \qquad (15.27)$$

$$(49 \text{ Btu}) (1/0.85 \text{ lb H}_2\text{O}) (1 \text{ °F/lb H}_2\text{O/Btu}) = 58 \text{ °F} \qquad (15.28)$$

Temperature increases in actual practice will be less, although they can be substantial. In some cases, they can be enough to contribute to pathogen destruction during lime stabilization.

Design of Alkaline Stabilization Systems

There are several different alkaline stabilization technologies available. Each system has advantages and disadvantages; therefore, it is important to evaluate and select a process that will produce the desired product on a site-specific basis. The following subsections describe several alkaline stabilization processes, from traditional liquid lime stabilization to newer, emerging dry lime stabilization and advanced technologies.

Liquid Lime Stabilization

Liquid lime or prelime stabilization involves the addition of a lime slurry to liquid sludge to meet stabilization requirements. For wastewater treatment plants that practice land application of liquid sludge, such as subsurface injection on agricultural land, lime is added to the thickened solids. This practice has been limited to smaller facilities or those with short hauling distances to land application or use sites.

Solids or septage conditioning with lime before dewatering is a second method of liquid lime stabilization. Lime is combined with other conditioners, such as aluminum or iron salts, to achieve enhanced dewatering. This method has primarily been used with vacuum filters and recessed-plate filter presses. Stabilization is complementary in these situations because the lime dose for conditioning exceeds the dose required for stabilization. Table 15.8 shows the range of liquid lime stabilization dosages required to maintain pH 12 for 30 minutes.

Table 15.8 Lime dosage required for liquid lime stabilization[a] (U.S. EPA, 1979).

Type of sludge	Average solids concentration, %	Average lime dosage, lb calcium hydroxide/lb[b] dry solids	Average pH Initial	Final
Primary sludge[c]	4.3	0.12	6.7	12.7
Waste activated sludge	1.3	0.30	7.1	12.6
Anaerobically digested combined	5.5	0.19	7.2	12.4

[a]Dose required to maintain pH 12 for 30 minutes.
[b]lb/lb × 1000 = g/kg.
[c]Includes waste activated sludge.

Dry Lime Stabilization

Dry lime or postlime stabilization involves the addition of dry quicklime or hydrated lime to dewatered cake. The lime is mixed with the cake using a pug mill, plow blender, paddle mixer, ribbon blender, screw conveyor, or similar device.

Either quicklime, hydrated lime, or other dry alkaline materials can be used for dry lime stabilization, although the use of hydrated lime is limited to smaller installations. Quicklime is less expensive and easier to handle than hydrated lime. Additionally, the heat of hydrolysis released in slaking the quicklime when it is added to dewatered cake can enhance pathogen destruction.

Sludge Combustion

Sludge combustion is either the total or partial conversion of organic solids to oxidized products (primarily carbon dioxide and water), or the partial oxidation and volatilization of organic solids by starved air combustion to products with a residual caloric value. The primary objective of thermal destruction is to reduce the quantity of solids or sludge requiring disposal. The term *incineration* refers to high-temperature thermal destruction in the presence of excess air. *Starved air combustion* is thermal destruction that restricts airflow (oxygen) to produce three potentially energy-rich products: gas, oil or tar, and a char.

Although wastewater sludges generally are organic, they will sustain combustion without auxiliary fuel addition only if sufficient quantities of water have been removed. Hence, dewatering should always precede thermal destruction. Dewatered cake with 20 to 30% solids can be burned with auxiliary fuels. Dried cake with 30 to 50% or higher total solids is *autogenous* (i.e., it can be burned without auxiliary fuel).

The term *combustion*, or *burning*, is the rapid combination of oxygen with fuel resulting in the release of heat. The normal combustible elements of wastewater sludge—carbon, hydrogen, and sulfur—chemically combine in the organic sludge as grease, carbohydrates, and protein. The combustible portion of the sludge has a heat content approximately equal to that of lignite coal. The products of complete combustion are carbon dioxide, water vapor, sulfur dioxide, and inert ash. Good combustion requires proper proportioning and thorough mixing of fuel and air and the initial and sustained ignition of the mixture.

Gaseous products from the combustion of wet organic sludges consist of dry combustion gases, excess air, and water vapor from the moisture in the sludge and the oxidation of hydrogen. The heat content of dry combustion gases is determined by multiplying the individual weight of each gas by its specific heat content at the exit temperature. The weight of moisture in the exit gases resulting from the combustion of hydrogen usually is combined with the weight of the moisture in the wet feed to calculate the total heat content of the total water vapor exiting in the combustion gases. If supplemental fuel is used, the volumes and heat contents of the stack gases resulting from fuel and sludge combustion must be separately calculated.

Sludge Heat Values

The composition and quantity of fuels to be burned are fundamental inputs for the design of any combustion system. The composition determines the fuel's heating value, and the fuel's quantity determines the required size of the unit.

> **Note on Terminology**
>
> One of the earliest measures of heat energy, still widely used by American engineers, is the *British thermal unit*, or Btu, defined as the amount of energy necessary to heat one pound of water one degree Fahrenheit. The internationally accepted unit of energy is the *Joule*. Other common units for energy are the *calorie*, and *kilowatt-hour*. These are all interchangeable and are used by different groups of scientists and engineers.
>
> The most accurate measure of heat value is obtained with the use of the *calorimeter*, a stainless steel bomb immersed in water in which the sample of the material is combusted under 30 atmospheres of pure oxygen. The rise in temperature of the surrounding water is then measured and the heat value calculated.
>
> An important aspect of calorimetric heat values is the distinction between *higher heat value* (HHV) and *lower heat value* (LHV). The higher heat value is also called *gross calorific heat value* or sometimes the *gross heating value* while the lower heat value is also called the *net heating value*. The distinction is important in engineering combustion work.
>
> In a calorimeter, as organic matter combusts, the products of combustion are (ideally) carbon dioxide and water. The water produced is in vapor form. As the calorimeter bomb cools in the water jacket, the water vapor in the steel bomb condenses, yielding heat that is measured as part of the temperature rise. The higher heat value is calculated by including the contribution due to this latent heat of vaporization, or the heat required to produce steam from water. But such condensation does not occur in large furnaces such as sludge combustors. The hot flue gases carry the water vapor outside the furnace, and condensation cannot take place. The lower heat value is the higher heat value minus the latent heat of vaporization that has occurred in the bomb calorimeter. For design purposes, the lower heat value is a much more realistic number, but American engineers typically express heat values in terms of higher heat value, HHV. When heat values of sludge or other fuels are specified, such as in Table 15.9, these are (at least in the United States) higher heat values unless otherwise noted.

The heat value of the feed sludge represents the quantity of heat that can be released per unit mass of solids. The fuel's heat value, the prime indicator of combustion potential, depends on its carbon, hydrogen, and sulfur fractions. Carbon burned to carbon dioxide has a heat value of 3.4×10^4 kJ/kg (1.46×10^4 Btu/lb), and hydrogen has a gross heat value of 14.4×10^4 kJ/kg (6.2×10^4 Btu/lb); sulfur has a gross heat value of 1.0×10^4 kJ/kg (0.45×10^4 Btu/lb). Consequently, any changes in the carbon, hydrogen, or sulfur content of a fuel will raise or lower the sludge heat value. Table 15.9 lists some heat values of representative residuals from wastewater treatment. Note that the highest heat values are for grease and scum. Also note the significant drop in heat value once a sludge has been stabilized by digestion. It makes no sense therefore to digest sludge ahead of combustion.

Process Alternatives

The five different types of thermal destruction processes discussed in this section include the following types of furnaces: multiple hearth, fluidized bed, infrared, combination multiple-hearth fluidized bed, and slagging.

Table 15.9 Representative heat values of some wastewater residuals.

Material	Combustibles, %	High heating value, Btu/lb[a] of dry solids
Grease and scum	88	16 700
Raw wastewater solids	74	10 300
Fine screenings	86	9000
Ground garbage	85	8200
Digested sludge	60	5300
Chemical precipitated solids	57	7500
Grit	33	4000

[a]Btu/lb × 2.326 = kJ/kg.

Multiple-Hearth Furnace

A cross section of a typical multiple-hearth furnace is shown in Figure 15.15. The multiple-hearth furnace comprises a vertically oriented, cylindrically shaped, refractory-lined steel shell containing a series of stacked horizontal refractory shelves or hearths. A hollow cast-iron rotating shaft runs through the center of the hearths. Rabble arms are attached to the center shaft above each hearth and may have air flowing through them for cooling. Teeth on each arm may be angled in the direction of rotation (forward rabbling) or the teeth may be reversed (back rabbling). Back rabbling increases the detention time of the cake and improves drying. Combustion air enters at the bottom of the hearth. The combustion air circulates upward through the drop holes in the hearths, countercurrent to the sludge flow. The air exhausts through the top hearth to the waste heat exchanger (if present) and then to the air pollution control equipment.

Flow of solids generally is downward through the furnace; however, the movement is a combination of outward and inward radially descending travel on successive hearths that is interspersed by drops of elevation through either peripheral holes in the hearth or an annular opening at the center. The radial travel is caused by the sweeping action of metal blades (rabble teeth) set at an angle in the horizontal arms mounted in the center shaft. The movement occurs in a series of discrete steps, each tooth in turn acting on a given parcel of solids. Thus, the solids are pushed forward and radially several times on each hearth and come to rest after each plowing action. In addition to moving the sludge, the rabbling process cuts, furrows, turns, and opens the sludge, thereby exposing new surfaces to the drying process. The effective surface area for drying is estimated to be 130% of the plan area of the hearth (Niessen, 1988). Dry ash discharged from the bottom hearth of the furnace is handled in either a wet or dry form.

The multiple-hearth furnace includes three zones: the drying zone, combustion zone, and cooling zone. Most of the sludge water evaporates in the upper hearths or drying zone. The heat-transfer mechanisms for the drying hearths are approximately 15% convection and 85% radiation (Lewis and Lundberg, 1988). The temperature rises through the drying zone from 430 to 760 °C (800 to 1400 °F). The dry product combustibles are burned at temperatures of 760 to 930 °C (1400 to 1700 °F) in the central hearths or combustion zone. The volatile gases and solids burn in the upper middle hearths, and the fixed carbon burns in the lower middle hearths. Incoming combustion air cools ash to 90 to 200 °C (200 to 400 °F) in the lowest hearths or ash cooling zone; ash exits the bottom of the furnace.

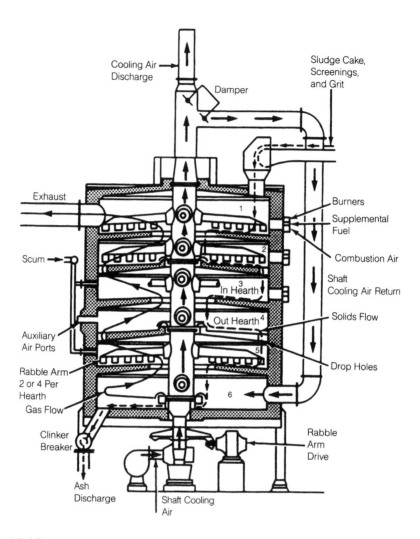

Figure 15.15 Typical multiple-hearth furnace.

Multiple-hearth furnaces have been successfully used for thermal destruction of various sludges. Advantages reported for multiple-hearth furnaces include minimum space requirements, reliability, ease of operation, effectiveness, reduction of sludge volume, and production of a sterile ash. Disadvantages reported for multiple-hearth furnaces include high maintenance and labor requirements, high energy use, air pollution problems, odor problems, high capital cost, and system complexity.

Fluidized-Bed Furnace

In a fluidized-bed combustion furnace the sludge solids are set in fluid motion (fluidized) in an enclosed space (fluidized-bed zone) by passing combustion air through the fluidized-bed zone in such a way as to set all particles in that zone in a homogeneous boiling motion (Figure 15.16). In this state, the particles are separated from each other by an envelope of fluidizing gas (air for combustion), thereby presenting an extended surface for a gas-to-solid reaction such as air to carbon or hydrogen. It is this

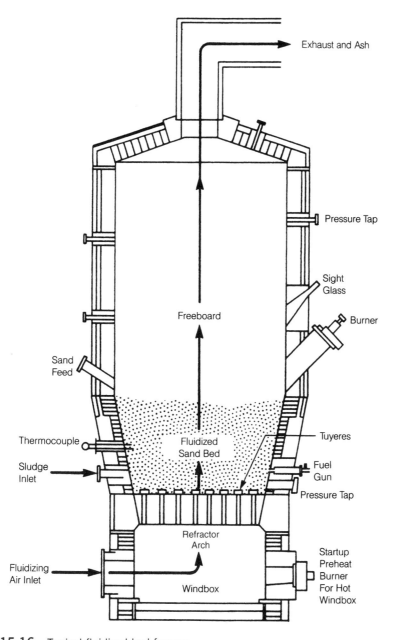

Figure 15.16 Typical fluidized-bed furnace.

extended combustion surface that makes the high thermal efficiency found in most fluidized-bed combustors possible. Capacity is, therefore, a function of the total bed area reactor, although it is usually expressed as either fluidized-bed surface area or freeboard area and the maximum gas velocity permitted because of particle entrainment. At combustion equilibrium, the fluidized bed resembles a boiling liquid and obeys most hydraulic laws. The dispersion of fluidizing gas throughout the fluidized-bed zone by the specially designed orifice plate ensures complete mixing. Temperature variations from any one spot in the fluidized bed to any other spot typically will not exceed 6 to 8 °C (10 to 15 °F).

Design of Sludge Combustion Systems

Selection and design of a thermal destruction system for municipal sludge is a complex, technical, and highly specialized task. Design procedures include empirical methods that rely on extensive pilot-plant kinetic data. Such information is generally regarded as proprietary and may be patented by the manufacturers who will perform the design calculations after receiving input data, conditions, and functional specifications from the operating agency.

A multiple-hearth combustion furnace should be able to achieve satisfactory combustion of dewatered cake and concentrated scum within an exhaust gas temperature range of approximately 430 to 760 °C (800 to 1400 °F) and burn dewatered cake and concentrated scum to a sterile ash with no more than 5% combustibles in the ash. The combustion will be efficient so as to limit emissions with an opacity that does not exceed 20%, particulates that will not exceed 0.65 g/kg (1.3 lb/ton) of dry solids input, and the temperature of the exhaust gases that will not exceed 67 °C (120 °F) at an oxygen content of the flue gas in the range of 6 to 10%.

A fluidized-bed furnace typically is feasible for plants with flows greater than 38,000 m^3/d (10 mgd); a multiple-hearth furnace typically is economical at plant flows greater than 76,000 m^3/d (20 mgd). Departures from these general rules may result from site-specific factors such as plant location, land availability for other reuse or disposal options, and air emission regulations. Fluidized-bed furnaces generally are more economical than multiple-hearth furnaces at installations that require high-temperature combustion of exhaust gases because multiple-hearth furnaces require afterburners to achieve high temperatures.

Emissions Control

Inadequately designed or operated thermal processing facilities for wastewater sludge can significantly contribute to air pollution. Two categories of air contaminants associated with thermal processing are odors and combustion emissions.

Odors, the most noticeable form of air pollution, result in approximately half of all citizen complaints to local air pollution control agencies. Odors from thermal processing are particularly offensive. The quantity and quality of combustion emissions from the thermal processing depend on the thermal processing method used, sludge composition, and auxiliary fuel composition.

Pollution Control Technology

Particulates, the predominant air contaminants from thermal processing, include both solid particles and liquid droplets (excluding uncombined water) that are swept along by the gas stream or formed through condensation of the flue gas. Particulates from sludge combustion are enriched with volatile trace metals such as cadmium, lead, and zinc. Particle sizes are mostly smaller than 2 µm, and volatile elements can be found primarily in the submicron particles. Technologies for controlling particulates from gas streams include mechanical collectors, wet scrubbers, fabric filters, and electrostatic precipitators. Selection of a particulate collection system depends on the nature of the particulate matter, conditions of the gas stream, and emission limits.

Conclusions

Engineers who design wastewater treatment plants are often viewed by the public they serve as the bad guys. Why is it that technology does not have a higher standing in the eyes of the public?

Consider this example: engineer Kurt Prufer used his engineering skill and common sense to build the ovens at Auschwitz, Germany, eventually expanding them and the associated ancillary processes to be able to gas and incinerate over 2000 people a day (Vesilind, 1994). Prufer did a technically excellent job, and certainly he did nothing illegal. He simply followed the dictates of his superiors, his clients, and society. He trusted "others to assess those consequences ... beyond the engineering task" (deRubertis, 1994). If nothing else, Prufer was politically correct by the standards of his society which he felt he was serving.

We do not need such an extreme example to illustrate that engineers, by virtue of their training and skill, can and often do cause much more direct and irreparable harm to people, cultures, and the environment. Nothing gets built without engineers. Engineers are an indispensable component of a civilized society, so it can be argued that they have the most clear and compelling responsibility to do "the right thing"—which may not be what the client wants.

Engineers who take this view might say, "Yes, if all of us *did* decide to undertake only engineering projects that we felt were the right thing to do, then some things would *not* get built." What would have happened if Adolf Hitler could not have found a single engineer to design and build the ovens at Auschwitz?

References

Avallone, E. A.; Baumeister, T. (Eds.) (1978) *Marks' Standard Handbook for Mechanical Engineers*, 8th ed.; McGraw-Hill: New York.

Benefield, L. D.; Randall, C. W. (1980) *Biological Process Design for Wastewater Treatment*; Prentice-Hall, Inc.: Englewood Cliffs, New Jersey.

Burd, R. S. (1968) *A Study of Sludge Handling and Disposal*; Publication No. WP-20-4; Federal Water Pollution Control Administration, U.S. Department of the Interior: Washington, D.C.

Buswell, A. M.; Neave, S. L. (1939) Laboratory Studies of Sludge Digestion—Illinois. *State Water Surv. Bull.*, **30**.

Camp, T. R.; Stein, P. C. (1943) Velocity Gradients and Internal Work in Fluid Motion. *J. Boston Soc. Civ. Eng.*, **203**.

Chang, Y. (1967) The Fungi of Wheat Straw Compost: Part II—Biochemical and Physiological Studies. *Trans. Br. Mycol. Soc.*, **50**, 667.

Christensen, G. L. (1982) Dealing with the Never-Ending Sludge Output. *Water Eng. Manage.*, **129**, 25.

Daigger, G. T.; Bailey, E. (2000) Improving Aerobic Digestion by Prethickening, Staged Operation, and Aerobic-Anoxic Operation: Four Full-Scale Demonstrations. *Water Environ. Res.*, **72**, 260.

deRubertis, K. (1994) Discussion of *Why Do Engineers Wear Black Hats?* by P. A. Vesilind. *J. Prof. Issues Eng. Educ. Pract.*, ASCE, **119** (1), 330.

Eastman, J. A.; Ferguson, J. F. (1981) Solubilization of Particulate Organic Carbon during the Acid Phase of Anaerobic Digestion. *J. Water Pollut. Control Fed.*, **53**, 352.

Farrell, J. B.; et al. (1986) *Microbial Destructions Achieved by Full-Scale Anaerobic Digestion*; EPA-600/D-86-1031; U.S. Environmental Protection Agency, Cincinnati, Ohio.

Garber, W. F. (1982) Operating Experience with Thermophilic Anaerobic Digestion. *J. Water Pollut. Control Fed.,* **54,** 1170.

Golueke, C. G. (1977) *Biological Reclamation of Solid Waste;* Rodale Press: Emmaus, Pennsylvania.

Handbook of Chemistry and Physics (1974) 55th ed.; CRC Press: Boca Raton, Florida.

Haug, R. T. (1980) *Compost Engineering;* Ann Arbor Science Publishers: Ann Arbor, Michigan.

Jewell, W. J.; Kabrick, M. (1978) Autoheated Aerobic Thermophilic Digestion with Air Aeration. *Proceedings of the 51st Annual Conference of the Water Pollution Control Federation,* Anaheim, California, October 1–6.

Kelly, H. G. (1991) Autothermal Thermophilic Aerobic Digestion: A Two Year Appraisal of Canadian Facilities. *Proceedings of the Environmental Engineering Specialty Conference of the American Society of Civil Engineers,* Reno, Nevada.

Kelly, H. G.; Wong, V. (1997) Stability in Autothermal Thermophilic Aerobic Digestion. *Proceedings of the Water Environment Federation and American Water Works Association Specialty Conference,* Philadelphia, Pennsylvania, August.

Koers, D. A.; Mavinic, D. S. (1977) Aerobic Digestion of Waste Activated Sludge at Low Temperatures. *J. Water Pollut. Control Fed.,* **49,** 460.

Lee, K. M.; Brunner, C. A.; Farrell, J. B.; Erlap, A. E. (1989) Destruction of Enteric Bacteria and Virus During Two-Phase Digestion. *J. Water Pollut. Control Fed.,* **61,** 1421.

Lewis, F. M.; Lundberg, L. A. (1988) Modifying Existing Multiple-Hearth Incinerators to Reduce Emissions. *Proceedings of the National Conference of Municipal Sewage Treatment Plant Sludge Management,* Palm Beach, Florida.

Mavinic, D. S.; Koers, D. A. (1979) Performance and Kinetics of Low Temperature, Aerobic Sludge Digestion. *J. Water Pollut. Control Fed.,* **51,** 2088.

Niessen, W. R. (1988) Thermal Processing of Wastewater Treatment Plant Sludges. *Proceedings of the National Conference of Municipal Sewage Treatment Plant Sludge Management,* Palm Beach, Florida.

Parkin, G. F.; Owen, W. F. (1986) Fundamentals of Anaerobic Digestion of Wastewater Sludge. *J. Environ. Eng.,* **112,** 5.

Storey, G. W. (1987) Survival of Tapeworm Eggs, Free and in Progottids, during Simulated Sewage Treatment Processes. *Water Res.,* **2,** 199.

Stukenberg, J. R.; Clark, J. H.; Shimp, G. F.; Sandino, J.; Crosse, J. T. (1992) Compliance Outlook: Meeting the 503 Pathogen Reduction Criteria with Anaerobic Digestion. *Proceedings of the Water Environment Federation 65th Annual Conference and Exposition,* New Orleans, Louisiana, September 20–24.

U.S. Environmental Protection Agency (1979) *Process Design Manual for Sludge Treatment and Disposal;* EPA-625/4-78-012; Office of Technology Transfer: Washington, D.C.

Vesilind, P. A. (1994) Closure to *Why Do Engineers Wear Black Hats?* by P. A. Vesilind. *J. Prof. Issues Eng. Educ. Pract.,* ASCE, **119** (1), 331.

Symbols Used in this Chapter

A = surface area of digester over which losses occur (m²)

b = endogenous decay coefficient (d^{-1})

C_p = specific heat of water, 4.2 kJ/kg·°C (1.0 Btu/lb/°F)

G = root-mean-square velocity gradient (s^{-1})

G_m = volume of methane produced, m³/d (ft³/d)

G_v = volume of total gas produced, m³ (ft³)

G_{sgp} = specific gas production, taken as 0.8 to 1.1 m³/kg (13 to 18 ft³/lb) of volatile solids destroyed

k	= maximum specific substrate use rate, g COD/g VSS·d
K_c	= half saturation concentration of biodegradable substrate in feed sludge (g COD/L)
K_d	= reaction rate constant (d^{-1})
M_{sgp}	= specific methane production per mass of organic material (BOD or COD) removed, m^3/kg (ft^3/lb)
N	= number of reactors in series, assuming reactors are of equal volumes
P_v	= fraction of VSS in the digester (% of suspended solids)
P_1	= absolute pressure at surface of liquid, Pa (psi)
P_2	= absolute pressure at depth of gas injection, Pa (psi)
Q	= daily sludge feed rate (m^3/d)
Q	= gas flow, m^3/s (ft^3/min)
Q_i	= influent average flowrate (L/d)
Q_1	= heat required, kJ/d (Btu/d)
Q_2	= rate of heat addition required to compensate for losses from the digester (kg cal/h)
S_{eff}	= concentration of biodegradable substrate in digested sludge (mg/L)
S_i	= influent digester BOD$_5$ (mg/L)
T	= temperature (°C)
T_1	= temperature of sludge entering digester, °C (°F)
T_1	= temperature of surroundings (°C)
T_2	= temperature of product leaving digester, °C (°F)
T_2	= temperature of sludge in digester (°C)
U	= heat-transfer coefficient, kg cal/m^2·h in °C
U_e	= effective heat-transfer coefficient
$U_{1,2,...}$	= effective heat-transfer coefficients of individual materials
V	= tank volume, m^3 (ft^3)
V	= volume of the aerobic digester (L)
V_e	= effective volume of each reactor (m^3)
V_s	= volatile solids destroyed, kg (lb)
W	= power dissipated per unit volume, Pa·s (lb-s/ft^2)
X	= digester suspended solids (mg/L)
X_{out}	= COD or VSS out of the reactors (mg/L)
X_{in}	= COD or VSS into the first reactor (mg/L)
X_i	= influent suspended solids (mg/L)
Y	= portion of the influent BOD consisting of raw primary solids (%)
Y	= yield of anaerobic organisms resulting from growth (g VSS/g COD destroyed)
Z	= power
β	= temperature coefficient
ΔCOD_R	= COD removed across digester
ΔOR	= organics (COD or BOD) removed daily, kg/d (lb/d)
ΔX	= biomass produced kg/d (lb/d)
μ	= absolute viscosity, Pa·s (l-s/ft^2) (for water, 720 Pa·s at 35 °C [1.5×10^{-5} lb-s/ft^2 at 95 °F])
θ_c	= solids retention time (d)
θ_{min}	= limiting solids retention time for required digester performance

16

Beneficial Use and Ultimate Disposal

Introduction

While the effluent from wastewater treatment is discharged to the most appropriate watercourse, there is no such outlet for the sludge, and the disposal of these residues of wastewater treatment has been one of the most difficult aspects of plant operation. In this chapter, a number of options are discussed for sludge disposal or biosolids use. (Recall that *sludge* becomes *biosolids* if it is treated in a way to achieve compliance with federal regulations)

Designing sludge disposal as a part of the overall plant design is difficult. Not only do the regulations change, but the people who agree to accept the sludge products may change. It is important for engineers and operators to work together to develop several options for sludge and biosolids disposal, using the one that is of the greatest benefit at any given time.

In conjunction with the trend toward beneficial use of wastewater solids compared with disposal, the design of any sludge or biosolids project must take into account the quality of the sludge to be disposed of and the prevailing local and state regulations as well as the federal regulations.

In most cases, the selection of both sludge treatment and biosolids disposal is governed by disposal regulations. The regulation governing sludge disposal in the United States is the 40 CFR 503 regulations promulgated by the U.S. Environmental Protection Agency (U.S. EPA).

The 40 CFR Part 503 Regulations

The 40 CFR Part 503 regulations (U.S. EPA, 1994) address the use and disposal of wastewater solids generated from the treatment of domestic wastewater and septage in the United States. Wastewater sludge and other wastewater solids disposed in municipal solid waste landfills or used as landfill cover material must comply with the requirements of the 40 CFR Part 258 municipal solid waste landfill regulations, although they do not fall under the Part 503 regulations. For the most part, the rule was written to be self implementing, which means that citizen suits or the U.S. EPA can enforce the regulations before permits are issued. The standards are incorporated into National Pollutant Discharge Elimination System (NPDES) permits issued by the U.S. EPA or permits issued by states with approved solids management programs.

Part 503 regulations do not address ocean disposal of wastewater solids. Although ocean disposal at one time was an acceptable practice in the United States and was practiced widely by communities on the Atlantic coast, the U.S. Congress passed the Ocean Dumping Ban Act in 1988 making dumping an unlawful practice after 1991. While there has been some controversy regarding the scientific bases and environmental effects of ocean disposal, this practice has ceased to be an option in the United States. In other countries, ocean disposal may be an option if studies indicate that the environmental effects are negligible or beneficial. However, ocean disposal is not discussed in this book because it is not allowed in the United States.

Land Application

Land application includes all forms of applying bulk or bagged biosolids to land for beneficial uses at agronomic rates (that is, rates designed to provide the amount of nutrients including nitrogen, phosphorus, potash, and so on needed by the crop or vegetation grown on the land while minimizing the amount that passes below the root zone). These beneficial-use practices include application to the following: agricultural land, such as fields used for the production of food, feed, and fiber crops; pasture and range land; nonagricultural land such as forests; public-contact sites such as parks and golf courses; disturbed lands such as mine spoils, construction sites, and gravel pits; and home lawns and gardens.

The two most important characteristics of sludge that both limit its use as well as make its disposal on land beneficial are the heavy metals and the nutrients. Results from a study of sludges in Ontario, Canada, are shown in Tables 16.1 and 16.2. Note the high variability of metals and nutrients, emphasizing the point that in the design of solids disposal, the characteristics of the sludges must be known. In the absence of such information, conservative design parameters must be estimated.

Responsibility for complying with the Part 503 rule rests with the person who prepares biosolids for land application or applies biosolids to the land. The regulation establishes two levels of quality with respect to heavy-metal concentrations: *pollutant ceiling concentrations* and *pollutant concentrations* (high-quality biosolids), and two levels of quality with respect to pathogen densities—*Class A* and *Class B*. In addition, the regulation establishes two types of approaches for meeting vector (i.e., flies, rodents, and birds)-attraction reduction: processing or the use of physical barriers. Under the Part 503 regulations, fewer restrictions are imposed on the use of higher quality biosolids.

To qualify for land application, biosolids or wastewater-sludge-derived material must meet at least the pollutant ceiling concentrations, Class B requirements for pathogens,

Table 16.1 Heavy metal concentration in Ontario sludges (Antonic et al., 1974).

Component	Primary plants[a] Anaerobically digested sludges			Secondary plants[b] anaerobically and aerobically digested sludges		
	Range, mg/L^{-1}	Mean[c], mg/L^{-1}	Standard deviation, mg/L^{-1}	Range, mg/L^{-1}	Mean, mg/L^{-1}	Standard, deviation mg/L^{-1}
Zinc	2.8–130	74.3	48.3	4–225	55.5	57.4
Copper	4.6–150	54.5	46.9	7–148	34.8	30.2
Nickel	0.7–15	4.4	4.7	0.26–16.8	6.5	14.9
Chromium	2–68	16.5	20.9	2–430	41.6	51.7
Lead	11–86	40.9	30.3	3.7–60	21.8	25.1
Cadmium	0.2–2.6	0.7	0.7	0.1–8.7	1.4	2.0
Cobalt	<0.6–1.4	1.0	0.3	0.3–3.6	0.8	0.8

[a]Number of plants = 10.
[b]Number of plants = 30.
[c]Arithmetic mean.

and vector-attraction reduction requirements. A number of general requirements and management practices apply to biosolids that are land applied unless they meet criteria for pollutant concentration limits, Class A pathogen requirements, and vector-attraction reduction requirements. In all cases, however, the minimum frequency of monitoring, recordkeeping, and reporting requirements must be met.

Because of the continuing concern with sludge disposal, many farmers are insisting on accepting only Class A sludge, forcing wastewater treatment plant operators to modify their sludge treatment schemes.

Table 16.2 Nutrient characteristics of Ontario sludges (Antonic et al., 1974).

Constituent	Primary plants[a] Anaerobically digested sludge			Secondary plants[b] Anaerobically digested sludge			Secondary plants[c] Aerobically digested or waste activated sludge		
	Range, mg/L^{-1}	Mean, mg/L^{-1}	Standard deviation, mg/L^{-1}	Range, mg/L^{-1}	Mean, mg/L^{-1}	Standard deviation, mg/L^{-1}	Range, mg/L^{-1}	Mean, mg/L^{-1}	Standard deviation, mg/L^{-1}
Total solids—TS%	2.8–12.5	8.8	2.9	2.0–12	4.1	1.8	2.2–4.5	2.75	0.95
Volatile solids—VS%	24–61	43.4	10.5	36–70	41	8.5	41–69	55.8	9.9
Ammonia nitrogen—N	100–590	326	78	250–1200	628	245	20–180	110	24.8
Total Kjeldahl nitrogen—N	950–2900	1736	913	1600–3000	2114	495	650–2300	1358	576
Total phosphorus—P	240–2600	713	399	390–2900	975	603	440–1200	730	303

[a]Number of plants = 10.
[b]Number of plants = 25.
[c]Number of plants = 8.

Table 16.3 Land application pollutant limits (dry-weight basis).

Pollutant	Ceiling concentration limits,[a] mg/kg	Cumulative pollutant loading rates, kg/ha	High-quality pollutant concentration limits,[b] mg/kg	Annual pollutant loading rates, kg/ha·a
Arsenic	75	41	41	2.0
Cadmium	85	39	39	1.9
Copper	4300	1500	1500	75
Lead	840	300	300	15
Mercury	57	17	17	0.85
Molybdenum	75	—	—	—
Nickel	420	420	420	21
Selenium	100	100	36	5.0
Zinc	7500	2800	2800	140

[a]Absolute values.
[b]Monthly averages.

Pollutant Limits

To be land applied, biosolids must meet the pollutant ceiling concentrations and cumulative pollutant loading rates or pollutant concentration limits. Bulk biosolids applied to lawns and home gardens must meet the pollutant concentration limits. Biosolids sold or given away in bags or other containers must meet the pollutant concentration limits or the pollutant ceiling concentrations and be applied at an annual product application rate that is based on the annual pollutant loading rates. Pollutant limits for land application are listed in Table 16.3.

Pathogen and Vector-Attraction Reduction

Biosolids are classified into two categories, Class A and Class B, based on pathogen reduction criteria. Bulk biosolids applied to agricultural and nonagricultural land (for example, forests, public-contact sites, and reclamation sites) must meet at least Class B requirements and one of the several available vector-attraction reduction options. Bulk biosolids applied to lawns and home gardens and biosolids sold or given away in bags or other containers must meet the Class A criteria and one of the several available vector-attraction reduction processing options. One of the ten vector-attraction reduction options described later in this section also must be met when bulk biosolids are applied to agricultural or nonagricultural land.

Management Practices

The following management practices apply to land-applied biosolids that have not met the criteria of pollutant concentration limits, Class A pathogen requirements, and vector-attraction reduction:

- Bulk biosolids shall not be applied to flooded, frozen, or snow-covered ground so that the material enters wetlands or other waters of the United States unless authorized by the permitting authority;
- Bulk biosolids shall not be applied at rates above agronomic rates, with the exception of reclamation projects when authorized by the permitting authority;

- Bulk biosolids shall not be applied if likely to adversely affect a threatened or endangered species;
- Bulk biosolids shall not be applied less than or equal to 10 m from waters of the United States unless authorized by the permitting authority; and
- Biosolids sold or given away in a bag or other container shall have either a label affixed to the bag or container or an information sheet provided to the person who receives the biosolids for land application that provides the name and address of the person who prepared the analysis information on proper use, including the annual application rate that does not cause any of the annual pollutant loading rates to be exceeded.

Further, when biosolids that meet Class B pathogen reduction requirements—but not Class A requirements—are applied to the land, the following site restrictions must be met:

- Food crops with harvested parts that touch the biosolids–soil mixture and are completely above the soil surface (such as melons, cucumbers, squash, and so on) shall not be harvested for 14 months after application;
- Food crops with harvested parts below the soil surface (for example, root crops such as potatoes, carrots, and radishes) shall not be harvested for 20 months after application if the biosolids remain on the land surfaces for at least 4 months before incorporation to the soil;
- Food crops with harvested parts below the soil surface (for example, root crops such as potatoes, carrots, and radishes) shall not be harvested for 38 months after application if the biosolids remain on the land surfaces less than 4 months before incorporation to the soil;
- Food crops, feed crops, and fiber crops shall not be harvested for 30 days after biosolids application;
- Animals shall not graze on a site for 30 days after biosolids application;
- Turf shall not be harvested for 1 year after biosolids application if it is placed on lawn or land with a high potential for public exposure unless otherwise specified by the permitting authority;
- Public access to land with high potential for public exposure shall be restricted for 1 year after biosolids application; and
- Public access to land with a low potential for public exposure shall be restricted for 30 days after biosolids application.

Distribution and Marketing

The regulation of products that are distributed and marketed is addressed as a part of land application rather than as a separate practice under Part 503 regulations. Bulk biosolids are frequently applied to farmland, forests, and reclamation sites in liquid or dewatered cake forms at little or no cost to the land owner. At a minimum, these materials must meet the pollutant ceiling concentrations, Class B pathogen reduction requirements, and vector-attraction reduction requirements and should be applied using the cumulative pollutant loading rates if they do not meet the pollutant concentration limits.

Conversely, biosolids or wastewater-sludge-derived material that is considered suitable for distribution and marketing either in bulk or bags and other containers for uses on lawns and home gardens must meet the Class A pathogen reduction requirements, a vector-attraction reduction processing option, and the pollutant concentration limits (unless labeled, application rates are prescribed that do not exceed the annual pollutant loading rates).

If biosolids or wastewater-sludge-derived products meet the pollutant concentration limits, Class A pathogen reduction requirements, and a vector-attraction reduction processing option, they are typically not subject to the general requirements and management practices applicable to land application practices.

Composting

Composting can achieve compliance with Class A pathogen reduction requirements by monitoring for regrowth and operating under the following processes to further reduce pathogens:

- With in-vessel composting or static aerated pile systems, the temperature of the biosolids is maintained at 55 °C or higher for 3 days and
- With windrow composting, the temperature of the biosolids is maintained at 55 °C or higher for at least 15 days, during which time the windrow will be turned a minimum of five times.

Other operating conditions may be able to meet Class A pathogen reduction requirements based on meeting time and temperature or pathogen testing requirements. For vector-attraction reduction, composting must achieve temperatures of greater than 40 °C for 14 days and achieve an average temperature during that period of higher than 45 °C.

Heat Drying

There are few aspects of the Part 503 rule that will lead to changes in established heat-dried biosolids programs. Temperatures used in drying systems are in excess of 70 °C, and retention times in the dryer are 30 minutes or longer, which meets the requirements for Class A pathogen reduction. Drying processes must also meet the requirements of the thermal equation given in the Part 503 regulation. Similarly, vector-attraction reduction can easily be met by dryers that produce a product for marketing. The degree of dryness required is greater than or equal to 75% solids if the product does not contain unstabilized primary sludge, and greater than or equal to 90% solids if the product does contain unstabilized primary sludge.

Lime Stabilization

One patented process for alkaline stabilization complies with the Class A pathogen reduction requirements by elevating pH to above pH 12 for 72 hours while at the same time elevating temperature to above 52 °C for 12 hours or longer followed by air drying to greater than 50% solids. Other alkaline stabilization approaches may qualify the biosolids as Class A based on meeting the pasteurization on time and temperature criteria.

Surface Disposal

The surface disposal subpart of the Part 503 regulation applies to solids disposal operations such as the following:

- Dedicated disposal surface application sites, that is, application of sludge nitrogen at higher-than-acceptable agronomic rates on lands owned or leased by the wastewater authority that are highly controlled for access and operations;
- Monofills (sludge-only landfills);
- Piles or mounds; and
- Impoundments or lagoons.

There are many wastewater sludge lagoons or places where solids have been piled that are no longer active. In most cases, these entities would not be regulated under Part 503 regulations, particularly if they have been closed in a proper manner (U.S. EPA, 1994, 1995). However, if these sites or operations were still active in 1993, the date they become inactive could be critical in determining whether they are regulated under Part 503 regulations. If inactive sites have old wastewater sludge removed from them in the future, the use or disposal of the solids at that time would likely fall within the jurisdiction of Part 503 regulations if the material is used or disposed of through a practice covered by Part 503 regulations.

Part 503 regulations allow for storage up to 2 years without any restrictions or control. However, if the solids remain on the land beyond 2 years, the U.S. EPA may consider this disposal and regulate it as a surface disposal site.

In cases where surface disposal sites use liners and leachate collection systems, there are no pollutant concentration limits because pollutants leaching from the solids mass will be collected in the leachate and treated as necessary to avoid a pollution problem. For the site liner to qualify, it must have a hydraulic conductivity of less than or equal to 1×10^{-7} cm/s.

For surface disposal sites with no liner and leachate collection system, limits on three pollutants—arsenic, chromium, and nickel—are established in the Part 503 rule. While these vary, based on the distance of the active surface disposal site boundary from the site property line, the most extreme values allowed are listed in Table 16.4.

Different limits for these three pollutants can be developed through a site-specific assessment, as specified by the permitting authority, that shows the site has different parameters from those used by the U.S. EPA in establishing the maximum allowable concentration limits.

As a management practice, the Part 503 rule requires that surface disposal operations not cause the groundwater maximum contaminant level for nitrate to be exceeded or the existing concentration to be exceeded if it already exceeds this level.

Pathogen and Vector-Attraction Reduction

The pathogen and vector-attraction reduction requirements of the 40 CFR Part 503 regulations are complex. The pathogen reduction requirements for surface disposal and land application are operational standards for two classes of pathogen reduction: Class A and Class B. All biosolids that are sold or given away in a bag or other container for application to land, lawns, or home gardens must meet Class A pathogen requirements. All solids that are land applied or placed on surface disposal sites must meet at least the Class B pathogen requirements, except wastewater sludge placed on a surface disposal site that is covered daily with soil or other material. Specific requirements for the two classes of pathogen reduction and the rationale for these requirements are noted in the following subsections.

Table 16.4 Maximum allowable pollutant concentrations in wastewater sludge for disposal in active landfills without a liner and leachate collection systems (U.S. EPA, 1994).

Location in the Part 503 rule	Distance from the boundary of active biosolids unit to surface disposal site property line, m	Pollutant concentration[a]		
		Arsenic, mg/kg	Chromium, mg/kg	Nickel, mg/kg
Table 2 of section 503.23	0 to less than 25	30	200	210
	25 to less than 50	34	220	240
	50 to less than 75	39	260	270
	75 to less than 100	46	300	320
	100 to less than 125	53	360	390
	125 to less than 150	62	450	420
Table 1 of section 503.23	Equal to or greater than 150	73	600	420

[a]Dry-weight basis (basically, 100% solids content).

Class A Pathogen Requirements

When prepared for sale or give-away for land application or producing other products, Class A biosolids must meet one of the following criteria at the time of use or disposal:

(1) A fecal coliform density of less than 1000 most probable number (MPN)/g total dry solids or
(2) A *Salmonella* sp. density of less than 3 MPN per 4 g total dry solids.

In addition, the requirements of one of the following alternatives must be met (the item numbers given here are the same as those in the Part 503 regulations):

(1) *Time and temperature*—an increased temperature should be maintained for a prescribed period of time according to the guidelines listed in Table 16.5.
(2) *Alkaline treatment*—the pH of the material is raised to greater than pH 12 for at least 72 hours. During this time, the temperature of the biosolids should be greater than 52 °C for at least 12 hours. In addition, after the 72-hour period, the biosolids should be air dried to at least 50% total solids.
(3) *Prior testing for enteric virus and viable helminth ova*—the solids are analyzed for the presence of enteric virus (plaque-forming units [PFU]) and viable helminth ova. If the wastewater sludge is analyzed before the pathogen reduction process and is found to have densities of enteric virus less than 1 PFU/4 g total solids and viable helminth ova less than 1 ova/4 g total solids, the processed biosolids are Class A with respect to enteric virus and viable helminth ova until the next monitoring episode. If the wastewater sludge is analyzed before the pathogen reduction process and found to have densities of enteric virus greater than 1 PFU/4 g total solids or viable helminth ova greater than 1 ova/4 g total solids, and tested again after processing and found to meet the same enteric virus and viable helminth ova levels (as listed under alternative 4), then the processed biosolids will be Class A. This applies with respect to enteric virus and viable helminth ova when the operating parameters for the pathogen reduction process are monitored and shown to be consistent with the values or ranges of values documented at all times.

Table 16.5 Time and temperature guidelines.

Total solids	Temperature (t)	Time (D)	Equation	Notes
≥7%	≥50 °C	≥20 min	$D = \dfrac{131{,}700{,}000}{10^{0.14t}}$	No heating of small particles by warmed gases or immersible liquid
≥7%	≥50 °C	>15 sec	$D = \dfrac{131{,}700{,}000}{10^{0.14t}}$	Small particles heated by warmed gases or immersible liquid
<7%	>50 °C	≥15 sec to <30 min	$D = \dfrac{131{,}700{,}000}{10^{0.14t}}$	
<7%	≥50 °C	≥30 min	$D = \dfrac{131{,}700{,}000}{10^{0.14t}}$	

(4) *No prior testing for enteric virus and viable helminth ova*—if the wastewater sludge is not analyzed before pathogen reduction processing for enteric virus and viable helminth ova, the density of enteric virus in the processed biosolids must be less than 1 PFU/4 g total solids and the density of viable helminth ova must be less than 1 ova/4 g total solids at the time the biosolids are used or wastewater sludge is disposed, prepared for sale, or given away in a bag or container for application to the land. The biosolids must also meet any of the eight solids processing options for meeting vector-attraction reduction requirements.

(5) Biosolids have been treated by *processes to further remove pathogens* (PFRP) or equivalent processes.

Class B Pathogen Requirements

Class B pathogen requirements are the minimum level of pathogen reduction for land application and surface disposal. The only exception to achieving at least Class B occurs when the solids are placed in a surface disposal unit that is covered daily. Biosolids that do not qualify as Class B cannot be land applied.

Class B biosolids must meet one of the following pathogen requirements:

(1) Treatment by processes to significantly reduce pathogens or equivalent processes or
(2) At least seven samples should be collected at the time of use or disposal and analyzed for fecal coliforms during each monitoring period. The geometric mean of the densities of these samples will be calculated and should meet the following criteria: less than 2.0×10^6 MPN/g total solids or less than 2.0×10^6 colony-forming units/g total solids.

In addition, for any land-applied biosolids and domestic septage that meet Class B pathogen reduction requirements—but not Class A—the site restrictions previously described must be met.

Pathogen Treatment Processes

Processes to significantly reduce pathogens (PSRP) are designed to reduce the pathogen concentration in sludges and are generally equivalent to standard processes presently used in wastewater treatment. A sludge is classified as having been treated to PSRP levels if it has been treated by any of the following means:

(1) *Aerobic digestion*—sludges are agitated with air or oxygen to maintain aerobic conditions for a solids retention time (SRT) and temperature between 40 days at 20 °C and 60 days at 15 °C.
(2) *Air drying*—sludges are dried on sand beds or on paved or unpaved basins for a minimum of 3 months. During 2 of the 3 months, the ambient average daily temperature exceeds 0 °C.
(3) *Anaerobic digestion*—sludges are treated in the absence of air between an SRT of 15 days at temperatures of 35 to 55 °C and an SRT of 60 days at a temperature of 20 °C. Times and temperatures between these endpoints calculated by linear interpolation have been approved by the U.S. EPA's Pathogen Equivalency Committee, and the practice is being followed in a number of U.S. EPA regions.
(4) *Composting*—using the within-vessel, static aerated pile, or windrow composting methods, the temperature of the biosolids is raised to 40 °C or higher for 5 days. For 4 hours during the 5 days, the temperature in the compost pile exceeds 55 °C.
(5) *Lime stabilization*—sufficient lime is added to raise the pH of the biosolids to pH 12 and maintain it there for 2 hours of contact.

The processes above do not sterilize sludge. Significant pathogen concentrations can still remain. If there is a chance that the sludge would come into human contact and diseases may be transmitted by the pathogenic organisms in the sludge, the use of PFRP may be necessary. None of these processes can be expected to result in complete sterilization. They will, however, produce sludges that are very unlikely to include infectious doses of pathogens and are therefore more acceptable to the public. Among the processes to further reduce pathogens are the following:

(1) *Composting*—using either within-vessel or static aerated pile composting, the temperature of the sludge is maintained at 55 °C or higher for 3 days. Using windrow composting, the temperature of the wastewater sludge is maintained at 55 °C or higher for 15 days or longer. During this period, a minimum of five windrow turnings is required.
(2) *Heat drying*—sludges are dried by direct or indirect contact with hot gases to reduce the moisture content of the wastewater sludge to 10% or lower. Either the temperature of particles exceeds 80 °C or the wet bulb temperature of the gas in contact with the biosolids as the biosolids leave the dryer exceeds 80 °C.
(3) *Heat treatment*—liquid sludges are heated to a temperature of 180 °C or higher for 30 minutes.
(4) *Thermophilic aerobic digestion*—liquid sludges are agitated with air or oxygen to maintain aerobic conditions, and the SRT is 10 days at 55 to 60 °C.
(5) *Beta ray irradiation*—sludges are irradiated with beta rays from an accelerator at dosages of at least 1.0 megarad at room temperature (approximately 20 °C).
(6) *Gamma ray irradiation*—sludges are irradiated with gamma rays from certain

isotopes, such as ^{60}Co and ^{137}Ce, at dosages of at least 1.0 megarad at room temperature (approximately 20 °C).

(7) *Pasteurization*—the temperature of the sludges is maintained at 70 °C or higher for at least 30 minutes.

Vector-Attraction Reduction Requirements

Vector-attraction reduction reduces the potential for spreading of infectious disease agents by vectors, such as flies, rodents, and birds. Alternative methods for meeting the vector-attraction reduction requirement imposed by Part 503 regulations include the following:

(1–2) *Aerobic or anaerobic digestion*—mass of volatile solids is reduced by 38% or more. Volatile solids reduction is measured between the raw wastewater sludge before stabilization and the point when the digested solids are ready for use or disposal. This criterion should be readily met by properly designed and operated anaerobic digesters, but is not as readily met by typical aerobic digesters. Publicly owned treatment works with aerobic digesters may need to meet vector-attraction reduction requirements through the alternative methods outlined in items 3 and 4 below.

(2) *Anaerobic digestion*—if 38% volatile solids cannot be achieved, vector-attraction reduction can be demonstrated by further digesting a portion of the digested solids in a bench-scale unit for an additional 40 days at 30 to 37 °C or higher, and achieving a further volatile solids reduction of less than 17%.

(3) *Aerobic digestion*—if 38% volatile solids reduction cannot be achieved, vector-attraction reduction can be demonstrated by further digesting a portion of the digested sludge with a solids content of 2% or less in a bench-scale unit for an additional 30 days at 20 °C, and achieving a further volatile solids reduction of less than 15%.

(4) *Aerobic digestion*—specific oxygen uptake rate is less than or equal to 1.5 mg O_2/h per gram of total solids at 20 °C. If unable to meet the specific oxygen uptake rate criteria, a wastewater treatment plant may be able to satisfy the alternative method outlined in item 3.

(5) *Aerobic processes*—for aerobic processes, such as composting, temperature is kept at greater than 40 °C for at least 14 days (the average temperature during this period is greater than 45 °C).

(6) *Alkaline stabilization*—pH is raised to at least pH 12 by alkali addition and, without the addition of more alkali, remains at pH 12 or higher for 2 hours and then at pH 11.5 or higher for an additional 22 hours.

(7–8) *Drying*—total solids is at least 75% when the biosolids do not contain unstabilized primary solids, and at least 90% when unstabilized primary solids are included. Blending with other materials is not allowed to achieve the total solids percentage.

(9) *Injection*—liquid sludges (now called biosolids) are injected beneath the surface with no significant amount of solids present on the surface after 1 hour; biosolids that are Class A for pathogen reduction must be injected within 8 hours of discharge from the pathogen reduction process. This alterna-

tive is applicable to the following: bulk biosolids land applied to agricultural land, forests, and public-contact or reclamation sites; domestic septage land applied to agricultural land, forests, or reclamation sites; and wastewater sludge or domestic septage placed in a surface disposal site.

(10) *Incorporation*—wastewater biosolids that are land applied or placed in a surface disposal site should be incorporated to the soil within 6 hours of application; biosolids that are Class A for pathogen reduction and are land applied must be applied to, or placed on, the land within 8 hours of discharge from the pathogen reduction process. This alternative is applicable to the following: bulk biosolids land applied to agricultural land, forests, and public-contact or reclamation sites; domestic septage land applied to agricultural land, forests, or reclamation sites; and wastewater sludge or domestic septage placed in a surface disposal site.

(11) *Surface disposal daily cover*—wastewater solids or domestic septage placed in a surface disposal site shall be covered with soil or other material at the end of each operating day.

(12) *Domestic septage treatment*— the pH of domestic septage is raised to pH 12 or higher by alkali addition and, without the addition of more alkali, remains at pH 12 or higher for 30 minutes. This alternative is applicable to domestic septage applied to agricultural land, forests, or reclamation sites or domestic septage placed in a surface disposal site.

At least one of the vector-attraction reduction alternatives 1 through 10 must be met when bulk biosolids are applied to agricultural land, forests, and public-contact or reclamation sites. One of the alternatives 1 through 8 must be met when bulk biosolids are applied to lawns or home gardens or when biosolids are sold or given away in a bag or other container for land application. One of the alternatives 1 through 11 must be met when wastewater solids are placed in a surface disposal site. Although domestic septage can be treated the same as wastewater sludge, when it is handled as domestic septage rather than wastewater sludge, alternatives 9, 10, or 12 must be met when septage is applied to agricultural land, forests, or reclamation sites; in addition, one of the alternatives 9 through 12 must be met when domestic septage is placed in a surface disposal site.

Combustion (Incineration)

Part 503 regulations establish requirements for wastewater sludge-only combustors (called *incinerators* in the regulations). The rule covers the incinerator feed solids, the furnace itself, the operation of the furnace, and the exhaust gases from the stack. It does not apply to facilities incinerating hazardous wastewater solids (as defined by 40 CFR Part 261) or sludges containing greater than 50-ppm concentrations of polychlorinated biphenyls. It also does not apply to facilities that co-fire wastewater sludge with other wastes, although up to 30% municipal solid waste as auxiliary fuel is not considered "other wastes". Furthermore, this rule does not apply to the ash produced by a wastewater sludge incinerator.

Pollutant Limits

Pollutant limits for wastewater solids fired in an incinerator were imposed for the following heavy metals: beryllium, mercury, lead, arsenic, cadmium, chromium, and nickel. The limits for beryllium and mercury are those that already exist under the

National Emission Standards for Hazardous Air Pollutants (40 CFR Part 261). Pollutant limits for the remaining metals will be determined using site-specific performance characteristics and emission dispersion modeling results.

Incinerators must also meet a monthly average limit of 100 ppm for tetrahydrocannibinol (THC) that is corrected for moisture level (for 0%) and oxygen content (to 7%). This limit is an indicator to control toxic organic compound emissions. The limit is based on the arithmetic mean of hourly readings for the month, with a requirement for at least two readings during each hour of operation. The THC measuring device used must be a flame ionization detector with a heated sample line maintained at 1500 °C or higher at all times and be calibrated at least once every 24-hour operating period using propane. Operating parameters, such as oxygen concentrations and information to determine moisture content, in the stack exhaust gases and furnace combustion temperature must be continuously monitored.

The rule specifically bans incineration if it is likely to adversely affect a threatened or endangered species listed under the Endangered Species Act, or its designated critical habitat. If threatened or endangered species are known to be present in the vicinity of the incinerator, an ecological risk assessment may be needed to verify lack of likely effect. Sludge incinerators also need to meet the requirements and be permitted under the Clean Air Act.

Design of Land Application Operations

Land application in this section refers to any beneficial use that applies biosolids to the land. Distribution and marketing of biosolids products are also discussed separately in this section. Land application includes agricultural programs that involve growing a variety of crops, applying biosolids to forest areas, or adding them in larger quantities to reclaim disturbed land (such as old mining sites). All of these uses derive benefit from the components of biosolids and may or may not involve crop and animal production that eventually become part of the human food chain.

When evaluating sites with a view to eliminating areas potentially unsuitable for land application, the following criteria may preclude the use of the land: steep areas with sharp relief; shallow soils; environmentally sensitive areas (for example, intermittent streams, ponds); rocky, nonarable land; wetlands and marshes; and areas bordered by surface water bodies without appropriate setbacks. Other fatal flaws would include culturally sensitive lands such as old cemeteries and burial grounds, as well as public recreational areas.

When sites are determined by the above procedures to be suitable, soil test data should be obtained for each field. For agricultural purposes, soil testing will include pH; cation-exchange capacity; nutrient status; and, as appropriate, background trace metals analysis (that is, metals regulated by federal and state requirements and important micronutrient trace elements). Physical properties of soil, such as profile depth and texture, may be helpful but are assumed suitable for agronomic purposes if the site is currently being used in agricultural production. For forest or reclamation sites, more extensive investigation of groundwater characteristics and underlying geologic features may also be appropriate. These physical properties are less critical for sites used for agricultural application at agronomic rates.

Final site selection is often a decision based on availability of the most suitable sites, particularly for small communities. For large-scale projects, it is imperative that many sites be selected and permitted by state agencies to provide a variety of sites to which biosolids can be applied throughout most, if not all, of the year.

Any biosolids to be land applied must meet criteria for maximum concentrations of trace metals and must be classified as nonhazardous. In addition to requiring a municipal source that is environmentally suitable, a land application program must be based primarily on the nutrient content of the biosolids. It is in this arena that agronomic input is essential. Biosolids are not complete fertilizers, and farming practices may need to be modified and adapted to realize the full benefits of biosolids application. The specifics of each municipal solids source and methods of application determine the best management within a particular cropping system.

Application Rates

For most biosolids produced and land applied in the United States, the limiting factor (at least on an annual basis) is the nutrient content of the biosolids when they are applied as a nitrogen source for a particular crop. Thus, the land area needed for an agricultural program will be determined by the crops to be grown and the amount of nitrogen those crops can use. Regular nutrient analyses provide the database for nitrogen-based (agronomic) application rates for each source of biosolids.

Application rates are established in dry tons per acre, based on the amounts of plant-available nitrogen per dry ton. Applying biosolids based on plant-available nitrogen provides an organic source of plant-available nitrogen that conforms more closely to the crop's growth pattern than does inorganic fertilizer. The nitrogen uptake requirement of the corn plant and the nitrogen release curve of biosolids are similar; soluble chemical fertilizer tends to provide more nitrogen initially than the crop can use and less as the crop matures and increases its demand for nitrogen.

Plant-available nitrogen can be estimated as

$$N_{PA} = 20\,[(NH_4^+)(K) + NO_3^- + (N_O)(f)] \qquad (16.1)$$

where
- N_{PA} = plant-available nitrogen in lb/dry ton of sludge;
- NH_4^+ = ammonia nitrogen (%);
- K = volatilization factor, estimated as 1.0 for subsurface injection and 0.5 for surface application;
- NO_3^- = nitrate (%);
- N_O = organic nitrogen (%); and
- f = mineralization factor, estimated as 0.2 to 0.4, depending on the type of biosolids.

For a sludge containing 0.5% ammonia nitrogen, 0.02% nitrate nitrogen, and 0.2% organic nitrogen, and if $K = 1.0$ and $f = 0.2$, the plant-available nitrogen of approximately

$$N_{PA} = 20\,[0.5(1.0) + 0.02 + (4)(0.2)] = 26 \text{ lb/dry ton} \qquad (16.2)$$

Methods of Application

Operating a land application program requires blending the needs of the treatment plant (which continuously generates sludge) with land-use patterns of the area where the product will be applied. The land requirement is more than a function of how much acreage has received a permit, it must also conform to what is happening on that land as a farming unit at the time biosolids are applied. The most common means of transport to the land is as a liquid in tanker trucks. The biosolids is either sprayed out the back of the tuck as it travels up and down the field, or it is injected to the ground.

Roads used for transporting biosolids to a farm site must be suitable for the type and volume of truck traffic required. Land application programs are often visible and, therefore, need to be particularly sensitive to public perception concerning transport and actual application. From a practical standpoint, it is imperative that the equipment used for a project be able to handle as efficiently as possible the type and amounts of biosolids generated. By transporting and applying biosolids to maximize mechanical efficiency, the needs of both the generator and the farmer (who needs to have the land available for planting) can be met.

Odor Control

One of the most important factors in a land application program is the prevention of aesthetically unacceptable or nuisance conditions. While these factors may have no environmental or health effects, they are critical to the acceptability of a program. Truck traffic, dust, and especially odors must be minimized in any land application program.

The most common aesthetic concern centers around the potential for odor from a land application program. Pathogen and vector-attraction reduction (disinfection) processes required for biosolids to be land applied also serve to reduce odor potential by decreasing the amount of organic material that will eventually be decomposed (biodegraded) by microbes in the soil. Odors from the decomposition of agronomic applications of treated biosolids are much less noticeable and persistent than odors from the decomposition of raw organic waste, such as animal manures. Applying biosolids at crop nutrient rates with the management practices required by regulations also reduces the potential for odor from a land application program.

Design of Landfilling Operations

This section presents information related to the planning, design, construction, operation, monitoring, and closure of landfills for wastewater sludges. To minimize project delays and expenditures, a carefully planned sequence of activities is needed to develop a wastewater residuals landfill.

A monofill site must meet siting criteria established by the U.S. EPA in Part 503 (U.S. EPA, 1994). The following are management practices related to siting:

- A landfill shall not be developed if it is likely to adversely affect a threatened or endangered species.
- A landfill cannot restrict the flow of a base flood; a 100-year flood event.
- The landfill must not be located in a geologically unstable area.
- The landfill must be located 60 m or more from a fault area that has experienced displacement in Holocene time.
- If the landfill is located in a seismic impact zone, it must be designed to resist seismic forces. (Seismic impact zone is defined as an area that has a 10% or greater probability that the horizontal ground level acceleration of the rock in the area exceeds 0.10 g once in 250 years.)
- The landfill cannot be located in a wetland unless a special permit is obtained.

When selecting a new site, the landfill footprint and site geometry must provide adequate landfill capacity to meet the planning period. Typically, landfill sites are selected to provide 10 to 20 years of operational capacity based on the municipal residuals' production rate for the subject wastewater treatment facility.

The operational life of a landfill is affected by many variables, including the production rate of the municipal solids to be wasted, the volume consumed by liners and capping systems, and the volume consumed by bulking soils and cover materials. The volume lost because of liner and cover systems is easy to calculate and is based on the thickness of these layers over the landfill area. Depending on the percent solids of the bulking material, a certain amount of bulking soil must be mixed with the municipal residuals to improve its handling and strength characteristics. The daily and intermediate cover material may use an additional 20% of the overall landfill volume. Actual capacity assessments need to be conducted on a site-specific basis, depending on the type of material to be disposed in the landfill, the proposed method of landfill operation, cover requirements, and other factors.

Part 503 does not require a liner system for all monofills. The need for a liner is based on the pollutant concentrations of the municipal residuals. The U.S. EPA allows landfilling without a liner if pollutant limits are below those established for arsenic, chromium, and nickel. As shown in Table 16.2, these pollutant concentrations are based on the distance between active landfill liner limits and the property line of the landfill site. If a liner is not used, a sampling and analysis program must be established in accordance with Part 503. In addition, a site-specific permit may be obtained from the permitting authority for the development of an unlined landfill even if pollutant concentrations are exceeded. Site-specific limits may be justified if site conditions vary significantly from criteria used by the U.S. EPA in deriving pollutant concentrations.

If pollutant concentrations limits are exceeded for a particular residual, the U.S. EPA requires the development of a lined landfill. Note that many states' regulations for monofills include more stringent requirements, including the need for a liner system.

Design of Dedicated Land Disposal Operations

Dedicated land disposal, a confusing term, means different things to different people. In this book, dedicated land disposal means the application of wastewater sludges to land for disposal purposes. It is similar in operation to agricultural or nonagricultural land application; however, the rates of application are beyond agronomic rates and no crop is grown. It is often regulated as if it were a landfill because the site is the final repository for large quantities and heavy loadings of municipal solids.

As discussed previously, general requirements apply to surface disposal of municipal sludges. These include compliance with all applicable Part 503 requirements; closure by 1994 of active units located within 60 m of a fault with displacement in Holocene time, in an unstable area, or in a wetland, unless authorized by the permitting authority; and the need for closure and postclosure plans at least 180 days before closing any active units. Also, site owners are required to provide written notification to the subsequent owner that municipal sludges were placed on the land.

Where surface disposal sites use liners and leachate collection systems, there are no pollutant concentration limits because pollutants leaching from the solids will be collected in the leachate and treated as necessary to avoid a pollution problem. For the site liner to qualify, it must have a hydraulic conductivity of $<1 \times 10^{-7}$ cm/s. For surface disposal sites with no liner or leachate collection system, limits on three pollutants are established in the rule. While these vary based on the distance of the active unit boundary from the site property line, the most extreme values allowed for arsenic, chromium, and nickel are listed in Table 16.4.

The three pollutants listed present the greatest threat of leaching to groundwater and exceeding the maximum contaminant level. Allowable concentrations of the three pollutants are reduced if the active unit boundary is less than 150 m from the site property line. The table shows the worst case limits if the site boundary is located from 0 to <25 m from the disposal site property line. Different limits for these three pollutants can be developed through a site-specific assessment, specified by the permitting authority, that shows the site has different parameters than the ones the U.S. EPA used in establishing maximum allowable concentration limits.

As a management practice, the rule requires that surface disposal operations not cause the groundwater maximum contaminant level for nitrate to be exceeded or the existing concentration to be exceeded if it already exceeds this level. Either results of groundwater monitoring or a statement from a qualified groundwater scientist must be used to demonstrate compliance.

Application Rates

All land application programs are aerobic systems at the surface incorporation layer. The situations evaluated here do not necessarily support crop or vegetation production. The goal is to maximize application rates to achieve maximum aerobic stabilization of the biosolids by soil bacteria. The rates of application are substantially beyond agronomic rates and are primarily limited by the evaporation rate of the moisture added by the biosolids to the soil.

Application rates range from 30 to 250 dry metric ton/ha (22 to 110 dry ton/ac/yr). Assuming a 50-year site operating life, total loading would be 1500 to 12,500 dry metric ton/ha. At these rates, and assuming that 25% of the biosolids are lost to long-term products of decomposition, biosolids addition adds a depth of approximately 0.25 to 1.3 cm/y (0.1 to 0.5 in./yr) to the soil. These rates of application are readily handled by bacteria in the soil. Because aerobic biological activity proceeds at rates of approximately one to two orders of magnitude faster than anaerobic biological activity, biosolids loading rates can be handled biologically as long as sufficient water is evaporated and percolated to allow the system to remain aerobic. The controlling factor of application rate is the evaporation rate, which affects how soon and how often equipment can be moved onto the site.

Assuming that a 6% solids slurry is applied at the above loading rates, the amount of water applied with the biosolids each year is 8 to 38 cm (0.25 to 1.25 ft). Evaporation of this much water can be accomplished from moist soil-biosolids surfaces in many but not all regions of the United States. High evaporation is associated with particular seasons of the year that have the necessary temperature, humidity, wind, and rainfall characteristics.

Methods of Application

Biosolids can be applied in slurry or dewatered form much like other land application methods. Achieving consistent surface application can be difficult with dewatered cake, although dewatered material has the advantage of allowing higher rates of application because of less moisture addition. The advantages of slurry application often outweigh increased moisture content and evaporation requirements because slurries can be pumped and it is much easier to achieve consistent application rates over the entire site. Slurries can be spread on the land or injected slightly beneath the surface (a few centimeters deep) with mobile equipment specifically designed for this purpose.

The high-rate surface application system is often referred to as *dedicated land disposal* because sites are dedicated for long-term biosolids disposal and do not have beneficial use aspects (U.S. EPA, 1995). The rates of application are almost always substantially greater than agronomic rates because the purpose is to minimize costs by keeping the site size as small as possible. Such sites require landfill or similar solid waste disposal permits from state regulatory agencies in addition to meeting federal requirements. These sites are truly dedicated as biosolids landfills, and the wastewater agency owns the site with the intent of forever using it as a disposal site. Conversion of such sites for other uses can only be accomplished after proper evaluation.

Conclusions

One of the most curious sections of the U.S. EPA Part 503 sludge regulations is the "vector attraction" requirement. Quite clearly, this has nothing to do with vectors. If a truly applicable vector-attraction measurement were to be formulated, a list of vectors would have to be identified (flies? mice? beetles?) and the probability of their carrying pathogens from sludge to humans established. The present "vector attraction" standard is based on volatile solids concentration. There is quite a gap there.

When the 40 CFR 503 regulations were being developed, the U.S. EPA scientists did not know how to measure the degree of "stabilization" and decided to package this under the title "vector attraction". Why is it that a sludge with volatile solids greater than 38% of total solids attracts vectors and one with less than 38% volatile solids does not? This limit appears to be merely an expedient way of stating the putrescence of a sludge. Typical anaerobic digesters achieve a volatile solids reduction of 38% and well-digested sludge does not smell too badly, so the limit has validity. But we have to recognize that volatile solids reduction has little to do with putrescence, and it certainly has little to do with how many vectors the sludge will attract. But because there is nothing better, this will have to do for now. There is still a lot of work to do in understanding the science, engineering, and policy of sludge disposal and the beneficial use of biosolids.

References

Antonic, M.; Hamoda, M. F.; Cohen, D. B.; Schmidtke, N. W. (1974) *A Survey of Ontario Sludge Disposal Practices*; Project No. 74-3-19; COA Research Report: Toronto, Ontario, Canada.

U.S. Environmental Protection Agency (1994) *A Plain English Guide to EPA Part 503 Biosolids Rule*; EPA-83Z/R-93/003; Washington, D.C.

U.S. Environmental Protection Agency (1995) *Process Design Manual for Suspended Solids Removal*; EPA-625/1-75-003a; Washington, D.C.

Symbols Used in this Chapter

- f = mineralization factor, estimated as 0.2 to 0.4, depending on the type of biosolids
- K = volatilization factor, estimated as 1.0 for subsurface injection and 0.5 for surface application
- N_{PA} = plant-available nitrogen in lb/dry ton of sludge
- NH_4 = ammonia nitrogen (%)
- NO_3^- = nitrate (%)
- N_O = organic nitrogen (%)

Index

A

A²/O process, 8-18
Acinetobacter, 8-3
Activated carbon
    addition, activated sludge, 6-27
    adsorption, 10-10
    characteristics, 10-12
Activated sludge
    biology, 6-5
    characteristics, 6-3
    load, 6-3
    pH, 6-17
    process, 6-1
    systems, sludge production, 6-14
    tank shape, 6-18
    toxics, 6-17
Administration, facilities, 2-25
Administrative offices, 2-27
Advanced integrated lagoon systems, 9-7
Advanced treatment, 1-15
Aerated grit chambers, 4-15
Aerated grit chambers, design criteria, 4-16
Aerated lagoon, 6-12, 6-21
Aeration tanks, hydraulics, 3-18
Aeration tanks, volume, 6-10
Aerobic digestion, 15-1, 15-26
Aftergrowth, 11-7
Air-lift pumps, 12-21
Air quality control, 2-19
Air-operated diaphragm pumps, 12-20
Alkaline stabilization, 15-1, 15-39
Alternative biological treatment, 9-1
Aluminum salts, 13-3
Americans with Disabilities Act, 2-13
Amoebic dysentery, 1-12
Anaerobic digester, heat-transfer
    coefficients, 15-17
Anaerobic digestion, 15-1, 15-2

concentration data for inorganics, 15-5
concentration data for organic materials, 15-6
Annual cost calculations, 1-21
Annual cost, 1-20
Application methods, biosolids, 16-17
Application rates, biosolids, 16-14, 16-17
Archimedes screw pumps, 12-22
Arrhenius equation, 11-4
Artificial soil, 15-38
Attached growth, treatment, 7-1
Attached-growth processes, nitrogen, 8-17
Autothermal thermophilic aerobic digestion, 15-30
Average annual flow, 2-11

B

Backwashing filters, 10-7
Backwashing, 10-15
Bacteria, testing, 1-11
Bar screens, 4-2
Bardenpho, 8-18
"Beaver disease", 1-12
Belt conveyors, 12-26
Belt filter dewatering, 14-23
Belt filter, performance, 14-24
Beneficial use, 16-1
Bernoulli equation, open-channel flow, 3-7
BHP, 3-20
Biochemical oxygen demand, measurement, 1-7
Biofilter–activated sludge process, 7-19
Biofilters, 7-4
Biological phosphorus removal options, 8-5
Biosolids, 16-1
Biotowers, 7-4
Bonds, 1-20
Brake horsepower, 3-20
Buchner funnel test, 13-9
Bucket elevators, 12-29

C

Capillary suction time test, 13-8
Capital costs, 1-19, 1-20
Capital recovery factor, 1-21
Carbon dosages, wastewater influent, 10-17
Carbon oxidation, 6-6
Cascade reoxygenation, 11-17
Catenary bar screen, 4-6
Centrifugal dewatering, 14-17
Centrifugal pumps, 12-17
Centrifugation, factors, 14-20
Centrifuges, 14-16
40 CFR Part 503 regulations, 16-2
Chain-and-bucket elevators, 4-20
Characteristic curves, 3-20
Characteristics, municipal wastewater treatment, 2-4
Chemical coagulation, settling tanks, 5-15
Chemical conditioning, 13-2
Chemical feed systems, 10-26
Chemical polishing, phosphorus, 8-7
Chemical sludge, 12-13
Chemical solids production, 12-11
Chemical treatment, 10-17
Chemical-feed pumps, 11-10
Chemistry, chlorine in water, 11-6
Chick's law, 11-3
Chlorinators, 11-10
Chlorine disinfection, 11-5
Chlorine disinfection, chemistry, 11-5
Chlorine toxicity, 11-7
Chlorine, odor, 11-8
Circular settling tanks, 5-5
Clamshell buckets, 4-20
Clarifier sizing, 6-13
Clarifier, hydraulics, 3-16
Class A criteria, 16-4
Class A pathogen requirements, 16-8
Class B pathogen requirements, 16-9
Coefficients, monod, 8-10
Coliform, testing, 1-13
Combined biological treatment, 7-18
Combined chemical and biological process, phosphorus, 8-5

Combustion, 15-1
Commercial wastewater flow, 2-6
Comminutors, 4-8
Complete mix, activated sludge, 6-18
Composition, domestic wastewater, 2-10
Composting, 15-1, 15-33, 16-6
Compound interest factors, 1-21
Concentration, domestic wastewater, 2-10
Conditioning, 13-1
Conduction drying systems, 14-33
Constructed wetlands, 9-17
Contaminants, domestic wastewater, 1-14
Contaminants, residential, 1-14
Continuous decanter scroll, 14-18
Control facilities, 2-26
Control strategies, activated sludge, 6-39
Convection drying systems, 14-33
Conveyors, 12-26
Costs, 1-19, 1-20
    calculations, 1-21
    operation and maintenance, 1-23
Countercurrent aeration, 6-31
Critical particle, 5-3
Critical scour velocity, 5-9
Cryptosporidium, 1-12, 11-4
Cysts, 1-12

D

DAF, 14-9
Darcy–Weisbach equation, 3-8
Dechlorination, 11-5, 11-8
Deficit ratio, 11-17
Denitrification filters, nitrogen, 8-17
Depth, activated sludge, 6-35
Depth, settling tanks, 5-7
Design criteria, 2-3
Design flows, 2-11
Design process flow, 2-11
Design process, 2-1
Design standards, 2-3
Design, trickling filters, 7-5
Detritus tanks, 4-17
Dewatering, 14-1
Diarrhea, 1-12

Die-off rate constant, 11-3
Diffused aeration, activated sludge, 6-28
Digester design parameters, 15-12
Direct dryers, 14-35
Direct–indirect dryers, 14-36
Disinfection kinetics, 11-2
Disinfection tanks, hydraulics, 3-18
Disinfection, 11-1
Disk dryers, 14-35
Dissolved air flotation thickening, 14-9
Dissolved oxygen, activated sludge, 6-16
Dissolved oxygen, measurement, 1-7
Domestic flow contributions, 2-4
Domestic wastewater, 2-8
Downflow activated carbon contactor, 10-15
Downward velocity, 14-5
Drag conveyors, 12-28
Dried sludge, transport, 12-27
Dry lime stabilization, 15-41
Drying bed dewatering, 14-27
Drying, 14-1
Dual-power multicellular aerated lagoons, 9-7
Dual-sludge processes, nitrogen, 8-15
Dysentery, 1-12

E

Economics, 1-19
Eductor, 13-6
Efficiency, 3-21
Effluent, standards, 1-13
Egg-shaped digester, 15-21
Elevation, 2-17
Emissions control, 15-46
Emulsions, 13-6
Energy balance, composting, 15-34
Entameoba histolytica, 1-12, 11-7
Environmental impact, 2-28
Equalization, 4-21
Escherichia coli H57, 1-12
Extended aeration, activated sludge, 6-22

F

f values (fraction of nitrogen lost), 9-13
Facility design, chlorine, 11-8
Facultative lagoons, 9-3
Facultative ponds, biochemical oxygen demand, 9-4
Facultative ponds, design criteria, 9-4
Federal sludge regulations, 16-1
Ferric chloride, 13-3
FGR–SGR, 8-19
Filter press dewatering, 14-25
Filters, backwashing, 10-7
Filtration theory, 10-3
Finding of No Significant Impact, 2-29
Fixed-growth reactor–suspended-growth reactor, 8-19
Flexibility, 2-24
Flights, 5-17
Floating aquatic systems, 9-17
Floating bed aerated filters, 7-19
Flow characteristics, 12-14
Flow distribution, 3-14
Flow per capita, 2-4, 2-5
Flow projections, 2-4
Flow variation, wastewater, 2-8
Flowrates, design, 3-2
Flow-through velocity, 5-9
Fluidized bed denitrification, 8-17
Fluidized-bed furnace, 15-46
Flux, 14-5
FONSI, 2-29
Free available chlorine, 11-6
Freeze–thaw conditioning, 13-11
Freezing bed, 13-12
Frictional head losses, 12-15
Funding, 1-19

G

Gas collection and storage, 15-14
Gas production, 15-13
Gas quality and quantity, 15-11
Gas-injection systems, 15-16
Gastrointestinal problems, 1-12
Gel polymers, 13-6
General obligation bonds, 1-20
Geology, 2-18

Giardia lamblia, 1-12, 11-7
Giardia, 11-4
Giardiasis, 1-12
Granular media filtration, 10-2
Gravity belt thickening, 14-12
Gravity settling, 5-2
Gravity thickening, 14-2
Gravity thickening data, 14-6
Great Lakes, standards, 2-5
Grinder pumps, 12-17
Grinders, 4-8
Grit disposal, 4-20
Grit removal, 4-14
Grit tanks, hydraulics, 3-17
Grit washing, 4-20
Grit, characteristics, 4-18
Grit, quantities, 4-18
Growth kinetics, 6-8, 8-10

H

Hazen–Williams equation, 3-9
Head loss, 3-5, 3-15
Head loss characteristics, 12-14
Heat drying, 16-6
Heat values, wastewater residuals, 15-42
Heavy metal concentration, Ontario sludges, 16-2
Heavy metal, land application pollutant limits, 16-4
Helical screw-conveyor centrifuge, 14-18
Hepatitis virus, 1-12
High rate, activated sludge, 6-23
Higher heat value, terminology, 15-42
High-purity oxygen aeration, 15-29
High-rate digestion, 15-8
History
   activated sludge, 6-3
   legislation/regulations, 1-3
   wastewater treatment, 1-2, 2-1
Hollow fine fiber configuration, membranes, 10-30
Hollow-flight dryers, 14-35
Horizontal-flow grit chambers, 4-17
Horizontal-flow grit chambers, design criteria, 4-18
Hybrid systems, activated sludge, 6-27

Hydraulic profile, 3-1
Hydraulic residence time, 5-8
Hydraulic retention time, definition, 6-4
Hydraulics, 2-24, 3-1
Hydrocyclones, 4-17
Hydrogen sulfide, 15-22
Hydrogeology, 2-18
Hypochlorite, 11-5

I

Imhoff tanks, 5-21
Impeller power requirements, 10-25
Incineration, 16-12
Inclined screw conveyors, 4-19
Indirect dryers, 14-35
Industrial wastewater, 2-9
Industrial wastewater flow, 2-6
Infiltration/inflow, 2-5
Infrared dryers, 14-36
Innovation, wastewater treatment plants, 1-4
Interface settling velocity, 14-4

J

Jar test, 13-8
Jet systems, 6-31

K

Kincannon/Stover model, trickling filters, 7-12
Kinematic wave/Velz model, trickling filters, 7-11
Kinetic constants, 6-13
Kinetic pumps, 12-16
Kinetics, 6-12
    disinfection, 11-2
    UV, 11-15
Kirshmer's values, 4-13

L

Laboratories, 2-27
Lagoon systems, 9-2
Lagoons, 14-31
Laminar flow, 3-8
Land application, 16-2
    design criteria, 6-13
    pollutant limits, 16-4
Land treatment processes, site requirements, 9-10
Land treatment systems, 9-7, 9-8
Land use, 2-14
Landfill disposal, 16-8
Landfill leachate, 2-7
Landfill, design criteria, 16-15
Launders, 3-14
Legislation/regulations, 1-3
Lime, 13-2
Lime precipitation, phosphorus removal, 10-19
Liquid lime stabilization, 15-40
Lime stabilization, 16-6
Loading rates, activated sludge, 6-21
Loading, trickling filters, 7-4
Logan model, trickling filters, 7-13
Low rate, activated sludge, 6-22
Lower heat value, terminology, 15-42
Low-rate digestion, 15-7
Low-temperature aerobic digestion, 15-29
Ludzack–Ettinger design, 8-13

M

Maintenance, 2-25, 2-26
Mannich polymer, 13-6
Manning equation, 3-11
Mass transfer, 1-16
Materials balance, 1-16
Maximum day flow, 2-11
Maximum hour flow, 2-11
Maximum velocity, 3-5
Maximum week flow, 2-11
Mean cell residence time, definition, 6-4
Mechanical stirring systems, 15-16
Mechanical surface aeration, 6-32

Membrane processes, 10-28
Microbiology, composting, 15-34
Minimum velocity, 3-5
Mixed liquor suspended solids, 6-2
Mixed liquor, 6-2
Mixing opportunity parameter, 10-25
Mixing, 10-23
MLSS, 6-2
Modeling, streams, 1-3
Molecular weight, polymer, 13-4
Monitoring facilities, 2-26
Monod kinetic coefficients, 8-10
Monofills, 16-15
Moody diagram, 3-9
Moving bed filters, 10-7
Multiple-hearth furnace, 15-43

N

National Fire Protection Association, 2-13
National Pollutant Discharge Elimination System, 1-13, 11-1, 16-2
National Research Council Formula, trickling filters, 7-10
Natural systems, 9-1
NFPA, 2-13
Nitrification, 6-9
   process design, 6-13
   trickling filters, 7-13
Nitrobacter, 6-15
Nitrogen concentration, wastewater, 2-9
Nitrogen removal processes, 8-7
Nitrogen removal, design, 8-22
Nitrogen, metabolism, 1-11
Nitrogen, plant available, 16-14
Nitrogen, testing, 1-10
Nitrosomonas, 6-15
Noise control, 2-19
Nomenclature, activated sludge, 6-6
Nonporous diffusers, activated sludge, 6-29
NPDES, 1-13, 16-2
Nutrient removal processes, design, 8-24
Nutrient removal, 8-1
Nutrients, activated sludge, 6-16

O

Occupation Safety and Health Act, 2-12
Odor control, 11-1, 11-19, 16-15
Odor, trickling filters, 7-9
Odorous compounds, wastewater treatment, 11-21
Open-channel flow, 3-11
Operating point, 3-23
Operation and maintenance costs, 1-23
Organic polymers, 13-3
Organics, domestic wastewater, 2-10
OSHA, 2-12
Outfalls, hydraulics, 3-19
Overflow rate, 5-4, 6-35
Overflow rates, primary sedimentation, 5-9
Overland flow systems, 9-13
Oxidation, 6-5
Oxidation ditch processes, nitrogen, 8-14
Oxidation ditch, 6-20
Oxidation towers, 7-4
Oxygen demand, 6-13
Oxygen transfer, activated sludge, 6-27
Ozonation, 11-23

P

Paddle dryers, 14-35
Part 503 regulations, 16-2
Partially mixed aerated lagoons, 9-5
Particle settling, 5-3
Pathogen reduction, 15-10, 16-4
Pathogen requirements, 16-8, 16-9
Pathogen treatment processes, 16-10
Pathogens, testing, 1-11
Paved drying beds, 14-30
Performance, trickling filters, 7-5
Peristaltic hose pumps, 12-21
pH adjustment, 10-22
Phased isolation ditches, 8-21
Phosphorus
   metabolism, 1-11
   precipitation, 10-17
   removal, 10-18
      design, 8-21
      processes, 8-2

testing, 1-11
PhoStrip II process, 8-19
Physical–chemical processes, 10-1
Pipe flow, 3-7
Pipelines, 12-23
Piping, hydraulics, 3-19
Plant expansion, 2-23
Plant layout, 2-21
Plant-available nitrogen, 16-14
Plug flow, activated sludge, 6-19
Plunger pumps, 12-18
Pneumatic conveyors, 12-29
Pneumatic ejectors, 12-21
Pollutant limits, incineration, 16-12
Pollutant limits, land application, 16-4
Polyelectrolytes, 13-1
Polymers, 13-1
   charge, 13-4
   dosage, 13-6
   molecular weight, 13-4
Population projections, 2-4
Porous diffuser systems, activated sludge, 6-28
Preaeration, settling tanks, 5-15
Preliminary treatment, 1-15, 4-1
Present worth factor, 1-22
Present worth, calculation, 1-22
Primary settling tanks, hydraulics, 3-17
Primary sludge characteristics, 12-12
Primary treatment, 1-15, 5-1
Process train, wastewater treatment plant, 1-15
Processes to significantly reduce pathogens, 16-10
Progressing cavity pumps, 12-19, 12-25
Project sequence, 2-3
PSRP, 16-10
Pulsed bed filters, 10-8
Pump characteristics, 3-19
Pump selection, 3-23
Pumping grit, 4-20
Pure oxygen, activated sludge, 6-25

Q

Quantity, municipal wastewater treatment, 2-4

R

Radiation drying systems, 14-34
Rapid infiltration systems, 9-15
Recessed-impeller pumps, 12-17
Reciprocating piston pumps, 12-21, 12-25
Reciprocating rake bar screen, 4-4
Recommended Standards for Wastewater Facilities, 2-5
Rectangular setting tanks, 5-5
Redundancy, 2-23
Reed beds, 14-30
References, 1-24, 2-30, 3-26, 4-25, 5-25, 6-42, 7-20, 8-25, 9-20, 10-33, 11-24, 12-30, 13-13, 14-38, 15-47, 16-18
Regrowth, 11-4
Regulations, 1-3, 16-2
   landfills, 16-16
   safety, 2-12
Reoxygenation, 11-1, 11-17
Return activated sludge, 6-13
Revenue bonds, 1-20
Reynolds number, 3-9
Rotary drum screen, 5-24
Rotary drum thickening, 14-15
Rotary lobe pumps, 12-21
Rotary screen thickener, 14-15
Rotating biological contactor, 7-1, 7-13
Roughing filter–activated sludge process, 7-18
Roughing filters, 7-4
Roughness coefficient, 3-10

S

Safety and health, 2-12, 11-8
Salmonella, 1-12
Sand drying beds, 14-27
Sand drying beds, design criteria, 14-29
Screening, 4-1
   characteristics, 4-10
   design criteria, 4-2, 4-11
   quantities, 4-10
Screens, 5-23
Screens, hydraulics, 3-17
Screw conveyors, 12-28

Screw–combination centrifugal pumps, 12-17
Scum collection, 5-19
Scum disposal, 5-19
Scum management, 5-20
Secondary clarification, trickling filters, 7-20
Secondary clarifier, definition, 5-2
Secondary settling tanks, hydraulics, 3-18
Secondary settling, 6-33
Secondary sludge characteristics, 12-12
Secondary solids production, 12-5
Secondary treatment, 1-15
Self-cleaning screens, 4-7
Septage, 4-21
Sequencing batch reactors, 6-26
Sequencing batch reactors, phosphorus, 8-7
Service area, characteristics, 2-4
Settling tank, definition, 5-2
Settling theory, 5-2
Settling velocity, 5-2
Settling zones, 5-3
Sewage, definition, 2-7
Sewer charge, 1-19
Sewer, definition, 2-7
Sewerage, definition, 2-7
Sewered population, 1-1
Shear test, 13-9
Shear-lift pump, 12-17
Shigella, 1-12
Side water depths, primary sedimentation, 5-9
Side-overflow formula, 3-14
Sidestream fermentation processes, phosphorus, 8-6
"Sigma concept", 14-19
Sinking funds, calculation, 1-22
Site access, 2-18
Site assessment, 2-14
Site requirements, land treatment, 9-9
Site selection, 2-14
Slow-rate systems, land treatment, 9-10
Sludge age, definition, 6-4
Sludge collection, 5-17
Sludge combustion, 15-41
Sludge compactability, 13-10
Sludge conditioning, 13-1
Sludge consistency, 13-10
Sludge generated, 6-12
Sludge volume index, 6-34
Sludge wasted, 6-12
Sludge yield coefficients, 12-7
Sludge, characteristics, 12-11
Sludge, properties, 5-18
Sludge, quantities, 5-18, 12-2
Sludge, storage, 12-13, 12-24
Sludge, transport, 12-13, 12-24
Soil absorption systems, 9-1
Soils, 2-18
Solid-bowl centrifuge, 14-16
Solid-bowl conveyor centrifuge, 14-18
Solids balance, 12-4
Solids disposal, 1-15
Solids flux curve, 14-4
Solids flux, 14-4
Solids loading rate, 6-34
Solids management, 12-1
Solids production, 12-1, 12-9
Solids retention time, definition, 6-4
Solids terminology, 1-10
Solids transport, 12-1
Solids treatment, 1-15
Solids, measurement, 1-9
Sources, municipal wastewater treatment, 2-4
Spiral wound, membranes, 10-30
Spiral wrap, membranes, 10-30
Square settling tanks, 5-6
Stabilization, 15-1
Stacked settling tanks, 5-6
Staff facilities, 2-27
Standard for Fire Protection in Wastewater Treatment and Collection Facilities, 2-13
Standard Methods, 1-6
Standards
   design, 2-3
   effluent, 1-13
   Great Lakes, 2-5
   streams, 1-13
   water quality, 1-13
Stefan–Boltzman constant, 14-34
Step feed, activated sludge, 6-24

Storage, 2-26
Streams, standards, 1-13
Streeter–Phelps, 1-3
Subnatant, 14-10
Subsided fixed-bed reactors, 7-19
Surface disposal, 16-6
Sulfonators, 11-10
Sulfur dioxide feeders, 11-11
Sulfur dioxide, 11-5, 11-6
Support facilities, 2-25
Suspended-growth biological treatment, 6-1
Suspended-growth processes, nitrogen, 8-12
SVI, 6-34
Symbols, 1-25, 2-31, 3-26, 4-25, 5-25, 6-43, 7-22, 8-27, 9-21, 10-33, 11-24, 12-31, 13-14, 14-39, 15-48, 16-19
Synthetic media filters, 7-4
System head curves, 3-21

T

Tank geometry, 2-23
Tapered aeration, activated sludge, 6-24
Temperature, activated sludge, 6-15
"Ten States Standards", 2-5
Terminology
   disinfection, 11-2
   mg/L, 14-2
   percent solids, 14-2
   residuals, 14-2
   solids, 1-10, 14-2
   sterilization, 11-2
Testing, water quality, 1-6
Thermal conditioning, 13-11
Thermal drying, 14-32
Thermophilic anaerobic digestion, 15-9
Thickening, 5-21, 14-1
Topography, 2-17
Torque flow pump, 12-17
Total organic carbon, measurement, 1-9
Total solids, measurement, 1-9
Toxicity, chlorine, 11-7

Transportation, 2-18
Trash racks, 4-12
Traveling bridge, 5-17
Trickling filter–activated sludge process, 7-19
Trickling filters, 7-2
   hydraulics, 7-7
   loadings, 7-7
Trickling filter–solids contact process, 7-18
Tubular conveyors, 4-19
Turbulent flow, 3-8
Two-stage anaerobic digestion, 15-9
Two-stage process, phosphorus, 8-5

U

Ultraviolet disinfection, 11-12
Ultraviolet inactivation kinetics, 11-15
University of Cape Town process, 8-18
Utility services, 2-19
U-tube systems, 6-31
UV, 11-12

V

Vacuum filters, 14-32
Vacuum-assisted drying beds, 14-30
Value engineering, 1-23, 2-3
Vector, trickling filters, 7-9
Vector-attraction reduction, 16-4, 16-11
Velz formula, trickling filters, 7-10
Ventilation, trickling filters, 7-8
V-notch weir equation, 3-13
Volatile solids loading, 15-13
Vortex grit removal, 4-17
Vortex pump, 12-17

W

WAS, 6-2
Waste activated sludge, 6-2
Wastewater, definition, 2-7
Water horsepower, 3-20

Water pollution control,
    legislation/regulations, 1-3
Water supply, 2-7
Weather conditions, settling tanks, 5-13
Wedge-wire drying beds, 14-30
Weir control, 3-12
Weir rate, 5-8
Wet scrubbers, 11-23
WET, 2-15
Whole effluent toxicity, 2-15
WHP, 3-20